S0-EJS-533

WORLD GEOGRAPHY OF PETROLEUM

AMERICAN GEOGRAPHICAL SOCIETY
SPECIAL PUBLICATION NO. 31

AMERICAN GEOGRAPHICAL SOCIETY
SPECIAL PUBLICATION NO. 31

CONTRIBUTORS

WALLACE E. PRATT
EUGENE STEBINGER
WILLIAM B. HEROY
GUILLERMO ZULOAGA
A. J. FREIE
OLIVER B. KNIGHT
GEOFFREY BARROW
FRANK B. NOTESTEIN
A. I. LEVORSEN
G. M. LEES
MAX STEINEKE
M. P. YACKEL
LYMAN C. REED
I. SWEMLE
D. DALE CONDIT
KIRTLEY F. MATHER
U.S. ARMY-NAVY PETROLEUM BOARD
JOHN W. FREY
ANASTASIA VAN BURKALOW
HERBERT FEIS

Editorial Adviser

S. HADEN GUEST

WORLD GEOGRAPHY OF PETROLEUM

EDITED BY

WALLACE E. PRATT AND DOROTHY GOOD

1950

PUBLISHED FOR THE AMERICAN GEOGRAPHICAL SOCIETY

BY PRINCETON UNIVERSITY PRESS

London: Geoffrey Cumberlege, Oxford University Press

Printed in the United States of America
by Vail-Ballou Press, Inc., Binghamton, N.Y.
Half tones by the Phoenix Engraving Company, New York.
Line cuts by the Beck Engraving Company, New York.
Colored fold map reproduced by A. Hoen and Company, Baltimore, Md.

PREFACE AND ACKNOWLEDGMENTS

GEOGRAPHY brings a new approach to the many-faceted study of petroleum. Briefly, scientific geography examines the earth relations of petroleum—where it is formed in the depths, how it is found and brought to the surface, moved over the face of the earth, and used in the air, on the land, and the seas—all this in the framework of spatial relations measured, qualified, and expressed in maps. In the present survey geography describes the world's petroleum-producing regions on a comparative basis and, in a world-wide view, synthesizes the broad problems of petroleum supply and utilization. Such a survey of the geography of petroleum opens out vital considerations of policy and power. The reader who will study the twenty-three chapters in this book, and with them the accompanying maps, diagrams, and tables, will have at his disposal much of the information—and fascinating some of it is—that will enable him to form his own judgements.

The statistics in the text and throughout the tables represent as far as possible conditions at the end of 1948 and the beginning of 1949, and this period is to be understood in statements where the present tense is used. In a few places later information is given, and in some others the only reliable available information was of earlier date.

The American Geographical Society expresses its grateful appreciation to the scientists who responded generously with their time and thought to the invitation to contribute to this volume. The Society also expresses its appreciation to the many organizations that contributed information and authorized their personnel to devote time and care to its requests.

The editors' task has been greatly facilitated by the cooperation of many governmental and research agencies, commercial concerns and institutions, and private individuals. Some of our informants have requested not to be referred to individually. These apart, we wish to record our thanks, for information and courtesies received, to the following (in alphabetical order): American Bureau of Shipping; American Gas Association; American Independent Oil Company; American Petroleum Institute; Anglo-Iranian Oil Company, Ltd.; Arabian American Oil Company; Professor Charles H. Behre, Jr., Columbia University; Professor Abram Bergson, Columbia University; the British Information Services (New York); Dr. W. M. Cadman; Cities Service Oil Company; Creole Petroleum Corporation; Mr. Maxim Elias; Professor W. B. Fisher, Aberdeen University; the French Information Services (New York); the Geological Society of America; Gulf Oil Corporation; Dr. Winthrop P. Haynes; Dr. Hans Heymann, Jr.; Institute of Petroleum (London); International Petroleum Company, Ltd.; Iraq Petroleum Company; the Israel Office of Information (New York); *National Petroleum News* (Cleveland); the Netherlands Information Bureau (New York); N.V. de Bataafsche Petroleum Maatschappij (The Hague); Petróleos Mexicanos; the Petroleum Press Bureau (London); Mr. Joseph E. Pogue, Chase National Bank; Shell Union Oil Company (New York); Sinclair Oil Corporation; Socony-Vacuum Oil Company; Standard Oil Company (N.J.); The Texas Company; Professor W. Taylor Thom, Jr., Princeton University; Trinidad Leaseholds (N.Y.), Inc.; Mr. K. S. Twichell; United Nations Department of Economic Affairs, Statistical Office and Division of Economic Stability and Development; United Nations Department of Social Affairs, Population Division; U.S. Bureau of Mines; U.S. Department of State; U.S. Economic Cooperation Administration; U.S. Federal Power Commission; U.S. Geological Survey; U.S. Maritime Commission, Division of Vessel Utilization and Planning; U.S. Department of the Interior, Oil and Gas Division; *The Westinghouse Engineer* (Pittsburgh); *World Oil* (Houston).

Maps and Diagrams. Of the sixty-one maps and diagrams in this book, forty-six have been prepared at the American Geographical Society from source maps and supplementary information in the Society's collections. The Society gratefully acknowledges the gift of source maps and diagrams, for use in designing some of the figures, from: the Anglo-Iranian Oil Company, Ltd., the Arabian American Oil Company, Mr. Geoffrey Barrow, Mr. A. J. Freie, the Geological Society of America, Dr. G. M. Lees, N. V. de Bataafsche Petroleum Maatschappij, Mr. Eugene Stebinger, the Standard Oil Company (New Jersey), the U.S. Army-Navy Petroleum Board, the U.S. Navy, and Dr. Guillermo Zuloaga. The remaining fifteen figures were provided by the contributors, either as individuals or through cooperating institutions, for direct reproduction after some slight emendations. The Society records its thanks for the gift of these drawings, which were used in the figures indicated: the Creole Petroleum Corporation, 7, 10; the Geological Society of America, 30, 31; Professor A. I. Levorsen, 23, 24; N. V. de Bataafsche Petroleum Maatschappij, 40, 41, 42, 43, 44, 46; Standard Oil Company (New Jersey), 39; the U.S. Maritime Commission, 54; the *Westinghouse Engineer,* 55.

Photographs. The editors wish to thank those whose courtesy made available the photographs used in this volume, as follows: Anglo-Iranian Oil Company, Ltd., Plates 16, 54, 55, 56, 57, 59; Arabian American Oil Company, Plates 62, 67, 68, 69; Brazilian Government Trade Bureau, Plates 43, 44; Chicago Bridge and Iron Company, Plate 14; Corporación de Fomento de la Producción (Chile), Plates 38, 39; Creole Petroleum Corporation, Plates 23, 29, 31; Fairchild Aerial Surveys, Inc., Plates 50, 51; Government of India Information Services, Washington, D.C., Plates 86, 87; Houdry Process Corporation, Plate 17; Hunting Aerosurveys, Ltd., Plates 60, 61; International Petroleum Company, Ltd., Plate 32; Iraq

Petroleum Company, Plate 58; McCarty Company and Emsco Derrick and Equipment Company, Plate 90; National Aerophotographic Service of Peru, Plate 1; N.V. de Bataafsche Petroleum Maatschappij, Plates 77, 79, 80, 81, 82, 83, 84, 85; F. B. Notestein, Plate 33; Sovfoto (New York), Plates 70, 71, 72, 73, 74, 75, 76; Standard Oil Company (N.J.), Plates 2, 3, 4, 5, 6, 7, 8, 9, 10, 11, 12, 13, 15, 18, 19, 24, 27, 28, 30, 32, 34, 35, 36, 37, 45, 46, 47, 48, 49, 52, 53, 78, 88, 91; Max Steineke, Plates 63, 64, 66; Trans-Arabian Pipeline Company, Plate 65; Union Tank Car Company, Plates 20, 21, 22; U.S. Army Signal Corps, Plates 92, 93, 94, 95; U.S. Maritime Commission, Plate 98; U.S. Navy, Plates 89, 96, 97; Yacimientos Petrolíferos Fiscales (Argentina), Plates 40, 41, 42; Dr. Zuloaga, Plates 25, 26.

DOROTHY GOOD
Editor, Special Publications

The American Geographical Society
New York
March 1950

A GEOLOGIST'S FOREWORD

THE distribution and the nature of the accumulations of petroleum in the crust of the earth are two prime subject matters of this book. It also portrays the character of the industry that has discovered and developed these mineral fuels at various places over the earth's surface, "reduced to possession" their fugacious liquid and gaseous components, and converted these raw materials into products currently useful to society; and it considers the impact of petroleum and the petroleum industry on our social and industrial economies.

While research on the *geology* of petroleum has expanded to proportions which now make it the dominant element in the training of young men who plan to enter the petroleum industry, yet in origin and distribution petroleum is intimately related to phenomena which fall mainly within the scope of the modern *geographer*. The winnowing and sifting action of tides and currents, in the shallow waters that surround the continents, on the sediments that are swept into them from the adjacent land-masses; the nourishment of marine life in near-shore waters, and the sea-bottom transformation of the organic remains of marine life; the relation of the continental masses to the profound oceanic deeps and to the intervening continental shelves; shore-line processes in general: all these are geographic phenomena, and they and other geographic agencies are main factors in the accepted theory of petroleum's origin and distribution.

The fields of inquiry of geography and geology overlap, and geology may properly claim for its own the past operation of earth processes which, in their present manifestations, are the interest of geographers. But geography touches petroleum and the petroleum industry over another wide area into which geology does not extend: geography, in its broad aspect, includes the study of the entire human environment; and the discovery of petroleum resources and the distribution of indispensable petroleum products, among the peoples of the world, are live problems in human geography, profoundly influencing human relationships over the whole earth.

It happens that the natural distribution of petroleum is admirably adapted to man's needs; great petroleum accumulations are conveniently situated with respect to transport routes serving the main centers of human population and culture. In a free, peaceful world, the widespread and equitable distribution of petroleum products should offer no formidable problem. But in a world in conflict, the control of these same petroleum resources becomes the object of vital military strategy.

Local geographical factors have operated to modify significantly the patterns of the petroleum industry in the various producing areas. The factor of sparse population in the surrounding regions has complicated the development of some of the earth's most prolific oil fields, and factors of climate have multiplied difficulties and hazards. But, in the present writer's opinion, the most formidable immediate barriers to the full development of the earth's petroleum resources lie in the serious restrictions of exploratory activity which have grown out of recent nationalization policies of many governments. These obstacles may be said to arise within the field of human geography: the attitude of mankind toward the mineral resources of the earth.

WALLACE E. PRATT

March 1950

CONTENTS

Preface and Acknowledgments v

A Geologist's Foreword. *By* Wallace E. Pratt ix

PART I. PETROLEUM IN THE GROUND. *By* Eugene Stebinger 1

Geological principles governing the occurrence of petroleum. The oil-forming process. Traps—the closed-pressure systems of oil occurrence. *The search for petroleum in the ground.* Geological exploration. Geophysical exploration. *The results obtained in the search for petroleum.* The estimation of reserves. The relative extent of exploration.

PART II. THE FUNCTIONAL ORGANIZATION OF THE PETROLEUM INDUSTRY. *By* William B. Heroy 25

Development. Production. Storage. Transportation. Refining. Distribution.

PART III. THE WORLD'S PETROLEUM REGIONS 43

THE WESTERN HEMISPHERE

1. The Caribbean Area as a Whole. *By* Wallace E. Pratt 45
2. Venezuela. *By* Guillermo Zuloaga 49
3. Trinidad. *By* A. J. Freie 80
4. Mexico. *By* Oliver B. Knight 95
5. Colombia. *By* Geoffrey Barrow 100
6. South America other than Caribbean. *By* Frank B. Notestein 120
7. North America. *By* A. I. Levorsen 130

THE EASTERN HEMISPHERE

8. The Middle East as a Whole. *By* G. M. Lees 159
9. Saudi Arabia and Bahrein. *By* Max Steineke and M. P. Yackel 203
10. The Union of Soviet Socialist Republics. *By* Eugene Stebinger 230
11. Europe West of the U.S.S.R. *By* Lyman C. Reed 240
12. Indonesia, British Borneo, and Burma. *By* I. Swemle 273
13. Other Areas in Africa, Asia, and Oceania. *By* D. Dale Condit 301

THE WORLD AS A WHOLE

14. Petroleum in the Polar Areas. *By* Wallace E. Pratt 308
15. Petroleum on the Continental Shelves. *By* Wallace E. Pratt 319
16. The Major Areas of Discovered and Prospective Oil. *By* Eugene Stebinger 325

PART IV. ASPECTS OF UTILIZATION 331

1. The Availability of Petroleum—Today and Tomorrow. *By* Kirtley F. Mather 333
2. Geographical Aspects of Petroleum Use in World War II. *By* The Office of the Army-Navy Petroleum Board of the Joint Chiefs of Staff 344
3. World Patterns of Civilian Utilization. *By* John W. Frey 354
4. A Statistical Survey. *By* Anastasia Van Burkalow 375
5. The Effect of the World Distribution of Petroleum on the Power and Policy of Nations. *By* Herbert Feis 392

Appendix. World Regions: Petroleum Production and Exports, 1938 and 1947 406

Bibliography 409

Abbreviations 433

Note on Glossaries 434

Conversion Factors 436

Index 438

MAPS AND DIAGRAMS

Facing Page
World Sedimentary Basins and Petroliferous Areas (fold map) 14
Page
Fig. 1. Idealized cross section of central portion of an oil basin 13
Figs. 2–5. The World's inter-regional trade in crude petroleum and in petroleum products, 1938 and 1947 40–41
Fig. 6. The Caribbean Petroleum Province 46
Fig. 7. Venezuela: annual production of crude petroleum 51
Fig. 8. Venezuela: physiographic regions and petroliferous basins 53
Fig. 9. The Maracaibo petroliferous basin 58
Fig. 10. Part of the Bolívar coastal field 59
Fig. 11. Petroliferous basins of eastern Venezuela 63
Fig. 12. Trinidad: oil fields and facilities 81
Fig. 13. Mexico: oil and gas fields and petroleum facilities 96
Fig. 14. Colombia: petroleum concessions 101
Fig. 15. The Mid-Magdalena fields 104
Fig. 16. South America: proved oil and gas fields, test drilling, and petroleum facilities 121
Fig. 17. Appalachian and north-central states: oil fields 132
Fig. 18. Appalachian and north-central states: natural gas fields 133
Fig. 19. Mid-Continent and western Gulf Coast States: oil fields 134
Fig. 20. Mid-Continent and western Gulf Coast States: natural gas fields 135
Fig. 21. California: oil fields 136
Fig. 22. California: natural gas fields 137
Fig. 23. East-central Oklahoma: oil pools 140
Fig. 24. East-central Oklahoma: detail of an area 141
Fig. 25. North America: natural gas pipe lines 144
Fig. 26. North America: oil pipe lines 145
Fig. 27. North America: population densities 147
Fig. 28. North America: oil refining centers 149
Fig. 29. North America: three types of prospective areas 155
Fig. 30. Generalized cross section of the East Texas field 156
Fig. 31. Southeastern United States: structural sketch 157

Fig. 32. The Near and Middle East: location map 165–167
Fig. 33. Parts of Persia and Iraq: oil-field development 173
Fig. 34. The Middle East: concessions, fields, and development 176
Fig. 35. Kuwait: oil-field development 184
Fig. 36. Saudi Arabia, Bahrein, Qatar: oil-field development 214
Fig. 37. U.S.S.R.: oil-field development and transport 231
Fig. 38. Europe: sedimentary basins, fields, and development 242–243
Fig. 39. Austria and adjacent countries: basins and fields 247
Fig. 40. Southeastern Asia: oil-field areas 275
Fig. 41. North Sumatra: oil-fields and facilities 285
Fig. 42. South Sumatra: oil-fields and facilities 286
Fig. 43. Borneo: oil-fields and facilities 288
Fig. 44. Netherlands New Guinea: exploration area 290
Fig. 45. North Sumatra: detail of tideland remapped after air surveys 292
Fig. 46. Burma: oil-fields and facilities 297
Fig. 47. Petroleum in the Arctic and adjoining areas 311
Fig. 48. Barrow, Alaska, monthly temperatures and hours of daylight 315
Fig. 49. The Continents and the Continental Shelves 320
Fig. 50. The share of the United States in the world output of crude petroleum 334
Fig. 51. Comparative trends of crude-oil production and of proved reserves 337
Fig. 52. Relation of reserves to production in eight nations 341
Fig. 53. European Theater of Operations: military petroleum distribution system 346
Fig. 54. Cross section of a "BT" tanker 349
Fig. 55. Components of aviation gasoline and how they are made 358–359
Fig. 56. World map, chief sources of energy, 1938 370
Fig. 57. Man-power years of energy produced, per capita by countries, in 1938 371
Fig. 58. Man-power years of energy produced per capita in 1938 and 1947 372
Fig. 59. Cartogram of petroleum utilization in 1938 381
Fig. 60. Index map, world's principal political units, January 1938 383

PHOTOGRAPHS

Plates	Illustrating	following pages
1– 7	Part I. Petroleum in the Ground	6
8–22	Part II. The Functional Organization of the Petroleum Industry	30, 46
23–91	Part III. The World's Petroleum Regions:	
23–35	The Caribbean Area	46, 62, 78, 126
36–44	South America Other than Caribbean	126, 142
45–53	North America	142, 150
54–61	The Middle East as a Whole	198
62–69	Saudi Arabia and Bahrein	214
70–76	The U.S.S.R.	214, 270
77–85	Indonesia and Burma	270, 286
86–87	India	302
88–89	The Arctic	302, 318
90-91	The Continental Shelves	318
92–98	Part IV. Aspects of Utilization:	
	Geographical Aspects of Petroleum Use in World War II	350, 366

TABLES

1. The World's discovered oil: production and reserves of crude petroleum, by principal producing countries, January 1, 1949 22–23
2. Venezuela: destination of petroleum exports in 1938, 1946, 1947, and 1948 50
3. Venezuelan oil-producing basins and districts: cumulative and average daily production at year-end, 1949 60
4. Trinidad and Tobago: exports of petroleum products in 1947, by quantity and destination 84
5. Trinidad: estimated deliveries of petroleum products into consumption annually in 1941 and in each year from 1944 through 1947 84
6. Trinidad: crude oil production, by fields, at end of December 1947 86
7. Trinidad: geological data on oil fields producing in 1947 87
8. Stratigraphic column of young Tertiary deposits in main oil fields, southwest Trinidad 88
9. Mexico: estimates of reserves and production of crude petroleum, by districts, as of January 1, 1949 98
10. The major oil companies operating in Colombia in 1948 114
11. United States: crude oil production during 1948, and proved reserves as of December 31, 1948, with principal states ranked in order of production 139
12. United States: natural gas net production during 1948, and proved reserves as of December 31, 1948, with principal states ranked in order of production 142
13. United States: total domestic requirements of all oils; and domestic requirements, production, imports, and exports, of leading petroleum products and of crude in 1947 and 1948 151
14. The Middle East's discovered oil: production and reserves of crude petroleum, by principal producing countries, January 1, 1949 162
15. Petroleum concessions in the Middle East, April 1949 177–179
16. The Middle Eastern countries: estimated areas and populations at midyear, 1948 196
17. Saudi Arabia: Pipe-line status at end of 1948 218

18. The U.S.S.R.: production of crude petroleum by area, 1939 and 1944 236
19. The principal sedimentary basins of Europe west of the U.S.S.R. 241
20. European countries, west of the U.S.S.R.: production of crude petroleum and crude-charge refining capacity in relation to requirements in 1948 250
21. Indonesia, British Borneo, and Burma: production of crude petroleum in 1948 and cumulative totals as of January 1, 1949 274
22. Netherlands Indies: trend of export trade and domestic consumption of certain classes of petroleum products, by weight, in selected years from 1913 to 1938 295
23. The chief types of petroleum products 356–357
24. Production of crude petroleum and estimated consumption of four leading products, by continents, in 1938 361
25. United States: estimated annual consumption of distillate and of residual fuel oils in 1941 and in 1945 363
26. The use of petroleum products, by countries, in relation to (1) number of persons per motor vehicle in each country, and (2) petroleum product most used in each country 385
27. Consumption of petroleum products in the United States, 1938 387
28. Comparative growth of production and consumption of petroleum in the United States and in the rest of the world, 1857 through 1948 388
29. Annual requirements of petroleum products in the United States and in the C.E.E.C. countries, 1938 and 1947 390
30. World Regions: production of crude petroleum and exports of crude petroleum and of petroleum products, in 1938 and 1947 406–408

PART I

PETROLEUM IN THE GROUND

PETROLEUM IN THE GROUND

BY EUGENE STEBINGER *

PETROLEUM, or, as its name implies, "rock oil," is a greasy liquid, with a unique and characteristic odor, occurring naturally at the surface of the earth and at depth. It is present in numerous surface showings on every continent. Petroleum is both simple and complex in its nature. It is simple in the sense that it is composed almost exclusively of only two of the elements, namely, hydrogen and carbon. It is complex in that it is essentially a mixture of closely related hydrocarbons, whose molecular structure varies step by step with the addition of atoms of either of the two components to atoms of the previous molecular setup. Thus an almost endless series of compounds comes into existence. These range from the light gas, methane, at one end, through an increasingly heavy succession of gases, liquids, semiliquids, and, finally, solids. Fortunately, the mixtures are readily separated into their useful components by distillation under gradually rising temperature, which effects a simple physical separation of the components by means of differences in their boiling points. Starting with the gases, some of which escape even without heating, the succession includes gasoline, kerosene, gas oil, lubricating oil, residual fuel oil, and, finally, asphalt. Careful use of words often makes it preferable to refer to "the petroleums," using the plural rather than the singular, and thus to imply the inclusion of all the variants, particularly the gases.

A summary of present-day concepts of the oil-forming process and of the nature of petroleum occurrence is herewith offered, as background for the detailed accounts of individual oil-bearing areas comprising later chapters. In addition, petroleum exploration as it stands today is portrayed, with the implications as to probable trends in the further expansion of petroleum enterprise throughout the world. A discussion of developed reserves worldwide is included, emphasizing their relative magnitude and distribution.

Geological Principles Governing the Occurrence of Petroleum

SOME BASIC CONTROLS

The great stores of petroleum in the earth are contained in stratified sedimentary rocks laid down in geological periods from Cambrian to Recent. The geologic time-scale from Cambrian upwards is divided into three major divisions, or eras, and these in turn into lesser divisions, or periods. The sequence, of world-wide application, from older to younger, or from

* Formerly Geologist, United States Geological Survey, and Chief Geologist, Standard Oil Company (N.J.).

bottom up in any normal occurrence of the formations in the field, is as follows: (1) Paleozoic Era, whose periods are Cambrian, Ordovician, Silurian, Devonian, Carboniferous, Permian; (2) Mesozoic Era, whose periods are Triassic, Jurassic, Cretaceous; (3) Tertiary Era, whose periods are Eocene, Oligocene, Miocene, Pliocene. Current estimates of the actual time intervals average 550 million years since the beginning of the Cambrian, 350 million years for Paleozoic time, 125 million for Mesozoic, and 74 million for Tertiary. A million years of Post-Tertiary glacial and Recent time is included in the overall figure.

The basic information for the geologic column and time-scale has been gradually assembled in the field by thousands of geologists throughout the world, since the beginnings of the science early in the last century. The vast assemblage of information on the changes of land and sea and the accompanying earth movements, as recorded in the sedimentary formations, and its interpretation in an orderly sequence applicable to all of the continents, has been an outstanding achievement of human endeavor in modern times. In recent years, first Triassic and then Cambrian rocks have been added to the list of the geologic periods productive of oil in important quantities. All of the others had already been proven, so that now the sequence is complete, thus establishing a remarkable continuity and universality in the oil-forming process since the Cambrian.

Throughout the reaches of geologic time the various land areas of the globe have undergone great changes. These changes have repeatedly reached the extreme, developing mountain chains where basins had previously existed in which sediments had accumulated in thicknesses of as much as thousands or even tens of thousands of feet. Conversely, these changes have reduced mountainous areas to low plains or plateaus, maintaining what appears to have been an approximate balance in the process, since as far back as Cambrian time.

Fundamentally these changes resulted from movements in the earth's crust, some reaching only comparatively shallow depths, measured in tens of thousands of feet, and others probably reaching deep into the earth's mass. All of these movements are included by geologists in the term "diastrophism." They are readily grouped in two classes: *orogenic* movements, the comparatively violent linear movements that have built up mountain chains on all of the continents, and *epeirogenic* movements, comprising much slower warping effects, either up or down, over great expanses of the earth.

These changes inevitably were accompanied by great invasions of the oceanic waters; and the resultant inundations came from first one direction and then another. Frequently they cut entirely across a continent, reducing it to an extensive group of islands surrounded by shallow seas. The marine sediments laid down around the margins of the continents, and in the repeated inundations reaching far inland, established the fundamental pattern of oil occurrence as it exists today. As both types of earth movements have repeatedly been of prime importance in causing invasions of the oceans over the continents, both have exerted a profound influence on the occurrence of petroleum.

The continued existence through these changes of great nuclear areas of Pre-Cambrian rocks—granites, gneisses, and schists—has influenced the continental framework since the Cambrian. They comprise the so-called "shields" of the continents, great, stable, positive areas that have been important sources of sediments through the ages.

There are six of these primary nuclear areas on the various continents and probably a seventh area of comparable proportions centered on Antarctica. Those that are well defined include: the Canadian Shield, centered on northeast North America; the Brazilian Shield, centered on northern South America; the African Shield, centered on north-central Africa; the Scandinavian Shield of northeast Europe; the Asiatic Shield of central Asia; the Australian Shield, centered on that continent. In a general way, there has been continuous sedimentation outward, first in one direction and then in another, from these nuclear areas since Cambrian time; the cumulative effects of sedimentary transport and retransport have been greatest in producing thick sections of the younger rocks, Cretaceous and Tertiary formations, up to distances thousands of miles from their centers. The ancient shield areas have, in this sense, profoundly influenced petroleum distribution.

THE OIL-FORMING PROCESS

Petroleum in its natural occurrence is closely associated with marine or semi-marine near-shore sediments. The relation is practically universal and has persisted since Cambrian time, very probably without material alteration. The study of the depositional zones of present-day seas and oceans therefore appears essential to an understanding of oil occurrence, particularly its initial stages. The findings of the oceanographers on the types and probable rates of accumulation of the sediments at various depths and bottom conditions, the determinations of the organic content of the muds and sands on the various bottoms in the shallower zones, and inferences as to the nature of the compaction process in the sands and muds after they have accumulated, all appear vital.

1. The Role of Sedimentation. Oceanographers, in their broad-scale studies of the oceans and seas, have found it convenient to classify the ocean-bottom environments on the basis of depth as: the *Littoral or Shore Zone,* the range from high to low tide; the *Neritic Zone,* from low tide to 100 fathoms, comprising the average range of the Continental Shelf; the *Bathyal Zone,* from 100 to 1,000 fathoms; and the *Abyssal Zone,* from 1,000 fathoms to the maximum depths. This last zone includes by far the greater part of the oceans. The bottom sediments of the Abyssal Zone are uniformly very fine characteristic oozes and very fine muds, with very slow rates of deposition. They appear to be of no interest from the petroleum standpoint. The sedimentary bottoms of the Neritic and Bathyal zones are composed of muds, sands, and gravels, in that order of extent. They are in general of types identical with those observed on land in exposures of the older marine formations. The zones with the maximum permanent accumulation of sediments appear to be the lower half of the Neritic and much of the Bathyal, thus including the depth range of 50 to well below 500 fathoms. Above 50 fathoms the Neritic bottoms are subject to strong tidal and shore currents as well as to wave action, and the sediments are highly oxidized and are constantly agitated by wave action. In addition, scavenger life in shallow waters ingests a large portion of the bottom sediments, greatly depleting their organic content. Below 50 fathoms, on out over the edge of the Continental Shelf, and well down the Continental Slope[1] permanent deposition increases rapidly, and the deposits are free from scavenger action, while oxidation practically ceases; thus there are created optimum conditions for the preservation of organic content.

The late Charles Schuchert, in the course of his classic studies of North American paleogeography, suggested that all marine strata laid down within 100 miles of former lands should be considered to be within the realm of likely petroleum sedimentation. This limit is in fair agreement with the above findings, except for areas with narrow continental shelves or excessively steep continental slopes. It is of interest as an estimate by a very able stratigrapher of how far seaward the initial stages of the oil-forming process are likely to have extended.

The rate of sedimentation doubtless varies widely through a range that as yet can be defined only in approximate terms. The conditions for a maximum rate include a rapidly sinking sea floor in proximity to rapidly elevating lands, the relief of which is being forced upwards to mountainous proportions. In other words, the earth movements involved are of the dramatic orogenic type. Under these conditions sediments totaling up to 30,000 or 40,000 feet in thickness have accumulated in a single geologic period. As the exploration for petroleum expands, the fact becomes increasingly evident that the great areas of prolific production are invariably associated with a setting of thick and rapidly accumulated sediments. These conditions evidently offer the optimum environment for petroleum accumulation.

On the other hand, conditions for minimum rates of sedimentation are those of the comparatively shallow seas that have often pushed in broadly and more or less aimlessly over the lowland areas of the continents. The earth movements involved in these transgressions of shallow seas are of the slow-moving, warping types (*epeirogenic*). Rates of marine sedimentation as low as 150 feet per million years, in contrast to possibly as much as 2,000 feet per million years in sites of rapid sedimentation, are indicated. In these areas of minimum rates of sedimentation the oil-forming process,

[1] The continental shelves and the continental slopes are discussed in Part Three, Chapter 15.

1. An aerial view toward the west of the Agua Caliente oil field in eastern Peru, on the Río Pachitea, south of Pucallpa (about 8°45′5″ S, 75°40′ W). This oblique mosaic gives an excellent idea of the part now played by aerial surveys as interpreted by the expert geologist. The geological structure is that of a closed anticlinal dome, asymmetrical on the east. The west flank dips 15 to 30 degrees; the east flank dips 40 to 80 degrees, toward the river. The rim-rock, which almost touches the two sides of the mosaic, is of Upper Cretaceous Rampart sandstone. The area enclosed by this rim-rock is 15 square miles, about 4½ miles in length, horizontally across the mosaic, and 3½ miles in width. The rampart formation (the rim-rock) is underlain by Cretaceous San Pedro shales, on which grows photographically light-colored vegetation. These shales, in turn, are underlain by Lower Cretaceous Campana sandstone, on which grows photographically dark-colored vegetation. The Campana sandstone has caused the central part of the structure to be topographically domal, and the upper part of the Campana is the oldest rock exposed at the crest. The lower part of the Campana contains 40 per cent of shales, and it unconformably overlies Permian limestone, in which there is no indication of petroleum. The oil production comes from the lower Campana, from saturated sands interbedded with the shales. (National Aerophotographic Service of Peru, photo by P. H. Blanchet, photogeologist for International Petroleum Company, Lima, Peru.)

2. Oklahoma. Folding in road cut on Black Knob Ridge, southeast of Atoka. (Carter Oil Company. Standard Oil Company, N.J., photo by Lofman.)

3. Surface geologist and instrument man working in field. (Carter Oil Company. Standard Oil Company, N.J., photo by Lofman.)

4. Arbuckle Mountains, Oklahoma. Two views of Arbuckle limestone. In the foreground, an exposed cross section of the outcrop, where cut was made for a highway. Behind it is the outcrop as seen from the undisturbed surface. These rocks, of Ordovician and later Cambrian age, produce oil and gas at Oklahoma City and other Oklahoma fields. (Standard Oil Company, N.J., photo by Corsini.)

5. Brunton Compass, an instrument used by surface geologist for measuring dip of beds. (Carter Oil Company. Standard Oil Company, N.J., photo by Lofman.)

6. The llanos, Colombia. Survey party. Dense jungle makes stream beds the only usable places for locating and mapping geological outcrops. Charte Anticline. (Standard Oil Company, N.J., photo by Collier.)

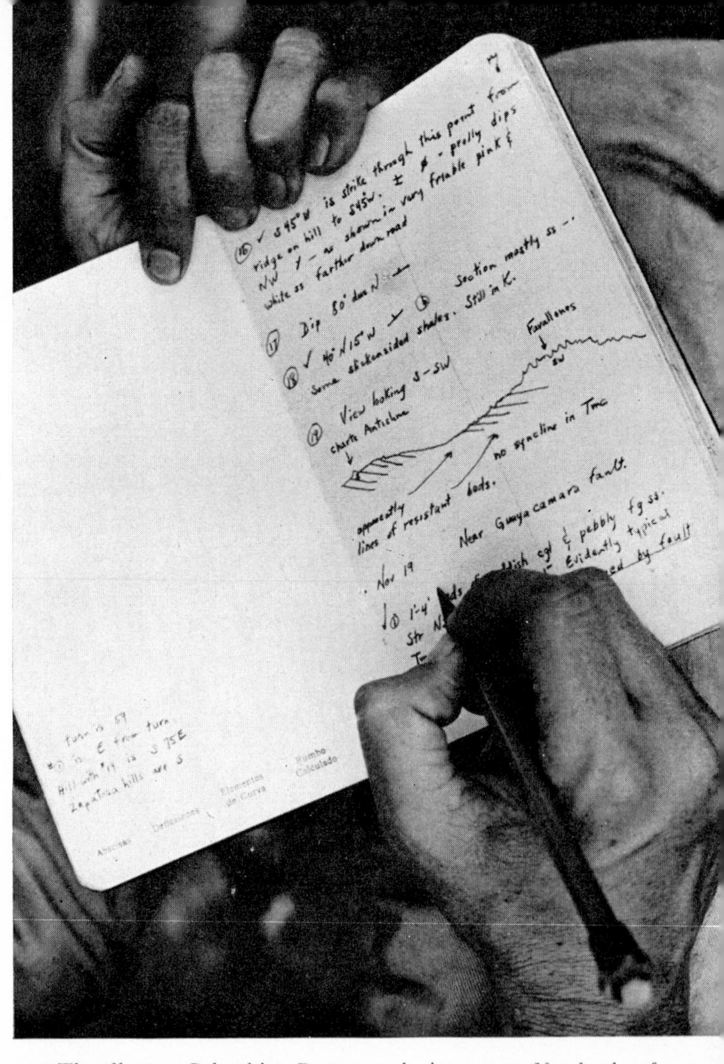

7. The llanos, Colombia. Paper work in camp. Notebook of Tropical Oil Company geologist covering the detailed survey of the Charte Anticline. (Standard Oil Company, N.J., photo by Collier.)

though still effective, has been slowed down appreciably. The resultant oil pools, often still large enough to be important, are generally of a mediocre character. The Mid-Continent and Eastern Interior regions of the United States, with thousands of small oil pools, are the outstanding examples of such a setting; nowhere else have such small pools been so intensively developed. In other parts of the world exploration has been restricted to areas of thick sediments and strong earth movements, which are the optimum settings for high yields of petroleum.

2. *The Role of Organic Matter.* A universal attribute of nearshore marine sediments is the presence of appreciable percentages of organic matter. The amounts are greatest in the fine-grained types: the muds and limy muds of the sea and their land equivalents, the clays, shales, and impure limestones. The average content of organic matter, as revealed by thousands of determinations from the sea-bottom samples, is 2.5%; that from the marine formations of the land averages 1.5%. The aggregate amount of organic content, if calculated from these percentages on the basis of the volumes of the sediments, is of very large proportions. Transformed into petroleum it would vastly exceed the stores in all the oil fields of the earth already found, or likely to be found in the future. Only a small fraction of the original source material survives to reach the final stage, where it takes on the useful form of free petroleum, trapped under high pressure, in a natural reservoir underground. It is all too apparent that the oil-forming process in its later stages, following the initial stage of the deposition of the sediments, is *highly selective.*

Why the selectivity? Whence the organic matter? Is it so varied in its nature that only a small fraction is of a type suited to the oil-forming process? Do sea-bottom environments vary so irregularly that they leave only a small fraction of their areas effective? And do bacteria, also, modify the oozes and leave but a small fraction of bottom area effective? Or is the selectivity induced only at a later stage in the compaction process, after hundreds of feet of oozes have been rid of excess sea water as the sediments tighten up? Do mineral catalysts become active at this stage under conditions that are highly selective?

These questions, typical of many others that could be propounded, serve to illustrate the present state of knowledge on this problem. That they can, and will, eventually be solved seems likely. Chemists have demonstrated ready ability to transform organic material of almost any type into petroleum, and then to reverse the process, and to stop it at will. The molecular combinations of the hydrocarbons can be varied by heat, pressure, catalysts, and other forms of persuasion, over a wide range, and fixed at any desired end-point. Then why are the geologists or, more particularly, the students of sedimentation, so slow in getting the answers? The answer to that query is not difficult. Nature is hiding an interesting natural phenomenon about as effectively as could be. Most of nature's terrestrial doings can be readily appraised: but here is a process that appears to be going on 30 to 100 miles out at sea on the sea bottoms or hundreds of feet, possibly well over a thousand, beneath them.

The bottoms have been patiently sampled, area after area. There certainly is no free oil in the few top feet of the sediments; there is only more organic matter like that from marine sediments on the land. How can we drill the numerous wells that would be required, 30 miles or more at sea, and recover suitable core-samples of the very soft, watery sediments? Furthermore, even this procedure might fail to yield the answers. The oil itself may not be generated until compaction has progressed much further and the rocks have been moderately stressed and strained, in the course of being uplifted to take their places among the other oil basins on the lands. In that event it would be necessary to search the earth for situations where uplift is part-way completed, and to try again with the drills.

After the above digression, it will be useful to record the observed data on some of the lines of investigation mentioned. Although the organic aggregate found in marine sediments is probably in part land-derived, the ocean waters appear to be by far the more important source. The vegetable and the animal debris from the lands has two very difficult hurdles to cross before it can move on to its final burial place, well out at sea in the critical zone of maximum *permanent* deposition. It must first avoid the active oxidation, as well as the scavenger action, which prevail in the river waters. Then it must escape the mud eaters of the littoral and upper Neritic zones —the teeming molluscan life responsible for the great shell beds of the geological sections. Apparently very little organic debris from the lands gets through to make a contribution to the petroleum-forming sediments. On the other hand, the surface and sub-surface organic content of the seas, floating and drifting about—the plankton of the marine biologists—has ready access to the zone of maximum permanent sedimentation without crossing either of these hurdles. The abundance of plankton to considerable depth is amazing, even in the clear waters of the sea, not to mention the more extreme conditions of colored seas and algal growths so dense that they hinder the navigation of small ships. The plankton assemblage, therefore, appears to make by far the larger contribution to the sediments.

There is an additional line of reasoning that appears conclusive as to the minor role that the land-derived organic aggregate has played in petroleum sediments, and that appears to point to the plankton of the oceans as the dominant source of organic matter in sediments since the Cambrian. It is large-scale plant life that makes the predominant contribution of organic debris to inland waters, from the swamps of the tropics to the muskegs of the subarctic regions. Such large-scale plant life is definitely dated as beginning in the lower Carboniferous, and promptly expanding to the lush coal-forming vegetation of that period, that, with numerous variations, has continued to the present. The beginning of such life in the Carboniferous doubtless greatly increased the land-derived organic matter available. Did this increase then radically alter the range in types of petroleum, or add any new types to those that had already been formed—in the numerous fields of each period ranging from Cambrian to Devonian inclusive, covering about 240 million years of geologic time? On the contrary, there appears to have been no marked change in crude types in the Carboniferous or in any later period, even though land life was increasing enormously in both the plant and the animal world. Already by Ordovician time the range in crude types created was broad enough to include the low-gravity asphaltic crude of the Trenton limestone, at one extreme, and high-gravity crudes from sands, at the other. Thus, by elimination, the marine plankton, mostly simple single-celled types of life, the source of a continuing and essentially unchanging organic aggregate since Cambrian time, seems to have made the major contribution of organic content to marine sediments.

In view of the important role that the plankton appears to play in our problem, a brief description may be useful. The marine biologists identify two main types: the permanent plankton, and the transitory or seasonal type. The permanent group are the more numerous; they include incredible multitudes of microscopic plants and animals which abound in all the seas, even in arctic latitudes. The most abundant are the single-celled algal plants. This plant life is the ultimate provider, through the photosynthesis of carbon dioxide into vegetable substance, for all the animal life of the seas. The total volume so produced annually is estimated at ten tons of carbohydrates for each sea acre; the crop greatly exceeds that of the earth's land areas. Transitory plankton includes innumerable species that have a brief seasonal drifting stage in their life history. The bulk of the transitory plankton includes the spawn, that is, the eggs and larvae, of countless species of fish, crustacea, many of the mollusca, most of the corals, and practically all of the echinoderms and sponges.

The range of the plankton is from the surface downwards to great depths: a great variety of bathy types are included in the total. Indispensable food materials, such as inorganic nitrogen compounds, phosphates, and silica, are supplied to the plankton by the upwelling action of sea currents, particularly from the deeps along the margins of the continents, in proximity to the areas of most rapid sedimentation. The net result of these processes is the production of an abundance of organic matter, a very small fraction of which could supply the entire organic aggregate found in marine sediments.

The Sargasso Sea of the North Atlantic is the extreme example. It occupies an oval area about 1,500 miles east-west and 600 miles north-south in driftless waters encircled by the clockwise rotation of the currents of the North Atlantic, including the Gulf Stream up the American coast, the southeast drift current off Europe and Africa, and the west-drifting equatorial current on the south. The west end of the Sargasso Sea lies 500 miles east of the Bahamas, and the north edge is 300 miles south of Bermuda. At the surface there is an accumulation of brown algae of the genus *Sar-*

gassum, comprising one of the greatest plankton assemblages of the oceans. Columbus became badly snarled up in it, in addition to the other difficulties of his first voyage. In the organic debris of sea bottoms the great accumulation of surface algae is augmented by similarly large volumes of low forms of animal life, making up a typical plankton assemblage. The annual crop of organic matter in the Sargasso Sea is stupendous. To the sedimentationist it is suggestive of the amounts of such drift that may be left by eddying effects along the landward side of the great ocean currents mentioned, not only at the surface but also at depth.

3. The Role of Compaction and Migration. Assured of an ample supply of organic matter, the inquirer naturally reverts to the zone of maximum permanent sedimentation, in the comparatively shallow marginal waters of the seas. Cyclic sedimentation there produces alternating layers of muds, limy muds, and sands, accumulating as watery oozes over the bottoms. With the continued sinking of the bottoms a great wedge of these sediments, amounting to hundreds and even thousands of feet in thickness, is gradually built up. There naturally follows a broad-scale compaction, due to the weight of the accumulated material. The weight is very slight, probably, in the first few hundred feet, but must increase uniformly with depth. As this compaction continues, the sea water, all-pervasive in the sediments in the early stages, is squeezed out, at least in part. In the sands some of it doubtless remains, even through the final stages of uplift to a land area, and it constitutes the "fossil" brines which under normal conditions are associated with oil pools. In the muds and limy muds, with their relatively high content of organic matter, the process of transforming the organic aggregate into petroleum is initiated, whether before compaction, during compaction, or later, after compaction. The petroleum-forming cycle is thus under way.

The details of the process, as has already been implied, remain obscure. The organic aggregate may enter a semi-soluble phase in the entrained sea water, which would thus facilitate movement from the muds to the sands. Bacteria present in great numbers in the original oozes, although they alone do not appear able to complete the process, may have played an important role in *conditioning* the organic content for the later changes. In a semi-soluble phase of the organic matter, catalysts, such as nickel, vanadium, lead, and iron, which are present in minute quantities in fine-grained sediments, may be effective in producing results similar to those obtained by the petroleum chemists in completely transforming the hydrocarbons. Furthermore, the pressure and moderate temperatures of the enclosing rocks, acting slowly over very long periods of time, may be as effective in completing the transformation as their more intensive application is in laboratory operations.

Petroleums in the gaseous forms are also, very probably, formed in the initial stages of compaction. These gases would facilitate the vertical movement of the liquids from the tight and fine-grained rocks, the original source materials, to adjacent porous sands where they would displace great stores of sea water, entrained in the course of the original sedimentation. It appears, therefore, that great accumulations of oil and gas may be built up by short-range migration from compacting muds into adjacent sands very gradually and in minute quantities. Presumably the process is largely complete before the final stages of the compaction process.

When oil and gas have arrived in a porous stratum, further migration would be essentially a lateral movement as the strata are tilted by differential elevation. If the sand be of considerable areal extent, the lateral movement in its stronger phases would probably include all the non-adherent liquid contents of the sand, including the large volume of salt water and the comparatively small amount of oil with its attendant gas. Even in very gently tilted strata such movement would probably be upward, in the direction of relief of pressure. If a trap—of any of the types to be described later—is encountered, it checks the mass movement of the fluids, and the inherent buoyancy of the oil and gas with respect to the heavier salt water will permit them to fill the trap from the top downward.

The above theoretical account of the oil-forming process from the compaction stage to the final pooling presents the prevailing view of a majority of geologists. The actual range of opinion is considerable. There is little difference of view regarding the character of source material or the validity of the compaction process, although everyone wishes to see more recorded specific observation. There is much debate, however, as to how widely migration has been effective: the minimum estimate of distances involved in migration is that movement results only from compaction, as a result of the fluids being squeezed out of the oozes and muds into the pores of immediately adjacent incompactible bodies of sand, whereas the maximum view would include movement of the fluids practically in artesian proportions, throughout the extent of the porous sands, and over many miles.

TRAPS, THE CLOSED PRESSURE SYSTEMS OF OIL OCCURRENCE

In the previous section reference was made to traps capable of checking the movement of fluids in a porous stratum, and permitting the oil and gas to fill the trap from the top down. The great volumes of oil pooled in the earth are confined as individual pressure systems in such traps, both large and small.

A trap, or closed-pressure system, capable of pooling oil and gas, exists wherever the configuration of an impermeable caprock and the underlying porous reservoir rock is such that oil and gas rising upward through the salt water, normally permeating the porous stratum, readily enter the reservoir and fill it from the top downward. The oil and gas are confined above the water against the impervious roof of the trap. They cannot escape until the trap is full, permitting them to "spill out" around its edges. This statement is applicable to oil pools in all of their considerable variations in form and size.[2]

Traps naturally group into two broad types. The first and more numerous are those created by local deformation of the strata, which produces a feature essentially structural in its nature, due to differential uplift or offsetting of the strata. The second group, occasionally very important because of the very large volumes of oil so trapped, are those that are essentially stratigraphic. These are produced by upward variations of the porous strata of the trap, which end in a complete loss of continuity and a complete sealing of the porous medium in the upward direction. In a very few places, elements of both main types are combined in a single trap. The fundamentals remain the same, and for present purposes these variants need not be considered.

The concepts can be more readily grasped by referring to Fig. 1. This presents an idealized cross section of a considerable portion of a single oil basin. It is greatly simplified, by limitation to a single oil zone and by omitting any considerable degree of unconformity and angularity in a thick section of formations. Vertical scale is greatly exaggerated. Immediate interest is centered in the various traps or reservoirs pooling oil across some 40 miles of the central portion of the basin. Traps *A, B,* and *C* belong to the structural group. *D* illustrates the most common type of the important group of stratigraphic traps. The element of local deformation of the strata is readily evident in *A, B,* and *C,* in contrast to the complete lack of that fundamental in *D,* the stratigraphic type. The deformation, it should be noticed, is strictly localized in each instance, being superimposed on, and subordinate to, the broad synclinal downsweep of the oil basin taken as a whole. In the diagram the exaggeration of vertical scale lessens the correctness of the impression gained in this respect.

In *A* the trap is in an anticline [3] in its simplest form. It is unfaulted, that is, without breaks offsetting the continuity of the strata, and it is symmetrical. One side is not noticeably steeper than the other, and there is no thinning and thickening of the softer units, namely, the shales, in the succession, such as often characterizes overturned or asymmetric folds. The anticline shown affects only one oil zone, and therefore only one trap appears. Anticlinal folds often affect several porous zones, at higher and lower positions, each forming its own trap, varying in width across the anticline, and comprising a vertical succession of traps whose petroleums may be very different, and whose production rates equally various. The simple setting presented in Fig. 1, therefore, might be vastly complicated. It would still remain an

[2] "Trap" and "reservoir" are essentially synonymous. "Trap" is the more inclusive term, however, in that it includes the entire rock assemblage effective in pooling the oil. "Reservoir" in the more limited sense, as used by petroleum engineers, comprises solely the portion of the porous strata holding oil and gas.

[3] An upfold or arch of stratified rock in which the beds or layers dip in opposite directions from the crest.

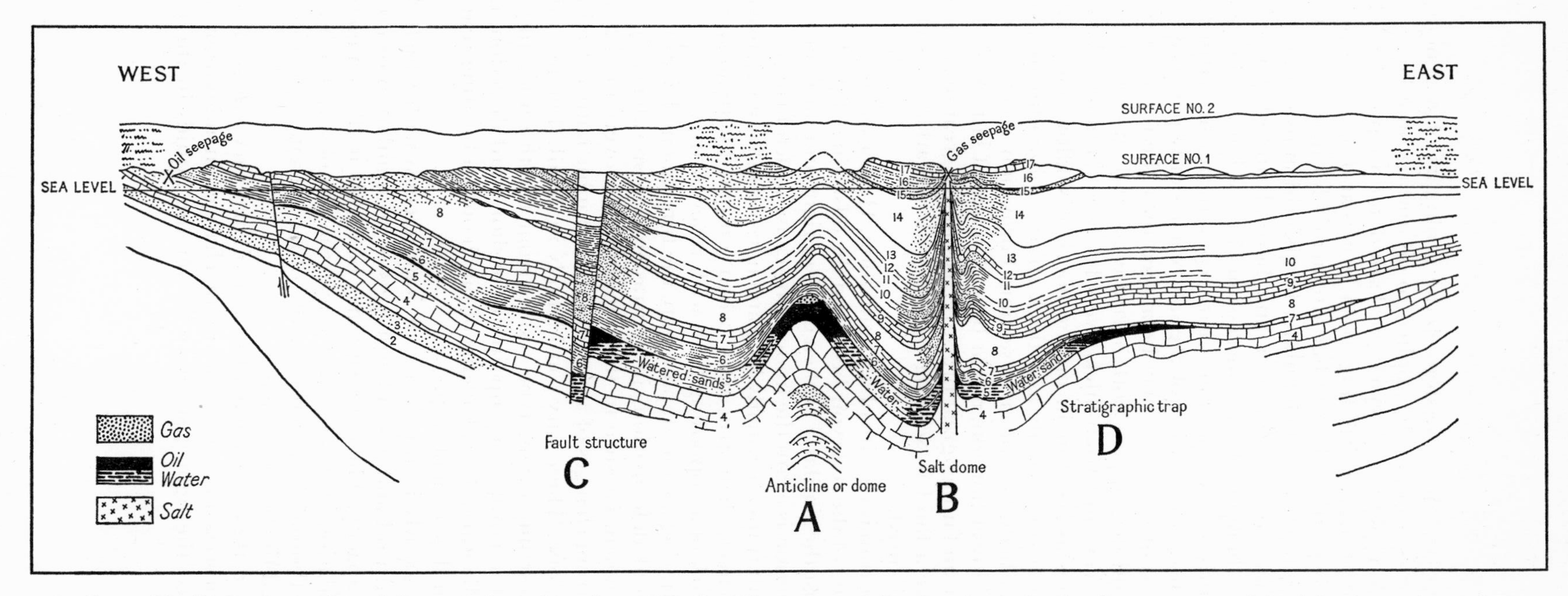

Fig. 1. Idealized cross section of the central portion of an oil basin. The structural features shown in the diagram are explained in the sections of Part I that are entitled "Traps—The Closed Pressure System of Oil Occurrence," and "The Interpretation of Readings."

anticline, however, the obsession of the geologists of a generation ago, when geology first came to be applied to oil-finding on a large scale.

At *B* various traps formed on a typical salt dome are presented. Deformation and uplift of the surrounding strata, just as at *A*, remain fundamental. The salt core is pictured in a rather attenuated form, piercing to the surface. Variants often show much bulkier forms, with mushrooming and consequent overhang of the salt over and beyond underlying sedimentary rocks. As at *A*, there might be a succession of traps both higher and lower, as well as faulting, which would greatly complicate the setting. In many instances the upward piercing of the salt reaches only part way to the surface. In that event the strata may be arched over the top of the salt mass, and traps for oil may appear both over the salt and on the flanks. Whence the salt? Quite certainly from thick-layered deposits at much greater depths. Salt often forms as a sediment on the floors of shallow, evaporating seas, and it is present occasionally as a thick stratigraphic unit in a succession of sedimentary beds. On the cross section, Formation 1 in the sequence, or possibly a still lower horizon, might be such a bed of salt. At *B* it would lie at great depth, and under great pressure. Salt under great pressure responds practically like a fluid in seeking escape, with remarkable results similar to the form pictured.

At *C* a fault trap is the site of accumulation. Here a seal has been formed by the breaking of the reservoir beds, and the offsetting of the broken edges upward against the similarly broken edges of tight and impermeable strata. Local deformation is again the underlying cause which brings the trap into existence. Fault traps are usually long and narrow and not extensive areally. They are, in effect, "half structures" when compared with an anticline, because of the lack of an opposing limb with reversed dip.

At *D*, illustrating a stratigraphic trap, the fundamental distinction is the absence of local deformation, in contrast to its prevalence in the types already mentioned. Closure in the trap under its overlying impermeable cover results from changes of an up-dip stratigraphic nature, limited to the porous zone alone. These vary considerably in detail. The prevailing, and by far the most important, type results from the thinning and eventual pinch-out of the porous-zone updip in the shoreward direction, i.e. to the right on the diagram. The elements of pinch-out and sealing remain in all the variants of this form of trap.

Traps vary greatly in the degree of completeness with which they have been filled with oil and gas. The process of accumulation often stops far short of full capacity. The net result may be no oil at all, or a mere bubble of an accumulation over a few acres near the top of a trap whose closure extends over thousands of acres. At the other extreme, the accumulation may have filled the trap completely.

Furthermore, the proportions of gas and oil may vary widely. The original oil-forming process in the sediments or changes in migration may have greatly favored the light gaseous ends over the liquids in the petroleum mixture, and have resulted solely in gas accumulation. In other variants the trap may be filled with gas only part way down from the top, or there may be no gas cap whatever over the oil, the moderate proportion of gas present being entirely in solution in the oil.

Even though the original conditions of sedimentation may have been ideal for supplying petroleum abundantly, the effectiveness of accumulation in a trap may have been greatly curtailed, or eliminated, by limited porosity in the reservoir rocks. More often than not, these are sandstone, a rock type that varies widely in the features that control its porosity, namely, the fineness and roundness of the grains of sand and the nature of their cementation. Frequently the reservoir rocks are limestones and dolomites, the carbonate rocks that form the reservoirs effective in many of the world's largest oil fields. Porosity and permeability vary in the calcareous rocks as in the sandstones, but to a much greater degree. Pure limestones are dense and impervious. Others range through a variety of types from low-porosity, fragmental accumulations, essentially like fairly tight sands, through oolites and dolomites, to highly porous reef deposits of algal and coralline origin. In addition, the porosity induced in limestones by waters either at the surface of an unconformity, or well below the surface, may be very effective in affording voids for pooling. Effects induced by solution below an unconformity are likely to be highly unpredictable in form and extent. In reef deposits the algal and coralline growths, although often of considerable thickness, are also prone to diverge greatly from the regularity to be expected in normal sediments.

The Search for Petroleum in the Ground

GEOLOGICAL EXPLORATION

The application of geology to the problem of oil-finding, gradually recognized by the industry as an important function in its operations, has during the past thirty years placed exploration on a sound basis. Within a single generation the search for oil has been rationalized by a considerable degree of scientific control, closely paralleling similar recognition of new methods and techniques in other industries. These methods and techniques, some only refinements of methods long used in the science, others radically new in their applications, require a brief description.

As a preliminary to a succession of costly surveys using the different techniques required, it is essential that the results of a regional study be at hand, organizing information available from all useful sources. Careful planning is required. The choice of a single basin for operations may be necessary from among several contiguous basins. In that event, the variety of information on geology, as well as on the practical needs involved in planning and carrying out the succession of surveys, may be considerable. Maps of practically every type are eagerly sought out, and aerial photography of at least parts of the area may be available. A decision as to the ultility of supplementing these by further photographic surveys from the air may be necessary: aerial photographs constitute a tool useful in every type of the succession of surveys required.

Let us assume that the regional studies mentioned above have resulted in a decision to make an exhaustive study of one of the several basins in a region, the choice having been determined by trustworthy reports of a gas seepage near the center of one basin, in addition to otherwise equal promise in comparison with the other basins. The decision requires the study of surface outcrops, to the extent permitted by the nature of the terrane, and the application of geophysics in the following order: (a) survey with magnetometer, (b) survey with gravimeter, (c) survey with seismic equipment. In addition, a core drill may well be made available for coring to moderate depths. The equipment and methods, and their limitations, as well as the results obtained by each survey, are discussed briefly below. The reasons for making the surveys in the sequence indicated will become apparent. For convenience of description and reference, the operations over the area of the basin profile, Fig. 1, are chosen as typical of the basin as a whole.

Study of Surface Outcrops: Case I. Terrane conditions are those of Surface No. 1 of the profile, in a semi-arid region, with nearly perfect outcrops and moderate vegetation. Geologists make rapid progress working from west to east. Complete stratigraphic section from Formation No. 4 upward is gradually built up, and thickness is impressive. Oil seepage on west identified. Definitely not a live occurrence. No sign of gas. Only weathered oil staining in a "dirty" sand unit, No. 5 of the sequence. Faults of structure *C* not recognized. All dips uniformly eastward in sandy shale. Scattered fossils from Formations Nos. 13 and 14 appear to maintain a steady sequence upward from older to younger. Anticline *A* promptly identified. Plane-table survey on Formation No. 15 outlines extensive closed trap, symmetrical and unfaulted. Gas seepage *B* appears highly anomalous. Plane-table survey on limestone, Formation No. 7, appears to place gas occurrence in bottom of syncline to east of Anticline *A*. Seepage accompanied by sulphur water, all on a mud flat with poor exposures. At the close of three months' work in their sector, geologists are advised by the paleontologists working from east to west that all of Formations Nos. 5, 6, and 7 on the west, including the horizon of the seepage-showing, are missing (despite good exposures) on the east margin of the basin (off the edge of diagram) where the succession is Nos. 4, 8, and 9, with a large time-break at the top of No. 4. Geologists become cautious and remain non-committal as to value of Anticline *A*, pending suggestions of basinal depth and form from the geophysical surveys.

Study of Surface Outcrops: Case II. Here terrane conditions of Surface No. 1 are those of moderately dense tropical forest in hilly lowlands. Four- to five-fold more time has been consumed than was required for Case I. After trenching and pitting over a considerable area, the existence and stratigraphic setting of the oil seepage with relation to Formations Nos.

4, 5, and 6 has been identified. Arduous traversing eastward has afforded scanty information on the exposed section. Thickness likely to be considerable, but results purely qualitative. No evidence of fault structure *C*. Vague suggestions of Anticline *A*. Pitting very slow, because of lack of labor. Geologists recommend use of core drill, to follow further confirmation of the structure by geophysical surveys.

After over a year of effort in a tropical setting, always arduous and sometimes dangerous, only a beginning has been made in deciphering the detailed geology of the basin, so necessary in arriving at the objectives *A, B, C,* and *D* of Fig. 1.

Study of Subsurface Formations: Case III. In the average lowland areas where most oil is found, the proportion of the area blind to surface study is about seventy-five per cent. Surface studies, effective over only twenty-five per cent of the various terranes, although fundamental in building up the understanding of the framework of the basins, can make only a modest contribution. Let us consider that Surface No. 2 is the actual topographic surface of the region, underlain by a few hundred feet of flat-lying clay-shales and sandstones, which evidently were laid down long after the tectonic disturbances in the underlying oil-bearing strata had ceased, and long after erosion had etched out the characteristic profile of Surface No. 1. Under these circumstances study of the geology of the surface outcrops has very little utility in efforts to solve the oil-finding problem. Are there other methods for solving the problem? The major contribution comes from two sources, each more or less independent of the thickness of cover which blankets the geology. These are (1) geophysics, which embodies the application of certain principles of physics to the study of the earth's crust, and (2) sub-surface geological studies, which by means of exploratory drilling extend our perception of stratigraphy and geologic structure into the third dimension. Of these two methods, geophysics carries the greater part of the exploratory burden, but they overlap, and the sub-surface studies also are frequently effective in the finding effort.

GEOPHYSICAL EXPLORATION

Only the physical principles embodied in the design and operation of the gravity meter, the magnetometer, and the seismograph are of present proven utility in the field of geophysical exploration for the petroleums. Research has led to the design and trial of various other types of equipment, but their performance is uncertain, and their description will not be undertaken.

The Gravity Meter. The intensity of gravity on the earth's surface varies from station to station as the combined result of four main factors: the centrifugal force of the earth's rotation; the elevation of the station with reference to sea level; the flattened form of the earth at the poles; and the varying densities of the rocks in the earth's crust below the station. Such variations in the earth's gravitational field, though small, are readily measured by a gravimeter, which is able therefore to detect any local dense mass at depth, such as a buried granite ridge or an anticline which lifts a section of older and denser rocks above its normal position. Conversely, the gravity meter detects any local mass of less density than its surroundings, for example, a salt intrusion uplifting heavier sedimentary rocks in the typical salt dome. The quantitative physical measurements can therefore be interpreted in terms of definite geological concepts at depth.

The gravity meter is essentially a precise spring balance, with mirrors and a telescope to permit refined readings. It is enclosed in an insulated case, inside which a constant temperature is maintained. Stations in the field are usually spaced at half-mile to mile intervals. The network of station readings over the salt dome *B*, Fig. 1, when contoured, showed a gravity minimum of roughly circular outline. This comprised the first suggestion that salt in large amounts might be present in the area. Stations over the Anticline *A* showed nothing worthy of note. It was suggested, however, that the density effects of a salt mass at great depth might be neutralized by the relative uplift of heavier materials nearer the surface. Fault structure *C* gave no significant change in the gravitational field. The meter thus had suggested the existence of one of the three features with local deformation of note in this part of the basin.

In addition, interpretation of the overall survey indicated asymmetry in the synclinal setting, the steeper limb being on the west. A deep sedimentary section over the interior of the basin seemed assured.

The Magnetometer. The magnetometer is a precise modernized version of the miner's dip needle, long used in the search for iron ore. It measures the earth's magnetic field from station to station. Wider station spacing is the rule than for the gravity meter, but general procedure, including contouring, is similar. Interest centers in local irregularities. These are generally caused by differences in the magnetic permeability of the rock types, and particularly, in oil-basin geology, by variations in the basement crystalline rocks beneath the sedimentary strata. In places structural trends in the basement have a definite relation to those in the sedimentary section, and a magnetic survey proves useful. The overall survey of the earth's magnetic field in conjunction with that of the gravity meter is useful in determining the relative depths and symmetry of the basin as a whole.

The Seismograph. The seismic method of prospecting is based primarily on the fact that rocks vary in the speed at which they transmit earth waves, the speed of transmission being much greater in hard dense rocks than in the softer types. The rates of transmission vary from around 4,000 feet per second in soft sedimentary formations to as much as 23,000 feet per second in hard and dense igneous rocks. In practice, high-frequency seismic waves are generated in the area under survey by using explosives planted in shallow boreholes. A basic feature of the technique is a photographic record, made automatically on a single moving strip of film, of the instant of the shot explosion and the arrival times of the seismic waves at a series of stations whose positions with reference to the shot point have been determined. The time intervals are recorded to the nearest thousandth of a second. From the records of a series of explosions at different points, the structural attitude of the formations, particularly the harder ones, in a basin can be determined to great depths. The procedure, under favorable conditions, is very effective in identifying the local deformational features, so important in the pooling of oil.

The Interpretation of Readings. If we refer again to Fig. 1, it will be recalled that in Case II the geologists working on Surface No. 1, under terrane conditions of a moderately dense woodland in a low hilly area, had noted suggestions of structure on *A*, but nothing on *B* or *C*. Let us suppose that the gravity meter had reported a gravity minimum at *B*. Then, with leads at *A* and *B*, the seismic equipment was moved in, and it was able to define the structure accurately in four to six months' operations in the difficult terrane. In Case III, with a complete blanket formation under Surface No. 2, although conditions for seismic work were more difficult, the structural settings at *A, B,* and lastly *C,* were finally determined.

The oil pools on Fig. 1 have been lettered in the order of their discovery. It is noteworthy that no comment, even the most speculative, has been offered suggesting the existence of a prospect at *D,* the stratigraphic trap of the basin under consideration. Unfortunately, no survey of the sequence described has any resolving power over such an occurrence. Notwithstanding the great variety of means available, none is effective on this very important target among the group which comprises our objective. About 40 per cent of the oil discovered to date in the United States has been pooled in stratigraphic traps. Discoveries of traps of this character were first made by random drilling in the early days of the industry in Pennsylvania and Ohio, and by the same method have since been repeated in practically every producing area.

The Results Obtained in the Search for Petroleum

THE ESTIMATION OF RESERVES

Estimates of the recoverable oil in place in the porous reservoir rocks of a trap involve the measurement of the volume of porous rock saturated with oil, to which a recovery factor, determined by analyses of cores of the reservoir rock obtained in the course of drilling, is applied. The volume of the reservoir is calculated from the thickness of the oil-saturated part of the porous rock, as determined from the cores, and from its area, as defined by geological and geophysical information. The determination of the recovery factor is based on techniques in core analysis that have been greatly refined by petroleum engineers in the past ten years. These techniques determine the porosity, permeability, and water content of the sands, as well as shrinkage of the oil due to loss of gas content on reaching the surface. The techniques also involve calculations of the degree to which recovery is controlled by the viscosity of the oil and by its dissolved gas. With careful coring and careful analyses of the cores, it is possible to estimate the recoverable oil content within fairly close limits. The above procedure is applied only to that part of the reservoir that has been proved by drilling. The resultant totals can reasonably be called proved reserves, since it is possible to demonstrate their accuracy from close-in data in hand, without long-range assumptions as to continuity and porosity of the sand. Upward revisions are required as drilling proceeds, until the complete extent of the reservoir is proved.

For many years oil producers have cooperated, through committees of technical associations or under government auspices, in issuing annual estimates of reserves, by fields and by the larger political divisions, and most of them are carefully prepared. Errors, possibly as great as 25 per cent in either direction, may be present in the figures for individual reservoirs. When applied, however, to entire states and countries with hundreds of fields, the errors tend to compensate. Unfortunately, the total of these reserve figures, which are widely published in the United States and elsewhere, are often assumed to represent estimates of the total ultimate resources of oil available. This is particularly so when emphasis is placed on the merely ten to twenty years' supply, at the current consuming rate, represented by the proved reserves reported. The public, and frequently leaders in public affairs, thus assume that the estimates include the oil to be discovered in new fields for years to come; and the resultant confusion may inspire ill-advised proposals to meet an assumed impending oil famine.

World proved reserves as of January 1st, 1949, amounted to about 78 billion barrels, according to the estimates set forth in Table 1. Cumulative production since the beginning of the industry to the same date amounted to 58 billion barrels, making total discoveries of petroleum to that date 136 billion barrels. Of this world total of the oil discovered, the discoveries in the United States were 64 billion barrels, those in the Soviet Union 11, in Venezuela 13, and in the Middle East, 35.

THE RELATIVE EXTENT OF EXPLORATION

A clear understanding of the relative extent of world-wide exploration, compared with that in the United States—the present example of maximum effort in oil-finding—is essential for a comprehension of many problems regarding petroleum. As a measure of the effort made to find oil, total capital expenditures might be summarized, or expenditures and area covered by geological and geophysical surveys might be cited; but complete data on these activities are not readily available even on a country-wide basis, much less on a world-wide. Figures on the number of wells drilled are available, however, even for remote areas of the earth. The totals, merely, of the wells drilled are recorded, indicating the number that were productive and the number that were failures. (Failures are commonly designated as "dry holes" in the industry, regardless of the large volumes of water, usually salt water, that may have been encountered before they were abandoned.) The total of productive wells drilled is of little use as an index of the extent of exploration in a given area, for they reflect only economic considerations, such as demand for the oil and a ready market. The number of purely exploratory wells drilled would, of course, be useful, but this figure is not generally available. Even in the United States the figures extend back only a decade, and their reliability is questionable. For instance, it is difficult to get agreement as to how far from proved ground a well must be if it is to be called an exploratory well.

The most readily available index of the *effort* made to find oil is the number of dry holes drilled in a given area, whether within a proved field, or around its margins in attempts to extend the productive limits, or far out in exploratory tests. By citing dry holes drilled we may make comparisons of the *effort* to find oil, to date, in the various countries of the earth. By this measure the contrast between the oil-finding effort in the United States and that in other parts of the world is astounding. The approximate numbers of dry holes drilled since the beginning of the industry for important political entities, as of January 1st, 1946, were as follows: United States, 252,540; U.S.S.R. (prewar area), 2,650; British Empire (prewar area), 5,750; South America, 2,900.[4]

If the number of dry holes drilled in each country, or group of countries, be divided into the square miles of favorable sedimentary area in each, the resultant ratios offer further comparisons of the relative intensity of exploration. The sedimentary area favorable, to greater or less degree, for exploration in the continental area of the United States amounts to 2,392,000 square miles. This is not much over half that of each of the other three units cited: the U.S.S.R. has an estimated 4,338,000 square miles of favorable area; the British Empire, 4,402,000 square miles; and all of South America 4,123,000 square miles. The United States as a whole would therefore have been tested with one dry hole to each 9.4 square miles of favorable sedimentary area; the U.S.S.R., one dry hole to each 1,633 square miles; the British Empire, one to each 765 square miles; and South America, one to each 1,421 square miles. Other countries or regional groupings would show similar or, in many instances, much lower degrees of exploratory effort.

It is beyond the scope of the present chapter to analyze why the effort expended in exploring for so highly prized a commodity as petroleum has been extremely variable, from place to place. Doubtless a complex of human values is involved, including national temperament, mental and social habits, and the resultant social order. The results of exploration in the United States appear to indicate that petroleum is abundant in the sedimentary formations of the earth; and these results may be a criterion of the degree of effort required for effective exploration elsewhere. In other words,

[4] The world-wide figures on dry holes drilled, and the corresponding sedimentary areas involved, are estimates taken from unpublished studies of oil-finding experience. They were in part cited by Wallace E. Pratt in his *Oil in the Earth,* 1942, pp. 40, 65.

Table 1. The world's discovered oil: production and reserves of crude petroleum, by principal producing countries, January 1, 1949.

	Cumulative Production from beginning to January 1, 1949		Annual Production in 1948	Estimated Reserves on January 1, 1949
	Beginning Year	Crude Oil Produced		
A. THE WESTERN HEMISPHERE				
		(In millions of barrels)		
The Caribbean Area				
Venezuela	1909	4,521.6	489.985	9,000.0
Trinidad	1897	371.9	20.202	250.0
Colombia	1921	443.3	24.371	300.0
Mexico	1901	2,348.9	58.370	850.0
U.S.A., Caribbean part [a]	1896	9,746.9	708.079	9,986.0
Cuba	1933	1.9	0.112	3.0
Total, Caribbean Area		17,434.5	1,301.119	20,389.0
South America, except Caribbean Area				
Ecuador	1917	43.8	2.611	30.0
Peru	1896	371.7	14.069	160.0
Bolivia	1930	4.3	0.464	15.0
Argentina	1908	411.5	22.992	250.0
Chile		—	—	10.0
Brazil	1940	0.5	0.143	15.0
Total, South America except Caribbean Area		831.8	40.279	480.0
North America, except Caribbean Area				
Canada	1862	141.1	12.361	500.0
U.S.A., except Caribbean	1859	27,465.0	1,343.354	17,339.9
Total, North America except Caribbean Area		27,606.1	1,355.715	17,839.0
Total, Western Hemisphere		45,872.4	2,697.113	38,708.0
B. THE EASTERN HEMISPHERE				
The Middle East [b]				
The Persian Gulf Oil Province				
Iran	1913	1,938.2	190.395	7,000.0
Iraq	1927	411.6	26.466	5,000.0
Kuwait	1946	68.7	46.547	10,950.0
Qatar	1940	26.0	—	500.0
Saudi Arabia	1936	345.0	142.853	9,000.0
Bahrein	1933	98.9	10.915	170.0
Total, Persian Gulf		2,888.4	417.176	32,620.0
Egypt	1911	120.3	13.173	120.0
Turkey	1948	—	0.025	1.0
Total, Middle East		3,008.7	430.374	32,741.0
The U.S.S.R. [c]	1863	6,127.7	213.600	4,275.0

[a] The parts of the United States here included in the Caribbean Area are Florida, Alabama, Mississippi, Louisiana, East Texas, and the Texas Gulf Coast.

[b] The "mediterranean" area of the Middle East, broadly conceived, includes also the producing areas of the Black and Caspian sea borders in Rumania and the U.S.S.R.

[c] Including Sakhalin.

I. PETROLEUM IN THE GROUND

	Cumulative Production from beginning to January 1, 1949		Annual Production in 1948	Estimated Reserves on January 1, 1949
	Beginning Year	Crude Oil Produced		
		(In millions of barrels)		
Europe, except the U.S.S.R.				
Albania	1933	11.8	0.358	9.0
Austria	1935	51.4	5.711	75.0
Czechoslovakia	1919	4.1	0.187	2.0
France	1918	14.0	0.362	4.0
Germany	1880	92.8	4.447	45.0
Great Britain	1919	4.1	0.323	4.0
Hungary	1937	42.1	3.581	40.0
Italy	1860	3.4	0.065	1.0
Netherlands	1944	5.4	3.444	50.0
Poland	1874	277.4	0.989	20.0
Rumania	1857	1,187.4	29.500	350.0
Yugoslavia	1935	2.1	0.366	4.0
Total, Europe, except the U.S.S.R.		1,696.0	49.333	604.0
Indonesia, British Borneo, and Burma				
British Borneo	1913	157.2	18.588	150.0
Burma	1889	288.0	0.300	50.0
Indonesia	1893	1,174.6	31.906	1,000.0
Total, Indonesia, British Borneo, and Burma		1,619.8	50.794	1,200.0
Other Oil-Producing Countries				
Algeria	1914	0.3	—	1.0
Morocco	1918	0.4	0.096	1.0
India	1888	43.0	1.930	32.0
Pakistan	1914	11.0	0.560	
China	1939	3.3	0.532	20.0
Formosa	1895	1.0	0.017	
Japan	1875	95.2	1.291	15.0
Australia and New Zealand	1935	—	0.002	—
New Guinea (Australian)	—	—	—	50.0
Total, Other Oil-producing Countries		154.2	4.428	119.0
Total, Eastern Hemisphere		12,606.4	748.529	38,939.0
Total, Western Hemisphere		45,872.4	2,697.113	38,708.0
Total, World		58,478.8	3,445.642	77,647.0

Sources: Production figures are from *World Oil,* Vol. 129, No. 4, July 15, 1949, and reserve figures are from DeGolyer and MacNaughton's *Twentieth Century Petroleum Statistics 1949,* with the following exceptions: production and reserves by states, used in the separation of the Caribbean region of the United States from the remainder, are taken from the *Oil and Gas Journal,* Vol. 47, No. 39, Jan. 27, 1949; production in Qatar, Turkey, Burma, Formosa, India, and Pakistan are from Mr. L. G. Weeks' table, accompanying his article, "Developments in Foreign Fields in 1948," *Am. Assn. Pet. Geol. Bull.,* Vol. 33, No. 6, June 1949, p. 1124.

a density of exploration represented by a ratio of not more than 50 square miles of favorable sedimentary area per dry hole drilled would appear necessary throughout the world for even an approximate appraisal of the world's ultimate petroleum resources.

PART II

THE FUNCTIONAL ORGANIZATION OF THE PETROLEUM INDUSTRY

THE FUNCTIONAL ORGANIZATION OF THE PETROLEUM INDUSTRY

BY WILLIAM B. HEROY *

CONCURRENTLY with the growth of the petroleum industry from its primitive origins to its present commanding importance, a highly functional industrial organization has been perfected, extending from the search for the raw materials to the distribution of the products. Each of these functional operations may be carried on by individual units of the industry in widely separated portions of the globe, but most of these units are interdependent to such an extent that the industry as a whole is highly integrated. However separated in space such individual units may be, experience and technologic development have created a striking similarity of pattern in their methods of operation. While local activities may vary because of climate and topography, and individual companies may develop special methods that they regard as more workable and efficient, there is so high a degree of similarity in the way in which the work of the industry is performed that it can be described in terms that are quite generally applicable.

Prospecting has been dealt with in Part I, and we may therefore proceed to the next step—development.

Development

Petroleum and natural gas have accumulated in underground reservoirs, in which these fluids are confined within the porous spaces of the rocks, generally under considerable pressure. The reservoir pressure normally increases with depth, so that in wells 10,000 feet or more in depth, pressures in excess of 5,000 pounds per square inch may be present. As the depth to which wells are drilled has gradually increased, the equipment used for drilling has become heavier, both to increase the effectiveness of the operations and to control the heavy pressures that are encountered when the reservoirs are penetrated.

Two principal methods of drilling oil and gas wells have evolved from nearly a century of industrial experience. Both these methods depend upon the cutting of a circular passage downward through the rocks by some type of drilling bit.

In the percussion, or cable-tool, method (Plates 10 and 11) the bit is suspended at the end of a rope and is alternately lifted and dropped to crush the rock, which is removed by bailing. In the rotary method (Plates 8 and

* Consulting Geologist; formerly Chief, Foreign Production Division, United States Petroleum Administration for War; Past President, American Association of Petroleum Geologists.

9) the bit is fastened to the bottom of a shaft of pipe and cuts its way through the rock by rotation. A heavy fluid, generally composed chiefly of mud, is circulated through the drill pipe and bit to the bottom of the hole, and thence back to the surface through the annular space between the wall of the hole and the pipe. This fluid serves to lubricate the bit and also to remove the cuttings from the hole. Both methods of drilling require the use of heavy equipment installed at the surface, driven by powerful engines.

The cable-tool method of drilling doubtless had its origin in the spring pole. That simple device consisted of a bit hung by a rope or chain from the end of a pole which was so supported that it would move up and down when pulled by a rope. The downward pull on the rope would cause the bit to drop and pound the rock at the bottom of the hole, while the spring of the pole would lift the bit for another blow. From this simple tool, invention and experience have developed the modern cable-tool drilling rig.

The essential components of such a rig are: (1) the derrick; (2) the drilling line and tools; (3) the walking beam, which raises and lowers the tools; (4) the bailing equipment, for removing cuttings; (5) the equipment for handling casing; (6) the equipment for pressure control; and (7) the power plant. Cable-tool methods are used today chiefly for drilling shallow wells. In some areas, where the rocks are hard and the walls of the hole will stand firmly, or where accurate sampling of the rocks penetrated is desired, they are still used for drilling deep wildcat wells, but less than one-tenth of the oil and gas wells now completed are drilled with cable tools.

The principal elements of a rotary drilling rig are: (1) the derrick; (2) the drilling string, consisting of the swivel, kelly, drill pipe, drill collar, and bit; (3) the rotary table, which turns the drilling string; (4) the hoisting equipment, consisting of the draw works, a wire line, the crown and traveling blocks, a large hook from which the swivel is suspended, and elevators for lifting the drilling string; (5) the pumps and hose for circulating the drilling fluid; (6) the valves and other equipment for controlling pressure; and (7) the power plant. In addition, equipment is available for taking cores of the formation, for testing fluids found in the underground reservoirs, for cementing and perforating casing, and for other special purposes.

Through the use of such equipment wells are drilled not only to find new fields but, once they are discovered, to extend the productive area step by step, until the boundaries of the field have been reached and determined. In some fields several productive reservoirs may be found at different depths, each requiring a separate group of wells for its development.

In the early days of the industry wells were spaced very closely, as it was thought that more production was obtained thereby. It was, of course, true that the oil in the reservoir was more quickly obtained: but at the cost of less ultimate production. A better knowledge of the physical laws that control the underground movement of fluids has led to the adoption of wider well spacing (Plate 51). Where a field is under a single ownership, or the owners have combined their interests for unit operation, the optimum spacing may be selected, and the field developed with comparatively few wells. In other cases, where the ownership is divided among many interests, more wells are usually drilled than are necessary to obtain the maximum ultimate production of the field. The East Texas field is an outstanding example of excessive drilling, with an area of about 125,000 acres and over 26,000 wells. Under present practice and with average conditions, a spacing of not less than 20 to 40 acres per well is favored by conservative operators.

As the shallower fields have been found and developed, the demand for more oil has led to deeper drilling, and each year the average depth of the wells completed increases, as fields are developed at greater depths. The average depth of all wells drilled in the United States in 1948 was 3,434 feet. Not only has the average depth of drilling increased, but the greatest depth reached in drilling, and the greatest depth at which oil and gas are found in commercial quantities, have also increased. New depth records are achieved almost every year. As this is written, the deepest producing well is one in the West Poison Spider field, Wyoming, which reaches a depth of 14,309 feet. The deepest hole which has been drilled is in Ventura County, California, with a total depth of 18,734 feet. (Later, in 1949, a hole was drilled to the depth of 20,521 feet, in Sublette County, Wyoming.—Ed.)

Production

It is the function of the producing branch of the industry to operate the wells so as to extract and save the oil and gas present in the underground reservoirs. The natural gas which is associated underground with the oil is held in solution in the petroleum by the reservoir pressure. In some fields, where the amount of natural gas is in excess of the volume that the oil can hold in solution, it may occupy a separate portion of the reservoir overlying the oil zone. A wide variety of conditions are found, ranging from fields where very little gas is present to those where no liquid hydrocarbons are present, and the reservoir is occupied exclusively by gas.

The natural gas occurring with the oil plays an important role in the production of oil. Not only does the dissolved gas tend to make the oil flow more freely, but the pressure of the gas forces the oil upward through the well to the surface. This lifting power of the gas is of great economic importance, because it saves the cost of pumping the wells. It is therefore advantageous to the producer to conserve the gas pressure so that as much as possible of the oil can be produced by flowing. It is good producing practice to return to the reservoir all or part of the gas produced with the oil, in order to maintain the pressure in the reservoir at something approaching the original. This is usually accomplished by drilling some wells especially for the purpose of pressure maintenance.

Where the pressure is sufficient to cause the oil to flow, a string of pipe is set within the casing, which extends downward in the well to a point near the reservoir to be produced. This pipe is called the tubing, and the oil and gas flows through it to the surface. When the well has reached more than one oil or gas reservoir, the space between the casing and the tubing is sometimes used to flow the fluids from a second reservior. As the oil and gas flow upward toward the surface, the pressure decreases, and the gas in the oil gradually comes out of solution. At the surface, the fluids pass from the well into a separator, which acts to free the oil from the gas, so that it may be flowed to the "gathering tanks" located near the well. (Plate 12 illustrates a more centralized variant.)

The natural gas, after separation, is disposed of according to the local circumstances. Where the field is in some remote area, or the volume of gas is small, the gas may be burned at the well. In other fields, the gas is gathered through a special system of pipe lines to a central point, where it is processed to remove and liquify some of the heavier constituents, such as propane and butane. The gas thus dried may be returned to the reservoir to maintain pressure, may be manufactured into carbon black or other products, or may be flowed to main gas-transmission lines and disposed of for public-utility purposes.

In most of the oil fields presently producing no provision has been made for the maintenance of the reservoir pressure, and the wells have, in the course of time, ceased to flow naturally. Under these circumstances, the oil must be lifted to the surface by some artificial means. In the early stages, while the reservoir still has some pressure, gas or air may be introduced into the well under pressure to flow the oil. Eventually, however, some type of deep-well pump is required to bring the oil to the surface. The barrel of the pump is placed within the casing near the bottom of the well. It operates on the principle of the force pump, and is actuated by a long string of rods which are connected at the surface to a walking beam which alternately lifts and lowers them. Power for operating the pump may be supplied from a central plant, or by individual engines or motors.

In the absence of repressuring, approximately half of the total production of an oil field will be obtained by flowing, and the rest by pumping. Some fields have proved to be extremely long lived. However, the production is eventually reduced to a point where the amount of oil produced no longer compensates for the operating costs. The equipment is then salvaged, and the wells abandoned. In some fields it has been found possible, after the primary production of the field has been recovered, to obtain a second crop of oil from the reservoir by the injection of gas or water; this process is called secondary recovery, and may result in the production of a large amount of additional oil.

Storage

In the oil-field landscape a conspicuous feature is the presence of the numerous tanks in which the oil is collected after it flows from the wells. These fields or gathering tanks are used to measure the production of the wells they serve, and to hold the oil in storage until it is convenient to run it through gathering lines to a central group of tanks that may serve one or more fields.

8. Rotary drill. Washington Pool, McClain County, Oklahoma. Looking up derrick from drill floor. In order are: kelly, swivel with mud hose attached, elevators, and traveling block. (Standard Oil Company, N.J., photo by Corsini.)

9. Rotary drill. Ida Stowers No. 1, Natchez, Mississippi. "Going in the hole." Derrick man on "monkey board" loops guide rope around pipe rack, preparatory to guiding pipe to center of derrick. The pipe now in elevator will be lowered into hole as soon as the joint is made up at rotary below. The unlatched elevator will then come up for this new length of pipe. This process continues until the whole "string" is in the hole, and then the kelly joint is "stabbed" into the string and set down in the rotary, and drilling is resumed. (Humble Oil and Refining Company. Standard Oil Company, N.J., photo by Libsohn.)

10. Cable-tool drilling. Cut Bank, Montana. Unthreading the bit with wrenches and circle. (Carter Oil Company. Standard Oil Company, N.J., photo by Rosskam.)

11. Cable-tool drilling. Cut Bank, Montana. Spudding in. Lowering bailer at right is the big spudding bit. (Carter Oil Company. Standard Oil Company, N.J., photo by Rosskam.)

12. Talara, Peru. Gathering lines of producing wells, which carry the oil to a battery station where oil and gas are separated, then pumped to gas plants and refinery. (Standard Oil Company, N.J., photo by Collier.)

13. Aruba. Elevation view of refinery and tank farm of Lago Oil and Transport Company, Ltd. (Standard Oil Company, N.J., photo by Morris.)

14. Storage tanks of 55,000-barrel, 10,000-barrel, and 20,000-barrel capacities. The large tank shown at the left is equipped with a Horton Lifter Roof. The vapor spaces of the two fixed roof tanks, directly to the right of the lifter roof tank, are connected to the vapor space of the Horton Lifter Roof, which serves as central storage unit for all three. Hydrocarbon vapors that would otherwise be forced out of the two cone-roof tanks, or displaced by unbalanced filling and emptying, flow through interconnecting vapor lines into the lifter roof, later returning to the cone-roof tanks. The lifter roof has a 5-foot lift and 39,500 cubic foot capacity. (Chicago Bridge & Iron Company.)

15. Storage. Aruba Refinery of Lago Oil and Transport Company, Ltd. Part of spheroid tank farm, the largest in the world. (Standard Oil Company, N.J., photo by N. Morris.)

16. Abadan refinery, Persia. Aerial view of part of Abadan Island, showing pipe lines entering from the mainland, storage tanks, refinery installations, administration buildings, and loading jetties. (Anglo-Iranian Oil Company, Ltd., photo by Hunting Aerosurveys, Ltd.)

Field tanks are usually small, with capacities ranging from a few hundred to several thousand barrels, depending on the production of the wells. The tanks of the central groups, or tank-farms, have much larger capacities (Plates 14 and 15). In the American practice tanks of 55,000 and 80,000 barrels capacity are most common, and a single tank-farm may provide storage for several million barrels.

Other large assemblies of tanks for the storage of crude petroleum are located at strategic points along main pipe lines, at ocean terminals where oil is delivered for loading on tankers, and at the terminals of pipe lines at refineries (Plates 16 and 18).

In the aggregate the amount of crude petroleum thus held in storage at all times throughout the world is very large. At some periods the amount held in storage in the United States has exceeded 400 million barrels. In recent years United States stocks have ranged between 200 and 250 million barrels. Stocks in other countries are generally lower than before World War II, and a figure of 350 million barrels may represent roughly the total amount of present world crude storage. The storage of these large quantities of oil is required to provide working stocks at refineries, and to take care of seasonal demands. In comparison with current requirements, the quantity of oil normally held in storage is about two months' supply.

Transportation

The great distances over which the bulk of the world's oil production is moved from the fields to the points of ultimate consumption have brought about the development of a system of transportation for the industry which in extent surpasses that of any other industry, and in tonnage is exceeded by that of only one other commodity, coal.

The petroleum industry has adapted itself to geographic conditions by developing four specialized forms of transportation. The consumer of petroleum products is most familiar with the tank trucks that speed over the highways and the tank cars that are hauled by the railroads (Plates 20, 21 and 22). He sees but little of the principal means of transportation: the far-flung pipe lines and the great fleets of tank ships and barges.

In the early decades of the industry, centers of refining were not far from the oil fields, and the products were transported to market in barrels by wagons, by barges, and, later, by rail. The growth of large centers of consumption and the shifting of production to more distant regions have brought about the establishment of large refineries at points convenient to such markets, to which the crude petroleum is moved either by pipe line or by tanker, as circumstances may determine.

The transportation of petroleum commences at the wells, from which the oil is collected by gathering lines (Plate 12), usually of small diameter (2 to 4 inches), and delivered to field storage. This is a group of large tanks located near a pumping station equipped with powerful pumps for moving the oil through the first section of the pipe line. Some fields may be served by a single pumping station, but large fields may be served by several pipe lines, each with its own pumping station and tank farm. If the pipe line is short, up to 50 miles in length, one pump station may be sufficient to move the oil through its entire length. On longer lines intermediate stations are installed at intervals usually of about 40 miles. At each intermediate station the oil is received under pressure, and moved forward by the pumps. One or more tanks are usually installed to hold the oil temporarily, for maintenance uses and in case of line breakage. The usual design of station calls for three pumps, two to be used for regular operation and one as a relief, so that repairs may be made. The pumps are sometimes operated by Diesel engines, obtaining their fuel from the pipe line, but where electric power is available it is often cheaper to operate the station by electric motors. The largest pipe lines operate at pressures up to 800 pounds per square inch. The greatest diameter thus far used is 30 inches, with a capacity of 326,000 barrels per day, in the line from the Abqaiq field to the Ras Tanura refinery, in Saudi Arabia. In the United States crude-oil gathering lines reported to the Interstate Commerce Commission at the end of 1948 totaled about 47,000 miles and trunk pipe lines about 63,000 miles, a total of about 110,000 miles of line. The Soviet Union has the second largest pipe-line system, with a total mileage of oil pipe lines of about 3,500 miles.

Some oil fields are not reached by pipe lines but depend on railway transportation. The oil is gathered and moved to a tank farm convenient to the railroad, at which a long siding is installed for convenience in filling the tank cars. A system of overhead pipes is provided so that a number of cars can be loaded at the same time. Such an arrangement is called a "loading rack." The cars are then moved in trains to the refinery, where similar racks are installed for emptying the cars. The tank cars vary in size in different countries; those used in the United States average about 200 barrels in capacity. Similar facilities are installed at refineries for loading tank cars with refined products (Plate 20). Most of the tank cars which may be seen moving over the rails are transporting such products as gasoline and fuel oil to bulk marketing stations.

Some tank trucks are used for transporting crude oil from a few small and remotely situated oil fields, but they are used mostly for distributing petroleum products from bulk stations to service stations, houses, and industrial plants.

The transportation facilities mentioned above are those used for the movement of petroleum and its products on land, but transportation by water is also a major function in the work of the industry. The types of equipment range from small barges used on rivers and other inland waters to deep-water tank ships of great size (Plates 96 and 98). Some of the smaller vessels are self-propelled, but they are usually moved by tugs, which may haul a string of several barges. Barges for oil were probably first used on the tributaries of the Ohio River in western Pennsylvania. They are still in use in large numbers on the principal rivers and canals, although the amount of petroleum so transported is small, in comparison with that hauled in tankers.

The first vessel constructed especially as a tank ship is reputed to have been built in England in 1886. Its hull was divided by bulkheads into compartments, so that different kinds of oils could be separated and so that the ship could be stabilized. Experience through the years has greatly increased the size and efficiency of tankers. The modern vessel has comfortable quarters for officers and crew, and is provided with numerous safety devices to prevent and control fires and explosions of its inflammable cargo. It is equipped with powerful pumps for the rapid discharge of cargo, and some, even of the largest capacity, can be emptied in less than twelve hours. Tankers today are being equipped with the latest navigation aids, such as radar, gyroscopic compasses, and sonic depth finders.

The largest ships now in service have a capacity of about 240,000 barrels, a displacement of 28,000 tons dead weight, and a speed of better than 16 knots, and larger are building. The average tanker is somewhat smaller, with a displacement of about 16,000 dead weight tons.

A parallel function to the transportation of crude petroleum is that of natural gas. This branch of the industry has reached its greatest development in the United States, where a system of natural-gas transmission lines has been developed that, with those now under construction, covers the country from coast to coast. The shorter lines may be operated by taking advantage of the pressure at which the gas is produced from the well, but the larger and longer lines have repressuring stations at intervals.

Altogether, over 259,000 miles of natural-gas pipe lines are in operation in the United States (figures of December 31, 1948), of which field and gathering lines account for about 29,000 miles. By the conversion of the "Big Inch" pipe line, built during World War II for oil transportation, to natural-gas transmission, continuous movement of natural gas from Texas to New Jersey, and to Staten Island, New York City, is now in effect. The completion of a new pipe line, the largest line yet undertaken—30 inches in diameter in the portion from New Mexico to the California border, and 34 inches in diameter from there on to San Francisco—will connect the gas-producing areas of the mid-continent area with the Pacific Coast.

Outside the United States the largest natural-gas transmission operation is that of the Soviet Government. The longest line is believed to be the line from Saratov to Moscow, with a diameter of 12¾ inches and a length of about 500 miles (Plate 76).

Refining

That part of the petroleum industry which relates to the conversion of crude petroleum into the hundreds of products so important to our modern way of life is essentially a manufacturing process. The great majority of refineries are relatively small, capable of treating less than 10,000 barrels of crude oil daily, and making a few well known products, such as gasoline, kerosene, and fuel oil. But some are immense installations, with capacities of several hundred thousand barrels per day, and, in addition to the major products, manufacture a wide range of lubricants, waxes, solvents, and special chemicals.

What a refiner can produce is basically dependent on the characteristics of the crude petroleum available to him. While we speak of petroleum as though it were a simple substance, it is actually a highly complex mixture of components, most of them liquid at ordinary temperatures. So numerous and intricately intermingled are these components that many have never been isolated in pure form. The crude oils from adjoining fields, and even from wells in the same field, may differ appreciably; and the difference between the crude oils of different districts or provinces may be much greater.

In practice, the crude petroleums from many fields in the same district, having generally the same characteristics, are mingled and transported to the refineries through the same pipe lines. Crude oils of the principal types, the paraffinic and the asphaltic, are usually segregated, and oils having special characteristics are run and processed separately.

What a refiner produces is also dependent on the requirements of his market. In some areas the greatest demand may be for gasoline, and the refining processes will be designed to obtain the highest yield of this product. Other refiners may seek to obtain the greatest production of lubricants. The particular processes that may be utilized are thus determined both by the nature of the crude supply and by the product demand, as well as the available refining technology.

The refining of petroleum is now largely a chemical industry, and the processes in use represent one of the greatest achievements of applied chemistry. In the beginning, however, the industry was developed on the basis of a very ancient industrial process, distillation. This process had been extensively used in a predecessor industry, the manufacture of "coal-oil," and was adapted by the earlier refiners to the treatment of petroleum. Commencing with the simple still, the industry has gradually improved its equipment and processes until, after a century of effort, the catalytic cracking unit, with its great size and complex controls, marks the present high level of achievement in this branch of petroleum technology.

The basic refining processes are distillation and cracking. The former, a simple physical separation of the oil into lighter and heavier fractions, is accomplished by heating the crude oil in a shell or pipe still. The shell still is simply a boiler, made in various sizes and shapes, in which the oil is heated. As the temperature rises, the more volatile portions of the crude change to vapor, which passes out of the still into a condenser, where it is cooled back to a liquid. By regulating the temperature of the still, the proportion of the crude that is driven off can be determined, and by running the oil through a succession of such stills operating at different temperatures, the crude oil is separated into a series of products. After all of the volatile products are driven off, petroleum coke remains.

The pipe still is a modern adaptation, in which the oil is flowed continuously in pipes around a heated chamber and thereby raised to the desired temperature, usually about 800°F. The oil then passes into a fractionator or "bubble tower," the function of which is to separate the heated oil into fractions or "cuts." This tower is a steel cylinder, set vertically, with a height ten to twenty times the diameter—many are more than 100 feet tall—and spaced vertically within the cylinder are horizontal perforated trays. As the petroleum vapors rise through the tower, they cool, and the resulting liquids collect selectively on the trays and are drawn off. This process is usually carried on at normal atmospheric pressure.

The heavier fractions drawn from the tower may be subjected to further treatment by another variant of the distillation process, in which the evaporation occurs under a partial vacuum. This process is used for the separation of products which normally vaporize at higher temperatures, and avoids their decomposition.

The cracking process of refining crude petroleum, on the other hand, is essentially one in which the fluid is intentionally subjected to high temperatures and pressures in order to cause its chemical decomposition. The motive for the development of this process was the desire to increase the proportion of gasoline that could be recovered, by breaking or "cracking" the molecules in the heavier portion of the crude oil into the lighter and more volatile molecules required for gasoline. The cracking process, as originally developed, made use only of heat and pressure, and is now referred to as thermal cracking. It required the development of an important unit, the pressure still, in which the temperature may be raised to a range of from 850° to 1,200°F, and the pressure may range from a few pounds to as much as 1,000 pounds per square inch. Depending upon the pressures and temperatures used, the cracking may be carried on while the fluid is in either a liquid or a vaporous state: the first is called liquid-phase thermal cracking, and the second, vapor-phase thermal cracking. After cracking, the vapors are processed in a fractionator as though normal distillation had occurred. By these developments, the amount of gasoline recovered from crude petroleum has been increased from 20 to as much as 45 barrels of gasoline from 100 barrels of crude.

During the past few years a type of cracking process has been developed in which a catalyst is used to remove the coke formed during the cracking operation. In a broad sense, a catalyst is a substance which, while remaining itself unchanged, facilitates a chemical reaction. In the catalytic cracking of petroleum the substance used is generally a pulverized, clay-like material, or a specially manufactured bead or pellet. The cracking takes place in a large chamber, called the reactor. Several types of the catalytic process have been put into operation, the principal difference being the manner in which the catalyst is utilized. These sub-processes may be divided into the fixed-bed group and the moving-bed group.

In the fixed-bed group the catalyst is in the form of pellets, which are placed on trays in the reactor, a large vertical steel drum. Several reactors are used, each being operated intermittently, but with their operation timed so that a continuous flow of cracked vapors is sent to the fractionator. While the reactor is on stream, the vaporized charge stock is passed through it, and as the reaction takes place the excess carbon is deposited on the catalyst in the form of coke. In a short time the catalyst becomes fouled, and is then taken off stream and regenerated by burning off the coke. It is then ready for another charge.

In the moving-bed type the petroleum is vaporized, and the catalyst in the form of fine powder is poured into the vapor stream, both being at a temperature of from 1,000° to 1,200°F. Vapors and catalyst are violently agitated, and in the chemical reaction that takes place excess carbon is deposited on the fine particles of catalyst, which pass in suspension in the cracked vapor out of the reactor into a separator. Here the blackened catalyst is removed from the vapor, which passes to the fractionator, while the catalyst passes to a regenerator chamber, where it is heated to burn off the carbon, and is then cycled back to continue the process. The catalytic cracking unit is an immense chemical machine; some of the moving-bed type are more than two hundred feet high. The chambers and fluid passages are of unprecedented size, and hundreds of tons of catalyst are required for their operation.

Several variants of both fixed- and moving-bed types of catalytic cracking have been developed, and catalytic principles have been applied in other specialized processes for the treatment of petroleum and its products. In the hydrogenation process, for instance, hydrogen is chemically introduced into the charge, through the use of catalysts, to convert heavier oils into lighter products. Lower-grade gasolines are catalytically cracked or reformed to produce gasoline of higher quality. Bauxite and other catalysts are used to remove objectionable sulphur compounds from gasoline. In alkylation processes a catalyst, usually sulphuric or hydrofluoric acid, is used to combine paraffin and olefin to produce alkylate, an important constituent of aviation gasoline.

In the cracking processes that have been described or mentioned, the purpose is to convert heavier constituents of petroleum into lighter and economically more valuable products. In the other direction, processes have been developed whereby natural gas and petroleum-refinery gases may be treated to produce liquid hydrocarbons. This chemical change, called polymerization, consists in combining the small molecules of gases into larger ones. It is carried on in the presence of catalysts, under pressures ranging from 500 to 1,000 pounds per square inch, and at temperatures of 350° to 450°F.

In the early days the natural gas produced in oil fields was wasted, blown into the air. Many of the picturesque gushers are testimony of ignorance concerning the function of gas in oil production. But the knowledge gained during the past twenty-five years concerning the value of gas in oil production, the commercial success of long-distance gas transmission, and the advantages of gas as a raw material for manufacture, have created a great new sister industry.

To transmit natural gas under high pressure, say 700 to 800 pounds, through gas lines, it is first necessary to remove those gases which would become liquid under pressure. The common compressible gases are propane, butane, and natural gasoline. Propane and butane, known as Liquified Petroleum Gases, or LPG, are extensively employed in the manufacture of chemicals and as industrial and domestic fuels. Natural gasoline is blended with refinery gasoline to produce motor fuel.

After the various products have been separated in the fractionator, each may be subjected to further treatment. Some of those used to improve the quality of gasoline or to provide components for it have just been mentioned, and blending of these components to provide a motor fuel of the desired characteristics is an important step in refining operations (Fig. 55). The further treatment of the fractions that contain lubricants has been carried to a high degree of perfection; through vacuum distillation, dewaxing, acid treatment, solvent refining, and other special methods, a large number of lubricating oils have been developed. Waxes recovered from heavier fractions are used for the manufacture of paraffin, petroleum jellies, and vaselines. Some of these are used in compounding greases and other solids used as lubricants.

Most of the heavier fractions produced in refining are, however, manufactured into burning oils, such as kerosene, distillate fuel oil, and residual fuel oil. More than half of the world's petroleum is converted into these products. The residual fuel oil is the heaviest liquid fraction produced, and represents the heavy remnant after all the lighter fractions have been removed. A still heavier fraction, asphalt, is produced from some crudes, and from some refining processes a large amount of petroleum coke results.

During the war twelve plants were erected in the United States for the manufacture from petroleum of butadiene, the most important raw material for the manufacture of synthetic rubber; and many other special products were also manufactured at petroleum refineries for various military requirements.

A complete refinery, capable of producing most of the products discussed, is a large plant, covering hundreds of acres. The largest refinery now in operation is in Iran, and has a capacity of 500,000 barrels of crude oil per day, while the Netherlands West Indian islands of Aruba and Curaçao have refineries with crude-charging capacities of 400,000 and 356,000 b/d, respectively. In the United States there are about 375 refineries in operation (December 1948). The largest, in Louisiana, has a capacity of 235,000 b/d, and the three next largest, in Texas, have crude-charging capacities of 230,000, 206,000 and 190,000 b/d, respectively. The world's refining capacity at the end of 1948 has been estimated at just under 11 million barrels of crude oil per day, as compared with about 9 million in 1947.

Distribution

As in other industries, distribution is concerned with the movement of the various products from the processing plants to the ultimate consumer. In the case of the petroleum industry this is a stupendous task, requiring for the world as a whole the transportation and delivery of about 465 million metric tons in 1948. The volumes of the inter-regional movements of crude petroleum and of petroleum products in 1938 and in 1947 are indicated in Figs. 2–5.

The distribution of petroleum products is the branch of the industry with which people everywhere who consume petroleum products have contact. The users range from the *Queen Mary* to the *Sunset Limited,* from the owner of an expensive automobile to the humble worker who lights a candle before a remote shrine.

Compared to the complex facilities that produce and refine crude oil, those of the distribution system are comparatively simple. The process commences, for the major products, with movement in bulk from the refinery to primary distributing points. This may be accomplished by facilities that are essentially the same as those used for the transportation of crude petroleum. Tanker loads of gasoline and fuel oil move overseas from seaboard refineries to distant terminals near large centers of consumption. Such rivers as the Thames, the Scheldt, and the St. Lawrence have along their banks great docks equipped with a maze of piping that communicates with extensive tank farms for storage. Here products that may come from several distant sources are held in bulk and gradually distributed to smaller depots.

In recent years, especially in the United States, the construction of pipe lines built solely for the transportation of products has made rapid progress, and over 12,500 miles of line are now completed. These generally run from refineries near sources of production to bulk terminals near large marketing centers. Some of the product systems are of wide extent, reaching from the Gulf of Mexico to the Great Lakes, and thence to the eastern seaboard.

More important, however, than pipe lines for the distribution of products are tank cars. In Western Europe and in the United States much of the refined output is loaded directly into tank cars for railway shipment. In the United States more than 50 million tons of products are moved annually in tank cars, of which there are about 150,000 in petroleum service.

Canals and waterways, both in Europe and the United States, are also important avenues for moving petroleum products. The Volga, the Rhine, the Danube, and the Mississippi are outstanding water routes, on which large volumes of products are moved by fleets of barges.

By these various means, major products such as gasoline and kerosene move to marketing points. A large city will have numerous bulk terminals, operated by several units of the industry, each with groups of storage tanks assigned to different products. Depending on the means of transportation available, these storage depots will be located on railway sidings or water fronts. Smaller cities and towns will commonly have several bulk stations with smaller and less diversified tank equipment. In the United States even villages along railway lines will have a tank or two from which products are distributed locally.

In most areas where good highways have been constructed, tank trucks are used to transport products from bulk stations to retail outlets; and the local service station, with its installation of gasoline pumps, is generally kept supplied by periodic visits from such trucks. The oils used for heating buildings are delivered directly to the householder or the plant by similar equipment. Farms in rural areas may have their own storage tanks, which are supplied with tractor fuel by tank truck deliveries. Large tank trucks may operate over hundreds of miles, to transport products direct from refineries to bulk stations or large consumers; their place in the distribution system is largely determined by local economic conditions.

Parallel to the system of bulk transportation and distribution is another method of marketing petroleum products, the use of small containers or packages. The most commonly used containers are metal barrels or drums with a capacity of approximately 50 gallons, and cans with a capacity of about 5 gallons (Plates 92 and 95). In them gasoline and kerosene are shipped to areas where distribution by bulk methods is impractical. By small trucks or animal transport, and even on the backs of men, such barrels and cans are delivered into remote areas, so that the community is indeed inaccessible that does not today obtain some petroleum product for light or heat.

Barrels and cans are also used for the distribution of the more valuable products. Lubricating oils are generally packaged at the refinery, either in heavy drums or in smaller tins, to prevent contamination, and at the service station the motorist today usually obtains oil for his crankcase

---

FIGS. 2–5. The world's inter-regional trade in crude petroleum and in petroleum products, in 1938 and in 1947. The width of the arrows is proportionate to the volume of exports from each region, as indicated by the scale in the lower left-hand corner of Fig. 3. Bunker usage is treated as part of the local demand of each country, but the movements of bunker fuel from supply points to loading stations are included in the export tables. The actual quantities and their destinations are listed in Table 30 in the Appendix. The regions indicated by numerals in the cartograms are: (1) Alaska and Newfoundland; (2) Canada; (3) United States; (4) Mexico; (5) the Caribbean area except United States and Mexico; (6) other South America; (7) Europe, except U.S.S.R.; (8) U.S.S.R.; (9) west Africa; (10) north Africa; (11) south and east Africa; (12) the Middle East; (13) other Asia; (14) East Indies and Oceania.

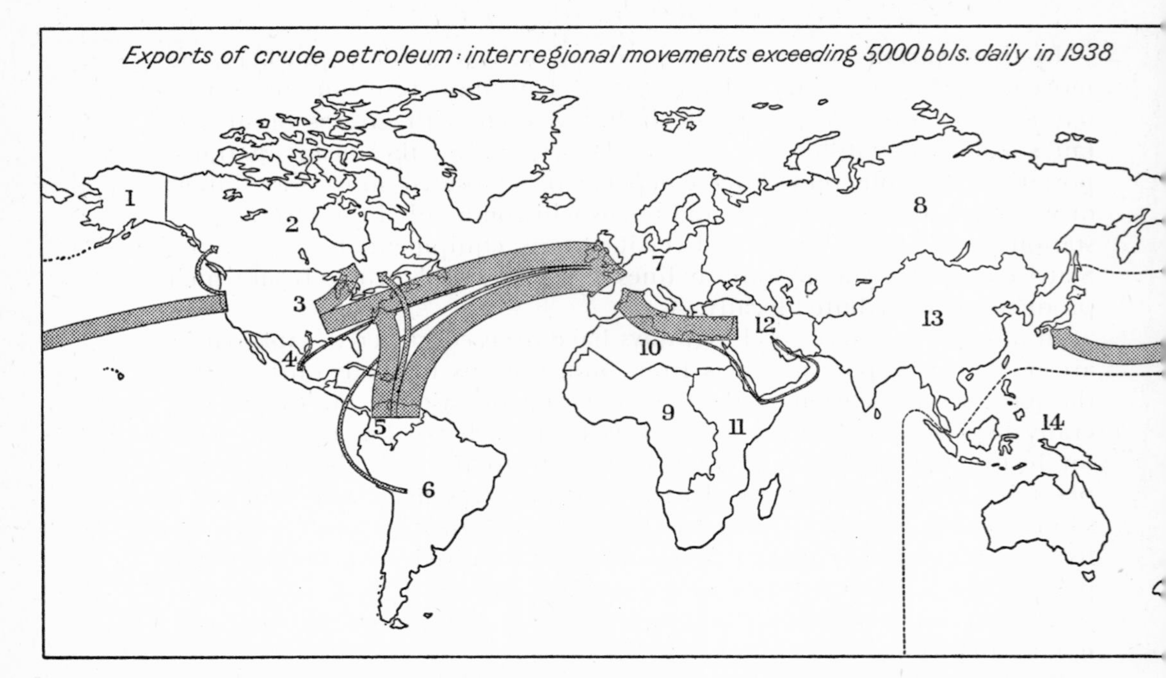

FIG. 2. World trade in crude petroleum, 1938. See page 39 for legend.

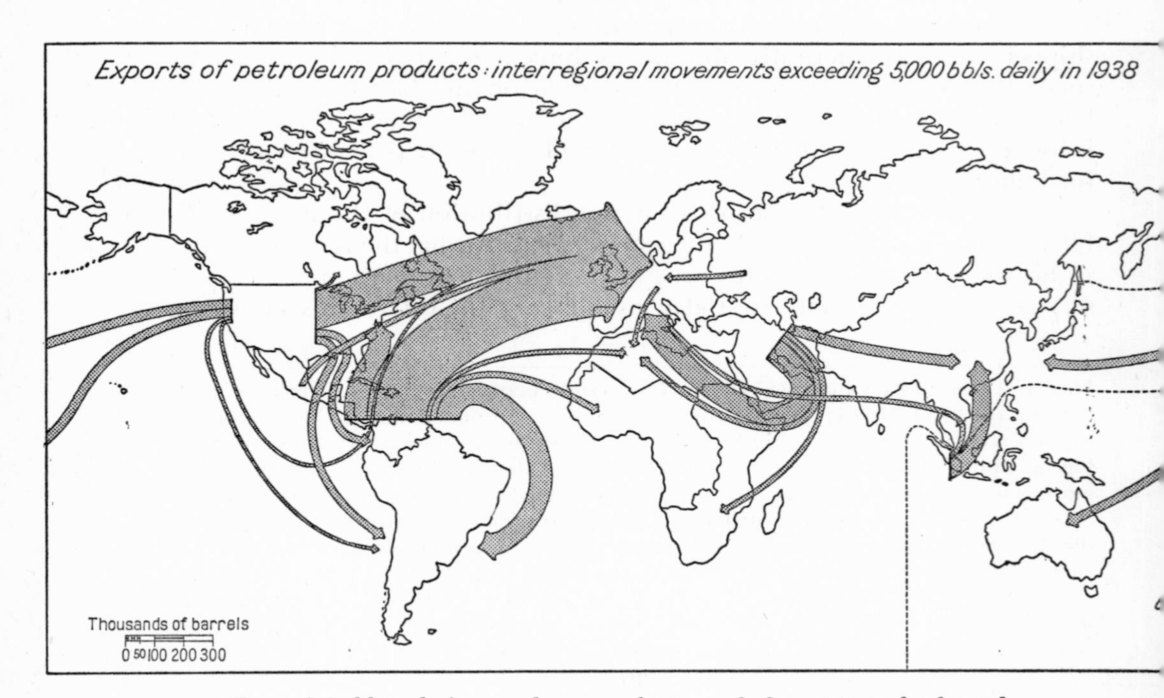

FIG. 3. World trade in petroleum products, 1938. See page 39 for legend.

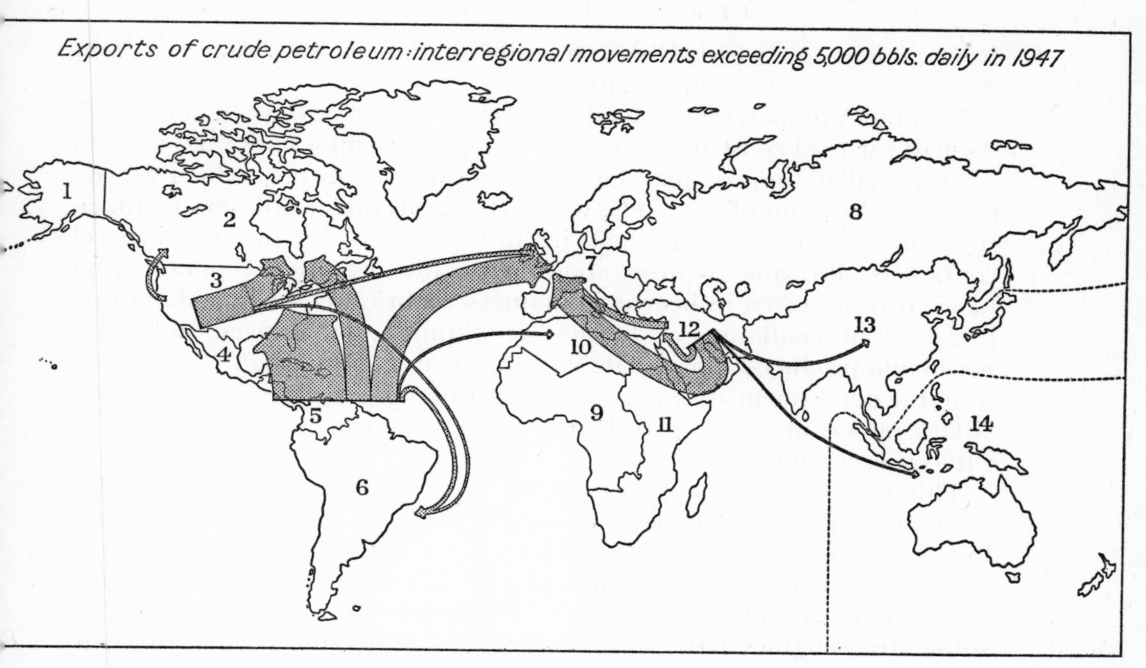

FIG. 4. World trade in crude petroleum, 1947. See page 39 for legend.

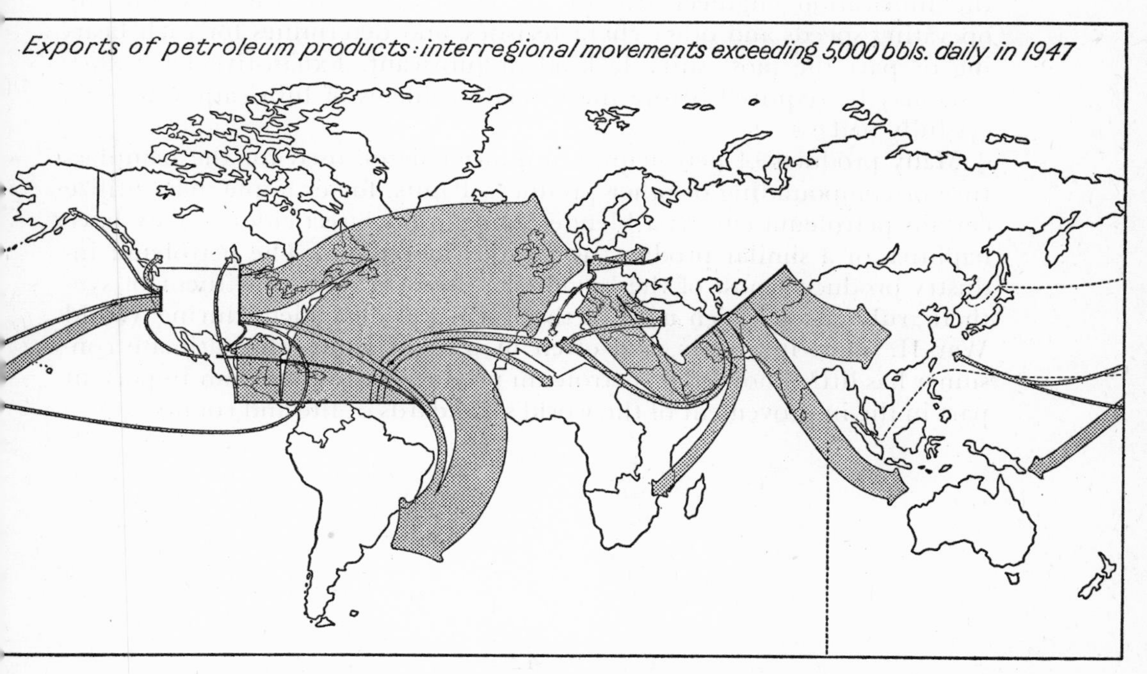

FIG. 5. World trade in petroleum products, 1947. See page 39 for legend.

from a quart can which has been sealed at the refinery and marked to show the exact grade and quality.

For other products special methods of distribution have been developed. Asphalt for road construction may be shipped in tank cars heated to keep it liquid (Plate 22), or may be poured into light drums, which are removed at the point of use. LPG, which is gaseous under ordinary temperatures and pressures, can be held in liquid form under pressure in steel containers. Propane requires about twice the pressure of butane, and hence is transported in heavier containers. Lubricants for household use, packaged in small cans or bottles containing a few ounces, and other petroleum products, appropriately wrapped, are sold through retail stores; from the department stores of the great cities to the shelves of the smallest village merchants, some petroleum products are carried in stock, almost without exception.

This vast system of distribution has come into existence in part because the world has need of the many products that petroleum can supply. But the enlargement of the demand for petroleum has resulted from the constant efforts of the industry to find out what the needs of the world are, and then to supply products that will better serve their purpose. As automotive engines have improved, for example, the refiner has constantly sought to improve the quality of gasoline, in order to give more mileage, smoother operation, and unobjectionable odor. Similarly, in the field of machinery, where each unit must be effectively lubricated to function properly and to operate with a minimum of wear and repair, the lubrication engineer studies the construction of the machine, its operating speeds and other characteristics, and determines for each bearing or part the most suitable kind of lubricant. Exhaustive laboratory tests may be required before the special problems of lubrication are successfully met.

Many products of petroleum are intermediates, used for the manufacture or compounding of other products. Paints, for example, may utilize certain petroleum cuts as a blender, and liquid insecticides usually have naphtha, or a similar product, as a major ingredient. The petroleum industry produced most of the butadiene, the principal constituent of synthetic rubber, on which most of the world was dependent during World War II. Thus indirectly, and often in ways of which the ultimate consumer has little knowledge, petroleum products have played an important part in the improvement of the world's standards of life and comfort.

PART III
THE WORLD'S PETROLEUM REGIONS

1. THE CARIBBEAN AREA AS A WHOLE

BY WALLACE E. PRATT *

FOR the purposes of this chapter, the Caribbean area may be defined to include the eastern part of Mexico, Central America, the West Indies, Trinidad, Venezuela, Colombia, and, in the United States, the states of Florida and Louisiana, together with the southern parts of Georgia, Alabama, Mississippi, and Arkansas, and the East Texas and the Gulf Coast districts of Texas. A prospective source of petroleum equally promising as the lands of the Caribbean area is the continental shelf that borders these lands beneath the waters of the Caribbean Sea and the Gulf of Mexico.

The Caribbean Sea and the Gulf of Mexico taken together form a single great land-locked body of water. The lands surrounding these seas and the shallow-water margins of the seas themselves constitute the greatest petroleum province in the Western Hemisphere, and, with the exception of the recently developed petroleum resources of the Middle East, they promise to yield more oil eventually than any other single province on earth.

Ever since the early years of the present century the Caribbean area has been the dominating factor in international commerce in oil; continuously during this period the bulk of the oil moving in world trade has originated in this area. In the earlier years Trinidad and the Gulf Coast region of the United States were the principal exporters. Later, Mexico took the leading position, which it, in turn, surrendered to Venezuela, the present leader, more than twenty years ago. Colombia became a significant exporter of petroleum in the early 1920's, and has consistently maintained this position ever since. All these countries possess large remaining petroleum resources, including, in the aggregate, an immense volume of proved reserves. The other lands of the Caribbean area—the Central American republics and the islands of the West Indies—are not as yet commercially productive (Cuba has produced a little oil for internal consumption for many years past), but they may become so. They are marked by a number of natural seepages and other favorable evidences of petroleum which have never been systematically explored.

The world's total consumption of petroleum had amounted to more than 58 billion barrels by the end of the year 1948. It is a remarkable circumstance that of this total consumption, about 46 billion barrels or

* Formerly Director, Geological and Geophysical Exploration Dept., Member, Executive Committee, and Vice President, Standard Oil Company (New Jersey); Past President, American Association of Petroleum Geologists.

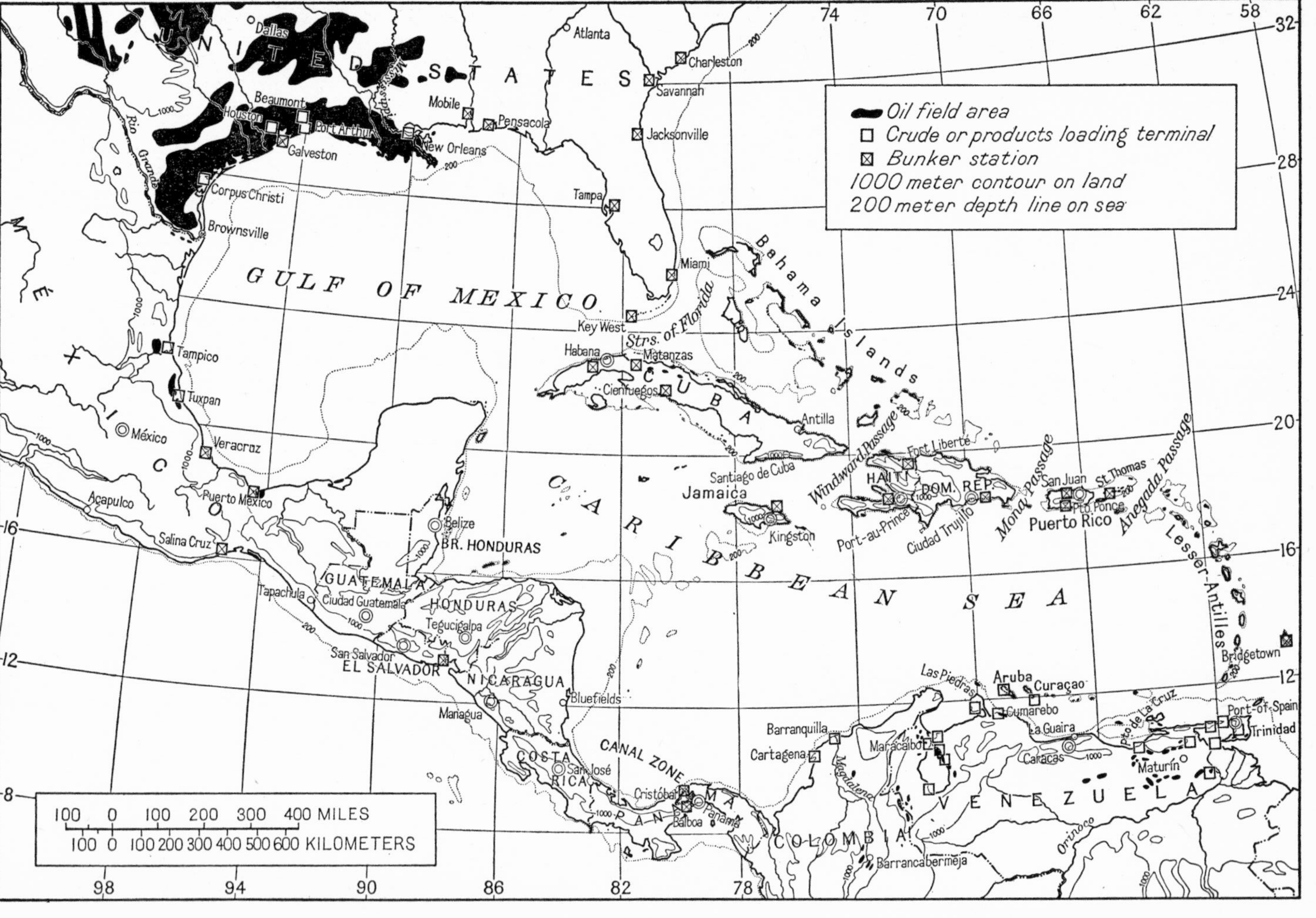

Fig. 6. The Caribbean petroleum province: oil-field areas, with marine loading terminals and bunker facilities. The bunker stations shown are those listed in Hurd's *Ports of the World* (London, 1948). The extensive oil-field areas shown in the United States portion of this map represent a broader definition of "oil-field area" than that employed for the remainder of the Caribbean petroleum province; difficulties of reproduction have made it impossible to show the very numerous United States fields as separate entities, as they are shown in Fig. 19.

17. Thermofor catalytic cracking plant built for the Leonard Refineries, Inc., at Alma, Michigan. Plant has 3,000 b/d capacity and went on stream March 18, 1947, marking a milestone in catalytic cracking progress. It was the first postwar cracking unit built to meet the peace-time competitive needs of the small refineries. (Houdry Process Corporation.)

18. Aruba refinery of Lago Oil and Transport Company, Ltd. General view of loading lines in tank field. (Standard Oil Company, N.J., photo by Morris.)

19. Aruba refinery of Lago Oil and Transport Company, Ltd. "Cat" plant and high-octane units. (Standard Oil Company, N.J., photo by N. Morris.)

20. Tank-car loading racks at night at an oil refinery. Cars are loaded by overhead pipes through the dome. (Union Tank Car Company.)

21. Tank car being steamed through the dome. The use of live steam for a minimum of two hours is required to free the car of any explosive gases or vapors. Before workmen enter to clean or repair, an explosimeter is used to determine whether the car is safe, and then flaming waste is lowered through the dome as a final precaution to test for possible explosions. (Union Tank Car Company.)

22. Blue Mounds, Wisconsin. Asphalt or road oil unloading through a Cleaver-Brooks Bituminous Pumping Booster to a tank truck. The tank truck will subsequently supply the asphalt for road building or similar work. The asphalt in the tank car is first softened to a flowing viscosity by steam-heater pipes permanently installed in the car. In the picture, the asphalt is being unloaded through the bottom outlet valve of the tank car into the Cleaver-Brooks Booster, where it is heated to a higher temperature than can be accomplished with the tank-car heater pipes, either for speedier unloading or for a particular temperature required for the work involved. The asphalt passes through the Booster back into the tank car through the dome, and this circulation is continued until the entire contents of the tank car has been heated to the desired temperature. (Union Tank Car Company.)

Monagas, Anzoátegui, and Guárico. Loading-line fixtures may be seen beside the waiting tanker on the right. Supplies of pipe for the oil fields and pipe lines are piled in the foreground. Before the development of the oil fields only a small fishing village existed here. There is now a rapidly growing city, and intensive development of agriculture and varied industries. (Creole Petroleum Corporation.)

24. Western Venezuela. Lake Maracaibo shoreline. (Standard Oil Company, N.J., photo by Vachon.)

25. Western Venezuela, Lake Maracaibo. Contrast between the old and the new. In the foreground the village of Pueblo Viejo, built on piles in the water of Lake Maracaibo. Such villages, found by the Spanish discoverers, originated the name of Venezuela (Little Venice). On shore, the Shell Oil Company's new industrial camp of Bachaquero. In the distance, the new development drilling in Bachaquero, part of the Bolívar Coastal Field. (Photo by G. Zuloaga.)

78% has been supplied by the Western Hemisphere. Of the Western Hemisphere supplies, about one-third has come from the Caribbean area, which has a record of past production amounting to about 17½ billion barrels. Its proved reserves on December 31, 1949, were estimated at over 20 billion barrels, out of a world-total of 78 billion. Of the proved reserves within the United States, which were estimated at about 28 billion barrels at the end of 1948, about 10 billion barrels, or a little more than one-third of the total, were situated in the territory adjacent to the Gulf Coast within the Caribbean area as defined above.

The Caribbean area is the Western-Hemisphere counterpart of the classic Mediterranean and associated inland seas in the Eastern Hemisphere. It is not without significance that the earth's two greatest mediterranean areas—of "seas in the midst of the land"—should also be the sites of the earth's greatest known petroleum provinces. Typically negative mobile sectors of the earth's crust, squeezed and depressed between the rigid positive buttresses that constitute the central framework of the adjoining continental masses, these areas of land-locked seas—major earth features persisting through much of geologic time—have provided an ideal environment for the generation of petroleum.

The Caribbean area is fortunately situated to serve as a source of liquid-fuel supply for the Western Hemisphere, and the petroleum it produces has been a principal article of commerce for the participating countries. Except for the activities of the petroleum industry, the tropical coastal lowlands of Mexico, Colombia, and Venezuela would scarcely have experienced the development and the improvement in health and living conditions that have come to them. In each of these countries the petroleum resources are owned by the central government, and in each the royalties and other taxes levied on the concessionaires who have carried out the exploration for and development of the petroleum have been major sources of national revenue. At the same time, the export of petroleum and petroleum products has made possible imports of much needed capital goods and manufactures, which have contributed materially to higher standards of living in the petroleum-exporting countries.

The development of the petroleum resources of the Caribbean area is far from complete; in many promising localities it has hardly been initiated. The enterprise is one that demands prodigious expenditures of capital, manpower, and technical skill, effectively and enthusiastically dedicated to the task. In the past these requisites have been supplied principally by the United States citizens who explored and developed the previously unrecognized petroleum resources of Mexico and Colombia, and by the British, who, in association with the Dutch, performed the same service in Trinidad and Venezuela. Each of these groups has remained active, and has contributed much to the further development of these resources in all of the producing countries.

The further development of the petroleum resources of the Caribbean area is of prime importance to all the peoples of the Western Hemisphere, and it would be fortunate if this enterprise could be carried forward more rapidly and expeditiously than in recent years. A principal impediment to its progress has been the divergent views of the governments, on the one hand, and the foreign concessionaires, on the other, as to their respective rights and interests in the joint venture. No people finds it easy to be tolerant of, or even fair to, the foreigner. Moreover, national pride in Latin America would be tremendously gratified if the sovereign state itself developed the resources of the nation. Yet governments seem ill-fitted for an undertaking of this character, and very few governments can afford to assume the financial risks involved.

The best interest of the peoples of the Western Hemisphere demands that this apparent conflict of interest be resolved. That there exists a large area where the true interests of both governments and foreign oil companies run parallel has been amply proved by the experience already gained in the Caribbean countries. This area of common interest must be defined and expanded as widely as possible, if the petroleum resources of the Western Hemisphere are to serve society most advantageously.

In the following pages the development of petroleum in Venezuela, Colombia, Mexico, and Trinidad will be discussed in separate chapters. Although that part of the United States bordering the Gulf of Mexico belongs geologically to the Caribbean area, the geography of the petroleum resources of this part will be discussed along with the rest of the United States in the section devoted to North America.

Exploration in Central America and the West Indies, aside from Trinidad, has been carried on fairly widely, though not intensively, in recent years. Test wells have been drilled in Costa Rica, Nicaragua, Panama, Haiti, the Dominican Republic, and Cuba; but commercial production has been obtained only in Cuba. Exploration by geological and gravimetric survey has been made in Guatemala and the Bahamas. In the latter an aerial magnetometer survey covering 84,358 square miles was carried out in 1947, and gravimetric mapping of shallow-water areas has been accomplished with the aid of diving bells and of tripods.

The surface evidence of the occurrence of petroleum on the island of Cuba is very impressive. Copious seepages of light oil, the distribution of which appears to be related to igneous intrusions which penetrate Jurassic or Cretaceous sediments, are to be observed in widely separated localities over the island. While several deep wells have been drilled in Cuba, exploration to date has been far from definitive. Recently new interest has been aroused by an ambitious project which has undertaken the drilling of additional deep tests in shallow water off shore along Cuba's northern coast.

2. VENEZUELA

BY GUILLERMO ZULOAGA *

VENEZUELA is the world's greatest exporter of petroleum and petroleum products, and is second only to the United States in current crude production. The volume of Venezuelan oil exports has increased approximately two and a half times since 1938, as indicated by Table 2. At the same time there has been an important shift in destination, occasioned by the increased consumption of Venezuelan production within the Western Hemisphere.

Domestic consumption in Venezuela, including local bunkering at Aruba and Curaçao, averaged 27,115 b/d in 1938, 81,190 b/d in 1946, 103,000 b/d in 1947, and 125,000 b/d in 1948. These figures amount to relatively small percentages of the total production in the same years: 531,604 b/d in 1938, 1,064,461 b/d in 1946, 1,190,794 b/d in 1947, and 1,338,759 b/d in 1948.

To evaluate export trends, it is desirable to divide Venezuelan crude-oil production into three categories. Approximately 40% of the production of 1,440,813 b/d in December, 1949, was made up of heavy crudes, principally valuable as a source of residual fuel oil. Much of the world's marine bunker requirements are so supplied. The fuel oil from these heavy Venezuelan crudes complements the United States crude oil, and leaves refiners in that country free to process their domestic crudes for more valuable light products. Other countries in the Western Hemisphere consume considerable quantities of petroleum fuel, and increasing long-range demands are expected to give impetus to this type of production.

The second basic type of Venezuelan crude production comprises the "specialty" crudes, including those desirable as sources of high-grade asphalts, waxes, and industrial lubricating oils. These crudes, representing about 15% of the total production, currently find their principal outlets in the United States, Canada, and Europe. Increasing demand for high-grade asphalt is anticipated, particularly in Latin America, where new road systems and industrial development are scheduled over the next several years.

* Member, Management Committee, Creole Petroleum Corporation. Formerly Petroleum Inspector for the State of Zulía; General Petroleum Inspector and Director of Geological and Mines Survey of Venezuela; Manager, Geology Department, Creole Petroleum Corporation.

Author's Note: In preparing this paper I have borrowed freely from reports and data in the files of the Creole Petroleum Corporation, and much cooperation has been given me by my fellow workers.

The remaining 45% of Venezuela's production is composed of light crude oils, having gravities similar to East Texas and West Texas crudes. This type of production finds its principal outlet in refineries in the Western Hemisphere outside the United States.

The two small Netherlands West Indies islands of Aruba and Curaçao, in the Caribbean Sea about forty miles off the northern coast of Venezuela, have long occupied a strategic position in relation to the petroleum industry of Venezuela. In a lesser degree these islands have shared also in the activities of the petroleum industry in other countries of the Caribbean area.

Table 2. Venezuela: destination of petroleum exports[a] in 1938, 1946, 1947, and 1948.

Area of Destination	Exports of Venezuelan Petroleum			
	1938	1946	1947	1948
	(In thousands of b/d)			
United States	136	346[b]	361[b]	403[b]
Canada	7	65	86	107
Other Latin America	18	180	204	235
Subtotal, Western Hemisphere	161	591	651	745
Europe	265	313	377	398
Africa	25	23	24	29
Other Eastern Hemisphere	11	8	7	9
Subtotal, Eastern Hemisphere	301	344	408	436
Total Exports	462	935	1,059	1,181

[a] Including re-exports of products from Aruba and Curaçao.

[b] Including deliveries to the U.S. Navy as follows: 12,000 b/d in 1946, 2,000 b/d in 1947, and 3,000 b/d in 1948.

Source: Data supplied by the Creole Petroleum Corporation.

When a subsidiary of the Royal Dutch Shell, in 1917, developed the first large oil field in Venezuela (the Mene Grande field), it was natural for the parent company to choose the adjacent island of Curaçao, a colonial possession of its own government, as the site for a refinery to process the rapidly mounting output of the new field. The Netherlands Government, in turn, welcomed this new industry as a prospective source of income for its people on these tiny, barren islands.

With this new activity, Curaçao soon grew into a thriving, prosperous community, and when, in 1928, an American company that had developed large oil production in Lake Maracaibo also sought a convenient refinery site, the Netherlands Government made it attractive to bring this refinery to the island of Aruba. Both these refineries have expanded their operations from time to time, as Venezuelan production increased. Their combined capacity now exceeds half a million barrels daily, and the Netherlands West Indies have thus become one of the world's chief sources of petroleum products, particularly of bunker and heavy fuel oil derived from the asphaltic crude oil of the Lake Maracaibo region.

For over twenty years more Venezuelan crude oil has been refined in the Netherlands West Indies than in any other country. Recently, Venezuela has adopted measures designed to stimulate petroleum refining within its own borders, and it is probable that refining of Venezuelan petroleum will in the future center more largely in Venezuela itself.

Various interests have announced the intended construction of new refinery capacity, totaling from 250,000 to 300,000 b/d, within Latin America. About 175,000 b/d capacity will be in Venezuela, and the remainder in countries which today import a considerable portion of their requirements from Venezuela. Appreciable quantities of the Venezuelan production of light crudes should find outlets through these new refineries.

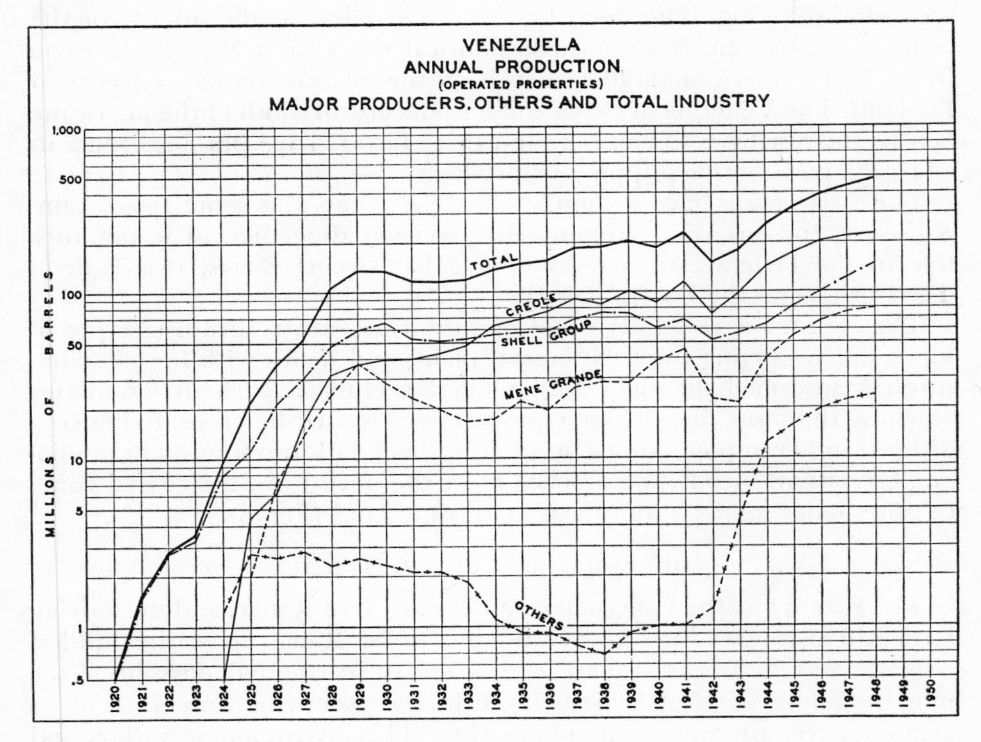

FIG. 7. Venezuela: annual production of crude petroleum from 1920 through 1948, in millions of barrels, distinguishing the shares of the Creole, Shell, Mene Grande, and "other enterprises" in the total output. This chart was prepared by the Creole Petroleum Corporation.

In summary, it appears that Venezuela's crude-oil production will play an increasingly important role in the development of Latin America and of the Western Hemisphere. While Venezuelan crude and products are currently moving to Europe as part of the overall program for rehabilitation of that area, it is foreseen that increasing requirements in this hemisphere will absorb nearly all the production, leaving very small quantities available for export to the Eastern Hemisphere. Requirements in that area are expected to be met by new production developed within its limits.

The Geographical Background

Venezuela, in the northern part of South America, has an area of 312,142 square miles and 1,740 miles of coast line, mostly on the Caribbean Sea, but some on the Atlantic. Of several navigable rivers, the Río Orinoco, with its great delta and tributary system, is the most important. Two partially enclosed bodies of water, the Golfo de Venezuela-Lago de Maracaibo unit in the west, and the Golfo de Paria in the east, give favorable access for shipping. The main harbor, La Guaira, in the middle of the country's Caribbean coast, is 1,800 nautical miles from New York, 1,000 from the Panama Canal, and 4,000 from the nearest commercial port in England. The Venezuelan coasts and harbors are all south of the hurricane belt of the northern Caribbean, and these factors have put Venezuela in a favored position for shipping its products.

The 1941 census gave a population of 3,850,000 plus some 100,000 unregistered Indians, or approximately 4,000,000 altogether, of whom 10% live in the Federal District. Venezuela has twenty States, two Federal Territories, and the Federal District.

The country is naturally divided into areas with contrasting types of topography, ranging from flat coastal plains and *llanos* with hot climates, through intermediate zones with moderate climate, to high mountains reaching the snow line. These regions, characterized by crops, living conditions, and even populations peculiar to them, fall into four main physiographic provinces: (1), the mountains of the north; (2), the coastal zone; (3), the llanos; and (4), the Guayanan hinterland (Fig. 8).

THE MOUNTAINS OF THE NORTH

These comprise the Venezuelan Andes, the Coast Range, and the Falcón-Lara hill country. The Andes and the Coast Range have the mildest climate, and as a consequence they have since colonial days been the most populous parts of Venezuela: although occupying only 12% of its area they contain 70% of its population. The intermontane valleys and gentle slopes are intensively cultivated; water power is largely used as a source of energy, and a young and active industrial development is in progress.

The Andes region is made up of the Sierra de Perijá and the Cordillera de los Andes, each a spur of the Colombian Andes. The Sierra de Perijá marks the western limit of Venezuela, and though the range itself is little known and virtually uninhabited, its foothills dipping to Lake Maracaibo are rapidly becoming one of the best cattle-raising areas. The Cordillera de los Andes, entering the western portion of Venezuela in the state of Táchira and continuing northeastward, consists of a narrow core of granite, schist, and gneiss, with flanks mostly of Cretaceous and Tertiary rocks.

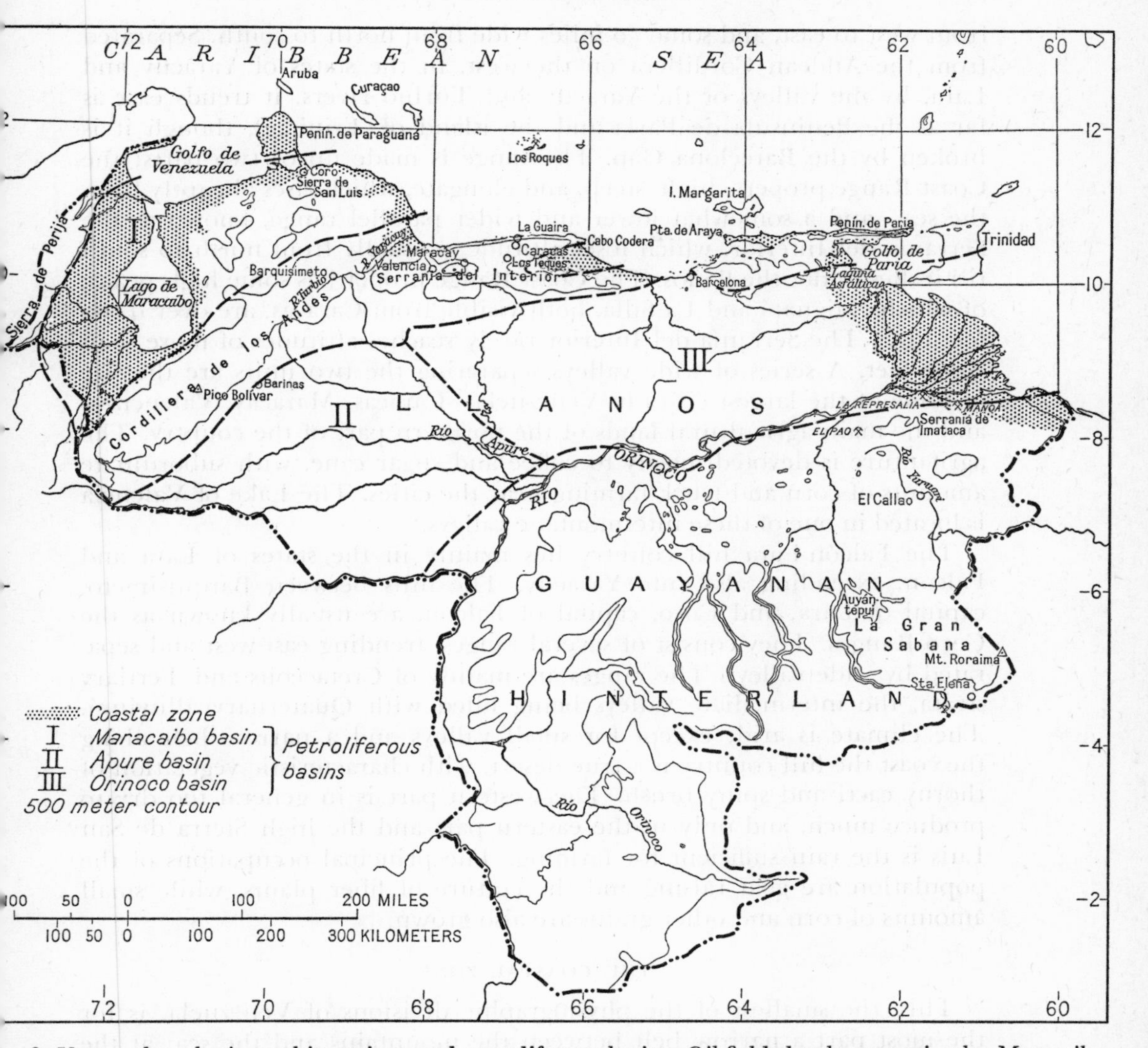

G. 8. Venezuela: physiographic regions and petroliferous basins. Oil-field development in the Maracaibo d Orinoco basins is shown in Figs. 9 and 11. In the Apure Basin a discovery well, Socony's San Silvestre o. 2, has been announced a few kilometers south of Barinas. The coastal zone continues eastward across the orthern part of the Orinoco petroliferous basin, to the point opposite Pta. de Araya.

Relatively speaking, these are young mountains, most of the present elevation dating from Post-Eocene time. The peaks culminate in the Pico Bolívar, 16,400 feet high. Narrow valleys and terraces are intensively cultivated, and here on the steep mountainsides above 4,000 feet is the only wheat-growing area in Venezuela; other crops are coffee, oats, peas, and corn.

The Coast Range is a continuation of the Andes, some 320 miles long from west to east, and some 50 miles wide from north to south. Separated from the Andean Cordillera on the west, in the states of Yaracuy and Lara, by the valleys of the Yaracuy and Turbio rivers, it trends east as far as the Península de Paria and the island of Trinidad, though it is broken by the Barcelona Gap. The range is made up of two parts: the Coast Range proper—high, steep, and elongate, which rises abruptly from the sea—and a somewhat lower and wider parallel range, known as the Serranía del Interior, which loses altitude gradually from north to south till it dies under the llanos. The Coast Range proper has some high peaks, of which Naiguatá and La Silla, both visible from Caracas, are over 9,000 feet high. The Serranía del Interior rarely reaches altitudes of more than 5,000 feet. A series of wide valleys separating the two units are the site of some of the largest cities in Venezuela—Caracas, Maracay, Valencia—and the main agricultural lands of the northern part of the country. The agriculture is devoted mostly to coffee and sugar cane, with subordinate amounts of corn and truck farming near the cities. The Lake of Valencia is located in one of these intermontane valleys.

The Falcón-Lara hill country lies mainly in the states of Lara and Falcón, extending also into Yaracuy. The hills between Barquisimeto, capital of Lara, and Coro, capital of Falcón, are usually known as the Coro Ranges. They consist of several ranges, trending east-west and separated by wide valleys. The ranges are mainly of Cretaceous and Tertiary strata, the intermediate valleys being filled with Quaternary alluvium. The climate is arid; except for small valleys and a narrow belt along the coast the hill country is a true desert, with characteristic vegetation of thorny cacti and spiny brush. The western part is in general too dry to produce much, and only in the eastern part and the high Sierra de San Luis is the rain sufficient for farming. The principal occupations of the population are goat raising and the culture of fiber plants, while small amounts of corn and other grains are also grown.

THE COASTAL ZONE

This, the smallest of the physiographic divisions of Venezuela, is for the most part a narrow belt between the mountains and the sea; at the western and eastern ends, however, it includes the Lake Maracaibo flats and the Orinoco Delta.

This zone has about 9% of the population of Venezuela, and is of prime importance both for its position as the site of the best harbors and ports and for its agricultural products. Cocoa, sugar cane, coconut, and bananas are produced on a large scale, and large-scale fisheries are moving rapidly into bigger production. Within this zone there lie the Lake Maracaibo oil fields, which constitute the greater part of the country's huge petroleum reserves.

The islands in the Caribbean Sea along the coast of Venezuela and the Península de Paraguaná may be included in the Coastal Zone, although their physiography does not necessarily conform to it. The waters off Margarita are rich in pearl oysters, and phosphate has been mined on Los Roques.

THE LLANOS

These great plains, between the mountains in the north and the Orinoco, are remarkably flat grass lands, mostly open, but with occasional patches of forest. The rivers, some of them very large, are sluggish because of the slight gradient. The lower reaches of the Orinoco fall only about 250 feet in 700 miles; hence it readily floods.

The llanos have a climate characterized by great contrast between wet and dry seasons. In the rainy season the rivers flood a considerable area, and cattle have to be moved to the higher points; malaria increases; and transportation becomes difficult. In exceptionally rainy years hundreds of square miles are flooded, and thousands of cattle drown. In the dry season (April to October) the rivers, except the largest, dry up. The ground becomes dusty and cracked, cattle have difficulty in finding drinking water, and most vegetation loses its leaves.

Although constituting 35% of the area of Venezuela, the llanos have only 18% of the population. This is cattle country, with practically no agriculture, except a few *conucos* along the streams close to the large towns. Game, including such animals as *chiguires* (large rodents), deer, and jaguars, is abundant. So also is bird life; at one time the exploitation of egret feathers was a thriving industry. Rivers are alive with alligators, *caribes* (small voracious fish), electric eels, poison rays, and other fish. Large areas in the llanos are covered with palms.

Geologically, the llanos are made up of Tertiary strata, capped here and there by Quaternary clays and gravels, with Recent muds and clays in the flooded areas. The eastern part of the llanos is characterized by *mesas*—flat, gravel-capped hills, usually separated from the lower surrounding country by steep cliffs or *farallones,* which may be as high as 1,000 feet.

The only mineral of economic importance in the llanos is oil; but oil is so abundant that it easily makes up for the absence of others. Its exploitation is bringing profound changes.

THE GUAYANAN HINTERLAND

The Guayanan Hinterland, or simply "Guayana," is the largest physiographic unit of Venezuela, occupying 45% of the area, but supporting only 2.5% of the population.

It is a territory of rich mines and poor people. The legends of the days of the Spanish *conquistadores* about El Dorado and Manoa have all been connected with this vast, mysterious territory. Fantastic as these legends sound, some have considerable basis in fact. Gold strikes of fabulous wealth have been made from time to time, to be followed by long periods of inactivity and general poverty. There is practically no agriculture, and cattle raising is on a very reduced scale. Rubber and *tonka* bean grow wild, and are harvested in periods of high prices. A large part of the scanty population is nomadic, living from hand-panning of gold and diamonds. Indian tribes, some wild and warlike, still inhabit large parts of this area.

The Guayana immediately south of the Orinoco is rather flat and in places looks like the llanos, but with low rounded hills. These are outcrops of hard rocks, which have resisted erosion since Paleozoic time. Farther south we find table mountains, or *tepuis,* like Roraima (9,000 feet high) and Auyán-tepuí, with vertical cliffs that make them practically unclimbable. It is only with the aid of ladders that a few scientific explorers and collectors have reached the tops. These mountains are of flat-lying sandstones and conglomerates, of continental origin and probably of Mesozoic age. Between the table mountains there are open plains, comprising the *Gran Sabana,* which is being considered for intensive colonization. Water is abundant, in swift streams and spectacular falls, some of several thousand feet, and the climate is healthful, in contrast to the warm area to the north, which is more susceptible to malarial infestation. Rubber and *sarrapia* (tonka beans) are the only agricultural products worthy of note.

The mineral resources of Guayana are three: gold, diamonds, and iron. The Callao mine, the most productive gold mine in the world during the latter part of the nineteenth century, lies about one hundred miles south of the Orinoco, in what is today the town of El Callao, on the Río Yuruari. The recorded production from this mine from 1871 to 1890 was valued at over $30,000,000, and during this boom period several other mines were discovered in the same area. A gold rush started, and a great number of young *Caraqueños* went into gold-prospecting, the majority losing all they possessed. Although the Callao mine has been exhausted, several of the others are still being worked. Every decade or so, a rich pocket is discovered, and the rush starts again.

Besides the lode or vein gold, a considerable amount of gold is produced from placers by panning. This is done with wooden *bateas,* by miners in groups following the rich sands from place to place.

Diamonds were discovered in Venezuela in the dry season of 1926, in the streams that drain the table mountains. Apparently the diamonds occur disseminated in the Roraima sandstones, from which they are eroded and concentrated by running water. The stones are of good quality, both for gem and industrial use, and a few very large ones have been found.

Before World War I some supposedly rich iron mines were developed along the Orinoco, at Manoa, in the Serranía de Imataca. A Canadian company shipped a few thousand tons of good ore, but the enterprise soon failed. In 1927 the Pao deposit was found, and a very large mine, the property of the Bethlehem Steel Company, is about to begin large-scale exploitation. The ore is of the highest grade known, equaled only by that of Brazil; it contains 69% of metallic iron. The Pao deposit contains some 60,000,000 tons of this ore—literally a mountain of iron. In recent years another high-grade deposit has been found at La Represalia, downstream. When these deposits, and others that will probably be discovered along the same trend, come to be exploited on a large scale, they will exert a strong influence on the iron economy of the world. In addition, there are billions of tons of lower grade (30%) iron ore, which may in the future influence the world market.

The Geology of the Venezuelan Oil Fields

There are three distinct petroliferous basins in Venezuela: the Maracaibo basin, the Orinoco basin, and the Apure basin (Fig. 8). They are of roughly equal area, some 20 million acres each.

THE MARACAIBO BASIN

This is by far the most important basin from the point of view of production and reserves. By the end of 1949 it had produced three and three-quarter billion barrels of oil, and proved reserves totaled seven billion barrels.

Geologically, the Maracaibo basin (Fig. 9) is an ideal example of a geosynclinal basin of sedimentation: almost completely surrounded by mountains, the basin has been slowly sinking and simultaneously accumulating a tremendous volume of sediments, while the mountains that border it have been rising at about the same rate. In the center of the basin is Lake Maracaibo, underlain by a total thickness of some 16,000 feet of sediments; while the snow peaks of the Sierra Nevada de Merida on the southeast have an equal elevation. From the mountains, streams bring down to the lake silts and sands which would soon fill it, were the bottom not subsiding at about the same rate that the sediments are being deposited. As the mountain rims have risen, the area of the lake has become smaller, and its edges have become exposed, and thus subject to erosion and to the analytical eye of the geologist.

Three main units of oil-bearing rocks underlie the surface of the Maracaibo basin. Lying on older and altered rocks are Cretaceous formations: sandstones, limestones, and shales some 6,000 feet thick. Above these, with only minor unconformity, Eocene rocks—mostly sands and silts—were laid down; later, after a short period of erosion, again with only minor unconformity, Miocene sands and shales were deposited. Prolific production has been found in the rocks of each of the three: Cretaceous, Eocene, and Miocene. That of the Miocene, the shallowest formation, was found first; later, those of the Eocene and the Cretaceous.

With the exception of a few small fields southwest of the lake, and the Cumarebo field and a few small fields to the northeast in Falcón, the producing fields of the lake basin are restricted to a broad belt, trending northwest from the discovery field of Mene Grande toward Maracaibo and beyond, and ending with the Mara field in the north. This is a re-

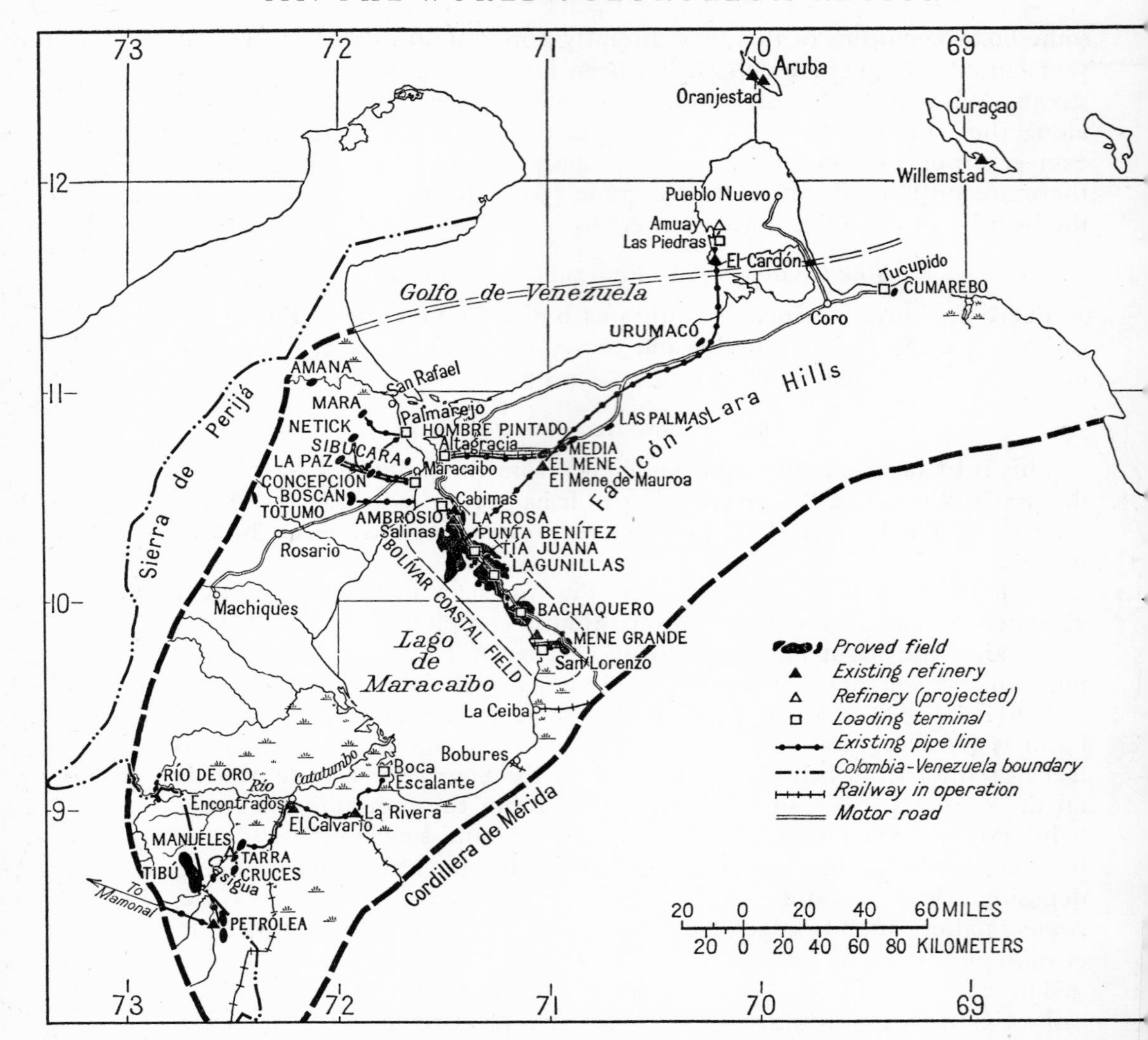

FIG. 9. The Maracaibo petroliferous basin: oil fields and facilities at the beginning of 1949. Names of oil fi are in Roman capitals. The size of the fields has been exaggerated about two-fold. In 1948 the Media, Palmas, and Urumaco fields were reported to be closed in and not producing, and the Totumo and Am fields were reported abandoned. The largest pipe line shown, that from the La Rosa field to Amuay, h diameter of 24/26 inches. The larger refineries are those at Aruba (400,000 b/d and 34,000 b/d), Cur (356,000 b/d), El Cardón (55,000 b/d), San Lorenzo (38,000 b/d), and Salinas (22,500 b/d).

markable trend; all the really large fields, whether in Miocene, Eocene, or Cretaceous rocks, fall within it. It may be divided into two parts: the Bolívar Coastal Field, on the northeast border of Lake Maracaibo, and the fields in Maracaibo and Mara districts on the northwest.

1. The Bolívar Coastal Field. In the early days it was not realized that many of the fields along the shore of Lake Maracaibo in the District

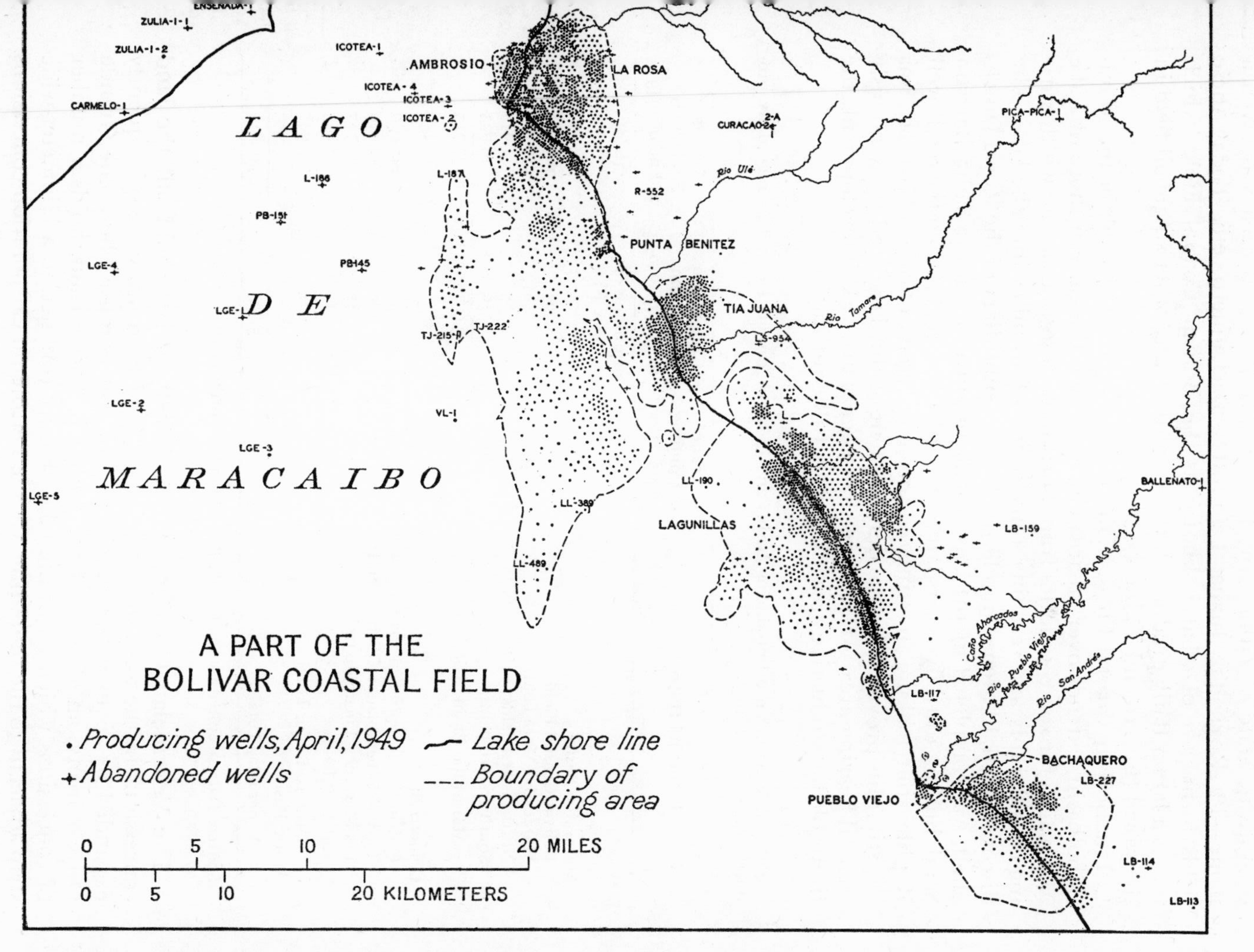

Fig. 10. Part of the Bolívar coastal field. Discovered in 1917, it has since been extended to a proved area of 178,000 acres, with more than 4,400 producing wells. The average daily production at the end of 1948 exceeded 735,000 barrels. This is one of the world's largest and most prolific fields.

of Bolívar, State of Zulía, would eventually join, and each was given a name—La Rosa, Tia Juana, Lagunillas, Bachaquero, etc. Today these fields form part of what is known as the Bolívar Coastal Field, which boasts a larger daily production potential than any other single oil field in the world (Fig. 10 and Table 3).

The Bolívar Coastal Field, high on the northeastern flank of the Maracaibo basin, extends over an area encompassing the eastern shore of Lake Maracaibo for some 43 miles from northwest to southeast, the width varying considerably and reaching a maximum of 10 miles into the lake. The discovery well, El Mene Grande, drilled by a subsidiary of the Royal Dutch-Shell, was completed in 1917, as a small producer from the Santa Barbara Sand, but not until five years later, December 14, 1922, when the Shell well R-4 blew out, making an estimated 100,000 barrels of heavy oil per day, was the stimulus provided for the development of this large and tremendously rich oil-bearing area. After more than twenty-five years' development, the limits of the productive area are not yet fully outlined.

Table 3. Venezuelan oil-producing basins and districts: cumulative and average daily production at year-end, 1949.

Basin and District	Year of Beginning Production	Cumulative Production to Jan. 1, 1950	Average Production at End of 1949
		(millions of barrels)	(b/d)
Maracaibo Basin			
Bolívar Coastal Field	1917	3,494	769,964
Maracaibo-Mara fields	1922	248	243,183
Northeastern fields	1921	73	4,943
Southwestern fields	1916	116	19,625
Maracaibo Basin, Total		3,931	1,037,715
Orinoco Basin			
Quiriquire-Jusepín-San Joaquín trend	1924	638	147,080
Tucupita-Temblador-Oficina trend	1933	409	216,416
Las Mercedes-Tucupido trend	1943	13	30,717
Scattered fields	1914	12	8,885
Orinoco Basin, Total		1,071	403,098
Apure Basin		—	—
Venezuela, Total		5,003	1,440,813

Source: Data supplied by the Creole Petroleum Corporation.

The Venezuelan Oil Concessions, a Shell affiliate, holds all the land concessions in the area, and its original land holdings were subsequently expanded by acquiring several large lake concessions. The Mene Grande Oil Company, an affiliate of the Gulf Oil Corporation, holds a number of concessions which lie in the lake, but are confined to a kilometer-wide strip adjacent to the shore line. Creole Petroleum Corporation concessions lie entirely within the lake.

Within the productive area the drill has penetrated sediments ranging in age from Recent to Eocene, Shell's VL-1 being the first well in the area to reach the Cretaceous. There are several unconformities within the section penetrated in the Bolívar Coastal Field, the most pronounced being that at the top of the Eocene. Approximate maximum thicknesses of the various formations making up the section are as follows:

Formation	*Age*	*Thickness*
Sediments	Recent	80 feet
El Milagro	Pleistocene	2,100
La Puerta	Miocene	2,200
Lagunillas	Miocene	1,000
La Rosa	Miocene	600
Icotea	Oligocene	460
Ambrosio	Eocene	360
Las Flores	Eocene	4,300
Potreritos	Eocene	2,880
Misoa	Eocene	3,400

Geologically, the Bolívar Coastal Field is a stratigraphic trap, and this applies to both Miocene and Eocene production. The geologic structure is monoclinal, the general dip being gently southwest towards the center of the basin. The post-Eocene sediments dip 2 to 10 degrees to the southwest, the essentially monoclinal structure being broken by several well-defined faults. The underlying Eocene erosional surface also dips to the southwest, but the Eocene beds themselves were folded and faulted before being subjected to truncation. These disturbances present a complex subsurface problem in the development of the Eocene reservoirs.

The surface of the productive areas consists principally of weathered sands and clays of Quaternary and late Tertiary age; east of Lagunillas and north of La Rosa, oil and asphalt seeps occur. At one place or another within the productive area of the Bolívar Coastal Field, the Lagunillas, La Rosa, Icotea, Las Flores, Potreritos, and Misoa formations have been found commercially productive, but by far the largest volume of oil to date has come from the Lagunillas and La Rosa sands. Production from the Potreritos sandstones has steadily increased during recent years; Eocene production began to receive serious consideration only some seven or eight years ago. Cretaceous production is also a future possibility.

Accumulations of oil and gas in the post-Eocene sediments are very largely in stratigraphic traps formed by the updip pinch-out of the Miocene sands, or by sealing of these sands at their outcrops by inspissated oil. Local structure plays a minor role; the Icotea sand (Oligocene) is found only in local lows of the Eocene erosion surface, and thus also contains only stratigraphic accumulations. In the underlying Eocene beds the accumulations are probably both structural and stratigraphic.

The oil produced in the Bolívar Coastal Field varies in gravity from

11° to 34° API. It is of asphaltic base with a wide range of characteristics, and is segregated into several grades. At the close of 1949 Creole's total daily production, 486,050 barrels, was being segregated as follows:

Type of Crude	*Grade*	*Production (b/d)*
La Rosa	Medium crude (25° API)	54,206 barrels
PB-32	Medium (22° API)	28,973
PB-101	Medium (high octane, 25° API)	9,143
TJ-102	Medium (25° API)	156,021
LL-370	Medium (26.4° API)	72,200
LL-453	Medium (29° API)	38,844
LL-261	Medium (25.5° API)	1,710
PB-Huron	Heavy (20° API)	1,899
Tia Juana	Heavy (17.7° API)	37,508
Lagunillas	Heavy (17.7° API)	57,401
Bachaquero	Heavy (14.2° API)	28,145

2. *The Fields in Maracaibo and Mara districts, State of Zulía.* In August 1944, the Shell Caribbean Company completed its well P-68 in Cretaceous limestone within the area of the older La Paz field, producing at the rate of over 5,000 b/d. This well was the first large-volume producer from beds below the Tertiary in this area. In 1945 well DM-2 in the Mara area was completed as the discovery well of the Mara field, in the same trend as the La Paz field. In all, 28 wells had been completed by March 1949, between La Paz field (Shell) and Mara (Shell-Creole-Texas), and they have a daily potential of about 200,000 b/d.

In the La Paz–Mara area there is a limestone section about 1,800 feet thick between the Upper Cretaceous Colón shales and the underlying basement rocks. This limestone is composed from top to bottom of: (1) a basal member of the Colón formation (100 feet); (2) La Luna formation (300 feet); (3) Cogollo limestone (1,400 feet). All portions of this section have been found productive.

These, and more recent Cretaceous discoveries, are of major importance, not only because of the volume and quality of the oil, but also because they open great possibilities for further developments in the Cretaceous elsewhere in Venezuela. Although it is rarely possible to make direct observations on the exact nature of the Cretaceous pay zones, it is believed that production is almost entirely from fractures in dense limestone. All portions of the section, from the top of the limestone to the beds overlying basement, are known to be productive.

The present Cretaceous fields lie in the western part of the Maracaibo basin, approaching its western margin, which is marked by the Perijá Mountains along the international boundary between Venezuela and Colombia. But it is quite possible that the Cretaceous rocks underlying the major part of the entire Maracaibo basin may show important local

26. Western Venezuela, Lake Maracaibo. Creole's storage and industrial development at La Salina, in the Bolívar Coastal Field. Oil-loading pier in foreground. (Photo by G. Zuloaga.)

27. Western Venezuela, Lake Maracaibo. Avenues of derricks of the Creole Corporation in the Lagunillas portion of the Bolívar Coastal Field. (Standard Oil Company, N.J., photo by Vachon.)

28. Western Venezuela, Lake Maracaibo. Weights used to sink caissons. (Standard Oil Company, N.J., photo by Vachon.)

29. Western Venezuela, Lake Maracaibo. A drilling well in the Bolívar Coastal Field. Derrick and drilling platform are mounted on caisson piles set in the lake bottom; boilers, engines, and mud tanks, on barges. (Creole Petroleum Corporation.)

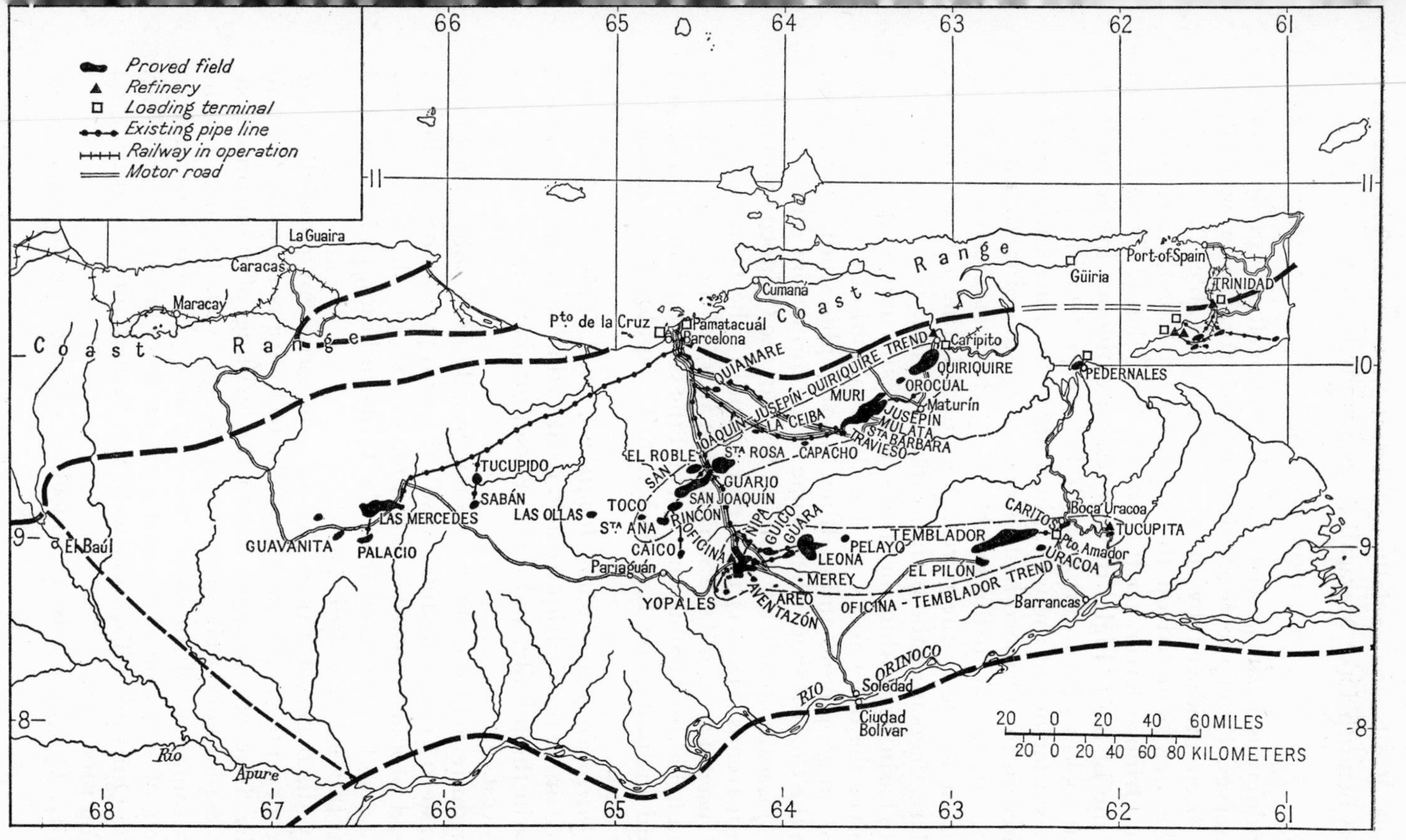

Fig. 11. Petroliferous basins of eastern Venezuela: oil fields and facilities at beginning of 1949. Names of oil fields are in Roman capitals, and the size of the fields has been exaggerated about two-fold. In 1948 the Pilón and Uracoa fields had no listed production. The largest pipe lines are of 16-inch diameter: these include double lines from the Oficina-Guico field area to Puerto de la Cruz, with a 16-inch loop from El Roble to the port, and the single line from Las Mercedes to Pamatacuál. The only refinery in eastern Venezuela with complete refining equipment is that at Caripito (55,000 b/d crude capacity, 1,200 b/d cracking). Two complete refineries, with capacities of 35,000 and 30,000 b/d respectively, are to be built at Puerto de la Cruz. Trinidad, with a refining capacity in excess of the island's output of crude, treats some 300,000 barrels monthly of Venezuelan crude, shipped in under bond. The oil fields and facilities of Trinidad are shown in more detail in Fig. 12.

accumulations of oil commercially significant in volume. The presence of these rocks beneath Lake Maracaibo has already been noted.

THE ORINOCO BASIN

This petroliferous basin is second to the Maracaibo basin in production and reserves. From the beginning of production, over one billion barrels had been produced by the end of 1949 (Table 3). The basin comprises the area bounded on the north by the Coast Range, on the south by the Orinoco River, on the east by the Atlantic Ocean, and on the west by the El Baúl "high" in Cojedes, an eminence in the underlying basement rocks which forms a saddle between the Orinoco basin and the Apure basin farther to the west. Since Mesozoic times sediments have been deposited in the Orinoco basin from the erosion of crystalline rocks in the Guayanan Shield to the south and of the metamorphic rocks in the rising Coast Range to the north. A gradual tilting towards the east has taken place, deepening the basin in that direction.

The whole basin lies within the llanos. Flat plains and mesas of Recent sediments cover the older rocks, so that the exploration for oil has been largely based on geophysics and drilling: stratigraphic studies have been restricted to the Coast Range, where some of the formations outcrop.

As in the Maracaibo basin, the main oil fields of the Orinoco basin occur in three main trends, viz., (1) the trend including the Quiriquire field, the Greater Jusepín area, and the San Joaquín area; (2) the trend including the Tucupita field, the Temblador area, and the Greater Oficina area; and (3) the trend including the Las Mercedes, Tucupido, and neighboring fields in the west.

1. The Quiriquire-Jusepín-San Joaquín Trend. This trend is composed of three units: the Quiriquire field itself to the northeast, the Greater Jusepín area in the middle, and the San Joaquín–Santa Rosa–Roble area to the southwest.

The Quiriquire field was the first commercially profitable oil field discovered in eastern Venezuela, and it is still the most productive. At the end of 1949 it had produced nearly 326 million barrels of oil, or nearly one-third the total so far recovered from the Orinoco basin, and it was then producing at the rate of 51,626 b/d.

The early discovery (1913) of the tar and gas seeps parallel to the mountain front on the northern edge of the tectonically unstable Maturin basin, in the State of Monagas, led to the drilling of numerous non-commercial exploratory wells between 1918 and 1927. In 1928, on the geological assumption of an oil migration from the southeast and an accumulation of seeps on the south nose of the Quiriquire anticline, Moneb-I, the Quiriquire discovery well, was drilled and brought in as a strong-flowing producer.

Structurally, Quiriquire is a monocline of Pliocene-Pleistocene age trending northeast-southwest, and dipping to the southeast at approximately 15 degrees. The Quiriquire formation which encloses the oil-reservoir rocks is an interfingered, overlapping, alluvial-fan series, continental in origin and deposited unconformably on truncated, folded beds ranging in age from Cretaceous to Miocene. The Quiriquire sediments consist of sandy clays interbedded with unsorted clastics, ranging in size from boulders to silty sands. In most areas an exact sand-by-sand correlation is impossible because of very rapid lateral lithologic variations; in such areas, points representing equal times of deposition are selected and utilized in correlative work.

The area of petroleum accumulation is an irregular oval about 6 by 3½ miles in extent, with the maximum thickness of net oil sand in the central part of the field (700 feet), and an average of 194 feet per well.

Major productive zones in the Quiriquire at present (1948) are basal members, Zeta, Eta, and Theta, which essentially contain light oil, but have numerous beds saturated with water and non-commerical heavy crudes. Members above Zeta are generally productive along the northern edge of the structure, but to the south (downdip) they largely contain water and tar oil.

A significant geological problem in the Quiriquire field concerns the location of the parent or source beds of petroleum, the continental Quiriquire formation being obviously a secondary reservoir. Attempts to locate the source beds have involved the penetration of the unconformity in key locations throughout the field. From information obtained through cores, the Cretaceous Santa Anita and Eocene Mundo Nuevo formations are the most likely of the possible sources; however, the possibilities of the Oligocene sediments have not been exhausted.

A gas cap is known to exist along the northern edge of the producing horizons; its absence in certain wells helped to locate the Moneb fault. Water is more pronounced at the extreme edges of the developed areas, but is found throughout the field at haphazard depths. There is no positive evidence of water drive, but such a condition may very well exist, since water flowed strongly from Quiriquire sediments in QGE-17, a well southwest of the field.

The irregularly varying gravity of the oil (13° to 36° API) in the productive section may be attributed to the lithology of the reservoir and the mode of accumulation. The crude produced is relatively heavy, with a sulphur content of 1.18%. The normal naphtha cuts have an exceptionally high octane rating, and the gasoline produced in cracking is also of high octane. The kerosene cut is very poor as a burning oil but excellent as a tractor fuel. Gas-oil cuts are of interest chiefly as cracking or hydrogenation feed stocks. The lube oils are of no major commercial interest. Asphalt suitable for warm climates can be produced, and bunker fuels from this crude are of exceptionally good quality. Quiriquire crude is best suited for producing aviation gasoline, normal and cracked motor gasolines, hydrogenation stocks, marine diesel fuel, and high quality bunker fuels.

The Greater Jusepín area had produced three-fourths as much oil as the Quiriquire field—230 million barrels by the end of 1949—but it was then producing 63,000 b/d, or more than 11,000 b/d more than the Quiriquire field was producing at that time. Production from the La Pica formation of Upper Miocene age was first obtained along the northern rim of the Maturin basin when Jusepín No. 1 was completed on October 11, 1938, by Creole Petroleum Corporation.

The Jusepín crude has a gravity range of 22° to 36° API, with an average of 31° API; it exhibits moderate sulphur, and is especially suited to the production of motor fuel of average quality, various grades of gas oils, diesel oil, and bunker fuels.

The San Joaquín–Roble area, in the southwestern part of the trend, had, by 1949, yielded 64 million barrels of oil since the beginning of production in 1936, and it was producing at the rate of 28,318 b/d at the end of 1949. Of this production, the greater part was coming from the San Joaquín–Guario field and from the El Roble field.

The San Joaquín field, discovered in 1939, in east-central Anzoategui, is situated 50 miles south of Barcelona, the state capital. The field lies 12 miles northeast of the Santa Ana field, which had begun production in 1936, and 12 miles southwest of the Santa Rosa field, discovered in 1941, all three fields being located along the axis of a regional uplift trending northeast-southwest. The discovery well of the San Joaquín field, JMN-1, was drilled to test a favorable structure revealed by surface geological mapping.

In the immediate area of the field the structure consists of an asymmetric anticlinal fold striking N 50° E. The crest of this fold is undulating and rises into three elongate domes, slightly en-echelon, which, from northeast to southwest, are called the Guario, Northeast, and Southwest domes, respectively. The northwest flank of the structure has an average dip of 10 degrees, while on the southeast flank dips range from 15 degrees in the upper beds to overturned in the lower beds, which also show thrust-faulting.

Pliocene, Upper Miocene, and Middle Miocene sediments are exposed in the area. The Pliocene beds outcrop over the extreme northeast plunge of the structure and overlap the older beds. The Upper and Middle Miocene beds outcrop along the crest of the structure where it rises to the southwest.

Accumulations of oil and gas are confined to the area of closure under each of the three domes. To date, reserves of oil have been found in 23 sands in the Naranja, Verde, Amarillo, and Colorado members of the Mio-Oligocene Oficina formation, and in sandstones in the upper 2,000 feet of the Oligocene Merecure formation. The thickest undisturbed

Oficina section found so far is 8,290 feet. The shallowest productive sand in this formation is the Naranja *E* at 3,840 to 3,903 feet in JM-15 (Northeast Dome), and the deepest is the sand now correlated as the Colorado *R* at 9,140 to 9,185 feet in JMN-7 (Southwest Dome). The total thickness of the Merecure formation is not yet known, only the upper 2,000 feet having been penetrated so far. The Merecure production obtained to date ranges in depth from 7,607–7,713 feet in JM-19 (Northeast Dome) to 10,660–10,705 feet in GU-8 (Guario Dome).

The great majority of the wells have been completed in one or more of the Oficina sands, with the Naranja *L* sand on the Guario Dome (Naranja *L*-2 reservoir) forming by far the most important reservoir known to date. In general, the reservoir sands of the Oficina formation are thin, ranging from a few feet to 40 feet, though some sands occasionally attain a thickness of 100 feet or more. They are in part massive, but usually contain intercalations of shale and carbonaceous material, and show rapid lateral variations in development. Some of the sands are coarse-grained and unconsolidated, but the great majority are very fine-grained and hard. Porosities usually range from 15 to 20%, occasionally reaching 25 to 30%. Permeabilities are generally low, though values in excess of one darcy have occasionally been obtained in some of the sands.

So far as is known, all the productive Oficina sands have gas caps, with the oil columns varying in height from 200 feet to more than 1,000 feet. The recovery mechanism is a combination of dissolved-gas and gas-cap drives. The oil produced is extremely waxy (average wax content 13%) and has a relatively high pour-point, although the gravity averages 43° API.

The 4 JM and 4 Guario wells in which the Merecure was tested or is producing indicate that this formation contains very large gas and distillate reserves, as well as some oil reserves, but further lateral and vertical exploration of the formation is necessary before its potentialities can be adequately assessed. There are not sufficient data at present to divide the penetrated section into members traceable from one well to another, but three separate reservoirs are recognized.

The Merecure section so far penetrated consists predominantly of very hard sandstones, with laminations of hard, black, silty shale, and is characterized by an almost continuous series of high-resistivity peaks on the electric log. The oil produced has similar characteristics to that from the Oficina sands, though it is about 10° API lower in gravity.

El Roble field, which lies on the northwest flank of the same regional uplift on which the San Joaquín field is located, is situated about 2½ miles northwest of the latter field in east-central Anzoategui. It was discovered in 1939 by well RPN-1, drilled to test a favorable structure indicated by surface geology. The structure of the field is a broad, gentle nose, the average dip being about 3 degrees. Small normal faults are known to exist, but have not yet been delineated.

Pliocene beds (Campo Santo formation) cover the entire surface area of

El Roble field. The Pliocene is less than 1,000 feet thick and is underlain by the Upper Miocene Freites formation, which is in turn underlain by the Mio-Oligocene Oficina formation.

Oil and gas have so far (1948) been found in 14 sands in the Amarillo and Colorado members of the Oficina formation, the upper 11 productive sands of the San Joaquín field (in the Naranja and Verde members of the Oficina formation) being saturated with water in all the Roble wells. The shallowest producing sand is the Amarillo *C* at 7,948–7,970 feet in RPN-6, and the deepest is the Colorado *R* at 10,867–10,970 feet in Anaco-1.

The Oficina section in El Roble is similar to that in San Joaquín, but is thicker and more indurated. Except for the Amarillo *C*, all the reservoir sands are hard, fine-grained, and somewhat cemented; they have low porosities and a permeability in excess of one darcy in some wells. The Amarillo member is well-developed in most of the wells, but appears to be shaling out to the southeast (thin and shaley in RPN-3 and RPN-6); it is the most important reservoir now known in the field and is open to production in 10 of the 11 wells drilled to date. However, the limits of production of most of the sands have not yet been defined, and it is thought that the Colorado *K* sand, which at present forms the second biggest reservoir, may contain important reserves west and southwest of its present proved area.

The crude produced by the Roble wells was originally similar to that from the Oficina sands of San Joaquín. Withdrawal, however, has resulted in rapid pressure declines, accompanied by an average 14° API rise in gravity to 58° API, a rapid rise in gas-oil ratios, and a drop in wax content to 5 per cent.

2. *The Tucupita-Temblador-Oficina Trend.* The daily production rate of this trend has considerably surpassed that of the Quiriquire-Jusepín-San Joaquín trend, to the north, although the latter still has the greater cumulative production (Table 3). The Greater Oficina area is the principal producing unit.

The Tucupita field in the eastern part of the trend has been in production only since 1945. At the end of 1949 the total oil produced amounted to six and one-quarter million barrels, and the current output was being produced at the rate of 7,865 b/d.

The Temblador area, in south-central Monagas, embraces several separate fields, but only the Temblador field was producing in 1949. The area's total production to the end of 1949 was slightly less than 32 million barrels, and the current rate of production was then 9,756 b/d.

The surface of the entire area is covered by the Quaternary Mesa formation. The discovery well of the area, Temblador-1, was located in 1936 on the north (basinward) side of a major regional fault (Temblador fault) revealed by a seismograph survey. The Temblador area is high on the southern limb of the Maturin basin, a part of the eastern Venezuelan geosyncline and structural basin. Structurally, the area is a homocline dipping northward at about 2 degrees and broken by a number of regional ENE-WSW normal faults, with smaller branch cross faults trending generally NW-SE. These two sets of faults are of primary economic importance because, in the absence of closed folds, they provided the necessary traps for oil and gas migrating up-dip from the central part of the basin, though lateral closure is in some cases achieved by sand pinch-outs, and not cross faults. The Temblador area fields are thus for the most part fault traps, and to a limited extent combination fault-stratigraphic traps.

The stratigraphic section encountered in the area consists of 300 to 500 feet of Quaternary sediments (Mesa formation), underlain unconformably by about 2,500 feet of Pliocene Las Piedras formation, which lies unconformably on the Miocene Freites formation, 1,900 feet thick. The Mio-Oligocene Oficina formation, about 500 feet thick, unconformably underlies the Freites, and is in turn unconformably underlain by about 600 to 700 feet of Cretaceous Temblador formation, which rests on the igneous-metamorphic Basement Complex (Guayanan shield). The formation thicknesses given apply to the Temblador field; in the Pilón and Uracoa fields the thicknesses are somewhat reduced, because of the more shoreward (southerly) position of these fields.

The oil-bearing reservoir beds in the area are confined to the Oficina and Temblador sands. In the Temblador field some free gas has been found in the Freites formation, and the Temblador formation contains only thick tar sands and heavy oil which have not been exploited; thus production is limited to the Oficina sands, which in this field are encountered throughout the entire thickness (500 feet) of the formation, though no single sand is productive over the whole area of the field.

In the Pilón field the upper part of the Oficina reservoir is gas-bearing; oil production is confined to the lower part of this reservoir and to the upper part of the Temblador formation. In the Uracoa field the upper part of the Oficina formation shows only a sandy-shale development; the lower part is well developed in Tabasca-1, but yielded only gas, and in Uracoa-1 (the only other well in this field) it failed to yield oil production, because of its poor development. The Temblador formation is productive in Tabasca-1.

The Oficina reservoirs consist of soft to hard, fine to medium grained sands, interbedded with thin shales and containing some thin coal seams. The productive Temblador sands are confined to the upper part of this formation, and consist essentially of soft, fine to coarse grained arkose, interbedded with hard shales. The Temblador field is producing under an active water drive, which, combined with the good porosities and permeabilities in the sands, is expected to result in a high ultimate recovery of oil. However, salt water is intimately associated with the oil, and has of necessity been produced with it, almost from the beginning. This contamination with salt requires an elaborate dewatering and desalting operation.

The oil produced from the Pilón and Uracoa fields is similar to but heavier than that from Temblador field. The Temblador crude, which is of naphthenic base with a moderate sulfur content, is best suited to the manufacture of aviation and motor gasolines, gas oils, and bunker fuels.

The Greater Oficina area in south-central Anzoategui embraces a number of oil fields in the vicinity of the Oficina field proper, which is the oldest of these fields. The area has produced over 371 million barrels since the beginning of production in 1933, and at the end of 1949 the average rate of output was 198,795 b/d. The discovery well of the area, Oficina No. 1, was located at a point where torsion balance and refraction seismograph data showed a structural disturbance in the general regional northward dip.

The bulk of the concessions in the area are held by the Mene Grande Oil Company, C.A., or jointly by Meneg and Creole. The remainder are held by the Socony Vacuum Oil Company of Venezuela. Meneg does the drilling in both its solely owned and its joint-interest concessions, and Socony drills in the concessions it holds.

The Greater Oficina fields are located on the southern limb of the eastern Venezuela geosyncline (the Orinoco basin). Outcropping in the area are the flat Quaternary Mesa formation and the Pliocene–Upper Miocene Sacacual group (Las Piedras and Algarrobo formations). These are underlain unconformably by the Miocene Freites formation and the Mio-Oligocene Oficina formation, neither of which outcrops. The latter, which is some 3,000 feet thick, lies unconformably on the Cretaceous Temblador formation, which in turn rests unconformably on the igneous-metamorphic basement (Guayana shield).

The axis of the southern limb of the eastern Venezuelan structural basin trends east-northeast and dips gently in that direction. The prevailing northward dip is of 2 to 4 degrees, increasing gradually towards the axis of the basin. There are a few very gentle noses and depressions, but the characteristic and economically important structural feature of the area is an intricate fault pattern. The larger faults trend parallel to the axis of the basin, and are invariably intersected by smaller faults trending northwest-southeast.

Accumulations of oil and gas in the Greater Oficina fields are found on the basinward (north) side of the longitudinal faults, lateral closure being provided by cross faults or, in some cases, by pinch-out of the sand. The number of sands varies from 50 to 70 in different parts of the area, with the maximum number of productive sands in any one field being 47 (West Guara field). The large number of sands, coupled with the many faults and some sand pinch-outs, has resulted in some outstanding examples of multiple-reservoir fields in the area, the biggest number of Oficina reservoirs in any one field being 100 (Oficina field proper). Oil staining, but no production, has frequently been found in the sands of the underlying Temblador formation.

The reservoir sands, which range in thickness up to a little more than

100 feet, are soft, compact, fairly clean, and of medium-grained texture. Porosities usually range from 20 to 25%, sometimes approaching 30%, and permeabilities usually range from 300 to 1,000 millidarcies, occasionally reaching several thousand millidarcies. A few of the sands carry only gas. Almost all the oil reservoirs have gas caps, with the oil columns varying in height up to 500 feet. Production depths vary from 4,000 feet to 7,000 feet, and, because of the many reservoirs, dual- and triple-zone completions are common.

The oil is usually of intermediate base, varying in gravity from 8° to 57° API, but no oil heavier than about 15° API has been exploited. Pour-points vary from below 50°F to 95°F, with the paraffin-wax content varying from 0.5% to more than 15%. The wax oils (more than 4% wax content) usually have a gravity lighter than 35° API. Sulphur content varies from less than 0.5% (lighter oils) through 1.2% (medium oils) to 2.3% (heavier oils).

3. The Las Mercedes–Tucupido Trend. In the total of 13.5 million barrels produced in this area up to 1950, since the beginning of production in 1943, 5.77 million came from the Las Mercedes field, discovered in 1943, and producing 11,009 b/d in December 1949. The Tucupido field, which began producing in 1945, had yielded 4.89 million barrels at the end of 1949 and was producing at the rate of 13,594 b/d. The Palacio field, discovered in 1947, was producing at the rate of 5,218 b/d. Two fields, Guavinita and Sabán, began a small production in 1948, and the Ruiz field, where production began in 1949, was producing 3,750 b/d in December.

THE APURE BASIN

Serious exploration in the Apure basin was begun only in 1947. Up to March 1949, two or three wildcats are reported to have found commercial production in both the Eocene and the Cretaceous, and future prospects for the basin are considered promising.

History of the Oil Industry

Oil was discovered in Venezuela by the Spanish *conquistadores,* who were impressed by the large asphalt seepages near the shore of Lake Maracaibo. In those early days asphalt became widely used for calking ships, and many English and French pirates of the sixteenth century forced an entry into Lake Maracaibo to get pitch.

The earliest commercial enterprise, however, was not launched until 1878, when a small Venezuelan company was started in the State of Táchira. It drilled a few wells some 60 feet deep in a seepage of high-gravity oil, and produced and refined some 15 b/d of crude, for local consumption as illuminating oil.

More serious attempts were directed to the production of asphalt. Early in the present century several North American and British companies became interested in exploiting the large asphalt deposits in both eastern and western Venezuela. In 1910 the New York and Bermudez Company started an ambitious exploitation of the large Guanoco asphalt lake in eastern Venezuela. It also drilled some wells, and discovered a heavy-oil field in 1913. However, the asphalt industry began to peter out with the first World War, and in later years the rapid development of large oil fields of asphaltic crudes supplied asphalt as a by-product.

The earliest commercial field of importance in Venezuela was Mene Grande, near the eastern shore of Lake Maracaibo, discovered by the Caribbean Petroleum Company, a subsidiary of the Royal Dutch-Shell group, in 1914, down-dip from one of the oil seepages. From 1917, when the field entered commercial production, to December 1949, it had produced 338 million barrels, and it was still producing at the rate of some 49,519 b/d.

In 1917 the Venezuelan Oil Concessions, Ltd., another subsidiary of the Royal Dutch-Shell group, discovered the La Rosa field. In 1922, when one of its wells started blowing wild for an estimated 100,000 b/d, it became world-famous, thus calling attention to the tremendous productive possibilities, and initiating an era of active prospecting, in both western and eastern Venezuela.

The history of the search for oil in Venezuela, and the development of its industry after the discovery of the Mene Grande field, has three distinct phases: (1) the wildcatting in the areas of surface indications of oil (seepages); (2) the search for oil in the llanos of eastern Venezuela by the extensive use of geophysical methods; and (3) the search for oil by deeper drilling in already productive areas.

The phase of wildcatting near the seepages was the heroic period in the history of Venezuelan oil, and by far the most successful. With scant knowledge of the geology and with primitive drilling tools, British and North American wildcatters went through Venezuela, "leaving a train of oil fields in their wake." The country yielded oil "more or less without regard to its geological constitution," [1] and the greatest Venezuelan fields were found at that time—Mene Grande, La Rosa, Lagunillas, La Paz, Quiriquire. The saga of this period has never been written, but it marks a turning point in Venezuelan history, for it changed a pastoral country into a major producer of one of the world's prime commodities.

When this first period of discovery ended in 1929, with the world crisis of that year, yearly production had increased from 2.2 million barrels in 1922 to 136 million in 1929.

During this first period there were several score companies scouring the country for oil, but only three were really successful: the Royal Dutch-Shell group, with its fields in the Districts of Sucre, Colón, Maracaibo, and Bolívar in the State of Zulía; the Lago Petroleum Corporation and the

[1] These phrases were first used by Wallace Pratt to describe the opening up of oil fields by wildcatters in the United States (*Oil in the Earth,* p. 47).

Standard Oil Company of Venezuela (now merged into the Creole Petroleum Corporation), with their fields in the Lake Maracaibo area in western Venezuela and at Quiriquire in eastern Venezuela; and the Venezuela Gulf Oil Company (today Mene Grande Oil Company), with its fields in the shoreline strip of Lake Maracaibo. Though many other companies have since searched intensively for oil in Venezuela, these three have continued to increase their reserves, both in the original concessions and in new ones.

The second phase, that of the search for oil in the llanos, was made possible by the development of geophysics, and by the increased knowledge of the geology of Venezuela. Over the surface of the great prairies there are no oil seeps, and rocks that the geologists knew were favorable for oil accumulation, which outcrop on the foothills of the Andes and Perijá ranges, are not exposed on the llanos. However, the geologists could project them from the outcrops hundreds of miles away and, with the aid of the new science of geophysics (the torsion balance first, and later the seismograph and gravimeter), which gave general suggestions of favorable places to test, they started discovering important fields in the eastern llanos. Oficina field, discovered in 1937 by the Gulf Oil Company in the midst of the Mesa de Guanipa in Anzoategui, marks the first success of the period, to be followed by the discovery of the Jusepín field in 1938 by the Standard Oil Company of Venezuela, and later the Roble-San Joaquín-Leona-Guara fields. The discovery of these fields initiated the production of light oils of the specialty class, suitable for such products as wax, etc. Of major significance to the country was the fact that the search for oil in the llanos brought highways, medical facilities, hospitals, and new cities.

The third phase, that of the search for oil beneath the known oil fields, got its really successful start with the discovery of production from Eocene rocks by the Lago Petroleum Corporation (Creole) in 1937, beneath their Miocene production in Lagunillas field. The discovery of that first well, LL-370, one of prime magnitude, together with the discoveries of the Mene Grande field in 1914, the Oficina field in 1937, and the Cretaceous production beneath the La Paz field in 1944, stand as four milestones in the development of the Venezuelan oil industry.

The Government and the Petroleum Industry

"The prime requisite to success in oil-finding," as Wallace Pratt has put it, "is the freedom to explore, and only slightly less imperative is freedom to develop and produce the oil once it is found." As Venezuelan petroleum legislation has evolved from the old *Codigo de Minas* of 1907 to the *Ley de Hidrocarburos* of 1943, taxes have been increased, and greater control has been introduced in various phases of the industry, but the provision that "any person, national or foreigner . . . may freely explore and carry on geological or geophysical investigations . . . in the national territory" has passed unchanged from law to law. By contrast, in Brazil and Colombia, countries where exploration has been restricted by antiquated or prohibitive legislation, the industry has not prospered at the same pace. A *résumé* of Venezuelan petroleum legislation from the beginning to 1948 is given at the end of this chapter.

Another factor of paramount importance has been the organization by the Venezuelan government of the department known as the Inspectoria Tecnica de Hidrocarburos, which was founded in 1930 under Minister of Fuel Gumersindo Torres to supervise the development of the industry. With a Venezuelan staff of petroleum engineers and geologists of first rank, it has maintained a standard of administration that has few equals in the world; and it has been able, during profound political changes, to maintain a continuity of action that has been the basis of the cordial relations of the industry and the government.

The capital required for the development of Venezuela's oil resources has, of necessity, been obtained abroad. As mentioned above, British and Dutch enterprises were the pioneers in the search for oil in Venezuela, followed promptly and aggressively by some of the major units in the oil industry of the United States. For many years considerably more than one-half of the oil production of Venezuela has resulted from the operations of the subsidiaries of three great international oil companies: the Royal Dutch-Shell (with a number of Venezuelan subsidiaries); the Standard Oil Company (N.J.), which is the controlling stockholder in Creole Petroleum Corporation, the largest producer in Venezuela at present; and Gulf Oil Corporation (Mene Grande Oil Company). Nevertheless, numerous other oil companies, Venezuelan, North American, and European, operate successfully in Venezuela, and many individuals, mostly Venezuelans, own substantial interests in the individual concessions and oil fields that are now merged to form the holdings of the larger producing companies.

A factor of prime importance in the economy of any country today is the price of gasoline and other petroleum products. In Venezuela the price of gasoline is the lowest in the world, partly because of the government's tax policy. In March 1949 the price of a gallon of gasoline in Caracas was the equivalent of 11.30 U.S. cents, or less than half the price in New York City, 25.10 cents, at the same date. Both price figures included tax. In Caracas the tax rate was equivalent to 1.22 cents a gallon, and that in New York to 5.50 cents.

Changes in the Economy of Venezuela Caused by the Industry

The petroleum industry has caused profound changes in the economy of Venezuela. One of the nation's leading economists thus describes what has happened:

"A nation in an initial stage of her development from a backward and disorganized agro-pastoral economy in which there suddenly appears, like an unnatural growth, the enormous creation of riches produced by the oil industry and which, in twenty years, changes from being the world's fifth or sixth exporter of coffee to its leading exporter of petroleum, finds herself faced with an upheaval so intense and so dangerous that out of it may come genuine and enduring greatness, or subjection to the fortunes of fuel oil.

"This, expressed summarily and even brutally, is the way we should plan in what we may call Venezuela's fundamental economic problem: we must use petroleum to develop Venezuela instead of permitting her future and the course she follows to be diverted and unnaturally displaced by the violent torrents of wealth produced by that industry and by all the special conditions that its existence brings about day after day." [2]

Before the development of the oil industry, thirty years ago, the country was not rich, but it was capable of producing sufficient food for its needs. The standard of living and the value of food consumed per person were, of course, very low, but Venezuela was self-sustaining.

Today, on the contrary, the situation appears paradoxical; the nation has a proved oil reserve of 8 billion barrels, worth potentially at least that many billion U.S. dollars, but its standard of living does not yet meet the expectations of its citizens, nor has it kept pace with the national income now being produced from oil. Of course, oil in the ground is not money in hand. But the existence of this stored potential provokes a feeling of money in the bank. There are unrealized riches on the one hand, and acute human needs on the other.

The oil industry has become the main source of the national income. In 1947–1948, when Venezuela adopted its first billion-bolivar budget (actually 1.5 billion bolivares, or approximately $450 million), the government's income from the oil industry was some $290 million. This is believed to be the highest per-capita budget in Latin America, and oil provides more than six of every ten bolivares of the income supporting this record expenditure.

The oil-produced government income has been and is being spent rapidly, in an attempt to give everyone a living more in keeping with the national resources. An intensive program of building is under way, including schools, hospitals, low-cost housing, sewers, roads, and waterways. Incidentally, by building roads for its own needs, the oil industry has given the nation 20% of its all-weather roads. Placement of immigrants, industrial production, food production, promotion of art, literature, music—all these betterments are under way. The needs are acute, but great progress is being made.

The effects of the capital invested in the oil industry in Venezuela may perhaps be summarized as follows. In 1914, the year of the discovery of

[2] Translation from the Spanish text (p. vii) of *Sumario de Economía Venezolana* (Caracas, 1945), by A. Uslar Pietri.

the Mene Grande field, Venezuela was growing most of its own food, raising cattle for its own needs (and even exporting some), and was exporting coffee and cocoa. In 1920, when oil production reached 1,426 b/d and proved reserves were estimated at 400 million barrels, Venezuela was still making a living by growing most of its own food (imports had started), and exporting coffee and cocoa. In 1930 oil production reached 370,000 b/d; then, after a slump, it climbed steadily to 562,000 b/d in 1939, hit a new peak in 1941, still another in 1944, passed the million mark in 1945, and exceeded 1.4 million b/d in 1949. Meanwhile, from 1914 onward Venezuela had steadily been losing the ability to feed and clothe itself. In 1946, although it continued to export coffee and cocoa, it imported 26% of the rice it ate, 21% of the corn, 93% of the wheat flour, and 38% of the sugar; and in 1947 it began to import meat.

The claim, however, by one section of opinion, that agriculture was ruined by the oil industry, when it drew men away from the farms, is not borne out by statistics; today, at the peak of its development, the oil industry employs some 59,000 men and women, only about 1.5% of the total population and only from 2 to 3% of the total working population.

A Résumé of Venezuelan Petroleum Legislation

THE OWNERSHIP OF THE OIL

In the majority of Latin-American countries the principle has been maintained, since colonial times, that ownership of all the subsoil minerals is vested in the nation, and that its control is effected through the government.

Early in the eleventh century there appeared the following Decree of the Royal Crown of Spain: "All mines of silver, gold, lead and any other kind of metal, or of any other thing whatsoever, which may be in our Royal Domains, belong to Us (the Royal Crown), wherefore no one shall dare to work said mines without our special license order."

The Laws of the Indies, promulgated in 1602, authorized the Royal Governors of the Spanish Colonies to apply the mining laws of Spain throughout the Colonies. The Mining Ordinances of New Spain, in 1783, made deposits of petroleum in the Spanish Colonies of the new world the property of the Royal Spanish Crown, and these ordinances were sanctioned, ratified, and made effective in the Gran Colombia (which included what is now Venezuela, Colombia, Panama, and Ecuador) by a decree of the Great Liberator, Simon Bolivar, issued at Quito, Ecuador, on October 24, 1829.

An article in the ordinances of 1783, later ratified by the Liberator, read: "Likewise I concede that there may be discovered, solicited, recorded and denounced in the manner aforesaid, not only the mines of gold and silver, but also mines of precious stones, copper, lead, tin, silver, antimony, calamine, bismuth, rock salt *and any fossil matters, whether they may be perfect minerals, bitumens, or juices of the earth,* and proper provision shall be made for the acquisition, enjoyment and development thereof."

The principle of the ownership of the subsoil minerals by the nation, and the control of their production by the government, has been maintained throughout the life of the Republic.

EXPLORATION AND EXPLOITATION

Petroleum was not of special interest to the Venezuelan government until the beginning of the year 1900. The activities of the few companies then active were limited to the exploitation of large asphalt deposits in the asphalt lakes of Guanoco and La Brea in eastern Venezuela, and at Inciarte at the northwestern corner of the State of Zulía, together with a primitive enterprise for collecting the oil from a seepage at Rio Paují in the mountain region of Trujillo in the Andes.

With the enactment in March 1918 of the first Petroleum Ordinance of Venezuela, it was provided that rights for the exploration and exploitation of asphalt, petroleum, and other hydrocarbons could be acquired only through special contracts granted by the Federal Executive. It was the intention of the government to give special attention to the production of petroleum by administering directly the granting of rights for its production. To this end the government adopted a procedure different from that governing the production of gold, copper, iron, coal, and other minerals, which had been and still are produced in Venezuela in large quantities.

All persons in Venezuela may freely undertake surface exploration, whether geological or geophysical, to look for oil, except on lands covered by valid petroleum concessions. This does not mean that an oil prospector upon finding a territory of oil-bearing possibilities, will be able to stake it off, apply for a concession, and be assured of obtaining it. The federal government has a discretionary power in the granting of concessions, and generally these are granted to responsible operating companies that have the means for exploring and developing the area promptly.

From the year 1918 to the present, various laws, decrees, and ordinances have been enacted, and each has been a definite forward step in Venezuela's petroleum legislation.

The 1920 Petroleum Law carried one of the most important provisions ever enacted in the mining history of the country: it grants to the landowner, *for a period of one year,* the right to obtain a permit for the exploration of oil, and the subsequent right to obtain an exploitation concession over the area covered by his property.

The 1920 Petroleum Law is the first and only legislation vesting in the landowner the exclusive right to obtain from the Federal Executive the right of exploration and exploitation of oil from the subsoil covered by his property. This provision of law tended to increase the value of private lands by giving the landowner an opportunity to obtain a petroleum concession and then sell it to an operating company, while maintaining intact his rights over the surface, with the probability of obtaining a reasonable royalty on the production of oil and a permanent income from rentals resulting from rights-of-way over his property for the necessary activities of the company.

Hundreds of landowners throughout the territory of the Republic appeared before the Land Registry Offices in the years 1920 and 1921 to file declarations of their decision to obtain a concession for the exploration and subsequent exploitation of their properties. These declarations, in accordance with the procedure established by the 1920 and successive laws, resulted in the granting of exploration and exploitation concessions covering more than 7,500,000 hectares. A great percentage of these concessions have been sold to operating companies, and incalculable wealth has thus been distributed among the farmers and cattle men of the country.

It was the 1922 Petroleum Law that established the basic principles of the present Venezuelan petroleum legislation. New petroleum laws were adopted in 1925, 1928, 1936, and 1938, but none of these radically modified the structure of the 1922 Law, and the basic principles of this law have made possible the development of the industry to the point where Venezuela has attained second place in world production. However, the fact that the petroleum concessions were governed by different petroleum laws, with varied provisions relative to inspection of production and ways of computing taxes, was confusing. A standardization of rights and obligations for all petroleum concessions was greatly to be desired.

An agreement was finally reached in 1942 between the government and the operating companies on the terms by which all active petroleum concessions would be converted to a uniform status. As a result of this cooperative effort, the 1943 Petroleum Law was enacted by the Venezuelan Congress, and the oil companies converted all their concessions in accordance with this law.

The 1943 Law authorizes the granting by the government of concessions for the following purposes:

1. The exploration of specified lots not exceeding an area of 24,710 acres, with the inherent right, after three years of exploration, to exploit an area up to a maximum of one-half the original exploration area. The other half of the exploration area reverts to the government as National Reserves.

2. The exploitation of specified lots not exceeding 1,235 acres. The exploitation concession does not carry the obligation to drill in search of oil, but an annual surface tax per hectare must be paid on it, and this tax is increased periodically.

3. The manufacture or refining of petroleum.

4. The transportation of oil, gas, etc., and of products derived therefrom.

30. Eastern Venezuela. Derrick under construction in jungle country, Quirequire field. (Standard Oil Company, N.J., photo by Vachon.)

31. Eastern Venezuela. The impact of the oil industry on the northern llanos. The recently completed pipe line and highway between the Jusepín area fields and Puerto de la Cruz. (Creole Petroleum Corporation.)

33. Eastern Colombia. Barco
cession. Mule transport of sup
during exploration, Catatu
basin. (Photo by C. W. Hubn

32. Colombia. Middle Magdalena basin. Double
of Andian National Pipe Line, near Barrancabern
(Standard Oil Company, N.J., photo by Collier.)

34. Colombia. Between Armenia and Ibague. A
ton truck loaded with structural steel climbing the
tral Cordillera. (Standard Oil Company, N.J., phot
Collier.)

The concessions mentioned in items *3* and *4* may be granted separately, but they shall always be considered as annexed to exploitation concessions.

TAXES

The most important *direct taxes* levied on the petroleum industry are the following:

1. Exploration tax of 2 bolivars per hectare per year during the three years of the exploration period on exploration concessions.

2. Initial exploitation tax (of 8 bolivars per hectare as a minimum) which may be considered as the actual purchase price of the concession. (Some acreage has been sold for as much as 2,000 bolivars per hectare, due to its proximity to a producing field.)

3. Surface tax on exploitation concessions. As already mentioned, this tax is increased periodically, during the 40 years' life of the concession, as follows:

First	10	years	= 5	bolivars	per	hectare	per	year
Next	5	"	= 10	"	"	"	"	"
Next	5	"	= 15	"	"	"	"	"
Next	5	"	= 20	"	"	"	"	"
Next	5	"	= 25	"	"	"	"	"
Last	10	"	= 30	"	"	"	"	"

4. Exploitation tax (government royalty). This tax is fixed at 16⅔% of the crude oil extracted, as measured in the producing field. This same tax applies to natural gas, even to that used as fuel in producing operations.

5. Tax on refined products used for consumption within the country, equivalent to 50% of the import duties which they would have yielded if they had been imported. This tax does not apply to products refined in the country for export.

In addition to the above direct taxes, the oil companies must pay all other general taxes, such as income tax, import duties on imports, consular duties, revenue stamps, etc.

An amended income tax law, published November 12, 1948, provides for an additional tax to be levied on oil companies and other industries holding concessions for the development of national resources whose net profits exceed taxes paid to the government, provided such companies have earned more than 10% on their invested capital. In arriving at the amount of the additional tax, the companies are allowed credit for other taxes paid, including import duty, concession surface tax, exploitation tax (royalty), normal income tax, and complementary income tax.

3. TRINIDAD

BY A. J. FREIE *

The General Framework

LOCATION AND AREA

TRINIDAD, the greatest oil-producing colony [1] in the British Empire, is an island lying immediately south of the Lesser Antilles in the West Indies, between 10° 2′ and 10° 50′ north latitude and 60° 55′ and 61° 56′ longitude west from Greenwich. It has an area of 1,862 square miles and is roughly rectangular in shape, with pronounced peninsulas at the northwest and southwest corners.

Trinidad is a prolongation of Venezuela, and at the north is separated from the mainland by a body of water six miles wide, known as the Dragon's Mouths. At the south there is a shallow channel, ten miles wide, known as the Serpent's Mouth, separating the island from the delta of the Orinoco River. The large volume of water discharged by the river into the southern end of the Gulf of Paria reduces the salinity of the Gulf's waters below that of normal sea water. The greater part of the Gulf is less than 100 feet in depth, with sudden deepening at the approach of the Dragon's Mouths. The Gulf of Paria has an area of approximately 2,000 square miles.

TOPOGRAPHY AND GEOLOGY

Topographically, Trinidad may be divided into five zones, running parallel to each other in a north-northeast direction across the island. It has three ranges of hills, separated by two belts of low and relatively flat land.

The highlands at the north, referred to as the Northern Range, consist of two parallel ridges having an average elevation of 1,500 feet. A few high peaks are present, of which the highest rises to 3,085 feet above sea level in the central portion of the southern ridge.

Immediately south of the Northern Range there is a lowland belt, known as the Northern Basin, extending across the island from the Atlantic Ocean to the Gulf of Paria. It has a maximum elevation of about 200 feet in the center to form a divide, from which the drainage goes westward via the Caroni River and eastward via the Oropuche River.

* Head, Producing Department, Creole Petroleum Corporation.

Author's Note: In the preparation of this paper the writer had access to the files of the Creole Petroleum Corporation, and also acknowledges the cooperation and assistance of his fellow workers. Trinidad Leaseholds, Inc., have kindly checked the map of petroleum facilities.

[1] Although Trinidad held the leading position in cumulative production up to the end of 1949, British Borneo, with an average rate of production of 67,750 b/d in 1949, was surpassing the Trinidad rate of 56,560 b/d.

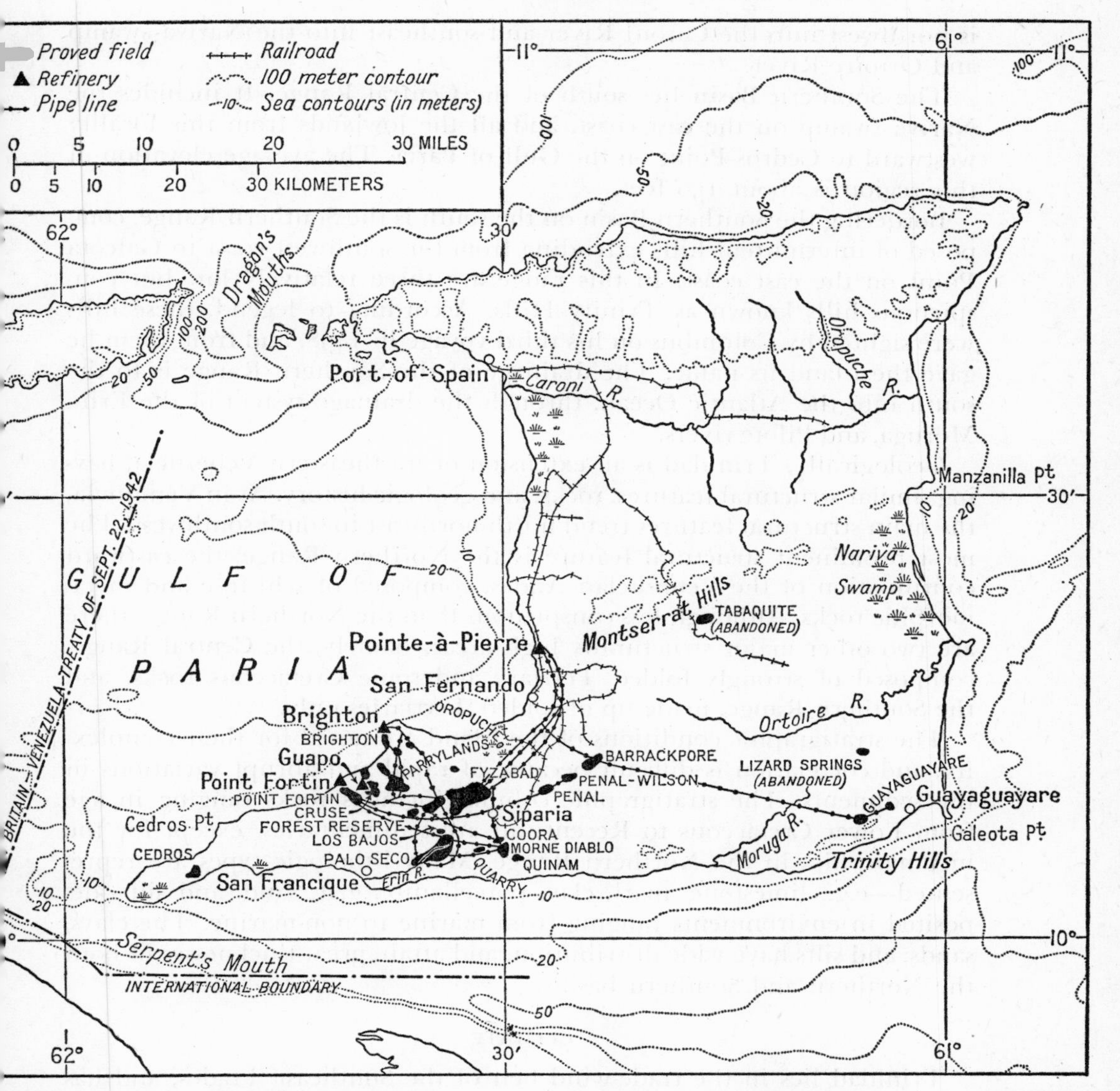

12. Trinidad: oil fields and facilities. Names of fields are in Roman capitals, except that the names of the po field and several lesser fields in the central area have been omitted for lack of space. The Cedros and Parrylands fields were not producing in 1948. The discovery field at Oropuche is indicated by cut-. Refineries and pipe lines are described in the text.

To the south of the Northern Basin there is a range of low-lying hills, about 7 miles wide, extending from the west coast at Pointe-à-Pierre to Manzanilla Point on the east coast, known as the Central Range. The most conspicuous topographic features in this range are found in its central part, where the Montserrat Hills and Tamana Hill attain an elevation of 1,000 feet. These hills form a drainage divide, and the main drainage is northwest into the Caroni River and southeast into the Nariva swamp and Ortoire River.

The Southern Basin lies south of the Central Range. It includes the Nariva swamp on the east coast, and all the lowlands from this locality westward to Cedros Point on the Gulf of Paria. The average elevation of this region is about 150 feet.

Bordering the Southern Basin on the south is the Southern Range, composed of intermittent hills extending from the southwest coast to Galeota Point on the east coast. In this range are three relatively low but conspicuous hills known as Trinity Hills. According to legend, these hills were sighted by Columbus on his third voyage in 1498, and from them he gave the island its name. The drainage of the Southern Range is to the south into the Atlantic Ocean, through the drainage system of the Erin, Moruga, and Pilote rivers.

Geologically, Trinidad is an extension of northeastern Venezuela, having similar structural features, rocks, and geologic history. As in Venezuela, the main structural features trend north-northeast to south-southwest. The most prominent structural feature is the Northern Range, the eastward continuation of the Venezuelan Andes, composed of schistose and metamorphic rocks. Although less conspicuous than the Northern Range, there are two other major structurally high areas: namely, the Central Range, composed of strongly folded Tertiary and some Cretaceous rocks, and the Southern Range, made up of folded Tertiaries only.

The stratigraphic conditions of the island are noted for their complexity, and correlation is difficult because of rapid and abrupt variations in the sediments. The stratigraphic column includes rocks ranging in age from Lower Cretaceous to Recent; all are sedimentaries, except for the metamorphics in the Northern Range. Many lithologic types are represented—e.g., limestone, marl, clay, porcellanite, grit, sand, and silt—deposited in environments ranging from marine to non-marine. The clays, sands, and silts have wide distribution, and attain great thicknesses in both the Northern and Southern basins.

CLIMATE

Trinidad lies in the trade-wind belt of the Southeast Trades, and has a warm tropical climate. The rainfall is heaviest in the east, where it averages about 100 inches per year, and diminishes westward to about 60 inches. As in most tropical countries, Trinidad has two seasons: the wet, from June to December, and the dry, from January to June. There is little variation in temperature between the two, but the dry season is slightly cooler than the wet. The mean annual temperature is 78°F (maximum 92° and minimum 67°) and is more temperate than the northern West Indian islands. Trinidad is not subjected to high winds, and lies outside the usual paths of the West Indian hurricanes.

3. TRINIDAD

POPULATION

The island's population at the census in 1946 totaled 530,809 [2] representing an increase of 35% over the 1931 census. The people are a heterogeneous group, one-third of them being East Indians, who constitute more than half of the rural population. The majority of the remaining two-thirds are negroes and mixed races of English, French, and Spanish extraction. The total number of Americans in 1946 was about 4,500, including personnel in the armed forces of the United States occupying military bases on the island.

The mixture of peoples found in Trinidad is directly related to its political history. The island was ceded by Spain to Great Britain in 1802 by the Treaty of Amiens. Prior to that date, it was raided by the British, Dutch, and French with resultant influences of these nationalities. A French immigration occurred near the close of the eighteenth century, when the Spanish governor offered special advantages for settlement, and many people came to the island, including a number of Frenchmen from Santo Domingo and elsewhere who sought to escape the persecution of the French Tribunals. The negro population was originally brought to the island as slaves. Shortly after 1834, when slavery was abolished, a labor shortage developed on the cocoa and coffee plantations. To remedy this situation, East Indian indentured servants were permitted to enter Trinidad in 1844. They came to Trinidad under contract to serve as laborers for a certain number of years, usually five, in return for wages and other benefits, plus passage. After expiration of the indenture contract, they could re-indenture themselves, seek other employment, or become peasant proprietors.

GOVERNMENT

Trinidad has enjoyed a stable government since it became a British Crown Colony in 1802. The colony is administered by an appointed Governor, a partially elected Executive Council, and a partially elected Legislative Council. The elected members constitute a majority in both councils. In 1945 universal adult franchise was granted.

ECONOMY

Trinidad is one of the most prosperous islands in the West Indies and owes its prosperity largely to the petroleum industry. Its principal economic activities are the producing and refining of petroleum, agriculture, and the transshipment of goods and materials to other West Indian islands and South America.

The colony has never been self-supporting in food; and, as a result,

[2] The island of Tobago, which is included administratively with Trinidad, had 27,161 people at the April, 1946, census, making a total of 557,970 in both islands. Their estimated population on December 31, 1947, was 586,700.

its principal imports are foodstuffs. Other imports consist mainly of crude petroleum for refining, agricultural and engineering machinery, textiles, vehicles, clothing, lumber, and construction materials. Its exports are mostly petroleum and petroleum products and tropical agricultural products.

Of the $82,262,232 worth of domestic commodities exported in 1947, petroleum and petroleum products were valued at approximately $61,-897,237, or 75% of the total, and asphalt and products at $2,925,183, or 3.5% of the total. The volume and destination of the petroleum exports during 1947 (excluding some 8,000,000 barrels of ships' bunkers included in Table 5: "Estimated Deliveries into Consumption Annually") is shown in Table 4.

Table 4. Trinidad and Tobago: exports of petroleum products in 1947, by quantity and destination.

Destination	Total	Quantities Exported Crude Oil	Motor Spirit	Kero-sene	Fuel Oil	Lubricat-ing Oil	Other [a]
	(In thousand barrels)						
United Kingdom	6,763	—	2,994	1,662	2,107	—	—
Brazil	1,727	—	936	126	653	—	12
Canada	1,088	1,088	—	—	—	—	—
United States	140	—	—	—	123	—	17
Cuba	135	—	—	—	135	—	—
Sweden	131	—	34	—	94	—	3
Italy	110	—	—	—	110	—	—
Argentina	79	—	—	—	79	—	—
Belgium	77	—	5	—	71	—	1
Venezuela	3	—	3	—	—	—	—
All other (military, unknown, etc.)	4,319	—	1,143	53	2,804	2	317
Total	14,572	1,088	5,115	1,841	6,176	2	350

[a] These represent asphalt and asphalt products only, except in the shipments to the United States, which are unspecified.

Table 5. Trinidad: estimated deliveries of petroleum products into consumption annually in 1941 and in each year from 1944 through 1947.

	Quantities Delivered in: 1941	1944	1945	1946	1947
	(In thousand barrels)				
Aviation Gasoline	49	417	176	55	81
Other Naphthas	227	278	264	287	345
Refined Kerosene	46	84	98	104	103
Distillates					
Land	73	175	136	114	114
Bunkers	699	832	796	632	1,011
Lubricants	23	21	14	18	20
Residuals					
Land	372	442	372	463	438
Bunkers	6,765	8,422	7,965	7,052	8,410
Total	8,254	10,671	9,821	8,725	10,522

3. TRINIDAD

Approximately 10,000,000 barrels of refined products, or the rough equivalent of 50% of the crude produced annually, is delivered into domestic consumption, including ships' bunkers and aviation refueling in Trinidad. Practically all the marketing is done by Shell, Trinidad Leaseholds, and Standard Oil Company of Trinidad. The volume of products delivered into all such consumption during 1947, and a few years prior thereto, is shown in Table 5. The figures in the table reflect abnormal consumption during the war years, when United States forces were highly active at the various military bases on the island. Since 1946 the consumption trend has been definitely upward for most products.

Petroleum

HISTORY

Trinidad's many surface indications of petroleum in the form of active oil seepages, asphalt deposits, and mud volcanoes aroused interest in the search for oil at an early date. The first well was drilled in 1866,[3] in the southern part of the island near La Brea, by the Trinidad Petroleum Company. It is reported that strong showings of heavy asphaltic oil were found, but not in commercial quantities. During the same year, Walter Darwent drilled a well to about 150 feet in the vicinity of San Fernando, but it, too, was non-commercial. Darwent was not discouraged by this failure, and in 1868 drilled another well located near an asphalt seepage at Aripero, about two and a half miles west of Oropuche village. The well was carried to a depth of 160 feet, and is reported to have struck oil near bottom. The records do not indicate that it was ever commercial. In about 1870, a well was drilled near the Pitch Lake at Brighton, but although strong showings of heavy oil were encountered, the venture was abandoned as a non-commercial proposition. During the next thirty years, no further drilling is reported, but around 1900 two wells were drilled near the site of the Darwent well, to a depth of about 500 feet. Both wells encountered showings of heavy oil, but neither obtained commercial production.

The disappointing results experienced at Aripero caused the shifting of interest to an area of asphalt seepages near Guayaguayare, on the east side of the island. Between 1902 and 1907 about thirteen wells were drilled at this locality; many encountered strong showings, but none was considered commercial.

At about this time the well-known British geologist, E. H. Cunningham Craig, reported on the oil possibilities of the island for the Trinidad Government. His reports revived interest in the search for oil, and the first commercial well was completed in 1907 near the Point Fortin anticline, in proximity to asphalt seepages.

[3] Gerald A. Waring, Geology of the Island of Trinidad, *The Johns Hopkins Univ. Studies in Geology, No. 7* (1926).

TRINIDAD OIL FIELDS

Since the completion of the discovery well, sixteen major fields, all operated by British companies, have been discovered. They have a combined proved acreage of some 18,000 acres. At the end of 1948 the total production of these fields and minor fields was about 56,000 b/d, and their total cumulative production up to that time was slightly over 370,400,000 barrels. Table 6 shows the production by fields at the end of 1947; and Table 7 the geological data of the fields.

GEOLOGICAL SUMMARY OF THE OIL FIELDS

The producing fields of Trinidad are located in the southern part of the island, in the relatively narrow Southern Basin. Accumulation is found in anticlines, domes, stratigraphic traps, synclines, and along faults.

It is not uncommon to find the crestal regions of the structural highs barren, and the oil concentrated on the flanks of the same structural features. In the case of the Fyzabad anticline, for example, the high north flank is practically barren, whereas good accumulation is found on the lower south flank in proximity to strike and dip faults. Apparently the oil migrated from older beds upward through disturbed zones and accumulated in sands along the faults, with migration to the crest of the structure blocked by sedimentary barriers. Trinidad operators are well aware of this condition and seldom overlook prospecting the structurally low areas, especially in regions of faulting.

Table 6. Trinidad: crude oil production, by fields, at end of December 1947.

Name of Field [a]	Number of Producing Wells	Crude Oil Produced: Daily Output	Crude Oil Produced: Total Cumulated
		(In barrels)	
Fyzabad	753	23,583	179,521,758
Guapo	353	9,317	66,072,085
Coora-Quarry	258	8,495	29,294,555
Palo Seco	186	1,455	23,696,100
Barrackpore-Penal	139	9,403	20,408,463
Point Fortin	78	1,827	10,517,844
Brighton	68	1,384	6,266,134
Los Bajos	29	392	3,360,812
Other fields	70	1,075	8,236,920
Abandoned fields	—	—	4,853,537
Totals	1,934	56,931	352,228,208

[a] Further data on most of these fields is given in Table 7.
Source: *World Oil Atlas 1948,* p. 125.

The production of Trinidad is obtained from the Miocene, with the exception of negligible quantities from the Cretaceous-Eocene at Lizard Springs. The main reservoirs are found in highly lenticular sands and silts of the Miocene Forest and Cruse formations (Table 8). The petrolif-

Table 7. Trinidad: geological data on oil fields producing in 1947.

Name of Field and Year Discovered	Proved Area	Producing Horizons [a]	Type of Accumulation	Depth Range of Sands: Upper	Depth Range of Sands: Lower	Gravity Range of Oil
	(Acres)			(Feet)		(Degrees API)
The Southwest						
Fyzabad (1920)	5,100	Forest & Cruse	Anticline	1,000	6,500	19.0 to 30.0
Guapo (1912)	3,050	" "	Plunging Nose	1,500	5,700	16.0 to 24.0
Parrylands (1911)	3,000	" "	Anticline	1,600	1,800	. . .
Forest Reserve (1913)	2,700	" "	"	300	8,500	14.0 to 33.0
Point Fortin (1908)	1,300	" "	Fault	1,500	7,500	19.5
Coora-Quarry (1936)	900	" "	Syncline	1,200	6,500	25.9
Brighton (1910)	500	Forest	Anticline	700	3,000	18.0
Los Bajos (1928)	400	Forest & Cruse	Fold-fault	1,000	6,000	27.5
Wilson-Penal (1908)	280	Wilson	Faulted Anticline	1,600	2,500	12.2 to 25.7
Cruse (1913)	250	Cruse	Stratigraphic	1,800	4,000	13.2 to 30.5
Barrackpore (1913)	120	Wilson	Faulted Anticline	500	3,500	17.3 to 30.6
Palo Seco (1929)	100	Forest & Cruse	Stratigraphic	1,500	5,500	17.8 to 33.4
Morne Diablo (1938)	50	Cruse	Fault	1,000	4,700	20.0 to 33.2
Quinam (1936)	40	"	"	1,000	4,300	25.9 to 34.1
The Southeast						
Guayaguayare (1907)	110	Forest & Cruse	Fault	500	5,000	15.3 to 49.9

[a] All the producing horizons named are in sands of Miocene age. In the Tabaquite field, discovered in 1913 and now abandoned, the Tabaquite producing horizon occurred in sands of Miocene and Oligocene age, the accumulation was of fault type, the depths of producing sands ranged from 300 to 1,800 feet, and the oil was of 42.0 degrees API.

Source: *World Oil Atlas 1948,* p. 125; and other sources.

erous section of these two formations has a thickness of about 2,100 feet, in which the oil sands are irregularly distributed, and frequently intercalated with salt-water sands. Some of the sands carry abnormally high pressures, apparently caused by avenues of communication with deeper horizons provided by faults and fractures. In the early days, the abnormal pressures caused many blowouts and presented other drilling problems, but improved mud techniques have overcome these difficulties.

REFINERIES AND PIPE LINES

The refineries of Trinidad have a total crude-charging capacity of 98,000 b/d. Practically all of this capacity is concentrated in the Trinidad Leaseholds refinery at Pointe-à-Pierre (65,000 b/d of crude and 50,000 b/d of cracking capacity) and in the United British Oilfields refinery at Point Fortin (30,000 b/d of crude). Both plants were completely modernized during the war years, to the point where they could supply the various types of products required by the Allied ships and planes operating in the general Trinidad area. After the close of the war the industry had excess capacity over that required for processing domestic production. To help this situation, approximately 300,000 barrels monthly are imported in bond from Venezuela and Colombia for processing at Pointe-à-Pierre and Point Fortin.

Table 8. Stratigraphic column of young Tertiary deposits in main oil fields, southwest Trinidad.[a]

Series	Formation		Suite	
Pliocene	La Brea		La Brea sands, silts, clays and porcellanite	Sheet sand deposits
			Unconformity	
	Morne l'Enfer	Upper	Upper Morne l'Enfer sands, silts and clays	
			Lot 7 silt (marker at base)	
		Lower	Lower Morne l'Enfer sands, silts and clays	
			Unconformity	
Miocene	Forest	Upper	Forest silts	Highly lenticular deposits
			Upper Forest clay	
		Lower	Forest sands	
			Lower Forest clay or intermediate clay (marker at base)	
	Cruse		Upper, middle and lower Cruse sands, silts and clays	
	Palo Seco		Clays, including Sphaeroidinella clay and Herrera conglomerate	
	Angular Unconformity			
Oligocene	Alley Creek-Princes Town marls			

[a] Compiled by K. Schmidt, paleontologist, Trinidad Leaseholds, Ltd.

All of the crude oil produced in Trinidad is transported from the fields to the refineries and terminals by pipe line. Since the fields are closely concentrated, and in close proximity to the refineries, there are only about 125 miles of crude-oil trunk line in the entire island. The longest line has an overall length of 23 miles, and the greatest diameter of any line is 10 inches. In addition a natural-gas line of 12-inch diameter extends the 16 miles from the Forest Reserve field to Point-à-Pierre, and a refined-products line of 2-inch diameter the 12 miles from the Fyzabad field to Point Fortin.

TRANSPORTATION FACILITIES

The petroleum industry has played an important part in promoting the improvement and growth of the transportation facilities of the island. Although the construction of hard-surfaced roads was begun prior to the discovery of oil, the highway system has been greatly expanded and improved since active exploitation was started around 1920, especially in the south, where the producing fields are located. Today, there are nearly 2,500 miles of all-weather roads, most of which are surfaced with natural asphalt obtained from the Pitch Lake. The South Trunk Road, running from Port-of-Spain south, and the Churchill-Roosevelt Highway from Port-of-Spain east to Piarco airport, are the leading highways on the island. The latter was built by the U.S. Government during World War II, and now serves as a connecting link with other roads in the north. This substantial mileage of first-class roads, plus the availability of an adequate supply of motor fuel manufactured by local refineries, has made the automobile the principal means of land transportation. In addition to the automobile, there is the Trinidad Government Railway which operates 118 route miles (153 track miles) of line connecting Port-of-Spain with the principal towns to the south. The operation of these lines, as in the case of the automobile, is facilitated by an adequate supply of fuel obtained from local sources.

Because of Trinidad's strategic location on steamship routes to South America and the Caribbean area, together with the fact that it has excellent ship-bunkering facilities and its own supplies of fuel oil, at an economic price, the island has developed into the most important shipping center in the West Indies. Trinidad is also the transshipping point for bauxite brought from the Guianas in small vessels for reshipment to the United States and Canada. No doubt, this point was selected not only because of its geographical location, but also because of the availability of fuel oil and bunkering facilities.

As in the case of the steamship lines, Trinidad is also at the crossroads of several international airlines operating between the United States, South America, Europe, and the West Indies. The planes of these lines are conveniently fueled with gasoline obtained from local refineries, whose aviation-gasoline manufacturing facilities were installed during the war years. The trends in local consumption of petroleum products and the consumption of aviation gasoline and bunker fuel are indicated in Table 5.

TARIFF STRUCTURE

The customs policy of Trinidad is to secure revenue and protect Empire trade. Duties are levied on the majority of goods imported, but, in general, goods from Empire sources are favored with a 50% tariff reduction. Practically all dutiable goods are subject also to a surtax of 15% on their assessed duty. All imports are subject to license, and must be obtained from sterling sources, unless they fall within the essential requirements of the colony. Certain oil-field equipment is either exempt from duty or enjoys a reduction, especially if obtained from Empire sources.

LABOR

Trade unionism, although still in its infancy, is more highly developed in Trinidad than in any other British West Indian territory except Jamaica. Interest in unionism was stimulated by the Trade Union Ordinance of 1932. The Oilfield Workers Union of Trinidad was organized under this ordinance, and has a collective labor contract with the oil industry through the Oilfield Employers Association.

EFFECT ON LOCAL ECONOMY

The petroleum industry is largely responsible for having developed and for maintaining a reasonably satisfactory economic status for the people of Trinidad. Without it, the island's economy would be reduced to the level of that found in many other West Indian islands. The greatest direct benefit derived from the industry is the setting up of foreign exchange, resulting from petroleum exports, to permit the importation of the necessary foodstuffs required to feed the island's dense population.

Of the many indirect benefits, the most important is the opportunity for employment of many workers in a progressive industry. In 1947, there were about 14,000 people employed by the oil industry (and 600 in the asphalt industry)—the greatest number in any industry except sugar, which employed an average of nearly 20,000. The oil industry has done much in promoting the welfare of its workers, especially as regards wages.[4] Wage rates in the oil industry under an agreement effective December 15, 1947, are: skilled workers, 34½ to 50½ cents an hour; semi-skilled, 32½ to 34½ cents an hour; unskilled, 29½ to 32½ cents an hour; an 8-hour day is observed. Under an agreement effective April 1, 1947, government manual workers receive $2.20 to $3.38 per 8-hour day for skilled work; from $1.77 to $2.12, for semi-skilled; and $1.72 for unskilled. (All rates are expressed in Trinidad currency.)

Other indirect benefits derived from the industry are, of course, the income from taxes and import duties; the supplying of the island with adequate fuel, and the promotion of shipping. The importance of the petroleum industry to the economy of the island and to the welfare of its people cannot be overemphasized and, no doubt, is fully appreciated by the government. In this connection, it should be stated that the Trinidad government has continued to maintain a friendly attitude toward the oil companies, all of which represent free and private enterprise, and has refrained from imposing unduly burdensome taxes and royalties. A summary of legislation on petroleum exploration and exploitation operations, and of the more important tax regulations, is given at the end of this chapter.

CRUDE OIL RESERVES

The proved crude oil reserves of Trinidad are estimated to be in the order of 300 million barrels, or almost the equivalent of the island's cumulative production up to the end of 1947. Prospects for future discoveries are believed to be reasonably good, and can be expected on the island itself, as well as in the submarine areas of the Gulf of Paria and on the continental shelf.

[4] Wage rates as cited in the *Annual Report on Trinidad & Tobago, B.W.I., for the year 1947,* London, Colonial Office, 1949.

Although the Tertiary of the land area has already been heavily prospected, additional discoveries are possible in the land portion of these sediments through extensions to existing fields, and by locating fault and stratigraphic traps outside the limits of presently proved reservoirs. Of the prospects on land, the Cretaceous also has possibilities of carrying oil accumulation. Sediments of this age are already productive in Venezuela, and, with their known extension across the Gulf of Paria into Trinidad, they cannot be overlooked when future exploration efforts are being considered.

The submerged areas have not yet been drilled and, except for a limited amount of geophysical work, are entirely unprospected. Taking into consideration that these areas, especially those underlying the Gulf of Paria, lie between the proved regions of eastern Venezuela and Trinidad, and, also, that similar sediments carry through, it would appear that the prospects for discovering additional reserves in these areas should not be discounted. Since the Gulf of Paria is relatively shallow, and its waters are sheltered and lie outside the hurricane belt, development costs should be sufficiently reasonable to permit carrying on profitable operations, providing accumulation is found in substantial quantities.

On the basis of the foregoing, it appears that future discoveries in Trinidad, including both the land and submerged areas, could at least equal, and perhaps exceed, the ultimate recovery expected from the acreage already proved. It would seem then, that a reasonable estimate for future discoveries would be in the order of 700 million barrels, of which 70% is estimated for the submerged areas and 30% for the land areas.

Notes on Petroleum Legislation

EXPLORATION AND EXPLOITATION OPERATIONS

With reference to oil operations, the classes of lands in Trinidad and Tobago may be summarized as follows:

1. Private lands, prior to 1889: Titles stemming from old Spanish grants or Royal grants, generally carrying with them all mineral rights.

2. Alienated lands, 1889 to January 1902: Titles granted by the Crown reserved for the Crown all gold, silver, and other precious metals or coals, excluding petroleum, underlying such lands, together with the right to exploit same.

3. Alienated lands, subsequent to 1902: Crown grants since 1902 reserved all precious metals, stones, *and petroleum* to the Crown, and the right to work and exploit same.

4. Unalienated Crown lands: lands still held by the Crown, both with respect to the surface and to the minerals lying thereunder.

5. Submarine areas: Areas lying under territorial waters and within the international boundary between Trinidad and Venezuela, as established by the Treaty of February 26, 1942, between Great Britain and

Venezuela. The precise location of this boundary is shown on Fig. 12. The area involved was officially annexed by Great Britain on October 29, 1942, and all minerals lying under these waters are owned by the Crown.

The basic regulations governing petroleum exploration and exploitation operations (usually referred to as mining operations) in Trinidad and Tobago are: (1) Regulations of December 7, 1934, as amended from time to time; (2) Submarine Regulations of May 22, 1945, as amended by Government Notices 99 of June 1, 1945, and 81 of January 22, 1946. In addition, there is certain incidental legislation relative to water control, safety measures, conservation, pollution, and workmen's compensation.

The provisions of the Land Regulations of 1934, as amended, may be summarized as follows:

1. Exploration license: The government may grant to a *bona fide* applicant an exclusive license to carry on geological and geophysical work on the licensed area for a period of one year. The period, however, may be extended at the option of the government over all or part of the area licensed.

2. *Area:* An exploration license may be granted on an area containing not less than 500 acres; the government reserves the right to fix the maximum area.

3. Fees: A filing fee [5] of £5 ($20 U.S.) is required.

4. Annual rental: Graduated upward from a minimum of £100 annually for tracts of from 500 to 1,000 acres, to £1,250 ($5,000) annually for tracts of from 50,000 to 100,000 acres.

5. Mining lease (exploitation lease): Any time during the term of the exploration license the holder thereof has the right to obtain a mining lease on any amount of the licensed area he chooses to select. The government has the right to issue a mining lease without first issuing an exploration license.

(a) Mining lease filing fees: Crown lands, £5 ($20 U.S.); Alienated Lands £20 ($80 U.S.).

(b) Term of lease: A mining lease is granted for a period of 30 years, with the right of renewal for an additional 30 years, and carries the provision that the government, at its option, may increase surface taxes up to 50%, and that royalty on the oil will be payable at rates then in effect or subsequently amended.

(c) Surface taxes: These taxes or rentals are offset by royalties for the current year and are graduated upward from 4 shillings ($0.80 U.S.) per acre annually for the first year to 15 shillings ($3.00 U.S.) per acre annually for the sixth and all subsequent years.

The principal features of the Submarine Regulations are as follows:

1. Combined exploration license and mining lease (exploitation lease). The government may grant, at its option, based on proper application,

[5] Sterling-dollar equivalents as of March 1949.

the exclusive right to explore for and exploit petroleum and natural gas under the area covered by such grant.

2. *Area.* A license may not be granted on a submarine area of less than 625 acres, unless special permission is obtained from the Government.

3. *Filing fees.* A filing fee of $24 Trinidad ($20 U.S.) is required.

4. *Term of lease.* A lease runs for a period of 30 years, with the right of renewal of up to 30 years, and carries the provision that the Government, at its option, may increase surface rentals up to 50% and that royalty will be paid at the rate then in effect or as established thereafter.

5. *Surface taxes:* As in the case of land licenses, the taxes or rentals are offset by royalties for the current year, and are graduated upward from $0.10 Trinidad ($0.085 U.S.) per acre annually during the first two years, to $1.80 Trinidad ($1.53 U.S.) per acre annually for the tenth and all subsequent years.

6. *Joint development obligation:* Under the Submarine (Oil Mining) Regulations, there is a provision that the Government has the power to require unit development of an area where various licensees have rights over a single geological structure. If the licensees cannot agree on a unit development scheme, the Government may submit a plan for consideration, and if this plan proves to be unacceptable, the matter will go to arbitration for final decision.

Private land leases are obtained through private contracts, and, though the lease contracts need not be uniform, they are usually for a term of 35 years, with the option to renew for a like period. Surface taxes, as in the case of Crown-land leases, may be offset by royalty payments. Royalties usually amount to 10% in cash, and are payable on the same basis as royalty payable on Crown lands.

In addition to the provisions already set forth under the 1934 Land Regulations and under the Submarine Regulations, there are a number of provisions that apply to both. Of these, the principal ones are summarized as follows:

1. *Royalty:* A minimum royalty of 10% of the real value, based on world prices, of crude oil saved in storage; a minimum of 10% of the real value of casing-head gasoline recovered; and 3 farthings, or approximately 1¼ cents U.S., per 1,000 cu. ft. of natural gas sold. (As a matter of interest, royalty rates are not spelled out in the lease itself, but are subject to Schedule II of the Land (Oil Mining) Regulations of 1934, as from time to time amended. These regulations provide that, commencing as of January 1, 1943, royalty rates may be changed at 3-year intervals at the option of the Government, or of lessees who produce at least 50% of the crude oil in Trinidad. Under these conditions the minimum royalty is fixed at 10%, but no maximum has been established. It has been reported, however, that a maximum of 12½% is being considered.)

2. *Refining obligations:* Both the 1934 Regulations and the Submarine

(Oil Mining) Regulations provide that, when a licensee's production reaches 100,000 tons per year, he is obliged to erect a refinery capable of processing 50% of the annual crude production. However, there is a provision in the Oil Refining and Mining Ordinance whereby the government may suspend the obligation under certain conditions, e.g. available capacity in other refineries, or if licensee has other refining capacity within the Empire.

3. Reciprocity: Neither the 1934 Regulations nor the Submarine (Oil Mining) Regulations permit the Government to grant mineral rights to an alien or to alien-controlled companies unless British subjects or British-controlled companies are afforded the same privilege by the government of the country of origin of such alien or alien-controlled companies.

4. Corporate and personnel requirements. Under the above conditions, a license may be granted on Crown lands to an alien-controlled company, providing that it is organized in Trinidad or in some other possession of the British Empire. Furthermore, at least two members of the Board of Directors, the Chief Local Representative, and a majority of the people engaged in the operation of the company, must be British subjects.

Other tax regulations. Some of the more important tax regulations, other than royalties and surface taxes, which are of particular interest to oil operations, are summarized as follows:

1. Companies are assessed an income tax of 40% on chargeable income derived from operations in Trinidad.

2. Deductions for the depletion of wasting assets, as permitted by the extractive industries in the United States, are not allowed.

3. Expenses incurred through the drilling of dry holes, except those items considered as fixed-capital investment, may be charged off in the same year in which they occur.

4. Drilling costs of productive wells, except fixed-capital investment items, are deductible from income over a period of seven years. In some instances, a shorter period may be granted, providing the Tax Authorities agree thereto.

5. A company suffering operational losses may carry such losses forward for a period of five years from the period in which they occurred, and deduct same up to a maximum of 50% from the chargeable income during the year of assessment.

6. Another form of tax, known as the Oil Operating Fee and Drilling Impost, is levied against licensees for the purpose of maintaining a Water Conservation Board. This board has its headquarters in Port-of-Spain, and maintains a technical staff whose primary function is to see that the producing areas are developed efficiently. To support this board, the licensees are required to pay $480 per year Trinidad currency ($400 U.S.), plus a drilling impost based on annual footage drilled. On current annual production, the impost comes to about 1/10 of a cent per barrel.

4. MEXICO: A BRIEF RÉSUMÉ

BY OLIVER B. KNIGHT *

MEXICO, although ranking sixth among the countries of the world in petroleum production, is less conspicuous as an exporting country. In 1938 about two-thirds of the petroleum produced was retained in the country, and imports added an amount equivalent to about one-seventeenth of production. In 1947 about three-quarters of the petroleum produced was retained in the country, and imports added an amount equivalent to one-eighth of Mexican production. In 1948, when the Mexican crude production totaled 58.4 million barrels, a still larger proportion was retained; exports totaled 13.1 million, or less than one-quarter of the oil produced. Imports of petroleum products contributed some 4.6 million barrels, equivalent to about one-twelfth of the whole Mexican production. In addition to liquid petroleum, substantial volumes of natural gas are imported from the United States.

Physiography, Population, Industries

Comprising an area of 764,000 square miles, Mexico is physiographically complex—high mountain ranges, intervening plateaus, and narrow coastal lowlands break up its surface. The central plateau extends north from Mexico City to the great central basin of the United States, being bounded on the east by the Sierra Madre Oriental, which roughly parallels the Gulf of Mexico, and on the west by the Sierra Madre Occidental, which parallels the Pacific Ocean. These two ranges, together with the Sierra Madre del Sur in southwestern Mexico and the zone of giant volcanos formed along a zone of cleavage marking the southern limit of the central plateau, form the mountain provinces. To the west and south the coastal plain is narrow, with the mountains rising, in places, abruptly from the ocean. The plain along the Gulf coast is more extensive, widening to over one hundred miles at certain points, and extending into the Yucatán lowlands which cover most of the northern part of the Yucatán Peninsula.

Mexico's population numbers about twenty million. The greatest density is in the central-plateau area near Mexico City, while the desert areas to the north and the jungle lowlands are sparsely populated.

Mining, agriculture, and stock raising are the principal industries, with manufacturing gaining in importance in recent years.

Early Discoveries

The history of petroleum in Mexico dates back to pre-conquest days,

* Economics Department, Creole Petroleum Corporation, Caracas, Venezuela.

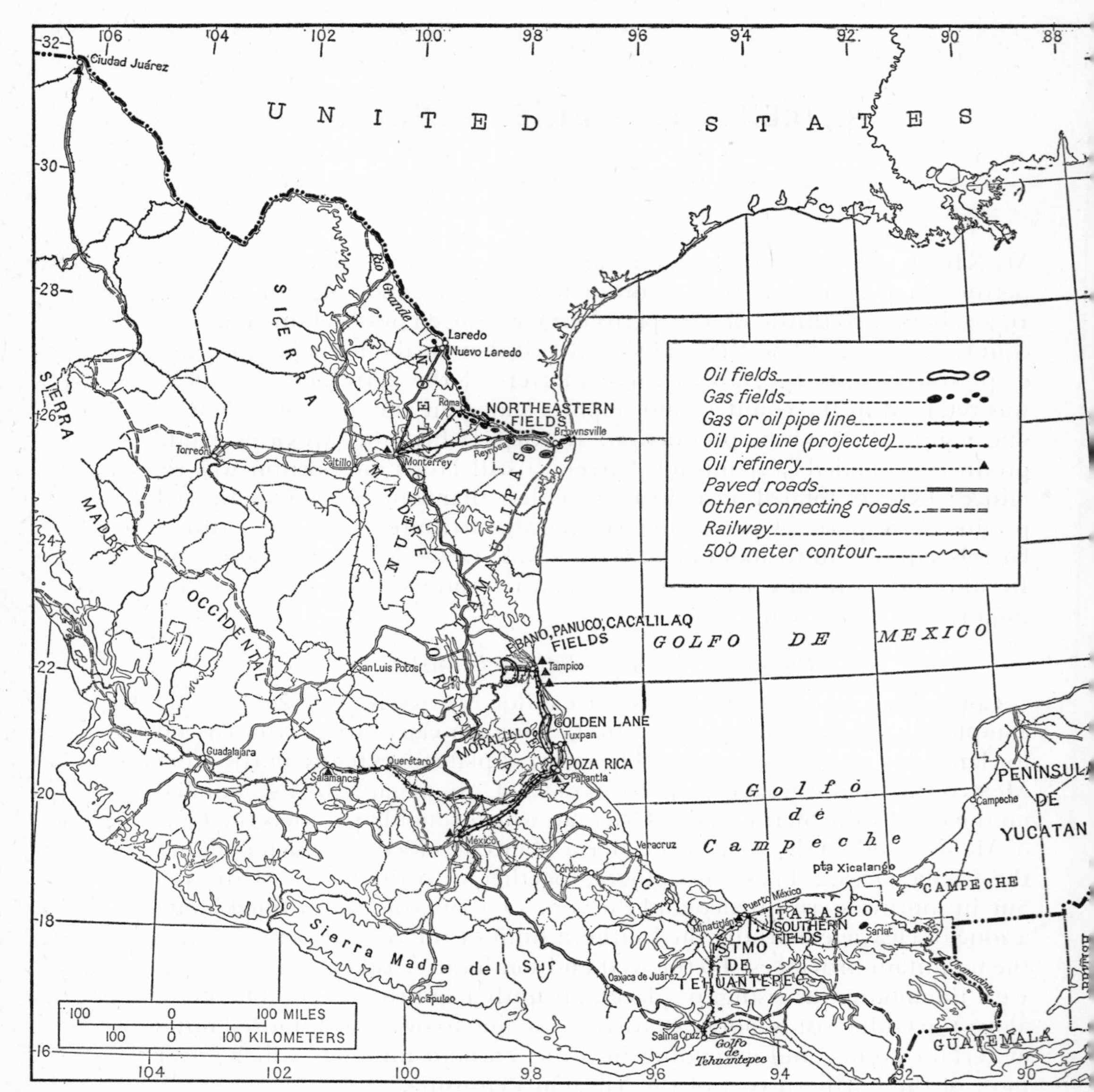

FIG. 13. Mexico: oil and gas fields and petroleum facilities. Tampico, with a crude refining capacity of 87, b/d, is the terminus of six crude-oil pipe lines capable of transporting 126,000 b/d. The oil port of Tux is the terminus of six crude lines with a total capacity of 200,000 b/d. Minatitlán, 15 miles south-southwest Puerto México and with a refining capacity of 27,500 b/d, is the terminus of four crude lines with a combi capacity of 30,000 b/d. Mexico City, with a refining capacity of 50,000 b/d, is the terminus of one crude l with a capacity of 50,000 b/d, from Poza Rica, and one gas line, also from Poza Rica. Monterrey is the ter nus of three gas lines from the northeastern fields. The refineries shown at Ciudad Juárez, Monterrey, a Nuevo Laredo are skimming plants, and the plant at Poza Rica is also for skimming. The refinery at Sa manca is under construction. Proposed pipe lines include, in addition to those shown on the map, a gas l from Monterrey to Mexico City and a crude-oil line from Poza Rica to Guadalajara. Data on pipe lines a refineries are from Pemex (Petróleos Mexicanos) and from *World Oil International Operations Iss* July 1949.

when asphaltic oil from numerous seepages along the eastern coastal plain was used by the Indian inhabitants for waterproofing dugout canoes, flooring their homes, and ceremonial fires. Attempts to exploit asphalt deposits commercially were made as early as 1864, and several unsuccessful sporadic attempts were made to drill for oil near seepages during the latter part of the nineteenth century. In 1880 some oil obtained from a tunnel driven into a hill near a seepage, and transported ten miles by burros, was refined into kerosene in Papantla and sold locally.

The first commercial oil discovery was made in the Northern District field west of Tampico in 1901, by the Mexican Petroleum Company of California. This was followed by discoveries in 1902 in the Isthmus of Tehuantepec and in the prolific Southern District field south of Tampico in 1908 by S. Pearson and Son, Ltd., of London, whose interests were acquired later by the Royal Dutch Shell. Within the next few years three of the world's most prolific oil wells were drilled in the Southern District field, each of which produced over 200,000 b/d initially, and which together have produced over one-fourth of the field's total production of one billion barrels, the other three-fourths having been produced by more than five hundred wells.

Production and Ownership

The increased demand for oil created by World War I resulted in the rapid development of Mexican oil fields, and by 1921 annual production reached its maximum rate of some 193 million barrels, placing Mexico second in importance to the United States as an oil producer.

Prior to 1921 subsoil rights, including petroleum, were vested in the landowners; this led to the fee purchase of large tracts by oil companies. The Querétaro Constitution of 1917, which was officially adopted in 1921, placed all petroleum rights under national ownership, giving landowners preferential rights to concessions of limited duration. The retroactive application of the new laws to rights held prior to 1921 resulted in lengthy litigation between the Government and the oil companies, which was terminated by the Morrow agreement that provided for the granting of perpetual concessions on properties and rights held prior to the promulgation of the Querétaro Constitution.

Subsequent to 1921 a rapid decline in producing rates followed an invasion of the principal fields by salt water and, in the same period, changes in the economic and social order resulted in a marked decrease in development and exploration activities. These factors, together with the depression starting in 1929, caused a production drop to 33 million barrels in 1932. This amounted to 2.3% of the world total, and placed Mexico in sixth rank as an oil-producing nation: which position it still held in 1948, with an annual production of some 58 million barrels.

The development of the Poza Rica field, located near Papantla in the southern part of the Tampico Embayment, which was discovered by Shell interests in 1930, made it possible to reverse the steady decline from 1921 to 1932, and in 1937 production had increased to 47 million barrels.

In March 1938 the activities of foreign oil companies in Mexico were terminated abruptly by the expropriation of their properties by the Mexican Government. Petróleos Mexicanos, a government-owned company, was formed to take over the operation of the expropriated properties and the development of Mexico's petroleum resources. Because of various difficulties, this company was unable to maintain production at the pre-expropriation level, and only by developing the prolific Poza Rica field was it able to produce an average of 40 million barrels annually from 1938 to 1945. A more active drilling campaign thereafter, together with increased outlets for crude oil, resulted in a steady increase in the producing rate from 45 million barrels in 1945 to 58 million barrels in 1948 (Table 9).

Table 9. Mexico: estimates of reserves and production of crude petroleum, by districts, as of January 1, 1949.

	Year of Discovery	Estimated Reserves Jan. 1, 1949	Annual Production 1948	Cumulative Production to Jan. 1, 1949
			(In thousands of barrels)	
Northeastern fields	1948	10,000	85	85
Central fields		930,000	52,228	2,218,272
Northern district	1901	60,000	10,205	801,341
Southern district	1908	70,000	7,715	1,089,103
Poza Rica	1930	800,000	34,308	327,828
Isthmus fields	1902	55,000	6,056	153,414
Total, Mexico		1,005,000	58,370	2,371,772

Sources: Estimated reserves from L. G. Weeks, "Developments in Foreign Fields in 1948," *Amer. Assn. Pet. Geol. Bull.,* Vol. 33, No. 6, June 1949, p. 1047. Production figures from *World Oil,* Vol. 129, No. 4, July 15, 1949, p. 89.

A new producing horizon in the El Plan field in the Isthmus of Tehuantepec, discovered in 1943, and the Misión field in northeastern Mexico, producing small quantities of gas and distillate, discovered in 1945, were the only commercial discoveries reported between 1938 and 1947. Since 1947 three discoveries of possible importance have been reported: one near Reynosa in northeastern Mexico, one near Tepetzintla, in the Southern District field area in northern Vera Cruz, and another at Agua Dulce, in the Isthmus of Tehuantepec.

Of the 2,372 million barrels of oil produced in Mexico to the end of 1948, 93% had come from fields in the Tampico Embayment, most of the remaining 7% having been produced from fields near Puerto México in the Isthmus of Tehuantepec. The gas fields in northeastern Mexico provide a source of fuel that has been an important factor in the recent industrial development in the Monterrey area.

4. MEXICO

Editors' Note on Operations of the American Independent Oil Company and the Cities Service Oil Company [1]

In March 1949, just eleven years after the expropriation of foreign oil properties by Mexico in 1938, a group of independent United States operators signed contracts with Petróleos Mexicanos (Pemex), the Mexican Government Oil Monopoly, to undertake geological and geophysical exploration and development drilling. The operators were the American Independent Oil Company, the Signal Oil and Gas Company, and Mr. Edwin W. Pauley, each originally holding a 33⅓ per cent ownership in the Mexican American Independent Company (C.I.M.A.), the operating company.

The American Independent Oil Company, of San Francisco, California, itself composed of ten independent United States oil operators, established in six states of the U.S.A., who in 1947 had joined forces to work in the international field, increased their initially owned 33⅓ per cent interest in C.I.M.A. to 53⅓ per cent by purchase of an additional 20% interest from Mr. Pauley. American Independent Oil Company de Mexico, S.A., a wholly-owned subsidiary, of the San Francisco company, was formed in April 1947.

C.I.M.A. has commenced active operations on the Gulf Coast side of the Isthmus of Tehuantepec in southeastern Mexico. This area includes both submerged lands and uplands, and is said to be highly promising. Extensive geophysical survey work has been carried out in the lagoon and undersea areas along the coast, and drilling operations have commenced at two locations, at Tortuguero, about nineteen miles east of the town of Coatzacoalcos, adjoining Puerto México in the State of Veracruz, and at Xicalango, near the town of Ciudad del Carmen at the entrance to the Laguna de Los Terminos in the state of Campeche.

Another type of agreement for foreign participation in the Mexican industry was reached in April 1948, by Petróleos Mexicanos and the Cities Service Oil Company, whereby the United States firm provides financial backing for exploratory drilling to be carried out by Pemex in the lands formerly held under concession by Cities Service subsidiaries. In 1949 test wells were being drilled under this arrangement in the area north of Tampico. None of these agreements grant any subsoil rights to the foreign company.

[1] Based on material supplied to the American Geographical Society by the American Independent Oil Company and the Cities Service Oil Company.

5. COLOMBIA

BY GEOFFREY BARROW *

IN THE South American part of the Caribbean area Colombia ranks second only to Venezuela as a petroleum-producing country. Since the beginning of commercial production in 1921 a total of 440 million barrels had been produced by the end of 1948 (Table 1). The average daily production of 64,880 b/d in 1948 represented a slight decrease from the average production of 70,700 b/d in 1947, the peak year.

The amount of crude petroleum exported annually from Colombia reached a peak figure of 22,498,000 barrels in 1940. Since then annual exports have diminished because of the greatly increased demand for refined petroleum products within the country. In 1948 the year's exports of crude petroleum amounted to 18,718,000 barrels. Crude petroleum ranks with coffee, gold, and bananas, as one of Colombia's four principal export items.

Domestic consumption in Colombia is increasing for almost all petroleum products, and this has been the trend for a number of years. To give specific examples, during the three years 1946–1948 inclusive, the consumption of motor gasoline went up 53%; aviation gasoline, 49%; diesel oil, 68%; and fuel oil, 11%.

The principal oil-producing areas of Colombia are found in two distinct petroliferous basins. In the Magdalena Basin are the producing fields of the De Mares concession, which produced about 9 million barrels of crude oil in 1948 and had a cumulative production of over 375 million barrels, and the fields of the Yondo concession, which produced about 5½ million barrels in 1948 and had a cumulative production of over 15 million barrels. In the Colombian portion of the Maracaibo Basin are the fields of the Barco concession, which produced around 8 million barrels in 1948 and had a cumulative production of nearly 45 million barrels. Each of these producing areas will be discussed in the following pages, and the lesser producing areas will be indicated briefly. Exploration in those parts of Colombia not draining to the Caribbean is discussed in Chapter 7 of Part III.

The Geographical Background

The Republic of Colombia, which lies in the extreme northwestern part of South America, is the only South American country with coast lines on both the Pacific and Atlantic oceans. Colombia has an area of

* Exploration Coordination Department, International Petroleum Company, Ltd., Toronto, Canada.

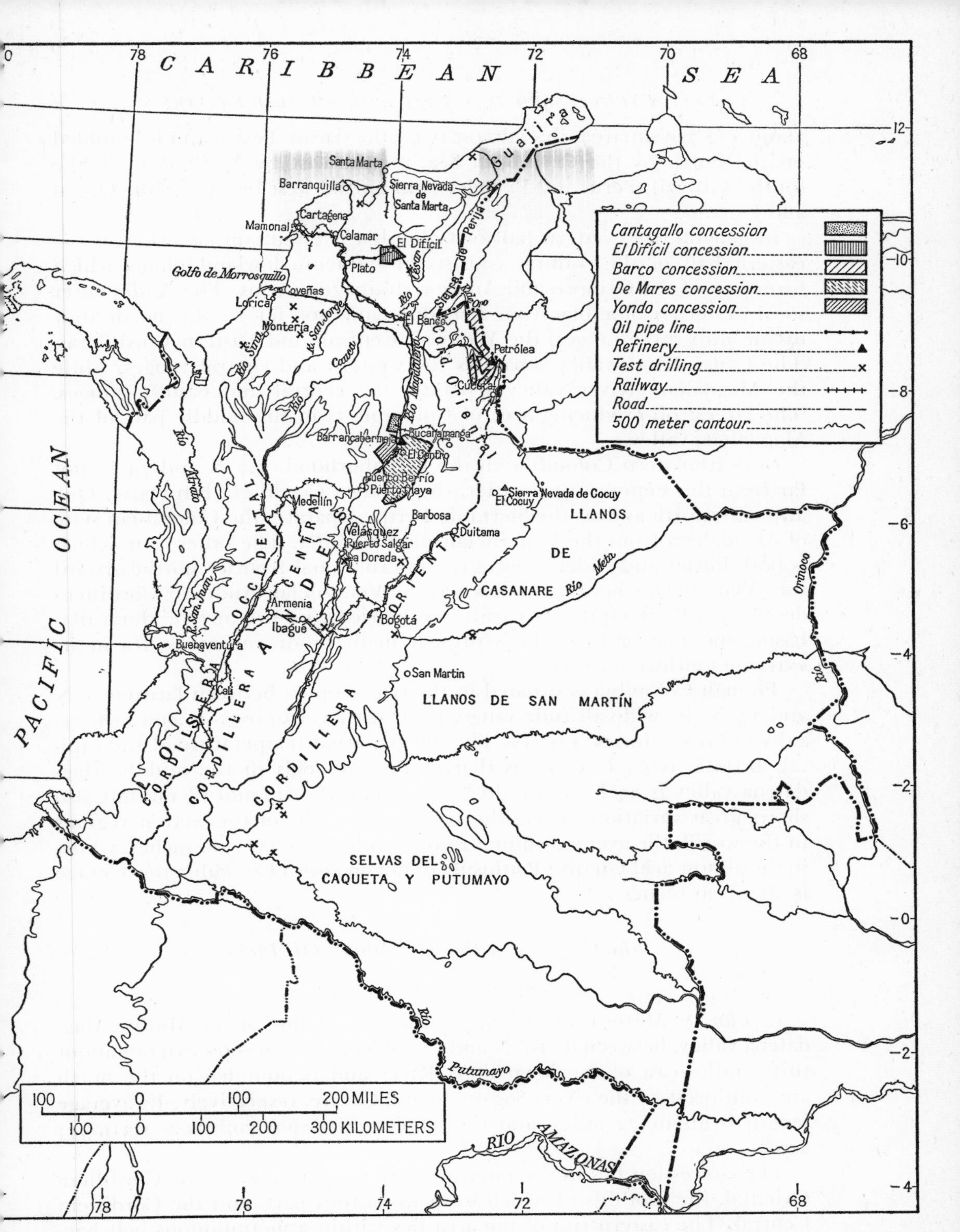

. 14. Colombia: petroleum concessions productive at the beginning of 1949 and areas being test drilled. e largest pipe line, of 12-inch diameter, carries crude from the Petrólea field to Coveñas. The longest, from rrancabermeja to Mamonal, consists of double lines of 10-inch diameter throughout the length of 334 les, and its branch from El Difícil to Plato has a combination of 6-inch and 8-inch pipe. These are used for de oil. The refinery at Petrólea has a capacity of 700 b/d, and that at Barrancabermeja, 17,000 b/d. A oducts line of 6-inch diameter leads from this refinery to Puerto Berrío. Additional products lines of simi- gauge are planned to serve Puerto Salgar, Medellín, and other parts of the plateau; and a products line m the port of Buenaventura to the plateau is also planned.

about 444,720 square miles (almost twice the size of Texas) and is bounded on the north by the Caribbean Sea, on the east by Venezuela, on the south by Brazil, Peru, and Ecuador, and on the west by the Pacific Ocean and Panamá.

In general, the western half of Colombia is mountainous, whereas the eastern half of the country consists of immense lowland plains which form part of the Orinoco and Amazon drainage systems. The Andes enter southwestern Colombia from Ecuador, and not far to the north they divide into three ranges, the Western, Central, and Eastern Cordilleras. The Cauca River valley separates the western and central ranges, while the Magdalena River valley separates the central and eastern ranges. The largest oil-producing area of Colombia is in the middle part of the Magdalena valley.

In northeastern Colombia, in the neighborhood of 7°N and 73°E, not far from the Venezuelan frontier, the Eastern Cordillera bifurcates. One arm runs north and, as the Sierra de Perijá, separates the Colombian state of Magdalena from the Venezuelan state of Zulía. The other arm, which is both loftier and wider, crosses the international frontier northeastward into Venezuela, where it is known as the Venezuelan Andes or Cordillera de Mérida. Between the two arms lies the richly oil-productive Maracaibo basin, and one of Colombia's present main producing areas lies in its extreme southwestern part.

Though Colombia is situated in the true tropics, between latitude 4°N and 13°N, its wide altitude range, from sea level to over 18,000 feet, results in great climatic variety. The mean annual temperature of the capital, Bogotá, is 55°F, whereas that of the oil fields in the middle Magdalena valley is 85°F. As would be expected, the amount of rainfall also shows great variation. Along the Pacific coast, one of the wettest regions in the world, the average annual rainfall is locally over 300 inches, whilst in the almost arid Guajira Peninsula of northeastern Colombia the average is about 30 inches.

The Geology of the Colombian Oil Fields

THE MAGDALENA BASIN

1. The De Mares Concession. On the eastern side of the Middle Magdalena valley, between 6° 10′ N and 7° 20′ N, the concession extends about thirty miles east of the Magdalena River and is bounded on the north and southwest by the rivers Sogamoso and Carare, respectively. Its average length is about 72 miles, and the whole concession embraces 1,319,344 acres.

The concession lies in the intermontane trough between the Cordillera Oriental, the foothills of which form its eastern part, and the Cordillera Central. The eastern rim of the area lies within a mountainous belt having a maximum local relief of some 2,500 feet and a maximum elevation above sea level of 3,500 feet. The main part of the concession lies in the low foothills of the Cordillera Oriental and in the lowlands bordering the Magdalena River, where the elevation averages less than 270 feet. Drainage of the area is in a general northwest direction, into the Magdalena River by its tributaries, the Sogamoso, Colorado, Opón, Carare, and their affluents. Except where artificial clearings have been made, the concession area is densely wooded.

The Middle Magdalena valley, wherein the De Mares concession lies, is generally considered a *graben,* or downthrown block, bounded on both east and west by normal faults. This structural depression was the depositional site of a great thickness of non-marine and brackish-water Tertiary sediments, which were laid down on Cretaceous marine sediments or on pre-Cretaceous basement rocks. The Tertiaries, from which all the oil produced on the concession is derived, are deformed into a series of strongly developed folds which are generally broken by closely related faults. The trend of the major folds is roughly parallel to that of the Cordillera Oriental.

Of all the structures that have so far been tested on the concession, only two have proved commercially productive. These two structures are faulted anticlines, the axes of which trend approximately 15°N–20°E, and are covered by the Infantas and La Cira fields.

In 1918 the first three wells of the Infantas field were drilled at the south end of the field near the Colorado River, and each well proved productive. Not until 1922 were any more wells drilled, but from that year on the development of the field progressed rapidly, reaching its peak in 1929, when 71 wells were drilled. The field is about 7 miles long and 1½ miles wide. It includes two producing horizons, the more important sands lying within a range of 2,200 feet between average top and bottom depths in lowest Tertiaries, with a lesser producing horizon at an average drilling depth of 1,500 feet in Oligocene sands. Some of the wells of this field have been drilled into the underlying Cretaceous, but none of them has yielded commercial production. The oil produced from the Infantas field is of asphaltic base, with an average API gravity of 26°. Up to the end of 1948, 466 producing wells had been drilled in the Infantas field, and had produced nearly 150,000,000 barrels.

The La Cira field was discovered in 1926, when three wells were drilled, and it has been actively exploited to the present time. Drilling activities reached their peak in 1938; during that year 107 wells were completed. Up to the end of 1948 a total of 821 producing wells had been drilled, with a total production of over 225,000,000 barrels.

The La Cira field is a little less than 5½ miles long and has a maximum width of 3¾ miles. Oil is produced from three horizons, the most important of which, as in the Infantas field, lies in the lowest Tertiaries and produces the bulk of the oil. The depth to the top of this zone varies from 2,200 to 4,100 feet; and the average penetration into it, of about 300 feet,

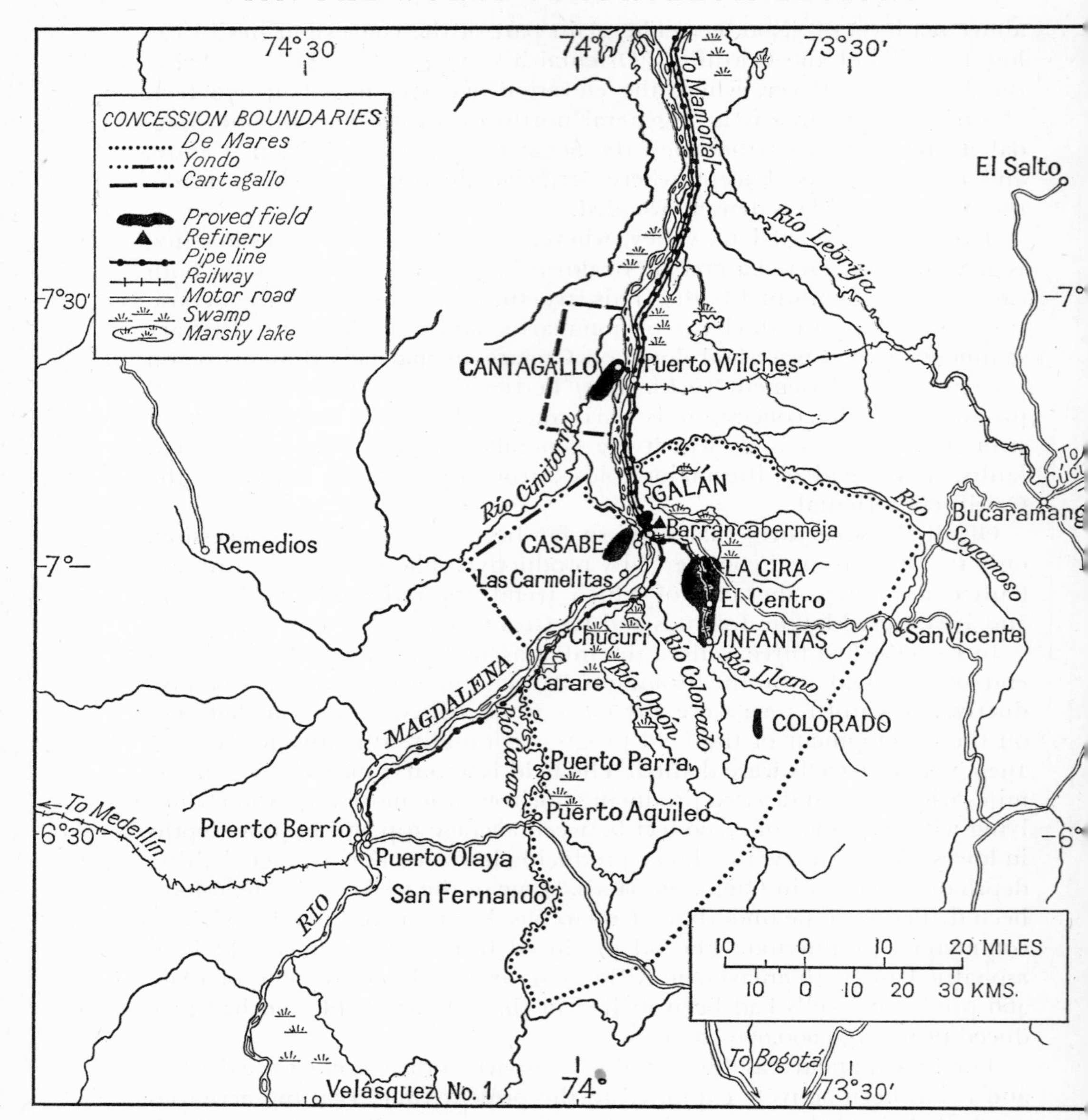

FIG. 15. The Mid-Magdalena fields: concession areas and development at the beginning of 1949. Name fields are in Roman capitals.

is made up of sand and shale in approximately equal amounts. There are two sand horizons in the Oligocene from which lesser production is obtained; the depth to the top of the lower of these zones ranges from 1,100 to 1,600 feet, and to the top of the upper zone from 600 to 1,000 feet. The overall thickness of the lower productive zone in the Oligocene averages 600 feet, whilst that of the upper zone varies between 30 and 350 feet.

Deeper drilling for the purpose of exploring the Cretaceous has not met with any success in the La Cira field. One of the wells drilled reached a depth of 8,051 feet, but encountered no commercial production. The average gravity of La Cira crude, which is asphaltic in base, is about 23° API. Since its discovery in 1926, the La Cira field has produced over 225,-000,000 barrels.

In addition to the Infantas and La Cira fields, two very much less productive fields, or pools, are located within the De Mares concession.

The Colorado field was discovered in 1932 after seven wells had been drilled, but results were not sufficiently encouraging to justify further attempts to develop it, and drilling operations were suspended. Improvements in drilling technique, together with a change in supply and demand conditions, led to a resumption of drilling in 1945, when an eighth well was drilled and completed as a producer. During the period 1946–1947 seven more wells were drilled, but overall results were disappointing, and drilling operations were again suspended.

The Colorado wells are drilled on a long narrow anticlinal structure located to the south-southeast of the Infantas field. Production is obtained from sands in two lower Tertiary zones, the average depths to the top of which are 3,000 and 4,000 feet. Colorado crude is of paraffinic base, with an average gravity of about 38° API. To the end of 1948, the Colorado field had produced a little over 250,000 barrels.

The Galan field, two miles northwest of Barrancabermeja on the east bank of the Magdalena River, is a northeastern extension of the Shell Oil Company's Casabe field, described in subsequent pages. Between June 1945 and June 1946, five wells were drilled and some production of heavy asphaltic oil, with a gravity of 19° API, was obtained from Oligocene sands, at depths of 3,000 to 4,000 feet. To the end of 1948 the Galan field had produced some 123,000 barrels of crude.

2. *The Yondo Concession.* On the western side of the Middle Magdalena valley on the 7° N parallel and immediately opposite the Tropical Oil Company's De Mares concession, the Yondo concession has an area of nearly 116,000 acres, and extends a maximum of about 19 miles west from the Magdalena River.

The Yondo concession lies in the same physiographic province as the western part of the De Mares concession, and the terrain consists mostly of riverside jungle and swamps, at an average elevation of 250 feet above the sea. The area covered by the concession rises very gradually to the west towards the Cordillera Central, but the maximum relief probably does not exceed 100 feet.

The regional geology of the Middle Magdalena valley, wherein the Yondo concession lies, has been outlined in the description already given of the De Mares concession. The stratigraphy and structure of the concession area itself are concealed beneath a blanket of undisturbed Quaternary sediments, and the so-called Casabe structure, along which the Casabe producing field is located, was discovered by geophysical exploratory methods. The structure is reported to be highly faulted and complex.

The Casabe field lies in the northeastern part of the Yondo concession, immediately opposite the De Mares concession, from which it is separated by the Magdalena River. The first well was drilled on the Casabe field in 1941 to a total depth of 8,280 feet, and was completed as a producer with an estimated initial production of 480 b/d of 19.5° API gravity oil. In the following year seven more wells were successfully completed, with initial productions varying between 150 and 1,230 b/d. Crude-oil production steadily increased from 5,700 barrels in 1941 to over 5,500,000 barrels in 1948, by the end of which year 99 wells had been completed.

3. The Cantagallo Concession. The Socony-Vacuum Oil Company's Cantagallo concession lies along the west bank of the Magdalena River some 280 miles from the Caribbean coast, and a few miles north of Shell's Yondo concession.

Drilling operations were commenced in 1941, and the first well drilled from young Tertiary sediments into igneous rocks at a depth of 1,153 feet. It was completed as a gas well. Cantagallo No. 2, located about 2½ miles south of the first well, was drilled to a total depth of 6,170 feet by March 1943, and later that year was brought in with an initial production of 286 b/d of 19° to 20° API crude through a ⅜-inch choke. Production is from Eocene sands at an approximate depth of 5,700 feet. By the middle of 1947, eleven wells had been drilled on the Cantagallo concession. Four were completed as oil producers, and two as gas wells. The average depth to the top of the producing zone is 5,860 feet, and its productive thickness is 160 feet.

During 1948, the annual production from the field was 385,000 barrels, while the cumulative production to the end of 1948 was nearly 550,000 barrels of heavy crude. Little drilling has been carried out on the Cantagallo concession since the No. 11 well was completed in 1947.

4. The Velasquez Private Lease. In 1946 the Texas Petroleum Company completed Velasquez No. 1, the discovery well on its Teran-Guaguaqui fee property, east of the middle Magdalena River and about ninety miles south of Tropical's De Mares concession. The well, drilled to a total depth of 8,455 feet, was brought in with an initial production of 383 b/d of 25.5 API gravity oil, from basal Tertiary sandstones. By the end of 1948, two of the five wells drilled had produced nearly 130,000 barrels of oil, most of which was consumed in development operations.

5. The El Difícil Concession. The Difícil field, discovered in 1943 by the Compañía Petróleo La Estrella de Colombia, a wholly-owned subsidiary of the Shell Company, is situated about one hundred miles east-southeast of Cartagena and forty-five miles east of the Magdalena River.

From the point of view of transportation of crude oil to seaboard, the Difícil field is better located than are other Colombian oil fields.

The discovery well, Difícil No. 1, bottomed in igneous basement at a total depth of 5,996 feet, and on completion flowed at the rate of 54 b/d of 45° API crude through a one-eighth-inch choke. This production was obtained from a porous coral-reef limestone of Upper Oligocene age immediately overlying basement rocks.

After this discovery, drilling was accelerated. By the end of 1948, twenty-four wells had been completed and the cumulative production of the field was over 750,000 barrels of high-gravity crude. The average drilling depth to the top of the producing zone is 5,700 feet, and its productive thickness is about 150 feet. The potential daily production of the Difícil field was believed to be about 5,000 b/d at the end of 1948.

THE MARACAIBO BASIN

The Barco concession is located in northeastern Colombia, immediately west of the Venezuelan boundary, between 8° 8′N and 9° 12′N (Fig. 14). The area of the concession originally granted to the Colombian Petroleum Company in 1926 was a little over one million acres, but in 1941 part was surrendered. There remained two non-contiguous lots, the larger one, Petrólea, adjacent to the Venezuelan border, having an area of 376,-000 acres, and the smaller one, Esperanza, in the southwestern part of the original concession, having an area of 85,000 acres.

The densely forested Barco concession area lies in the foothill zone between the Sierra de Perijá on the west and the Cordillera de Mérida on the southeast. This zone is characterized by wide river valleys, which form the upper part of the Catatumbo drainage system. The Catatumbo River flows southeast across the extreme northern part of the concession area, and continues, first east, then northeast, to empty into Lake Maracaibo. The Catatumbo River has many large tributaries, one of which, the Río de Oro, forms the northern boundary of the Barco concession.

Topography is varied. In the north a complex dendritic drainage system has cut the terrain into many narrow steep-sided ravines, separated by high winding ridges. To the southeast, in the Petrólea area, sandstone ridges have a relief of up to 1,500 feet; whilst in the southwestern part of the concession, outcropping massive sandstones give rise to a rugged topography, with a maximum relief of over 3,000 feet. A striking and not uncommon feature in the sandstone foothills is the occurrence of natural bridges over deep gorges. The maximum elevation of the concession area, near its western boundary, is about 5,500 feet, whilst the flood plains of the major rivers along its eastern side are not more than 160 feet above sea level.

The Barco concession area lies along the southwestern flank of the Maracaibo petroliferous basin, by far the greater part of which lies in Venezuela.

The sedimentary section underlying the concession ranges in age from Lower Cretaceous to Pliocene and Recent, and includes fourteen recognized formations. The Lower Cretaceous is of marine origin and averages 2,300 feet in thickness; the Upper Cretaceous is mostly marine, but includes some brackish water deposits, and it is around 3,100 feet thick; the overlying Tertiaries, which are characteristically non-marine but include a few marine horizons, have an average thickness within the concession of approximately 8,200 feet.

The concession area is folded into a series of anticlinal structures, which trend in a general north-south direction. Up to the present, by far the most productive structures have proved to be the Petrólea anticline in the southeastern part of the concession and the Tibú and Socuavo anticlines in the central part. Lesser production has also been obtained from the Río de Oro anticline in the extreme north of the concession, and from the Carbonera fault zone on the eastern flank of the Petrólea anticline.

The Río de Oro field, in the extreme north of the Barco concession, immediately south of the Venezuelan border, lies along the Río de Oro anticline in the easternmost foothills of the Perijá range. The structure is a narrow faulted asymmetrical anticline, the east flank of which is nearly vertical, whilst the west flank is comparatively gentle. Lower Eocene sandstones are exposed along the crest, which trends a little east of north, and extends from the Río Catatumbo north to the Río de Oro and into Venezuela.

The discovery well of this field was the first well drilled on the Barco concession. It was drilled by Doherty interests in 1920, as an offset to a small producing well which had been completed just over the Venezuelan border by the Colón Development Company. An oil-bearing sand in the Lower Eocene was penetrated at a depth of 449 feet, and the well began to flow strongly. Production soon dropped, however, to less than 50 b/d, and the sand was cased off. Drilling continued to a depth of 895 feet, where the hole bottomed in the Upper Cretaceous and was abandoned as non-commercial.

No further drilling on the field was undertaken until 1935, when the present concessionaires, the Colombian Petroleum Company, drilled Río de Oro No. 2 to a depth of 6,717 feet without obtaining production. Both this test and a third unsuccessful test drilled in 1936–1937 to a depth of 6,317 feet reached the Lower Cretaceous. The fourth test, drilled on the structure about three-quarters of a mile south of the discovery well, gave an initial production of 140 barrels of 38.6° API gravity oil. Río de Oro No. 5 proved to be the biggest well on the field, with an initial production of 3,800 b/d of 37.7° API gravity oil from an Upper Cretaceous sandstone.

By the end of 1940, thirteen wells had been completed, and the field, which never appeared to have a very large potential production, was considered to be drilled up. The cumulative production of the field at the close of 1941 was 64,708 barrels, and by the end of 1945 this amount had increased only to 66,419 barrels. No oil has been produced since 1945.

The Petrólea field is located in the southeastern part of the Barco concession along the Petrólea structure, which is a faulted asymmetrical anticline with a very steep to overturned west flank and a gently dipping east flank.

The discovery well was drilled is 1933. On penetrating the upper part of the Lower Cretaceous the well blew out, caught fire, and was not brought under control for 47 days. By the end of 1937 the productive area had been outlined by means of 18 scattered wells, and in 1938 intensive drilling of the Petrólea field was begun. By 1941, there had been drilled 138 wells, of which 126 are productive. These wells produce from the crushed limestones and calcareous sandstones of six different zones in the Cretaceous, and the gravity of the paraffin-base crude oil obtained varies between 36° and 48° API. The top of the producing zone in the Petrólea field varies between 130 and 1,180 feet, and the deepest well drilled in the field bottomed at 3,008 feet.

A freak well of unusual interest was Petrólea No. 200. On the morning of November 14, 1940, the well was spudded in, and on the afternoon of the same day, while fishing operations were being carried out at a depth of only 98 feet, the well blew out. Before being plugged, 5,200 barrels of oil had flowed in ten hours into a hastily built reservoir.

The Carbonera structure is a faulted thrust-fold, situated some six miles southwest of the structural high on which the Petrólea wells are drilled. It is a Tertiary prospect only, as the structural trap does not extend down into the Lower Cretaceous. Three producers have been drilled on the structure, to depths of 2,722, 2,234, and 3,359 feet, respectively, with initial productions of 46, 82, and 65 b/d of 19.6° to 22° API gravity oil. The wells produce from sands in the Middle and Lower Eocene.

By the end of 1945, the Petrólea and Carbonera structures had produced 20,488,303 barrels of oil. Production during 1946 was 3,462,972 barrels, during 1947 it was 2,867,359 barrels, and during 1948 it was 2,079,513 barrels, bringing the cumulative production to the end of that year up to 28,898,147 barrels.

The Tibú field, in the east-central part of the Barco concession, covers the Tibú and Socuavo structures. The Tibú structure is an anticline, which trends slightly west of north; it is slightly asymmetrical, the dip of the west flank averaging 15° and that of the east flank averaging 6°. The Socuavo anticline, which lies *en echelon* to the northwest of the Tibú anticline, trends slightly west of north in its southern half, north in its northern half. It, too, is asymmetrical, with a steep west flank and a comparatively gentle east flank. Production is obtained on both these structures from sandstone and limestone horizons in the Lower Eocene and Lower Cretaceous. In the Tibú field the average depth of the wells drilled is around 5,200 feet, and the deepest well drilled to the end of 1948 bottomed at 10,876 feet.

The discovery well on the Tibú structure was drilled in 1939–1940 to a depth of 5,399 feet, and was completed as a producer of 17 b/d of 31.2° API gravity oil in Lower Eocene sandstones. The first test was drilled on the Socuavo structure in 1941–1942 to a total depth of 9,850 feet, and penetrated 69 feet of basement complex. On completion, initial production was at the rate of 136 b/d of 49.7° API oil.

By the end of 1943, the Tibú field had produced 85,825 barrels of oil, but subsequent intensive development increased the annual production to 683,427 barrels during 1944, to 1,539,416 barrels during 1945, to 2,721,884 barrels during 1946, to 4,814,668 barrels during 1947, and to 6,086,490 barrels during 1948, by the end of which year the cumulative production of the field was 15,931,710 barrels.

EXPLORATORY DRILLING IN THE SINÚ AREA

In the Sinú Valley of northwestern Colombia, a considerable number of wells have been drilled by various companies in the past few years, but none of them has led to the discovery of any commercial oil fields.

On the Floresanto concession, which lies about twenty-five miles southwest of the Sinú river port of Montería, two deep tests and ten shallow wells have been drilled. Floresanto No. 1 was spudded in 1944 and drilled to a depth of 6,936 feet. It was later plugged back to 696 feet and completed in 1945 as a shallow Tertiary producer, with an initial production of 85 barrels per day of 51° API gravity oil. In the years 1945–1946, ten shallow wells with depths ranging from 1,203 to 2,175 feet were drilled in search of commercial production from this shallow horizon. This objective was not achieved, though from one or two of the wells some small production of high-gravity crude was obtained. In 1947, a second deep test, Floresanto No. 10, was drilled and bottomed in the Oligocene at a depth of 10,876 feet before it was abandoned as a dry hole.

The Floresanto wells were drilled by the Sinú Oil Company, a Socony-Tropical subsidiary. Before the Floresanto concession was surrendered in 1947 as non-commercial, it had produced a little more than 11,000 barrels of high-gravity crude, which was used in local operations. The average depth to the top of the producing horizon is 640 feet and the average productive sand thickness, 37 feet.

A few miles to the east of the Floresanto wells, the Sinú Oil Company drilled one well on its La Risa concession. This well was drilled in 1947 and abandoned as a dry hole after having reached a depth of 3,573 feet. The La Risa concession was surrendered without further testing.

In the period 1947–1948, the La Junta Petroleum Company, a subsidiary of the Texas and Socony companies, drilled three wells on its El Tablón concession. Tablón No. 1 was drilled to a total depth of 9,934 feet and abandoned as a dry hole; Jobo No. 1 to 6,944 feet and completed as a gas well; Jobo No. 2 to 7,504 feet and abandoned as a dry hole. Both the Jobo wells bottomed in igneous basement. The Tablón concession is located some fifty to sixty miles south-southeast of the Sagoc pipe-line terminal at Coveñas. Some twelve miles northwest of Tablón No. 1 well, another Texas-Socony joint-interest well, Sahagun No. 1, was spudded in 1948, and at the end of that year was drilling at 5,232 feet.

Further north, between the Sahagun concession and Coveñas, a test well was completed in 1948 by the Tripet Company, a subsidiary of the Cities Service, Richfield, and Sinclair companies. This well, San Andres No. 1, bottomed at 7,365 feet in the Cretaceous and was abandoned as a dry hole. A second San Andres well had been drilled to a depth of 5,793 feet by the end of 1948.

Refining and Transportation

At the river port of Barrancabermeja, in the De Mares concession, a Tropical Oil Company refinery with a present capacity of 23,000 b/d is located. This refinery was built in February 1922, during which year about 200,000 barrels of crude oil were refined to supply local demand. Since then it has been gradually expanded and modernized to keep pace with the ever-growing increase in this demand, until in 1946 over 5,000,000 barrels of crude were refined. The refined products include aviation fuel, motor fuel, tractor fuel, kerosene, lubricants, diesel oil, fuel oil, asphalts, and dry-cleaning fluids. For the transportation of these products, a fleet of river boats is owned and operated by the Tropical Oil Company. Delivery is made to bulk stations strategically located along the Magdalena River, whence the products are redistributed by rail and road to smaller bulk stations, or directly to dealer stations in the interior of the country.

The crude oil produced from the Infantas and La Cira fields in excess of that used to supply the Colombian demand is transported a distance of 335 miles, from Barrancabermeja in the De Mares concession to an ocean terminal at Mamonal, near Cartagena. In 1926 a 10-inch pipe line, with a capacity of 30,000 b/d, was completed between these two points by the Andian National Corporation, and in the following year rapidly increasing production made it necessary to increase the line's capacity. This was accomplished by the addition of some 285 miles of 10-inch loop line. In 1937 this was further augmented by 20 miles of 12-inch loop, and the line now has a capacity of 56,000 b/d, at 625 pounds operating pressure. Along the line there are ten pump stations, including that at Mamonal, where ocean-going tankers are loaded. During 1948, 4¾ million barrels of De Mares crude oil were exported through Mamonal, and at the end of that year the cumulative amount of crude exported was 320 million barrels.

The Casabe crude was used only for development purposes in the field until 1945, when deliveries were started to the Andian National Corporation's pipe line at Barrancabermeja for delivery to seaboard at Mamonal. From there oil is delivered by tanker to the Shell Oil Company's refinery in the Dutch island of Curaçao, or directly to other countries as required. No Casabe crude is refined in Colombia.

Towards the end of 1947, the construction of a 4½-inch pipe line across the bed of the Magdalena River was completed, thus tying the Cantagallo field to the Andian trunk line on the east bank.

In 1947 construction of a combination 6-inch and 8-inch pipe line from the Difícil field to the Magdalena River was begun, and by mid-1948 this had been tied into the Andian trunk-line system at Plato. On August 26, 1948, the Difícil-Plato pipe line was inaugurated, and the Difícil concession officially entered the exploitation stage.

In the Barco concession the Colombian Petroleum Company operates a small refinery, with a capacity of 500 b/d, at Petrólea, to supply local Colombian needs. Refined products, consisting primarily of motor fuel and diesel oil, are transported by railroad from Petrólea to Cúcuta, whence they are distributed to nearby dealer stations by road. In 1946 the crude oil which passed through the Petrólea refinery amounted to slightly more than 150,000 barrels.

The greater part of the crude oil produced from the Barco concession is piped through a 12-inch welded steel line to the Caribbean port of Coveñas on the Gulf of Morrosquillo. This line, which was completed in 1939, rises over a distance of 77 miles from an elevation of 216 feet at Petrólea to 5,285 feet at the Rieber Pass, where it crosses the Sierra de Perijá; thence it descends over a distance of 185 miles to cross the Magdalena valley on its way to the marine terminal at Coveñas. The present capacity of the pipe line is 25,000 b/d, but it is estimated that, increasing the number of pump stations along the line from four to eight, the capacity will be 70,000 b/d. During 1948, 7¾ million barrels of crude oil from the Barco concession were exported from Colombia through Coveñas, making a total of over 42 million barrels shipped from that port since the commencement of exportation from it in 1939.

In selecting the route to be followed by the pipe line, and in the actual construction of it, aerial transportation played an important role. To carry materials and personnel to the various construction camps built along the route, ten cargo planes were used, and at the completion of the project it was estimated that these planes had flown over 600,000 miles and carried some 5,500 tons of cargo.

For the Colombian market there are five principal distribution centers in the country: Bogotá, Cali, Barranquilla, Medellín, and Cúcuta. The Cúcuta market is served by the Colombian Petroleum Company refinery for its principal needs. The other four are served by the Tropical Oil Company, which has bulk and package stores in each of these areas. The larger storage plants in turn serve smaller bulk plants in the surrounding areas. For the importation of petroleum products, Tropical Oil Company has two stations, one in Barranquilla and the other in Buenaventura; the Tropical Oil Company finds it more economical to import motor gasoline, aviation gasoline, and diesel fuel into the Cali market from Talara, Peru, through the importation station in Buenaventura, than to attempt overland transportation from Barrancabermeja to this area.

The principal artery of distribution from the Barrancabermeja refinery is the Magdalena River. Since this river is only navigable about seven months out of the year, there is a considerable storage problem involved in the attempt to fill tanks during the "good river," especially for the Bogotá and Medellín areas.

During the wet season, when the river rises, there are frequent washouts on the railroads which carry products from the storage points to the consuming areas. This is particularly true on the Antioquia railroad from Puerto Berrío to Medellín, and on the Cundinamarca railroad from Puerto Salgar to Bogotá. The same is also true on the Pacific Railway, bringing imported products from Buenaventura to Cali.

During a great part of 1946 it was necessary for the Tropical Oil Company to maintain a very expensive overland transport by truck from Bucaramanga, near Barrancabermeja, to the railheads of Barbosa and Duitama, in order to supply the Bogotá area.

Frequently the Tropical Oil Company has to maintain overland transport, particularly of motor gasoline, on what is known as the Quindio road. This is between Ibagué and Armenia, the first being in the Bogotá marketing area, the second in that of Cali. The direction of this shipment varies with the necessity for keeping up stocks in either the Bogotá or Cali areas.

To help in solving some of these problems, the Andian National Corporation constructed a 6-inch pipe line for refined products from the refinery at Barrancabermeja up the Magdalena River to Puerto Olaya and Puerto Berrió, the supply point for Medellín. From Puerto Olaya it is possible to supply La Dorada and Puerto Salgar as well as Puerto Berrío; thus the line aids substantially in maintaining stocks both in the Medellín and Bogotá areas. This products pipe line, 68 miles in length, was placed in service toward the end of 1947.

Some Effects of the Development of the Petroleum Industry on the Colombian Economy

The petroleum industry has obviously had a very great effect in assisting the development of transportation facilities and local industrialization. To give some specific examples, there are now some 785 miles of highway in the country which are paved or surfaced with asphalt, and 1,875 miles more will be paved in the next few years. Total highway mileage in the country at present is over 11,000 miles.

The availability of aviation gasoline is, of course, essential to the development of commercial aviation, and the number of airplanes in service with domestic and international airlines is increasing rapidly. The number of cars, trucks, chassis, and buses is increasing with such rapidity as to create serious traffic problems in some of the cities. Annual imports of these types of vehicles rose from 2,036 in 1945 to around 12,000 in 1947. Due to a shortage of electrical energy in many parts of the country, industries and individual factories are bringing in various types of motors to operate their own generators; the majority of these motors are of diesel type. The number of tractors in use for farm cultivation has also shown a steady upward trend in the past several years; at the end of 1947 it was estimated that some 3,000 farm tractors were in operation in Colombia.

Table 10. The major oil companies operating in Colombia in 1948.

Local Name of Company	Parent Company	Nationality
Tropical Oil Company	International Petroleum Company, Limited	U.S.A.
Compañía de Petróleos Shell de Colombia	Anglo-Saxon Petroleum Company, Limited.	British-Dutch
Socony Vacuum Oil Company de Colombia	Socony-Vacuum Oil Company, Inc.	U.S.A.
Texas Petroleum Company	The Texas Company	U.S.A.
Colombian Gulf Oil Company	Gulf Oil Corporation	U.S.A.
Richmond Petroleum Company of Colombia	Standard Oil Company of California	U.S.A.
Colombian Petroleum Company	Socony-Vacuum Oil Company and The Texas Company	U.S.A.
Petróleos Ariguaní Sociedad Anonima	50% Tropical Oil Company, and 50% Colombian capital	Colombian
Superior Oil Company of Colombia	Superior Oil Company	U.S.A.
Colombian Phillips Petroleum Company	Phillips Petroleum Company	U.S.A.
Sinclair Oil Company of Colombia	Tri-Pet Corporation of Delaware. (Represents the combined interests of Sinclair Oil Company, Cities Service Company, and Richfield Petroleum Company)	U.S.A.
Stanolind Oil and Gas Company	Standard Oil Company of Indiana	U.S.A.

Climate and Health

In the middle Magdalena valley oil-field area there are two wet and two dry seasons, though of recent years the seasons have become less sharply defined. Annual rainfall has varied in the past twenty years between 92 and 209 inches, and averages 120 inches. The mean annual temperature is 85°F, with a mean difference between day and night temperatures of 10 degrees. The humidity is high, and averages 90%.

For an average distance of 12½ miles from the Magdalena River, swamps abound and supply an ideal environment for the prolific and sustained breeding of mosquitos. As was to be expected, malaria was a very common disease in the early days of development, and, if for no other reasons than economic ones, the undertaking of measures to reduce its prevalence was imperative.

In 1926, anti-malarial measures were undertaken systematically in the De Mares concession by the newly organized Sanitation Department of the Tropical Oil Company. Particular attention was paid to efficient screening of all dwellings, and to the installation of a piped water supply. Brush was cleared away over vast areas, water courses were cleaned, ravines opened up, earth ditches and lined ditches were made and kept in repair, many thousands of gallons of anti-larval oil were sprayed each year, and a routine system of sewage disposal was inaugurated.

The success of the anti-malarial campaign on the De Mares concession is well illustrated by the marked drop in the malaria frequency rate, reckoned by the number of disability cases occurring annually, from 845 per 1,000 employees in 1926 to 19 per 1,000 in 1942. Over the ten-year period 1932–1941, the average malaria frequency rate was 40.68 per 1,000.

The Barco concession area also has a wet equatorial climate, with a mean daily maximum temperature of 89°F and a mean daily minimum of 74°. The wettest period of the year is from October to mid-December, the driest from mid-December to the end of March. April and May are rainy months, while July, August, and September are usually showery.

In the Barco concession area malaria is the most prevalent disease, though this has been vigorously combated around the base camps by the screening of all living quarters, brush clearing, and careful sewage disposal.

The Petroleum Legislation of Colombia

THE HISTORY OF THE PETROLEUM CONCESSIONS

The history of the petroleum industry in Colombia has passed through various juridical phases since the year 1905, when the first two concessions were granted to De Mares in the Magdalena valley and to Barco in the Catatumbo region of northeastern Colombia.

In the year 1541 Gonzalo Fernandez de Oviedo y Valdes, a famous Spanish Royal Chronicler for the West Indies, wrote of his meeting with two of Don Gonzalo Jimenez de Quesada's captains:

"Among other things, they testified that one day beyond the town of Latora [Barrancabermeja], there is a bitumen formation or boiling well, which yields large quantities of a thick liquor that overflows the land. They said that Indians used the bituminous fluid in their homes for anointing purposes, that it fortified their legs and prevented them from being tired, and that Christians use the black, pitch-smelling liquor to calk their brigs."

But not until 1905, 364 years later, was any active interest displayed in the Barrancabermeja oil seeps. Then Señor Roberto de Mares filed, and four months later was granted, his request for the privilege of exploiting petroleum, coal, and asphalt for a period of fifty years in the areas bordering the Opón and Carare rivers. A contract was drawn up, defining the area of the De Mares concession, and the contractor undertook to "organize a company or syndicate with sufficient capital for the exploitation on a large scale of petroleum wells or petroleum sources" within that area.

One clause of the 1905 contract obligated the concessionaire, Roberto De Mares, to commence exploitation work within 18 months of its approval. This obligation was not fulfilled, neither in the original time allotted nor in the extension periods which were granted by the government in 1906 and 1907. On October 22, 1909, the contract was therefore declared to have lapsed.

For the next six years nothing was done, and then, in 1915, for reasons which are irrelevant to this discussion, the contract was again admitted. This time the concessionaire was given a twelve-month period, expiring June 25, 1916, in which to commence exploitation work. On May 17, 1916, a contract was signed between Señor De Mares and Messrs. Trees, Crawford, and Benedum of Pittsburgh, Pennsylvania, who undertook to exploit and develop the concession. In June 1916, these contractors started drilling operations and then organized the Tropical Oil Company, to whom the transfer of the concession was officially approved by the Ministry of Public Works on August 25, 1919, subject to acceptance by the newly formed company of certain modifications of the original 1905 De Mares contract. These modifications included:

(1) An obligation on the part of the Company to establish within two years a refinery of sufficient capacity to supply the country's needs.

(2) The concession to run for thirty years from the beginning of exploitation work.

(3) The concession to lapse if exploitation did not prove commercial.

(4) Royalty fixed at 10% of gross production (instead of 15% of net production).

The contract expires on August 25, 1951, when the De Mares concession reverts to the Colombian nation.

It was also in 1905 that a contract was signed between the Colombian government and General Virgilio Barco, granting the latter the right to exploit any petroleum or coal which might be discovered in a defined area of the Upper Catatumbo basin.

General Barco built a mule trail to the Petrólea valley, and in October 1906 was granted permission by the government to establish a provisional refinery in the city of Cúcuta. Not until years later was this refinery transferred from Cúcuta to the base camp at La Petrólea. Economically the venture met with indifferent success, and General Barco, with the government's approval, sold the concession to the "Compañía Colombiana de Petróleos," which was controlled by the Colombian Petroleum Company (Henry H. Doherty and Company). This company drilled the first producing well on the banks of the Río de Oro in 1920. Legal disputes followed, and after prolonged discussions the Barco concession was canceled by the government in February 1926. Appeals against this decision were made by Barco and his relatives and by the "Compañía Colombiana de Petróleos," but both appeals were dismissed.

In the meantime, private negotiations were carried on between the Colombian government, the Colombian Petroleum Company, and the South American Gulf Oil Company, and on March 3, 1931, the government granted to those two companies a new concession covering a large part of the area originally granted to General Barco. Systematic exploration of the concession area was commenced in July 1931 by the Colombian Petroleum Company under the control of the Gulf Oil Corporation, of which the South American Gulf Oil Company is a subsidiary. This arrangement continued until October 1936, when the management of the Colombian Petroleum Company was taken over by a partnership of the Texas Petroleum Company and the Socony Vacuum Oil Company.

Under the terms of the 1931 contract the Colombian Petroleum Company had the right to reserve up to one-half of the total concession area within the first ten years of that contract. They reserved approximately 460,000 acres for exploitation, the remainder reverting to the government. The exploitation period and the contract on the reserved area expires on August 25, 1981.

Between 1905 and 1931, in which year the petroleum laws of Colombia were completely revised and modernized, several other petroleum concessions were granted, though none of them have proved commercially productive.

The Law of 1931, which with only a few modifications has remained in force up to the present time,[1] regulates the exploration for and the exploitation of petroleum in both nationally and privately owned lands. In drawing up this petroleum law, the Colombian government sought the advice of experts from the U.S.A., Great Britain, Mexico, and Rumania; and before being ratified the law was debated in the Colombian parliament for four years.

NATIONAL LANDS

Under the existing Petroleum Laws of Colombia, the minimum area for which any one concession application on national lands may be made is 5,000 hectares (12,350 acres), and the maximum area that may be applied for is 50,000 hectares (123,500 acres); but exception is made as to this maximum limit in the less readily accessible plains and selvas of eastern Colombia, where applications may be made for areas as large as 200,000 hectares (494,000 acres). The maximum total area that any one applicant may be granted as a concession is 200,000 hectares (494,000 acres), so that the acquisition by major companies of concession areas in excess of this amount has to be accomplished by the formation and registration of subsidiary companies.

[1] By decree law published in January 1950, new petroleum legislation was proclaimed. This affects principally that part of earlier legislation that refers to rentals, royalties, and other matters of a fiscal nature.

The information pertaining to an area applied for, which must be submitted to the government through the Ministry of Mines and Petroleum, is clearly defined. It includes topographical and geological (or geophysical) maps of the area, a technical report leading up to the oil-producing potentialities of the area, the names of the technical men who have worked there, together with a statement regarding the exact time spent on exploration of the area, and proof that the applicant has the financial capacity deemed necessary to carry out his obligations as a concessionaire.

The Ministry studies the maps, technical data, and other documents; and provided that all is in order, and that the area applied for is free, the application is accepted. The length of time between the filing of an application and its acceptance has varied between nineteen days to over two years; recently this period has averaged eighteen months. The acceptance is published in the government's "Diario Oficial," and is publicly announced in the municipality or municipalities in which the application area lies. Within sixty days following this announcement any person may object to the proposed contract, and, provided that the objection raised is upheld by the courts, no contract will be signed by the government until in one way or another the objection is disposed of to the satisfaction of both the opposer and the would-be concessionaire. If the private ownership of the subsoil rights on an area partially or wholly within the national lands application can be proved before the courts by an individual or a group of individuals, then it will become necessary to revise the boundaries of the national lands applied for in such a way as to exclude the private property.

With certain reservations, an accepted national lands application may be desisted by the applicant or, if all oppositions have been disposed of, it may be formally contracted. The contractual period is divided into two parts, one for exploration and the other for exploitation. The exploration stage is divided into three periods: the initial period, ordinary extensions, and extraordinary extensions. According to the region in which the contracted area is situated, the initial exploration period is either three or five years. Subject to governmental approval, this can be extended ordinarily for three years, and extraordinarily for a further three years. Drilling on a contracted concession must begin six months before the end of the initial exploration period. The length of the exploitation period is thirty years, subject, at the option of the contractor, to an extension of ten years, provided he agrees to be governed by the petroleum laws in force at the time he exercises his option.

In order to surrender a contracted concession during the exploration period, or during any extensions of it which may have been granted, the concessionaire must prove to the satisfaction of the government that he has fulfilled all his legal obligations, that he has not found oil in commercial quantities, and that it is not technically justifiable to continue drilling in other parts of the concession.

5. *COLOMBIA*

PRIVATE LANDS

In order to lay claim to the petroleum which may underlie the privately owned lands of Colombia, the owner or owners of those lands must produce a title to them which emanated from the State prior to October 28, 1873, or, instead, official public documents, issued by a competent authority, accrediting its existence. Land titles granted to owners since that date do not carry with them the rights to hydrocarbons in the subsoil, which remain the property of the nation.

The time consumed in establishing a firm title to the subsurface rights of privately owned lands is very considerable, ranging from two to as much as eight years. Once the title is made good, however, the legal obligations inherent in the exploration and exploitation of the property are much less onerous than in the case of national lands. The average rental paid by a contractor to the owner of a property for subsurface rights is around U.S. $0.35 per acre per year, together with an agreed royalty, usually between 4% and 6%, payable if and when oil is commercially exploited.

ROYALTY

The government royalty on oil produced from national lands varies between 2% and 11% of the gross product exploited, according to the length of pipe line necessary to carry the products to seaboard. The minimum rate applies to producing fields over 900 kilometers (560 miles) from seaboard, the maximum to those within 100 kilometers (62 miles) of it.

The royalty collected by the government is distributed in the proportion of 45% to the national revenues, 50% to the department, and 5% to the municipality in which the producing wells are located.

The government royalty on oil produced from private lands varies between ½% and 7% of the gross product exploited, according to the date on which exploitation begins and according to the length of pipe line necessary to carry the products to seaboard. This royalty devolves in its entirety to the national revenues.

FORMATION OF NATIONAL OIL COMPANY

Having been passed by the Colombian congress and having received the presidential signature, a bill to authorize the organization of a National Oil Company became law on December 27, 1948. This company is authorized to engage in all phases of the petroleum industry, and one of its specific purposes will be to exploit oil fields, pipe lines, refineries, bulk stations, and movable and immovable property which, on the expiration of petroleum concession and other contracts, automatically revert to the nation. Provision is made whereby private capital, both national and foreign, may participate in the enterprise with the government.

6. SOUTH AMERICA OTHER THAN CARIBBEAN

BY FRANK B. NOTESTEIN *

THE oil fields of South America have strongly contrasting settings. The first of these—the inter-Andean valleys, such as the Maracaibo-Catatumbo basin of Venezuela and Colombia, and the Magdalena valley of Colombia —form part of the Caribbean area, and have been discussed in preceding chapters. In the rest of South America oil-producing areas are found along the Pacific coastal shelf of Colombia, Ecuador, and Peru, and in the great plains—part grassy prairie, part jungle—east of the Andes. This last province includes the prospective basins of eastern Colombia, eastern Ecuador, eastern Peru, Paraguay, southern Brazil, and Uruguay; and the Agua Caliente field in the *montaña* of Peru, the small foothill fields of eastern Bolivia, all the fields of Argentina, and a recently discovered field on the island of Tierra del Fuego, Chile. From a glance at the map, it is obvious that the great basin east of the Andes has been seriously developed only in the extreme south, where the plains, the narrowing of the continent, and the existence of a railway system have made the eastern slope relatively accessible.

Apart from the sedimentary basins associated with the Andes, there are three sedimentary areas in northern Brazil. These, from north to southeast, are: (1) the Amazon basin, extending for over 1,500 miles along the lower portion of the Amazon River, a great trough filled with a moderately thick section of Paleozoic formations overlain by a Tertiary fill; (2) the Piauhy-Maranhão basin, whose sediments contain some marine Devonian, which may be oil-bearing, overlain by continental Permo-Carboniferous and Mesozoic formations; and (3) the Bahia Basin, covering parts of the states of Bahia, Sorgipe, and Alagôas, filled with a 10,000-foot sequence of Upper Mesozoic and Tertiary rocks. Of these three areas, only the Bahia basin is currently productive.

The Pacific Coastal Zone

THE COLOMBIAN COAST

A considerable number of concession applications, submitted by both American and British companies, and covering much of the narrow zone of sediments on the Pacific coast of Colombia, suggests that this coast will be prospected in the near future. One drilling location is now being rigged in the Baudó valley. Results are, of course, problematical. That coast is

* Geologist, The Texas Company.

. 16. South America: proved oil and gas fields, test drilling, and petroleum facilities. The refineries and e lines shown include those under construction at the beginning of 1949. The quadrilateral boxes in the rthern part of the map indicate areas shown in more detail in Figs. 9, 10, 11, 12, 14, and 15.

a zone of extreme rainfall, and is completely jungle-covered. The topography is low-hilly, with the exceptions of the swampy San Juan delta and a belt of mangrove swamps about six miles wide along the immediate Pacific coast. The whole area is unhealthful in the extreme.

ECUADOR

The Ecuadorian oil fields are confined to a small area on and near Santa Elena Point, on the arid and windswept Pacific coastal shelf. The oil seepages here were well known in the colonial period. Thousands of shallow pits were dug, and seepage oil was collected and boiled down to tar for calking ships. Local tradition claims that the Spanish Armada was calked with tar from Santa Elena Point. The Point is almost on the Equator; yet it is swept by the cold Humboldt current, and the wise visitor takes along a leather jacket.

The coastal shelf of Ecuador, between the sea and the west foot of the Andes, is from 60 to 120 miles wide, and flat to moderately rolling. Arid at the southwest corner, it becomes better watered and better wooded as one proceeds north and east. The barren oil-field area proper is almost flat, and has apparently been subjected to marine planation at no distant date.

The oil fields of Santa Elena Point have been developed by British and Ecuadorian companies. The fields are not prolific, but they produce a high grade of crude oil. They are served by sea-loading lines and by two rather small refineries. The latter supply all the liquid-fuel requirements of the country. A railway connects the fields with the port of Guayaquil. Developed in a thinly populated area, which prior to the discovery of oil supported a few fishermen, the oil fields have had little effect on the population distribution or transportation routes of Ecuador, though the local supply of liquid fuel at moderate prices has stimulated the construction of highways in the populous plateau areas.

The coastal plain north of Santa Elena Point has been actively prospected by a Canadian firm, without success.

PERU

Like Ecuador, Peru has its petroleum industry largely concentrated on the Pacific coastal shelf. This part of the shelf is a zone of Tertiary sediments, 10 to 30 miles wide between the Pacific beaches and the foot of the Amotape Mountains, a spur range extending southwest from the Andes. The coastal zone is characterized by *tablazos,* or elevated marine terraces, which are deeply dissected by dry ravines. Farther inland there are hill ranges up to 1,000 feet high. The coastal shelf is one of the most desperately arid regions of the world, rains being spaced about thirty-five years apart. This coast is swept by the cold Humboldt current, and rainfall occurs only on those rare occasions when the warm *Niño* current from the north, shouldering the Humboldt current aside for a few weeks or months, extends southward along the coast. Heavy rains and destructive floods follow immediately. The periodicity of this warm-water invasion is one of the mysteries.

The oil fields, all on or close to the coast, were originally developed by two British companies and one small Peruvian operator. The larger of the British companies was later bought out by the International, a Canadian corporation affiliated with the Standard Oil Company (New Jersey), and the International has since then been the major operator. Within the past few years the Peruvian government has bought out the Peruvian company, and has expanded the operations. One new small field has been discovered at Punta Organos. As in the case of Ecuador, these oil fields of the Pacific coast are moderately prolific of very high-grade crude oil.

In opening the fields there have been no jungles to clear, no rivers to bridge, and no serious health problems to meet. No expensive pipeline outlets have been required. On the other hand, all the materials and manpower have had to be brought in. Water had to be distilled from the sea, or piped in from the Chira River. Again as in the case of Ecuador, the development of the oil fields has directly affected only a small area, but it has affected the rest of the country to the extent that an ample local supply of liquid fuel has been available, and it has avoided strain on the balance of trade and the exchange rate that would have resulted from importations.

The Plateau Area

Some years ago a very small oil field was developed on the high plateau near Lake Titicaca, but it proved short-lived and non-commercial. At an elevation of about 13,000 feet, this was the highest oil field in the world while it lasted. The Peruvian government has made efforts to develop additional oil fields in this bleak plateau area, so far without success.

The Great Plains and Foothills East of the Andes

THE EASTERN SLOPE OF COLOMBIA

The term *llanos* is applied to the great grassy plains extending from the eastern foothills of the northern Andes to the Orinoco River. Until a few years ago this prairie was a remote, somewhat legendary country devoted to grazing, primitive and untrammeled as West-of-the-Pecos in the days of Judge Bean. News traveled by word of mouth, usually in ballad-singer style to the air of "Oh Casanare." A wide cattle driveway extended five hundred miles from southern Venezuela to the foot of the mountains just east of Bogotá. But within the past few years the llanos have been given highway connection with the plateau, landing fields and air service have been established, and navigation has been opened on the Río Meta. Oil exploration has led to considerable road construction, and exploration parties have worked out many natural routes suitable for truck traffic in the dry season.

Several exploratory wells drilled by the Shell Oil Company and one by the Tropical Oil Company have shown a little heavy oil and the presence of tar sands—enough to indicate a petroliferous section—but no commercial oil has been discovered, and exploratory operations are at this writing (May 1949) at a standstill. The vast llanos of eastern Colombia remain unproven but still prospective, and numerous valid concessions are still held.

South of the plains area, geological and geophysical exploration is in progress in the dense jungles of southeastern Colombia, in the drainage systems of the Caquetá and Putumayo rivers, tributaries of the Amazon. Several shallow wells have been drilled for stratigraphic information. The answer in this area is several years away.

THE ORIENTE

The Shell Oil Company of Ecuador and the Standard Oil Company (New Jersey) are jointly prospecting a large concession in the vast forest region of the Oriente, the basin area east of the Ecuadorian Andes. Success in this area would open up eastern Ecuador generally, with marked effect on the economy of the whole country.

THE MONTAÑA

The Peruvians refer to the great forest country of the upper Amazon basin, east of the Andes, as the Montaña. (This word does not mean "mountainous": it is derived from *monte,* which means "forested.") The Montaña is a vast area of tropical rain forest drained by the Marañón, Huallaga, and Ucayali rivers, which combine to make the upper Amazon. There are a few river-port towns, of which Iquitos on the Amazon is the most important, and there are scattered agricultural and fishing settlements along the banks of the larger rivers. Away from the river banks the population consists largely of bush Indians, still entirely uncivilized. Much of the hinterland between the rivers is unexplored, except as it has been seen from the air. Topographically the area is characterized by a series of foothill ranges, of low mountains and high hills, paralleling the trend of the Andes. The intervening streams have the same northward trend until they join the eastward-flowing Marañón.

The Agua Caliente oil field (shown in Plate 1), on the Pachitea River in the foothills on the east side of the Peruvian Andes, is Peru's only field east of the Andes. It was discovered by aerial reconnaissance and developed by a California company, which still operates it. So far its production has been limited to meeting the demands of the local market in the Montaña. A small refinery supplies refined products, sufficient to supply the towns on the upper Amazon drainage and to fuel floating equipment on the rivers. Some products have been shipped to Iquitos and to Manaos, Brazil.

The economy of the upper Amazon, moribund since the collapse of the rubber boom about 1918, is showing signs of revival under the influence of improved river transportation. The discovery of the Agua Caliente field has undoubtedly influenced the Peruvian government in pushing its great Transandine highway project. The prospecting activities now in progress, on the part of the Peruvian government and various companies, may result in new discoveries of great importance, justifying a pipe-line outlet to the Pacific, or they may show that the oil resources of the Montaña are of only local significance.

BOLIVIA

This landlocked nation has several small oil fields in the eastern foothills of the Andes. The original petroleum discovery was made by the Standard Oil Company of Bolivia, and their holdings were later taken over by the Bolivian government, which is now the sole operator. The fields are on long narrow anticlinal ranges, clothed in dense but scrubby forest. The topography is rough, and the costs of development are high. Gradually the fields are being extended, and the search for new fields is in progress.

Since these fields are in the eastern foothill zone, whereas the industrial and population centers of Bolivia are on the plateau, the oil has been more accessible to northern Argentina than to the consuming centers of Bolivia. To meet this situation a highway has been built from Santa Cruz, at the eastern foot of the Andes, to Cochabamba, at the railhead on the plateau. In addition, a 350-mile 6-inch pipe line is being laid to connect the oil fields with a refinery at Cochabamba. A 45-mile 4-inch line branches off to supply the city of Sucre, also on the plateau. From the other direction, a railway is under construction from Corumbá, Brazil, to Santa Cruz. When these transportation arteries are completed, the Bolivian fields should be able to meet the requirements of Bolivia, and to supply some oil for export to southern Brazil. The oil fields are thus about to open up eastern Bolivia to settlement and development.

PARAGUAY

Union Oil Paraguay, a subsidiary of Union Oil Company of California, has been drilling carefully chosen deep wildcats since 1946. The sites are selected on the basis of exploration by the Western Geophysical Company, under contract to Union Oil. So far no showings of oil or gas have been found. Union Oil Paraguay holds a 23-million acre exploration concession in the Gran Chaco west of the Paraguay River, and it is in this large area that the seismograph work has been done and the test sites selected. In Paraguay Oriental, on the east side of the Paraguay River, exploration has been confined to minor geological studies.

URUGUAY

So far no prospect of oil production has been found in Uruguay, but in 1948 a United States firm was engaged by the government, which owns all mineral resources, to undertake extensive studies with a view to subsequent detailed geophysical surveys. It is estimated that several years will be required for completion of the present studies, which are concentrated in the northern part of the territory, where indications seem more promising.

At present Uruguay's petroleum industry is limited to one government-owned refinery in Montevideo that, with a capacity of some 14,000 b/d, is able to take care of the country's needs for all but some special products.

ARGENTINA

The Argentinian oil fields are in four groups, all along the eastern foot of the Andes: (1) the Salta fields in the extreme north of the country, on or close to the Bolivian frontier; (2) the Mendoza fields (Plates 41, 42) in the Province of Mendoza, central Argentina, only a few miles from the foot of the Andes; (3) the Neuquén fields in the Territory of Neuquén, also within sight of the Andes; and (4) the Comodoro Rivadavia fields in the Territory of Chubut, in the far south. The last named is much the most important. It is on the east coast, and farther from the mountains than the others.

The Salta fields, like those in Bolivia, are on long, narrow, tight anticlines, forming rugged foothill ranges at the eastern base of the Andes. These ridges are precipitous, densely wooded, and sparsely populated; the lowlands to the east support scrub forest, and are used for grazing and timber production. It cannot be said that these oil fields have contributed greatly to the local economy of the region, which is still dominated by cattle on the hoof. The oil produced is quite limited in amount, but of very high quality; part is refined locally in Salta and part is shipped by rail to the seaboard.

The Mendoza fields are in arid plains and badlands a few miles from the foot of the Andes. From these fields peaks over 22,000 feet high can be seen. Nearby there is the rich irrigated horticultural area of Mendoza, which was well served by rail and air lines long before oil was discovered. The oil produced is moved to seaboard by railway. The fields have thus contributed to the fuel supply of the nation, but have had little effect locally on population distribution, transportation systems, or industry.

The small Neuquén fields are in high, barren, sage-brush-covered plains, useful only for grazing—the area resembles the Red Desert country of Wyoming. The development of the fields has led to the extension of the rail-and-highway system into this area. No great influx of population has resulted, but the livestock industry has been given an improved outlet.

The Comodoro Rivadavia fields, in barren, windswept badlands suitable

35. Colombia. Middle Magdalena basin. Producing well in the jungle. El Centro field. (Standard Oil Company, N.J., photo by Collier.)

36. Peru. Lomitos repressuring station. (Standard Oil Company, N.J., photo by Collier.)

37 (opposite page). Peru. International Petroleum Company's refinery and the town of Talara, looking northwest.

38. Chile. Manantiales field. View from hill north and east of well 14, looking south and west toward camp and wells 14, 4, and 15. (Corporación de Fomento de la Producción.)

39. Chile. Straits of Magellan. View from the proposed pipe-line terminal site, looking south along shore of Bahía Gente Grande. (Corporación de Fomento de la Producción.)

only for sheep, front on the Gulf of San Jorge, and the products are shipped to refining and consuming centers by sea. A considerable industrial population has been attracted to the area and fair roads radiate to all parts of the interior. Small refineries, some port and sea-loading facilities, and a railway line extending about 125 miles inland are the chief items of permanent plant that have resulted. No important agriculture, aside from sheep grazing, has been or is ever likely to be developed. When the oil is exhausted, the area will revert to its primitive character as grazing land.

The Comodoro Rivadavia fields have for many years constituted the fuel mainstay of the Argentine. The oil is heavy and furnishes a high percentage of boiler fuel; the locomotives on the Argentine railways are fueled to a large extent on oil from these fields, and any interruption of this supply would cripple the nation's economy. An 1,100-mile 10-inch natural-gas line from Comodoro Rivadavia to Buenos Aires is now under construction. It should greatly alleviate the fuel situation at Buenos Aires.

Oil production in Argentina is in the hands of a government agency, Yacimientos Petrolíferos Fiscales, and of various foreign oil companies (British, American, and Swiss). The share of private companies in the industry is steadily declining because of the restrictive government policy, which effectively prevents the acquisition of new holdings by private operators. This policy has undoubtedly greatly retarded petroleum development.

CHILE

For several years the Corporación de Fomento of Chile has employed a North American contracting firm to make geological and geophysical studies on both sides of the Straits of Magellan, and in 1945 it announced the discovery of oil at Spring Hill, now called the Cerro Manantiales field, on the main island of Tierra del Fuego. The Andean axis in that region trends southeast, and this discovery is in the geosyncline northeast of the Andean axis. It is close to the Straits, with deep water open the year round. At the end of 1948 about sixteen wells had been drilled, nearly all of which proved productive of oil or gas. They are all shut in, however, until a pipe line can be constructed to a shipment point on the Straits.

The area is a bleak, windswept, subantarctic plain, which has no future except for sheep grazing and mineral extraction. The discovery will hardly do much for the local area, as it will not require railway or much road construction. If large production results, it will benefit the economy of Chile, which has in the past imported all its liquid fuel, to the detriment of its exchange rate. The potential production of the Cerro Manantiales field has been estimated at some 3,000 b/d, or the equivalent of about one-sixth of the current national consumption of petroleum products. A 42-mile 8-inch pipe line from this field to deep water is under construction. Recent dispatches announce the discovery of a new field in this area.

Brazil

Brazil has been economically handicapped by a paucity of mineral-fuel supplies, both solid and liquid, although the nation is otherwise blessed with great mineral resources. This situation springs chiefly from the fact that a large percentage of the nation's area is underlain by crystalline rocks.

For many years rather desultory prospecting for petroleum was conducted in various parts of the country, by national and state agencies and a few ill-equipped private operators, without success. Finally, in 1939, the Ministry of Agriculture drilled the discovery well of the Lobato oil field in the sedimentary basin of All Saints Bay, near the city of Salvador (Bahia).

Since 1939 the Conselho Nacional do Petróleo, organized in 1938, has been in complete charge of the search for petroleum. This agency has employed a North American drilling contractor and a North American geophysical company to continue prospecting the east coastal zone, where a narrow fringe of sedimentary rocks constitutes a coastal plain between the sea and the crystalline rocks of the Brazilian uplands, which rise in an abrupt escarpment to the west. Oil seepages and other indications of petroleum had long been known on the coastal zone, and by 1942 all efforts were concentrated in the basin adjacent to All Saints Bay. This is a down-faulted block of Tertiary to Jurassic sediments let into the eastern margin of the crystalline shield rocks. The relief is low and the climate tropical, with moderate rainfall. The area is well populated, the city of Salvador claiming 500,000 residents. As there is considerable local industry, there is a local demand for any petroleum products, including gas.

To date five rather small oil fields, one of which produces chiefly gas, have been proven in this basin, while four structural prospects in the same basin have been condemned by unsuccessful drilling. Efforts to find additional production are continuing, and by the end of 1949 potential production was rated at around 7,500 b/d.

The oil is of medium to light gravity and has a high wax content; it is derived from Lower Cretaceous to Upper Jurassic sandstones and limestones. In view of the very considerable number of wells drilled to secure this small production, it is doubtful that this operation would normally pay commercially, but the output is valuable to Brazil because all other petroleum has to be imported. These fields are all on low ground near the shores of the bay, and products not locally required can be conveniently shipped. Thus the accessibility of the area, as well as the exchange situation, justifies development, although the potentialities of the east coastal zone of Brazil do not appear great.

The Conselho Nacional do Petróleo has done geological and geophysical work in the Paraná basin of southern Brazil, so far without result, and is conducting geological and geophysical investigations in the lower Amazon trough. The difficulties inherent in exploratory work in the Amazonian forests are so great that much time and expense will be required to evaluate the prospects of that region. The early geological investigations by the Brazilian government in the Territory of Acre, extreme western Brazil, do not seem to have lent encouragement for further effort, but geological studies are under way in a systematic exploration of the entire country.

Brazil had three refineries in operation in the summer of 1949, with capacities of 150 b/d, 1,000 b/d, and 1,000 b/d, respectively. The first, operated by the Conselho Nacional do Petróleo near the Aratu (Bahia) field, the second by the Industria Matarazzo de Energia, S.A., in São Paulo, and the third by the Ipireango S.A. Companhia Brasileira do Petróleos in Rio Grande do Sul. The Conselho Nacional is building a refinery with 2,500 b/d capacity near the Candeias (Bahia) field, and another government refinery, with a capacity of 45,000 b/d, is planned for Belém. The Refinaria e Exploração de Petróleo União S.A. plans a 20,000 b/d refinery at São Paulo (to be served by pipe line from Santos), and the Refinaria do Petróleo Distrito Federal S.A. plans a 10,000 b/d refinery near Rio de Janeiro.

7. NORTH AMERICA

BY A. I. LEVORSEN *

THE earliest records of petroleum in North America antedate Columbus, and probably even the Indian. They are contained in the brea [1] or tar pits of southern California, near where Los Angeles now stands. Animals, many of species now extinct, waded into these naturally occurring basins in search of water and food, mired down, and finally succumbed to the black sticky substance, which they mistook for water in its reflected light. Thus these pits became cemeteries for the life of that time: great numbers of skeletons have been recovered for scientific study, and undoubtedly even greater numbers remain.

Other early records of petroleum are found in scattered reports, some legendary and others authenticated, of oil and gas seepages, oil springs, and brea and tar deposits. Some of these were well known to the Indians and early explorers, who often used the material either as a medicine or as a grease for their wagons. By the middle of the nineteenth century a number of wells dug for water had reported showings of oil and gas in many parts of the United States and Canada, and petroleum was not a complete stranger to the scientific thought of the time.

The Development of the Industry

The beginning of the petroleum industry, which has so revolutionized our mode of living, is generally taken to be the discovery of oil by drilling on Oil Creek, near Titusville in northwestern Pennsylvania, in 1859. This discovery was made near the site of an oil spring known to the Indians, and later the white man, as the source of small quantities of oil. Mr. Samuel M. Kier, a druggist of considerable enterprise, had bottled the waste oil from some of the salt wells at Tarentum, near Pittsburgh, and sold it as a medicine under the following exciting announcement:

> *The Healthful balm, from Nature's secret spring,*
> *The bloom of health and life to man will bring;*
> *As from her depths this magic liquid flows,*
> *To calm our sufferings and assuage our woes.*
>
> Kier's Petroleum or Rock Oil, Celebrated for its Wonderful Curative Powers. . . . Several who were blind were made to see. If you still have doubts, go and ask those who have been cured!

* Professor of Geology, School of Mineral Science, Stanford University; Past President, American Association of Petroleum Geologists.

[1] Brea is a viscous asphalt formed by the evaporation of petroleum from oil seeps.

> . . . It gets its ingredients from the beds of substances which it passes over in its secret channels. They are blended together in such a way as to defy all human competition. . . . Its discovery is a new era in medicine.

While looking at the derrick on Kier's advertisements, Mr. George H. Bissell, a lawyer from New York, conceived the idea of drilling for oil instead of salt. Professor Benjamin Silliman of Yale University had advised him that oil might be distilled, and the product burned in lamps. After considerable trouble in organizing and financing, the Seneca Oil Company was finally formed, and Mr. Edwin L. Drake, a retired railroad conductor and the owner of a small amount of the stock, was employed to go to Oil Creek to enlarge the spring and to drill a well in the hopes of finding oil. In order properly to impress the people living in the area with whom he had to deal, letters were addressed to him as "Colonel Drake," and the title has remained. The well was started from the Tarentum salt wells in June 1859 by a driller named William Smith and his two sons, and on August 27th it reached a depth of 69½ feet. While the drilling was shut down over the weekend, Smith saw some dark fluid in the hole and lowered a water can on a string. When the can came out it was full of oil, and when a pump was installed the well began producing at the rate of 25 barrels per day—the first discovery well drilled for oil.

The finding of oil in such quantities by drilling specifically for it set off a world-wide search for other deposits. The pattern of many future operations in the United States was set in these early days of the industry: first, in the manner in which the ownership of the oil was established and wells were drilled under lease rights, and, second, in the reasoning which prompted drilling near oil seepages. Many oil and gas fields have since been found by following similar reasoning. Seepages ordinarily occur at the surface of the ground, but a non-commercial showing of oil or gas in a well may be called a "subsurface seepage," and it is equally effective as an indicator of the presence of petroleum.

During the latter half of the nineteenth century the oil men moved west across the country. Searching the valleys that resembled those in which oil had been found in Pennsylvania, the drillers compared the formations with the rocks in West Virginia and Ohio, and the promoters bought their leases and drilled their wells near oil and gas seepages and springs. The "oil man" developed—typically American in nature, self-reliant, independent, capable, and aggressive. He has since added a large element of scientific observation and reasoning, which makes the present discoveries seem, outwardly, wholly the result of logical scientific thinking: but his scientific thinking does best when accompanied by a spark of native shrewdness, sometimes called "oil sense."

The continuous expansion of the petroleum industry since its beginning is one of the important factors to remember when thinking of its

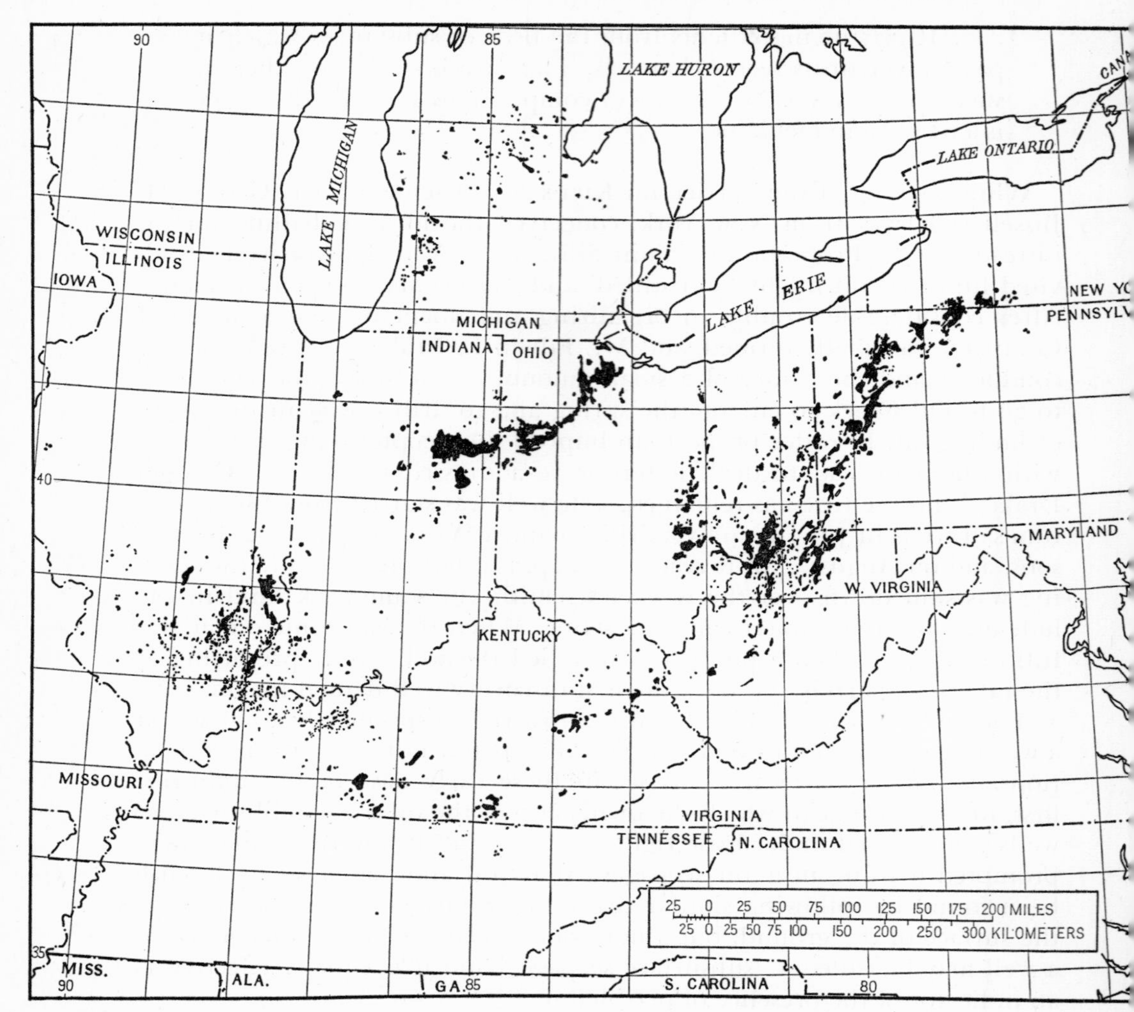

Fig. 17. United States, Appalachian and north-central states: oil fields, as of March 1946, drawn to s This map and those in Figs. 18 to 22 are derived from the map, "Oil and Gas Fields of the United Sta issued by the U.S. Geological Survey in 1946.

impact on the American scene. The industry is dynamic, and it is big business, a powerful combination which leaves a definite mark on the people of a country. The petroleum industry, with its huge capital investment, is exceeded in size in the United States only by agriculture, railroad transportation, and the public utilities. It has over 1,250,000 employees, and its wages are among the highest paid. It produces its raw material from over 4,000 separate oil and gas pools located throughout the continent, but concentrated in the north-central and Appalachian states, shown in Figs. 17 and 18; in the Gulf Coast and Mid-Continent

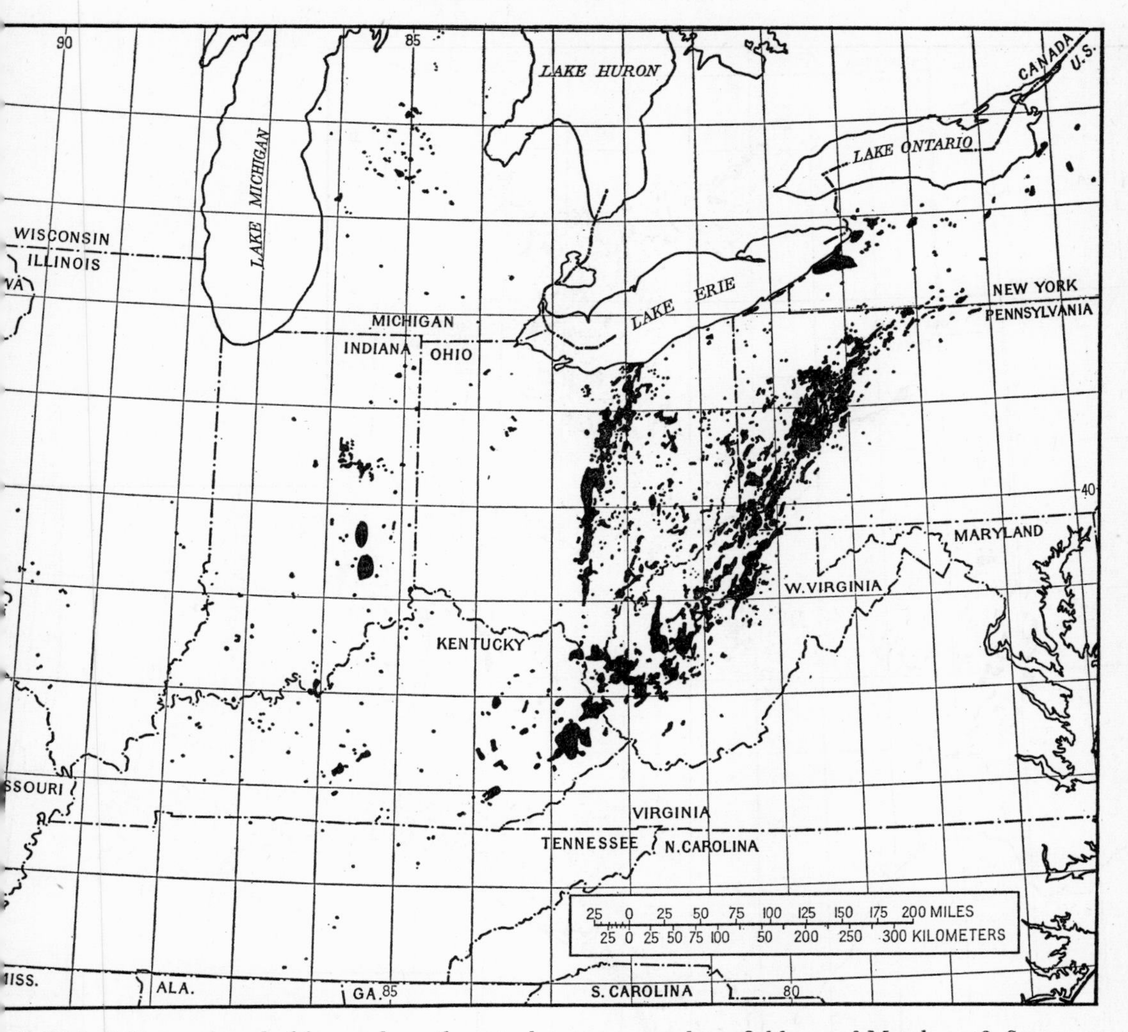

18. United States, Appalachian and north-central states: natural gas fields, as of March 1946. Source as Fig. 17.

states, shown in Figs. 19 and 20; and in California, shown in Figs. 21 and 22. From these widespread sources the crude oil is funneled through pipe lines to refineries, and the refined products are then distributed to every corner of the continent, and beyond it, from the arctic to the tropical jungles.

The broad pattern of the petroleum industry as developed in the United States is a sequence consisting of: (1) the production of crude oil and natural gas; (2) the transportation to refineries and markets; (3) the manufacturing or refining; and (4) the distribution to the consumer.

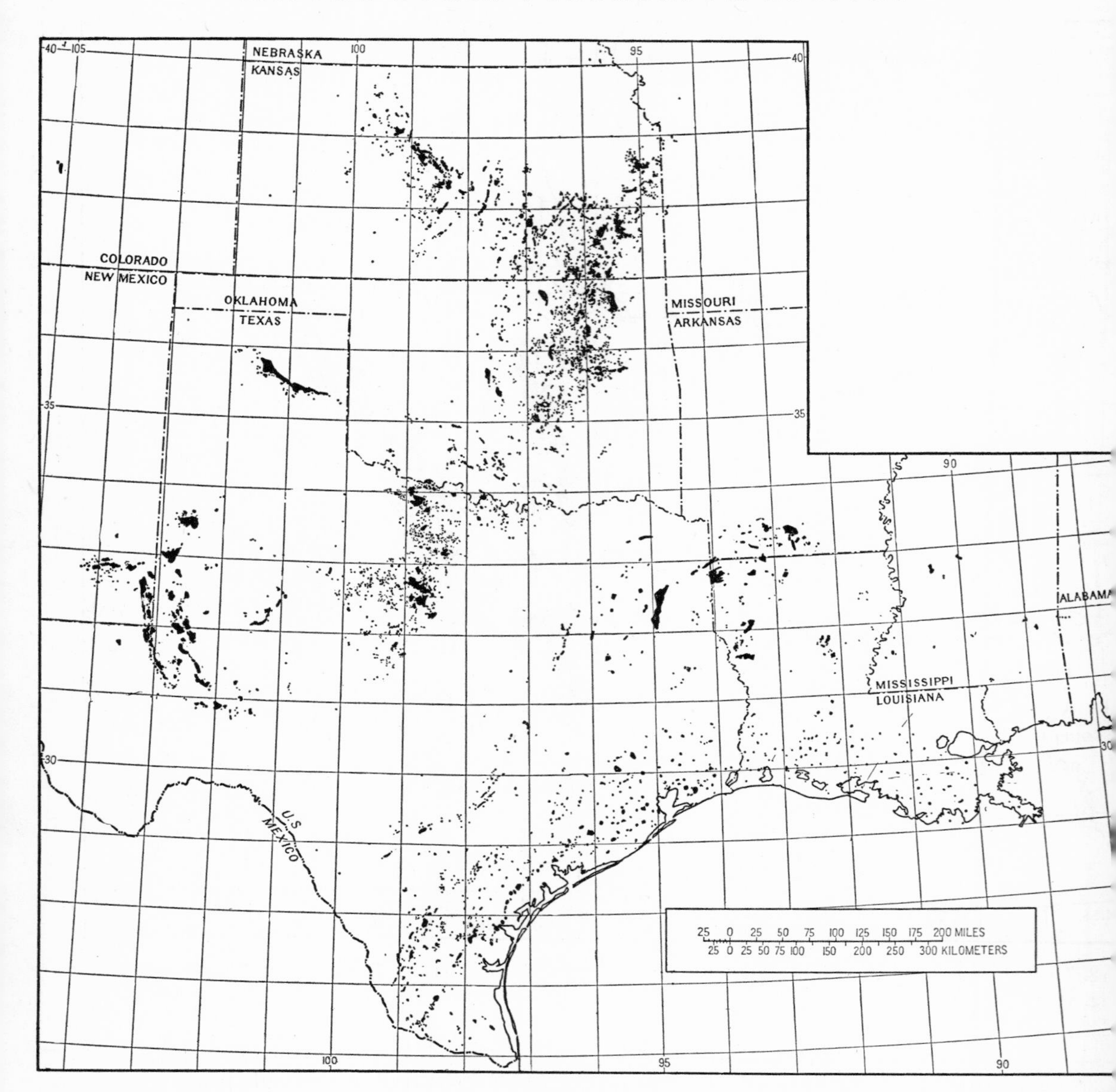

FIG. 19. United States, Mid-Continent and western Gulf Coast states: oil fields as of March 1946. Source as for Fig. 17.

Production

The production of the raw material of the industry consists of two parts: first, the exploration for and discovery of new pools; and second, the extraction of the oil and gas from these pools after they have been discovered. The heart of the oil industry in North America has been its ex-

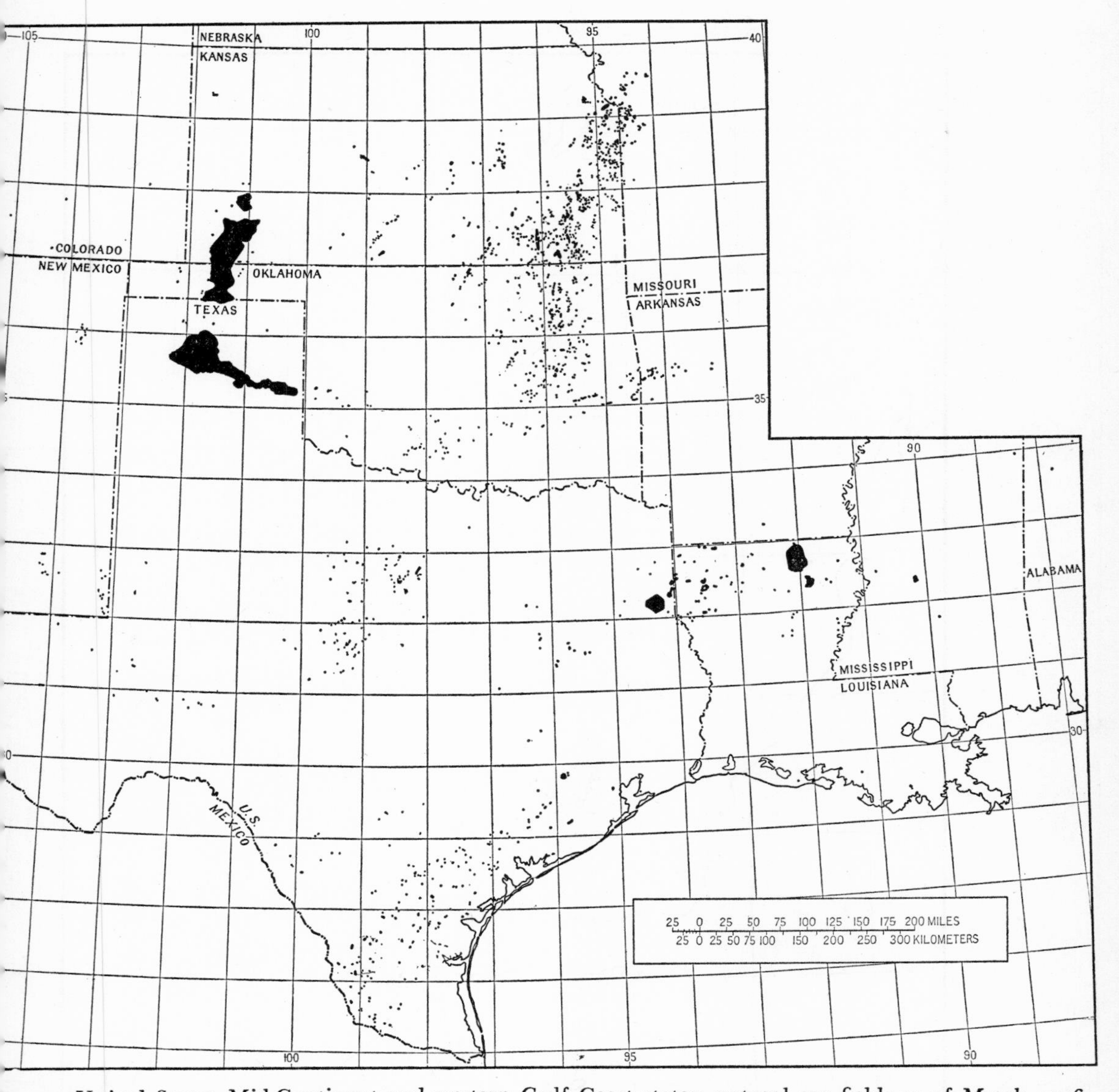

IG. 20. United States, Mid-Continent and western Gulf Coast states: natural gas fields as of March 1946. ource as for Fig. 17.

ploration program—the replacement, by new discoveries year after year, of the amounts of petroleum which are being consumed.

Discoveries are made only by drilling exploratory wells. In the decade since 1939, from 3,000 to 5,000 or more such wells have been drilled every year in the United States, and in 1948 nearly 7,000 were drilled. It has been estimated that one exploratory well has been drilled for every 12

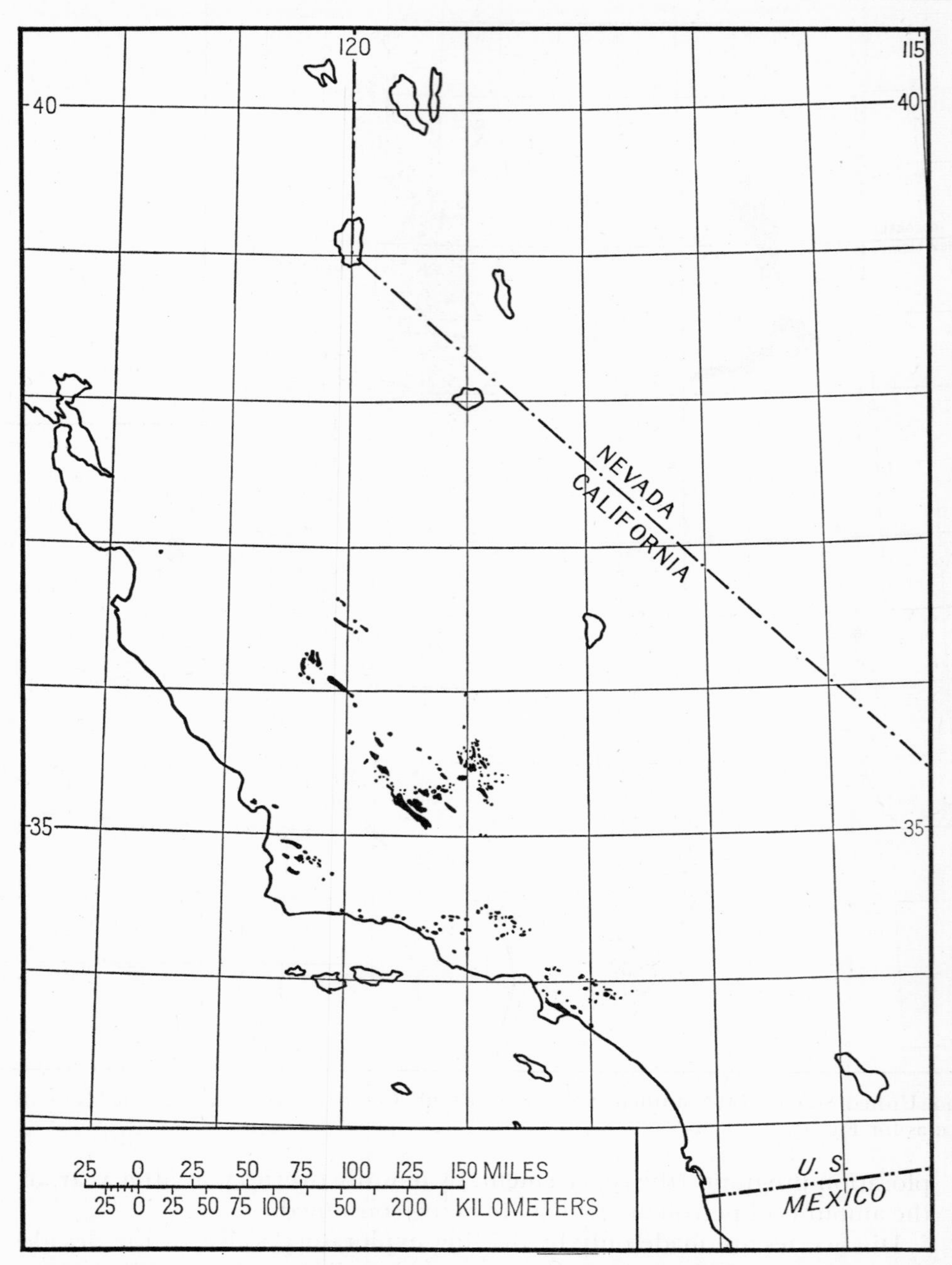

Fig. 21. United States, California: oil fields as of March 1946. Source as for Fig. 17.

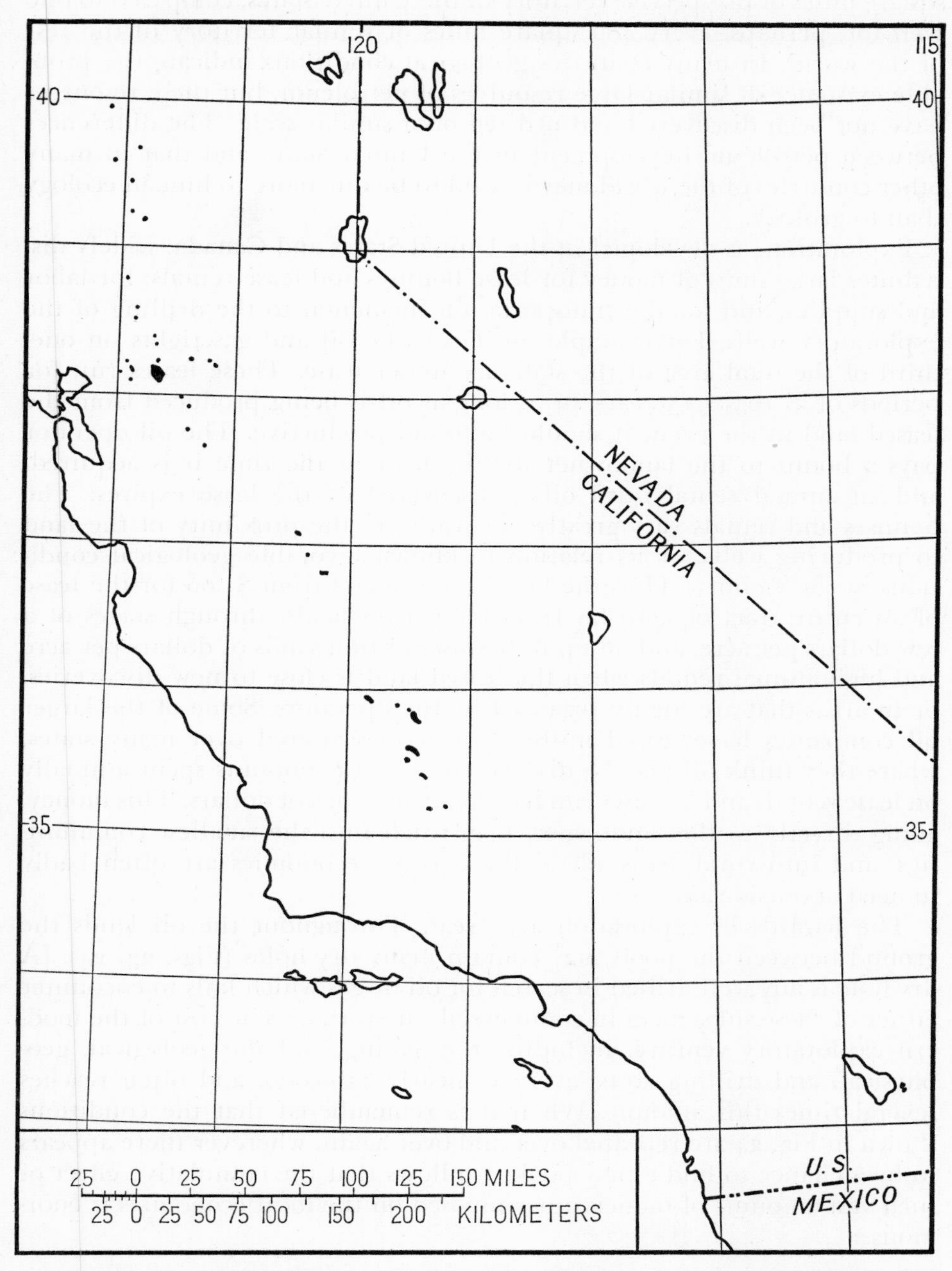

FIG. 22. United States, California: natural gas fields as of March 1946. Source as for Fig. 17.

square miles of prospective territory in the United States, compared to one well for, perhaps, every 480 square miles of similar territory in the rest of the world. In many countries geological conditions indicate the probable existence of similar large resources of petroleum, but these resources have not been discovered and utilized on a similar scale. The differences between petroleum development in the United States and that in many other countries of the world may be said to be due more to human ecology than to geology.

Exploration, as developed in the United States and Canada, widely distributes large sums of money for lease bonuses and lease rentals, for labor and supplies, and for the transportation incidental to the drilling of the exploratory wells. For example, in Texas the oil and gas rights on one-third of the total area of the state are under lease. These leases run for periods of 5, 10, or 15 years, or as long as oil is being produced from the leased land in the event it should be found productive. The oil operator pays a bonus to the landowner for the lease at the time it is acquired, and an annual rental until oil is discovered or the lease expires. The bonuses and rentals vary greatly, according to the proximity of the land to producing wells, or its relation to known favorable geological conditions in the vicinity. Thus the bonus rate ranges from $1.00 for the lease of an entire tract of land far from known oil fields, through stages of a few dollars per acre, and on up to bonuses of thousands of dollars per acre and high annual rentals when the leased land is close to new discoveries, or in areas that are highly regarded by the operators. Some of the larger oil companies have leased millions of acres scattered over many states, where they think oil may be discovered, and the amounts spent annually on lease rentals and bonuses run into many millions of dollars. This money, going directly to the landowner, may reach into the smallest communities, and into rural areas where the existing economies are often badly in need of outside aid.[2]

The hazards of exploration are great. Throughout the oil lands the ground between the pools may contain many dry holes (Figs. 23, 24). [A dry hole is any well drilled in search for oil or gas which fails to encounter either of these substances in commercial quantities.] The cost of the modern exploratory venture, including the leasing, and the geological, geophysical, and drilling costs, averages nearly $100,000, and often reaches several times this amount. When it is remembered that the conditions shown in Fig. 24 are repeated over and over again, wherever there appears to be a chance to find a new pool, it follows that the cumulative effect of such vast amounts of money superimposed on the local economies is enormous.

[2] In *Forbes Magazine* of October 1, 1949, the statement is made that the "farmers, ranchers and holders of royalty interests in the U.S. harvest a 'cash crop' of over a billion dollars yearly from the oil and gas industry. Last year their take was divided as follows: $400 million in lease and bonus payments, $800 million in royalties."

Not the least of the benefits which come to an area that is undergoing an oil boom is the improvement of land titles, since the ownership of every tract of land near a drilling well is carefully established. Another benefit, often important to marginal lands, results from the land taxes which must be paid in order to establish good titles to the leases. These taxes are in addition to the two billion or more dollars (1948) paid annually in other

Table 11. United States: crude oil[a] production during 1948,[b] and proved reserves as of December 31, 1948,[b] with principal states ranked in order of production.

Rank in Production	State	Production	Proved Reserves	Rank in Reserves
		(In million barrels)		
1	Texas	899.3	12,484.2	1
2	California	340.1	3,763.6	2
3	Louisiana	174.8	1,869.3	3
4	Oklahoma	155.2	1,250.4	4
5	Kansas	108.1	674.4	6
6	Illinois	62.8	393.4	8
7	Wyoming	56.0	715.8	5
8	New Mexico	47.4	552.2	7
9	Mississippi	43.6	365.4	10
10	Arkansas	30.1	300.0	11
11	Colorado	17.0	365.0	9
12	Michigan	16.9	69.4	14
13	Pennsylvania	12.9	109.8	13
14	Montana	9.5	118.9	12
15	Kentucky	8.6	59.2	16
16	Indiana	8.0	49.1	17
17	New York	4.4	67.1	15
18	Ohio	3.9	28.5	19
19	West Virginia	2.7	36.7	18
20	Alabama	0.5	3.5	20
	Miscellaneous[c]	0.6	3.3	
	Total, United States	2,002.4	23,280.4	

[a] Excluding condensate, natural gasoline, and liquified petroleum gases. The production of these in the United States during 1948, as reported in the source cited below, totaled 183,749,000 barrels, and reserves at the end of the year were estimated at 3,540,783,000 barrels (in accordance with the definitions used by the American Petroleum Institute's Committee on Petroleum Reserves). At the end of 1947 the proved reserves had been 3,253,975,000 barrels.

[b] Figures for 1948 based on eleven months' actual production, with an estimate for December.

[c] Including Nebraska, Florida, Missouri, Tennessee, Utah, and Virginia.

Source: *Reports on Proved Reserves of Crude Oil, Natural Gas Liquids, and Natural Gas,* Vol. 3, American Gas Association and American Petroleum Institute, New York, 1949.

taxes by the industry, and distributed wherever it operates. Thus the exploration for petroleum, even where the effort fails, brings not only the excitement of anticipated discovery into the community, but also financial as well as other advantages.

One productive wildcat[3] well constitutes a discovery, following which

[3] A wildcat is any well drilled for exploratory purposes, whether the site has been selected on a hunch, or as the result of the most elaborate geological, geophysical, and other surveys.

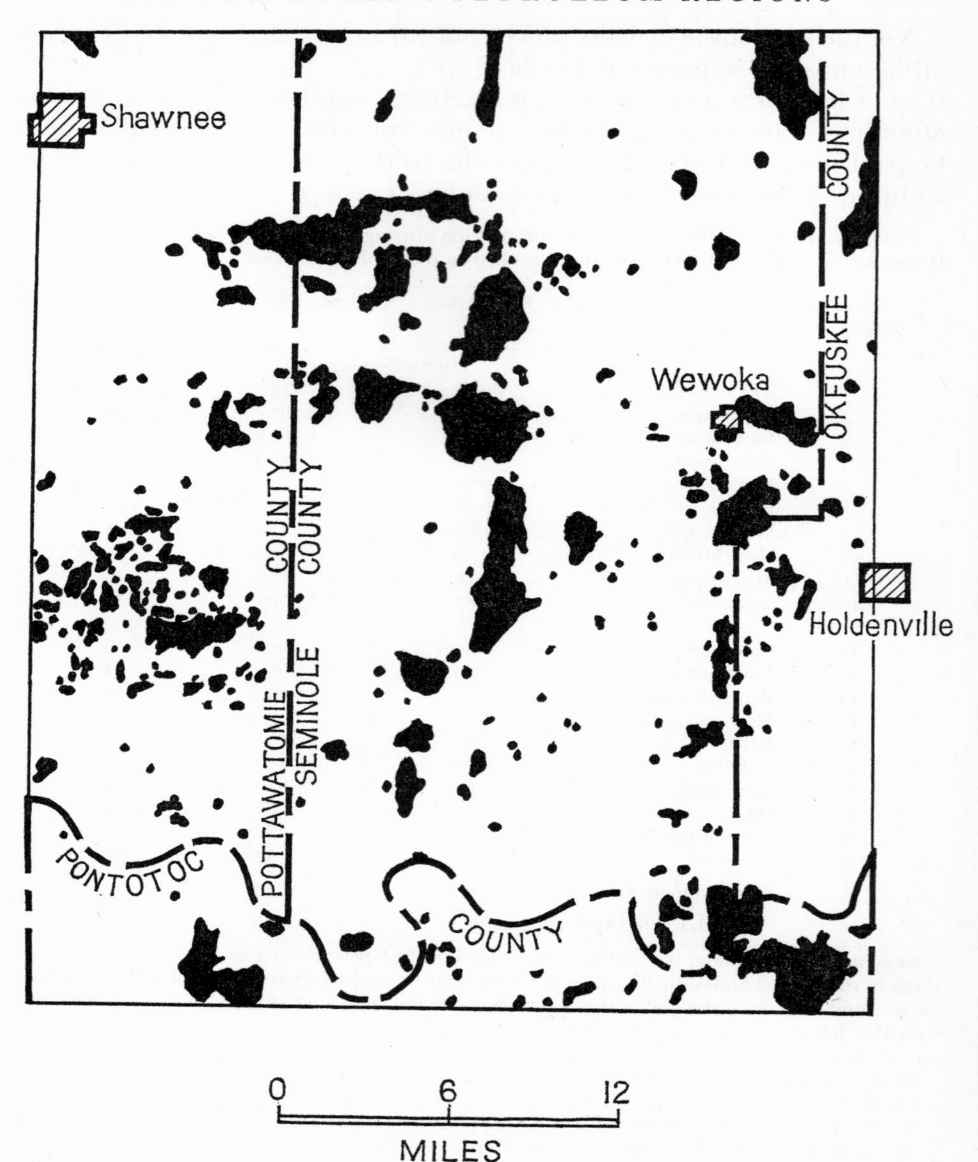

FIG. 23. East-central Oklahoma: oil pools (in black) in an area 30 by 36 miles. The map shows in some detail a distribution of oil pools typical of many in the producing regions of the United States. Numerous "dry holes," or failures, were drilled between the producing areas in the search for additional pools.

the engineers and drillers come in to develop the pool and produce the oil and gas. The depths of oil pools have increased, until now the average depth of all new wells is nearly 3,500 feet, and many pools, particularly in the Gulf Coast, are found at depths of 10,000 to 12,000 feet below the

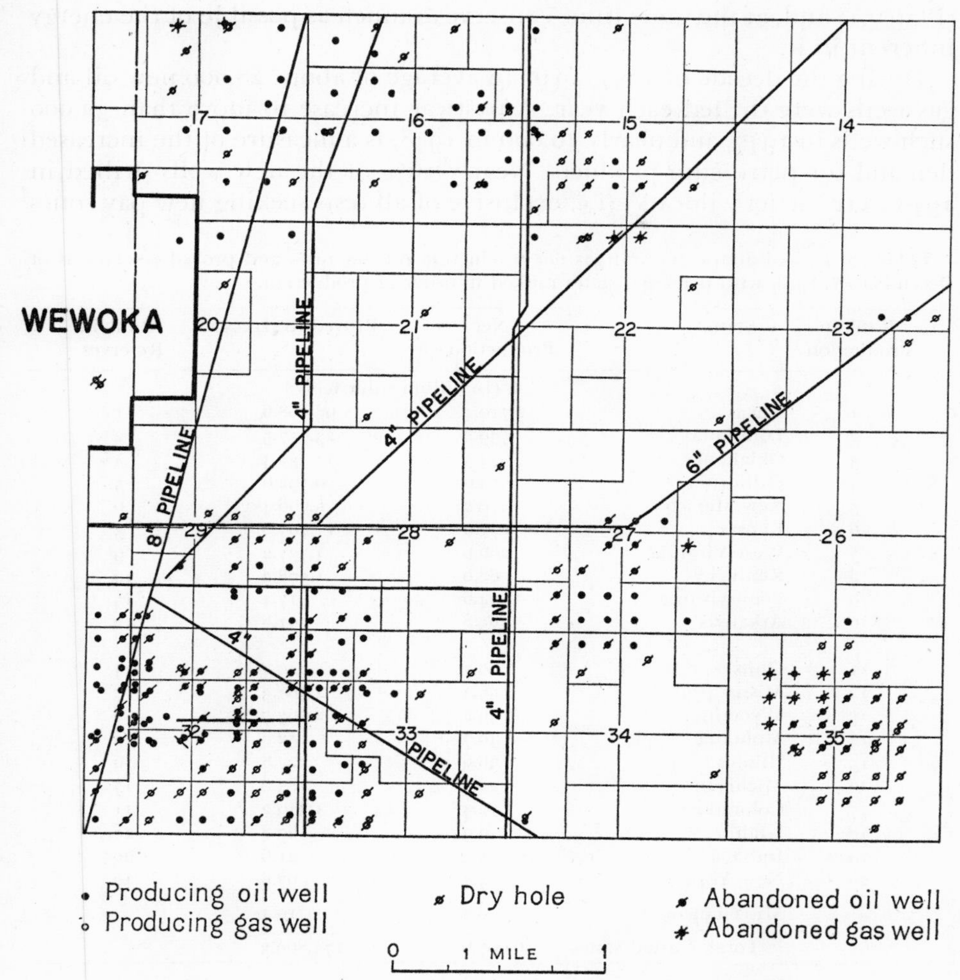

FIG. 24. East-central Oklahoma: detail of an area of 16 square miles east of Wewoka (see Fig. 23), showing oil and gas wells, dry holes, abandoned wells, land subdivisions, and the larger pipe lines. Not shown are the smaller pipe lines connecting each well to the larger lines. This area is characteristic of the oil-producing regions of the United States, and it gives an idea of the vast amount of effort that goes into the search for, and development of, petroleum resources.

surface. The deepest drilled well is more than 20,000 feet deep in Wyoming (1949). Wells in oil pools are generally drilled with spacings which range from one well to 10 acres up to one well to 160 acres. In some pools, however, notably the Long Beach field in California (Plate 50) and the East Texas field, the spacing is much less, and some wells are drilled on tracts as small as a fraction of an acre. The modern practice is to drill as few wells as are necessary to obtain the most oil and gas from the pool

(Plate 51) and, at the same time, conserve as much as possible of the energy inherent in it.

During the decade of 1937–1946 an average of about 28,000 new oil and gas wells were drilled each year. A postwar increase to more than 33,000 such wells in 1947, and nearly 40,000 in 1948, is a measure of the increased demand for petroleum products. Nearly 16% of the new wells drilled in 1948 were "strict wildcats" (i.e. exclusive of all tests seeking new pay zones

Table 12. United States: natural gas net production during 1948, and proved reserves as of December 31, 1948, with principal states ranked in order of production.

Rank in Production	State	Net Production [a] [b]	Proved Reserves [a] [c]	Rank in Reserves
		(In billion cubic feet) [d]		
1	Texas	2,710.4	95,708.6	1
2	Louisiana	759.2	23,977.5	2
3	Oklahoma	674.3	11,332.4	4
4	California	579.0	10,192.6	5
5	New Mexico	279.2	5,606.4	6
6	Kansas	277.0	14,407.8	3
7	West Virginia	206.0	1,737.2	9
8	Kentucky	88.0	1,378.2	10
9	Pennsylvania	74.6	617.4	15
10	Arkansas	65.8	901.8	12
11	Ohio	60.7	629.4	14
12	Mississippi	55.3	2,504.3	7
13	Wyoming	49.2	2,093.7	8
14	Montana	40.7	852.6	13
15	Illinois	36.6	227.8	16
16	Michigan	22.1	183.0	17
17	Colorado	12.4	1,349.2	11
18	Utah	6.9	69.8	18
19	Indiana	5.5	21.6	20
20	New York	4.4	67.6	19
	Miscellaneous [e]	0.3	10.4	
	Total, United States	6,007.6	173,869.3	

[a] Excludes shrinkage caused by natural gas liquids recovery.

[b] Net production equals gross withdrawals less gas injected into underground reservoirs; changes in underground storage are excluded.

[c] Proved reserves include non-associated gas, associated gas, dissolved gas, and gas in underground storage.

[d] 14.65 p.s.i.a., at 60°F.

[e] Includes Alabama, Florida, Missouri, Nebraska, and Virginia.

Source: *Proved Reserves of Crude Oil, Natural Gas Liquids, and Natural Gas,* Vol. 3, American Gas Association and American Petroleum Institute, New York, 1949.

in existing fields, or outposts attempting to extend known fields). Some 135 million feet of hole in new wells were drilled in the United States in 1948. Of the more than one million producing wells drilled in North America, about 443,000 were still producing at the end of 1948; of these around 3,000 were in Canada and Mexico.

These producing wells draw upon 4,000 or more pools, which range in

41. Argentina. Mendoza Province. Drilling rig set up by Yacimientos Petrolíferos Fiscales. (Photo by Y.P.F

42. Argentina. Mendoza Province. View of part of Y.P.F. refinery installations. (Photo by Y.P.F.)

43. Brazil. State of Bahía. Oil well C-56 in Candeias field. Depth 982.37 meters. (Brazilian Government Trade Bureau, N.Y.C.)

44. Brazil. State of Bahía. Six-inch oil pipe lines in Candeias field. (Brazilian Government Trade Bureau, N.Y.C.)

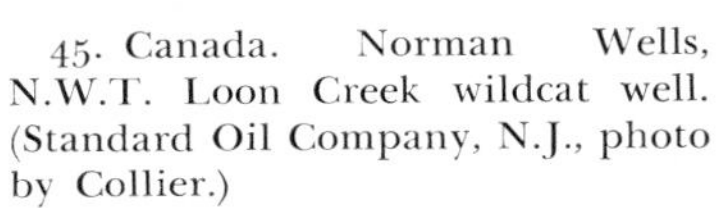

45. Canada. Norman Wells, N.W.T. Loon Creek wildcat well. (Standard Oil Company, N.J., photo by Collier.)

46. Canada. Norman Wells, N.W.T. Moving crew disassembling portable derrick. (Standard Oil Company, N.J., photo by Corsini.)

47. Canada. Norman Wells refinery, N.W.T. November 1944. (Standard Oil Company, N.J., photo by Collier

size from less than one million barrels up to the mammoth East Texas pool, which had a content of approximately 5 billion barrels. Only 130 pools originally contained more than 100 million barrels of petroleum each, or around 20 days' consumption in the United States alone, at our current rate. It is an alarming fact that wildcat drilling in the United States in recent years has discovered more pools (424 new oil pools in 1947), but smaller. The ownership of the producing pools is chiefly in the hands of some thirty major oil companies, which own over four-fifths of the proved reserves in the United States. On the other hand, the smaller operators—the small companies and individuals called the "independents" —are most active in discovering the new pools, and account for from two-thirds to three-fourths of those found annually.

Conservation and research are two key terms in modern oil and gas production practice. Waste, which characterized much of the early production methods, has been very largely eliminated. A few decades ago it was thought good practice if only about 20% of the recoverable oil in a new field was produced, whereas now 80% is not unusual. Many fields that were supposedly worked out and nearly exhausted have been rejuvenated by secondary recovery methods, such as gas repressuring or water flooding, and are again in profitable production.

Transportation

The transportation of petroleum from the wells to the refineries is chiefly accomplished through pipe lines and by barges and tankers. Relatively small amounts may be hauled by truck or tank car in the early development of an oil pool, but as soon as the production warrants, a pipe line is laid.

The first oil pipe line of any importance, constructed from Butler to Brilliant, on the Allegheny River near Pittsburgh, in 1875, came fifty years later than the first gas pipe line, for in 1825 gas was piped through hollowed logs into Fredonia, New York. Later, in 1875, a 5½-inch gas line into Titusville, Pennsylvania, was constructed, and by 1891 an 8-inch line carried gas 120 miles from the Indiana fields into Chicago. In 1946 there were over 81,000 miles of main gas-transmission pipe lines in the United States, and a total of both transit and distribution lines aggregating nearly 225,000 miles. As of December 31, 1948, this total had increased to 259,400 miles.

The first oil trunk line to the seaboard was a 6-inch line from McKean County in western Pennsylvania to Philadelphia, a distance of 235 miles, with a branch from Millway to Baltimore, a distance of 66 miles. From these beginnings the pipe-line system has expanded to the network shown in Fig. 26, which connects all the important oil pools on the continent with refineries or seaports.

The pipe-line system is divided into about 64,000 miles of gathering lines, generally less than six inches in diameter. These reach to individual leases

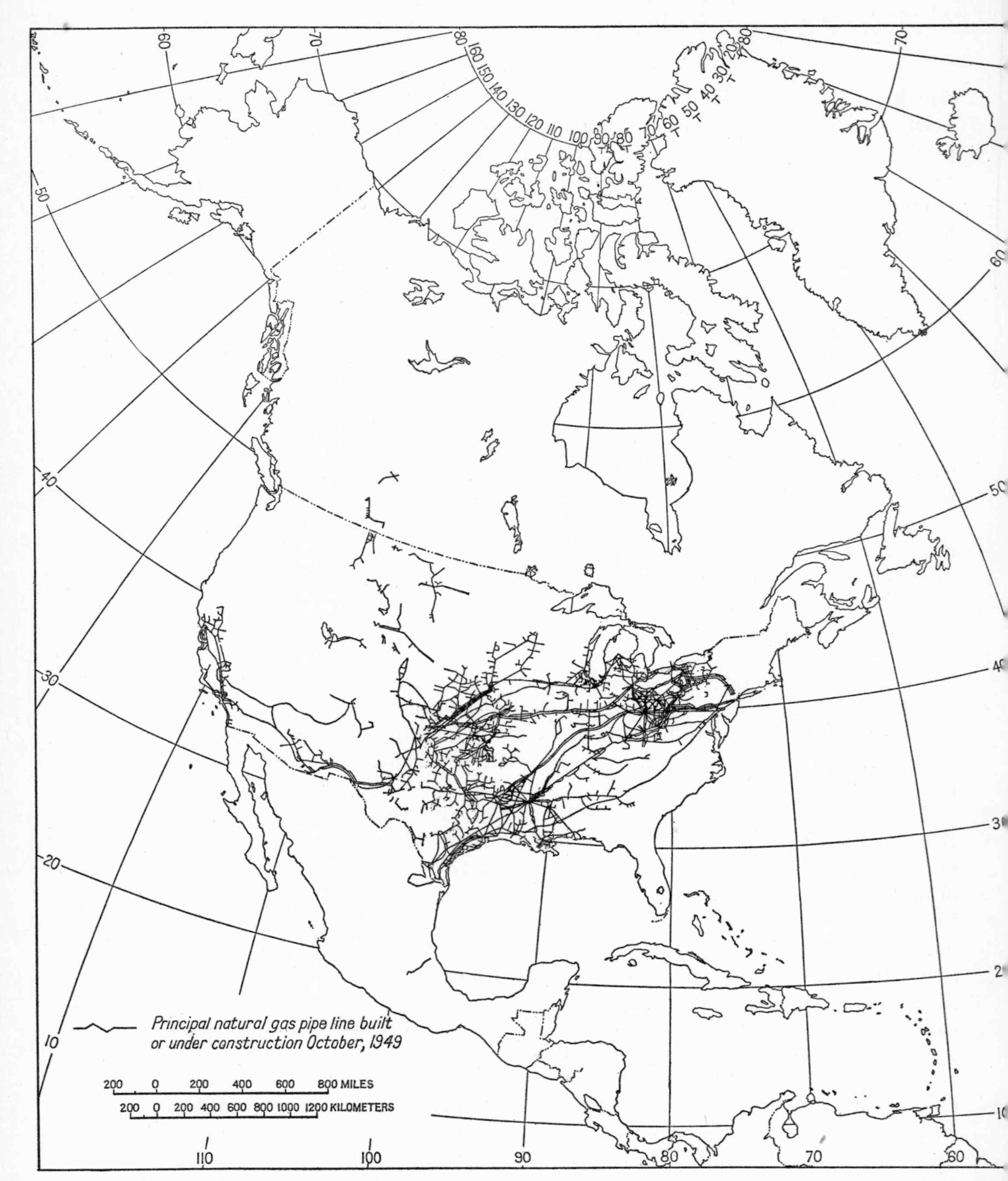

FIG. 25. North America: natural gas pipe line network about October 1949. Data for the United States from map of natural gas pipe lines published by U.S. Federal Power Commission in 1947 and map published by the *Oil and Gas Journal,* as a supplement to the October 6, 1949, issue. Data for Canada and Mexico from *World Oil,* July 15, 1949, and from other sources.

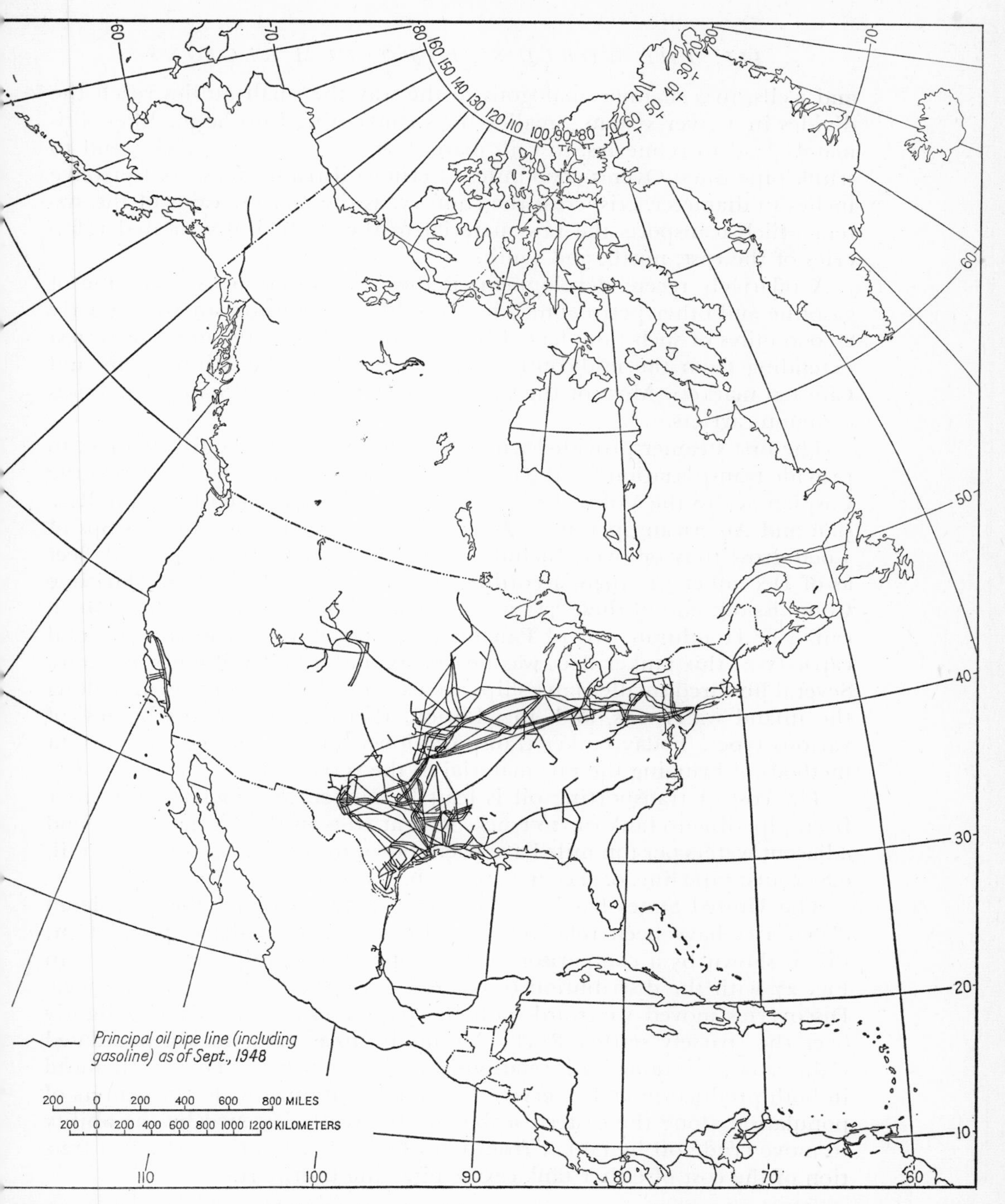

. 26. North America: network of pipe lines used for transporting crude petroleum and petroleum prod-
s about September 1948. Data for the United States from the map, "Oil and Gas Fields of the United
tes," published by U.S. Geological Survey in 1946, and from the map of crude-oil pipe lines, products lines,
l refineries issued by the *Oil and Gas Journal* as a supplement to its September 23, 1949, issue. Data for
er countries from *World Oil,* July 15, 1949, and from other sources.

and wells, in a manner analogous to the way the small gullies reach the divides in a river system. Smaller lines converging into larger lines ultimately lead to refineries, to rail-, barge-, or tanker-loading racks, and to trunk pipe lines. Over 73,000 miles of crude-oil trunk lines, six to twelve inches in diameter, criss-cross the country, as the main streams of the system which transports the raw material to the strategically located refineries of the east, north, and south.

A relatively recent development is the transportation by pipe line of gasoline and other petroleum liquid products. Beginning in 1931, nearly 20,000 miles of such lines have been laid in the United States, the largest extending from the mid-continent states north to the Minneapolis and Chicago markets. Most of the pipe lines of the United States operate as common carriers.

The first steamer provided with tanks to carry oil was built in 1879, to provide transportation for the oil from the Baku area in Russia across the Caspian Sea to the Volga. By 1900 many tankers had been built for Russian and American operators. There were 1,963 seagoing tanker ships of 1,000 gross tons or over, including whaling tankers, in the world fleet as of December 31, 1948, according to the records of the U.S. Maritime Commission; and of these 521 were of United States registry, 13 of Mexican, 4 of Honduran, 181 of Panamanian, and 27 of Canadian: the total capacity of this tanker fleet was nearly six and a half million gross tons. Several hundred smaller tank ships are used on the Great Lakes and along the inland waterways, and, in addition, there are numerous barges, of various types. Today tanker transportation is one of the most important methods of bringing the raw material to the refineries.

The cost of transporting oil is least by tanker, and increases in order from pipe line to tank car to truck. Recent costs in the United States and adjacent waters per ton mile have been shown to be: truck, 4.87 cents; rail, 0.83 cents; pipe line, 0.32 cents; tanker, 0.12 cents.

The United States has been fortunate in that most of the petroleum discoveries have been relatively close to the areas of dense population; this is shown by a comparison of the population-density chart shown in Fig. 27 with the distribution of oil and gas pools shown in Figs. 17–22. Discoveries moved westward with the population, and, skipping lightly over the sparsely settled Rocky Mountain areas into thickly populated California, maintained this relationship. Even now, with Texas well ahead in both production and reserves, the transportation costs to the centers of population along the eastern seaboard are greatly lessened by the ability to move crude oil by tanker from Gulf Coast to Atlantic ports, at a fraction of the cost of either tank-car or pipe-line transport.

Refining

There were 418 refineries in the United States at the beginning of 1949, with an aggregate crude-oil capacity of 6.7 million b/d, according to the

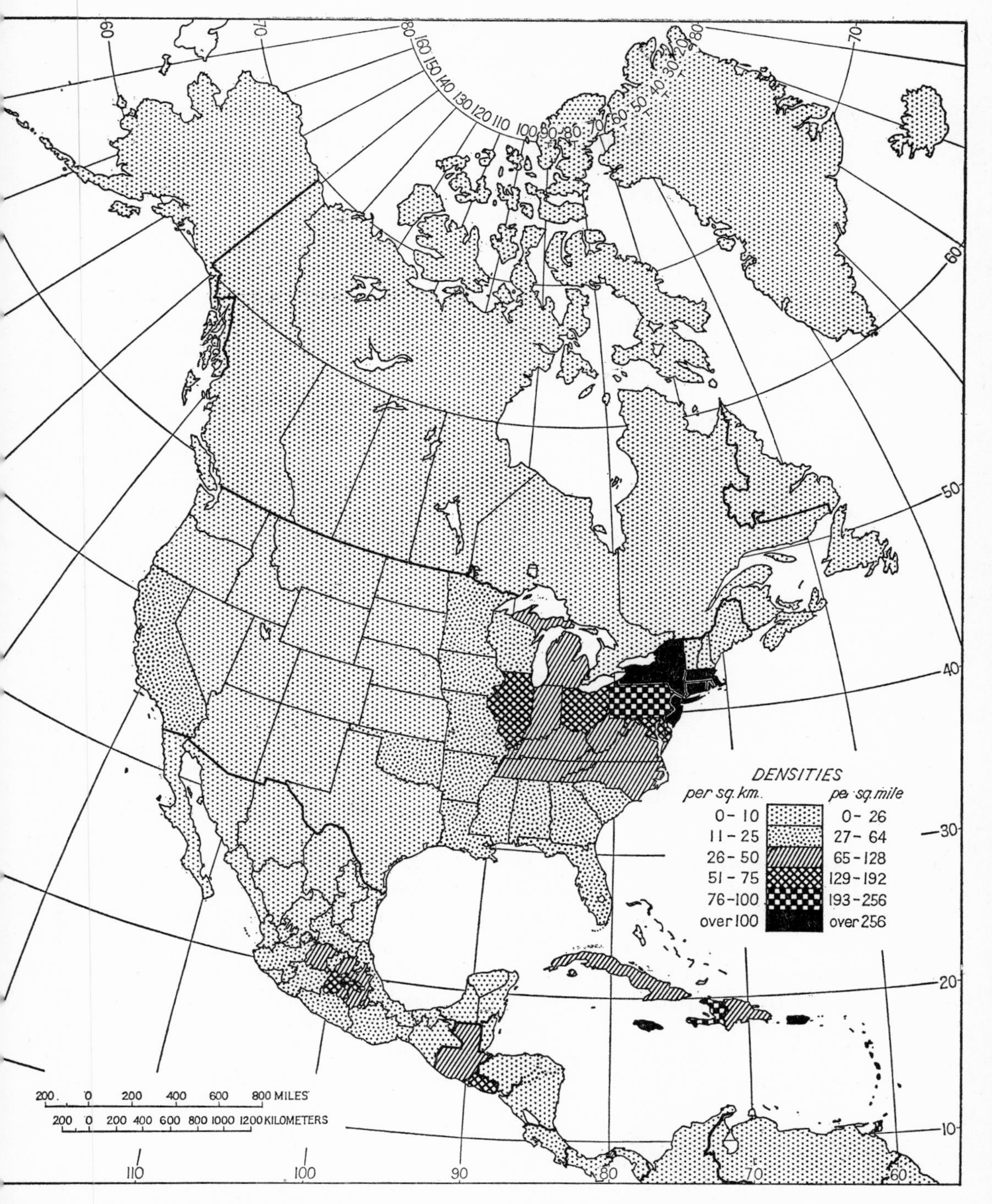

27. North America: population density, by states and provinces, about 1940. (United States and Mexican [c]ensuses of 1940; Canadian census of 1941.)

tabulation of the *Oil and Gas Journal.* Of these, 374 refineries were reported as operating, their aggregate crude capacity being 6.51 million b/d and their cracking capacity, 3.75 million b/d. In Canada there were at the same time 36 refineries with an aggregate crude capacity of 331,850 b/d, and cracking capacity of 167,400 b/d; and in Mexico, 7 plants with aggregate capacities of 160,350 b/d of crude oil, and 25,900 b/d of cracking capacity. The location of the refining centers, in the oil-producing areas, near large consuming centers, and alongside water terminals, is shown in Fig. 28.

The overall resources of petroleum are extended, not only by new pool discoveries and by improved engineering and production methods in producing fields, but also by higher refining efficiencies. In the case of the chief "money crop" of the refinery, gasoline, refinery technicians have increased the percentage of gasoline that it is possible to make from a barrel of crude oil: in 1915 it was about 18%; in 1940, 45%; and they promise nearly 60% in the near future by the use of catalytic-cracking methods, now in process of development. But to run all refineries for maximum yield of gasoline is not practical, because to do so reduces in proportion the yield of middle distillates, for which there is an insistent and increasing demand for household heating. As a result, the principal products of an average refinery in the United States in 1946 were reported by the Bureau of Mines as follows, in terms of a 42-gallon barrel of crude oil:

	Per Cent	*Gallons*
Gasoline	39.6	16.6
Kerosene	6.0	2.5
Gas oil and distillates	16.6	7.0
Residual fuel oil	24.9	10.5
Lubricating oils	2.6	1.1
Total	89.7	37.7

And in addition solid products—wax (0.2%), coke (0.6%), asphalt (2.6%), and certain other finished products (5.1%)—were produced.

Already the oil industry has helped develop over five thousand different products from crude oil, and its future use as a raw material for the manufacture of chemicals has been merely glimpsed. For example, while prior to 1940 petroleum and natural gas accounted for the raw material source of less than 5% of the total organic chemical production in the United States, in 1946 the ratio had jumped to nearly 25%, or over 3.5 billion pounds of raw materials for synthetic organic chemicals.

Marketing

The North American continent is the world's greatest market for petroleum products. In 1938 the average per-capita consumption of petroleum products ranged from 0.9 barrels in Mexico to 4.3 in Canada and 8.8 in the United States, the highest in the world. In 1947 the consumption rates

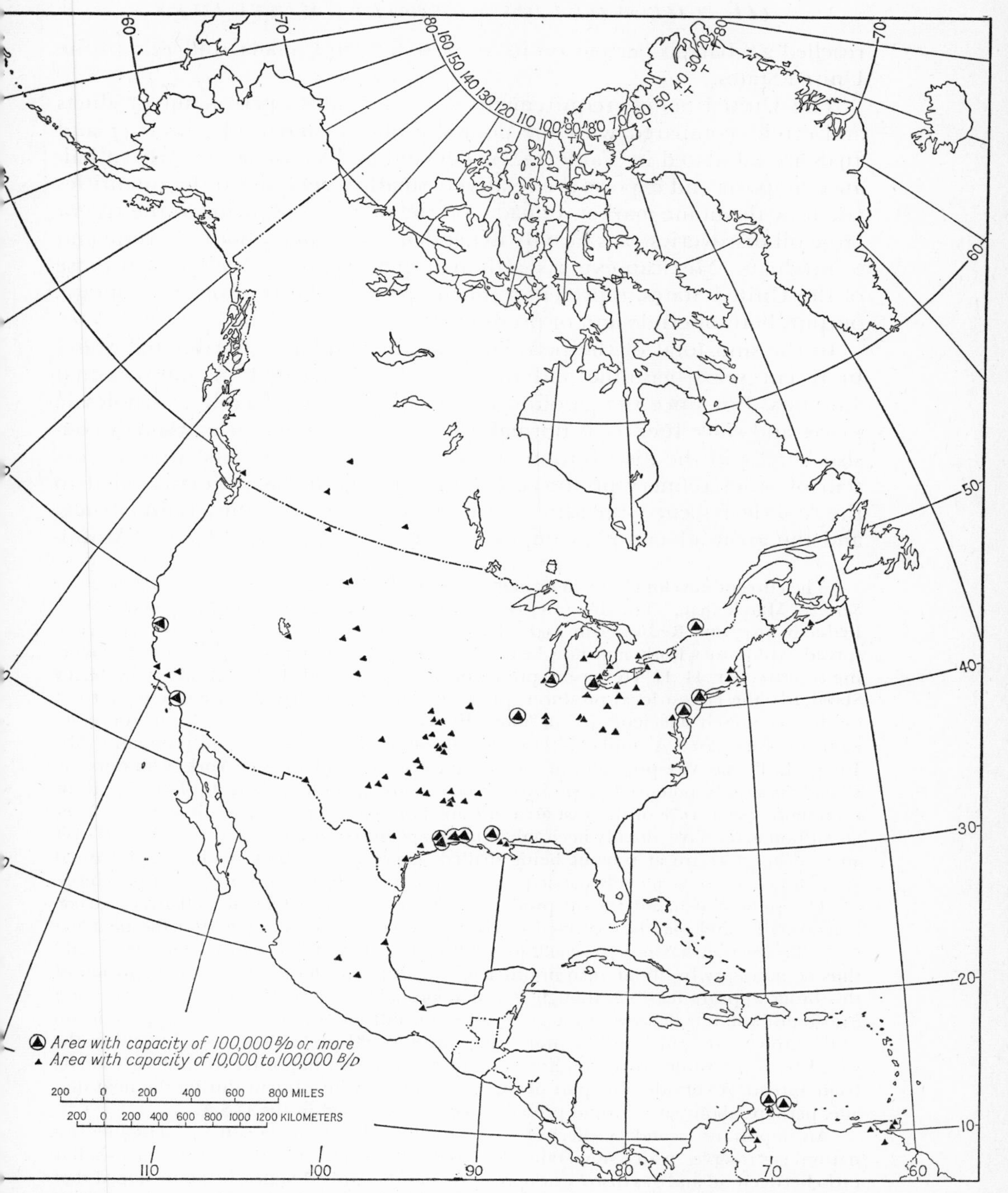

[FI]G. 28. North America: oil refining centers grouped by crude-oil charging capacity in use at the end of 1948. [D]ata for Canada, the United States, and Mexico from the tabulation in the *Oil and Gas Journal,* March 24, [19]49, pp. 269–285. Data for other countries from *World Oil,* July 15, 1949, and from other sources.

reached 2.1 barrels per person in Mexico, 6.75 in Canada, and 13.7 in the United States.

The United States' requirements for the leading petroleum products and crude, compared with the domestic output and trade, in 1947 and 1948 are tabulated in Table 13. As demand and supply are nearly in balance, imports and exports are relatively small in relation to the quantities taken by the home market. Canada, on the other hand, prior to the recent large oil discoveries, was heavily dependent on imports both of crude and of products.[4] Mexican exports, although quantitatively smaller than those of the United States, represent a much larger proportion of the domestic output, both of crude and of products.

In the sum total of the petroleum products used in the United States in recent years, gasoline and fuel oil are represented in almost equal amounts. But since the gasoline is almost twice as valuable at wholesale prices and since fuel oil is normally sold in bulk, modern marketing consists chiefly in the distribution and delivery of gasoline, and to a less extent of other refinery products. Gasoline marketing is intimately tied to the American scene—the filling station, the automobile, the gasoline truck, and the great advertising campaigns to make the names "Esso," "Mobil-

[4] The current developments in Canada were summarized as follows in *Petroleum Press Service,* March 1949. "The discovery of two major fields within a relatively short time—Leduc in 1947 and Redwater in 1948—has raised hopes that Canada, after years of widespread exploratory work, may now be on the verge of becoming an important oil-producing country. Mr. H. H. Hewetson, president of Imperial Oil, Ltd. (subsidiary of Jersey Standard), the largest local producer, predicted last month that the Dominion is at last on the road to self-sufficiency in oil, "along lines and at the pace of United States development in years past." A similar forecast was made by the well-known oil economist, Mr. Joseph E. Pogue, vice-president of the Chase National Bank of New York, who puts the Canadian supply potential at probably 5,000 to 10,000 million barrels of crude oil, or even more. Some 15% of the vast area of Canada ranks as prospective oil territory; over 50 million acres have already been taken up; 60 geophysical crews are at work; and over 40 wildcat wells are at present being drilled. Every important Canadian and United States oil company is actively participating in the search for oil in the western provinces.

"The present demand for oil products in Canada is at a rate of well over 250,000 barrels daily, and may be expected to rise to 400,000 barrels daily within the next ten years. Production has now reached 40,000 barrels a day, and a ten-fold increase would thus be necessary by about 1959 in order to catch up with local needs. An expansion of this order appears feasible, though it would probably require an investment of about $1,000 million, but in view of the geographical conditions of Canada a large proportion of the growing output from the prairie provinces would probably have to be sold in the world markets, while corresponding quantities would be imported to Eastern Canada from abroad. A considerable part of the $1,000 million investment during the next decade, perhaps about 50%, would probably have to be raised in the U.S.A. . . .

"An important secondary effect of the search for oil in Canada is the development of natural gas reserves, which now stand close to 4 trillion cubic feet. Provided that a similar ratio between oil and gas discoveries prevails as in the United States, it can be hoped that the expected finding of 2,000 to 3,000 million barrels of oil in the next 10 years will be accompanied by discoveries of an additional 5 to 10 trillion cubic feet of gas."—*Petrol. Press Serv.* (London), Vol. XVI, No. 3, pp. 66, 67.

Table 13. United States: total domestic requirements of all oils; and domestic requirements, production, imports, and exports, of leading petroleum products and of crude in 1947 and 1948.

	Domestic Requirements	Production	Imports	Exports
	(In million barrels)			
All Oils				
1947	1,989			
1948	2,106			
Gasoline				
1947	795	840	—	47
1948	867	920	—	38
Kerosene				
1947	102	110	—	7
1948	112	122	—	3
Fuel Oil				
1947	816	760	58	41
1948	839	847	56	35
Of which Distillate				
1947	298	312	4	30
1948	338	380	3	22
and Residual				
1947	518	448	54	11
1948	501	467	53	13
Lubricating Oil				
1947	36	52	—	14
1948	38	52	—	13
Crude Petroleum				
1947	1,953	1,856	97	46
1948	2,210	2,016	126	40

Source: *World Oil,* Vol. 128, No. 11, Feb. 15, 1949, p. 60, citing U.S. Bureau of Mines figures, with estimates for last two months of 1948.

gas," "Red Crown," "Fire Chief," "Koolmotor," "Phillips 66," and many other trade names household words. In the United States there are over 200,000 service or filling stations, and from 175,000 to 200,000 other outlets for gasoline and refined petroleum products used by the motorist. The filling-station lessee, attendant, or owner is as much a part of the urban or rural community as is the druggist or the grocer, and he is to be found wherever there are roads over which automobiles can be driven.

Refined products, in general, are carried from the refineries to jobber bulk stations by pipe line or by water transportation, and from there are distributed by tank truck to filling stations, commercial consumers, and farmers.

The result of the development of the petroleum industry in America has been that it gives the widest and most complete distribution of gasoline and refined products of any region on earth. It has been estimated that there is an average of one gasoline pump for every 21 automobiles in the United States, and one automobile for every four people. The industry has been developed as a mass production operation, and savings have been passed on to the consumer.

Resources and Reserves

As a result of its practices and methods, the American oil industry has developed a known underground producible reserve of crude petroleum and condensate of nearly 25 billion barrels, and an additional reserve of 2.5 billion barrels of other liquid hydrocarbons in the United States; 0.50 billion barrels in Canada; and 0.85 billion barrels in Mexico. In addition to the oil reserves, there is in the United States a reserve of 174 trillion cubic feet of gas, and a reserve from 5 to 10 trillion cubic feet in Canada. The distribution of the known reserves in the United States is shown by states in Tables 11 and 12. At our present rates of consumption, and without any further discoveries, this represents approximately twelve times our current annual oil consumption and somewhat over thirty times our current annual gas consumption.

It has been estimated that if the known natural-gas reserves were all synthesized into petroleum products there would be available an additional 15 billion barrels of oil, and plants are now being developed which will produce petroleum products commercially from natural gas; but, since gas is so valuable in its natural form, it is not believed that the bulk of our reserves will ever be used in this manner. However, by the use of modern chemical methods the liquid and gaseous hydrocarbons contained in our vast deposits of oil shale and coal can now become available for use. The cost is not yet competitive with the cost of petroleum products from naturally occurring crude oil and natural gas, but these secondary resources, which are known to be of the order of hundreds of times our current consumption, serve as a comforting backlog to meet the ever-increasing needs of the American citizen.

Today, as always since the beginning of the industry, the biggest petroleum resource is that which is as yet undiscovered. The present known reserves are but a working stock, and they reflect a balance between costs —of discovery, production, and carrying charges—and the incentive to discover new reserves. The fact that the present reserves represent only twelve times the current consumption is not necessarily alarming, since the average proved reserve has for many years fluctuated between twelve and fifteen times the annual consumption. Discovery in the past has more than kept pace with consumption, and reserves are currently the greatest recorded.

The industry's attention is at present focused on the possibilities of important discoveries in the continental-shelf area of the Gulf of Mexico, particularly off the coasts of Texas, Louisiana, and Mississippi, and to a lesser extent in the continental shelf along the Pacific states, especially that off California. The continental shelf is the submarine area adjoining the land which extends seaward to a depth of 100 fathoms (600 feet). It varies from a few miles in width in places off the California coast to more than 100 miles off the Texas and Louisiana shores. At present only the shallower portions are being prospected. Commercial oil and gas have been obtained at several places, and, as the exploration and drilling problems are solved, we may confidently expect many pools to be discovered within these areas.

Another region of great potential productiveness is the Alberta plains of western Canada, where several important oil fields have recently been found north and south of Edmonton. The wells are large, the drilling relatively shallow (less than 6,000 feet), and the potential area extends over thousands of square miles. It appears likely that, as a result of these developments, Canada will become an exporter of petroleum, rather than an importer, within a relatively short time.

The big problem of the petroleum industry in America is, evidently, how long it can continue to discover oil and gas to replace the steadily increasing amounts consumed annually. This is fundamentally a geological problem, and the answer is not simple, nor will it be uniform even among the best-informed petroleum geologists. The answer depends on the geologist's estimate of the volume of favorable unexplored rocks that can be explored under our present economic framework. If we think in terms of depths down to 20,000 feet, and of a price for petroleum that will provide risk money to drill holes to such depths, and if we think of the vast unexplored areas of western Canada (extending into the western United States), of the southern and southeastern United States, of the eastern coast of Mexico, and of the Yucatán Peninsula, then we may well look forward to continuing discoveries that will be adequate to supply the growing demands of the American peoples for many years to come.

Imagination is an essential factor. A geologist [5] does not physically "see" an oil or gas field, any more than a meteorologist, for example, sees a low- or high-pressure area, even though both commonly use contour lines to describe the ideas they intend to convey. Both are presenting mental concepts of the conditions as they are thought to exist. Until a discovery well has been drilled, any undiscovered oil or gas field at best exists only as an idea in the mind of the geologist. In a like manner, the basis for any undiscovered petroleum province—a petroleum province being a region in which there are a number of oil and gas pools having related geological conditions—the basis for such an undiscovered province exists only in geologic thought. It is rather sobering to remember that the future of this great natural resource, upon which many national and continental policies are based, industries planned, and defenses built, should rest in the mind and in the imagination of the geologist. Our petroleum resources will not be exhausted until after our imaginative powers have been exhausted. Our job is cut out for us!

Three of the prospective areas have been selected as typical of the

[5] The editors have secured the author's permission to quote directly, from this point to the end of the chapter, from his article, "Our Petroleum Resources," in the *Bull. Geol. Soc. Amer.*, Vol. 59, April 1948, pp. 283–300. Quotation marks have been omitted for the sake of readability.

kind and scale of geologic thought and imagination which, it seems, is necessary to analyze intelligently our undiscovered petroleum resources. . . . The data used are all taken from geologic articles and publications available to everyone, much of it in the publications of the Geological Society. It is the purpose here to give some indication of the enormous volumes of rocks favorable to petroleum production, but which are as yet essentially unexplored. Some of these areas and ideas are highly speculative at this time, but experience has shown over and over again that every producing area, whether an individual field or a great province, was at one time in a similar speculative position.

The areas selected are shown on the North American base map (Fig. 29). They are: (1) the overthrust fault belts; (2) the plains regions of western Canada and Alaska and of eastern Mexico and Guatemala; and (3) the wedge-out of Lower Cretaceous and Jurassic sediments in the southeastern states.

(*1*) *The Overthrust Fault Belts.* More than 2,500 linear miles of them, from 5 to 100 miles wide, are shown on the base map, in which older rocks are thrust over younger rocks, and in which the geology is so complicated as to make it extremely difficult to map and understand. Interpretation of geophysical measurements in these fault areas is generally impossible. The correct knowledge of the geology appears to lie in better and more accurate surface mapping, coupled with careful records of deep wells. Probably of equal importance is the need for geologists to learn more of the nature and mechanics of faulting.

Yet within these overthrust belts evidence of petroleum has been found at many places, and several oil fields discovered. One of these belts extends from British Colombia and Alberta south across western Montana and Wyoming and into Utah and Nevada; the Turner Valley pool of Alberta is located in it. Another belt is located in the western Ouachita Mountains of Oklahoma, and probably extends southwest across Texas. The South Mountain pool is one of several examples of oil pools in the Ventura region of California characterized by thrust faulting. The third belt is in the Appalachian Mountains, extending from New York into Alabama, where it passes under the overlapping Cretaceous and Tertiary rocks. The recently discovered Rose Hill pool in Virginia occurs in this belt. Apparently thrust faulting, and the sort of deformation which occurs in these belts, is not detrimental to the occurrence of petroleum. . . .

(2) *The Plains of Alaska and Western Canada, and of Eastern Mexico and Guatemala.* These two areas, which appear to be the northern and southern extremities of the western plains regions of the United States, much of which has been so productive of oil and gas, cover an area of over a million square miles in Alaska, Canada, and the northern United States, and 300,000 square miles in Mexico and Guatemala.

They are considered together because they have many common characteristics significant in the geology of petroleum. Both contain large vol-

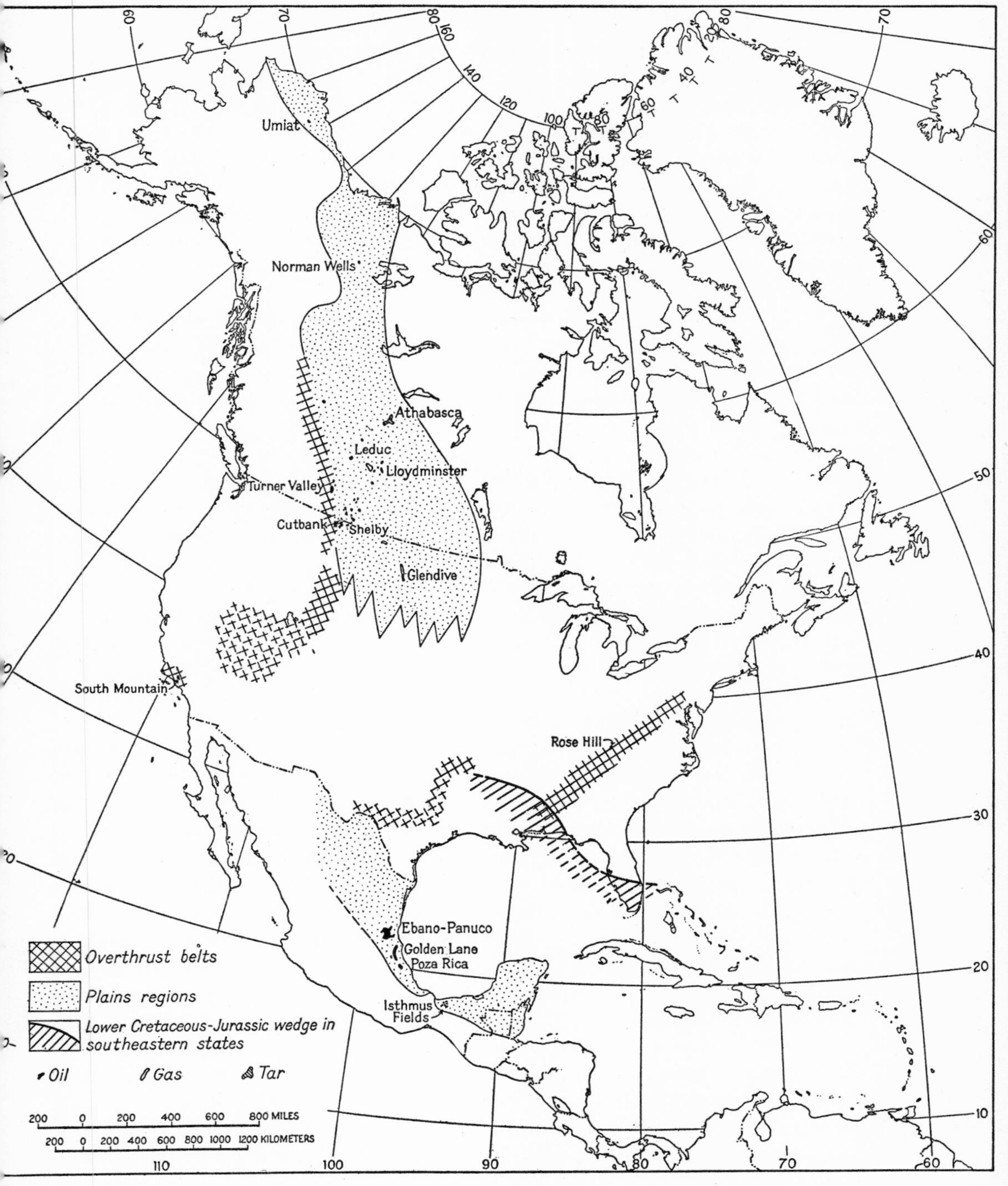

IG. 29. North America: diagrammatic map showing the three types of prospective areas selected for analy-
s. (Source: map prepared for the author's address, "Our Petroleum Resources," *Bull. Geol. Soc. Amer.*, Vol.
9, No. 4, April 1948.)

umes of sediments. Both also contain numerous and widespread evidences of petroleum in the form of seepages, asphalt and tar deposits, as well as proven oil and gas fields. There can be no question as to abundant source rocks. . . .

The average thickness of sediments in these two areas which might carry petroleum is of the order of 2 miles, which means a total volume of sediments of over 2½ million cubic miles—a truly large volume of prospective material which is today essentially unexplored! It is inconceivable that such a large volume of rocks, in which every prerequisite of a petro-

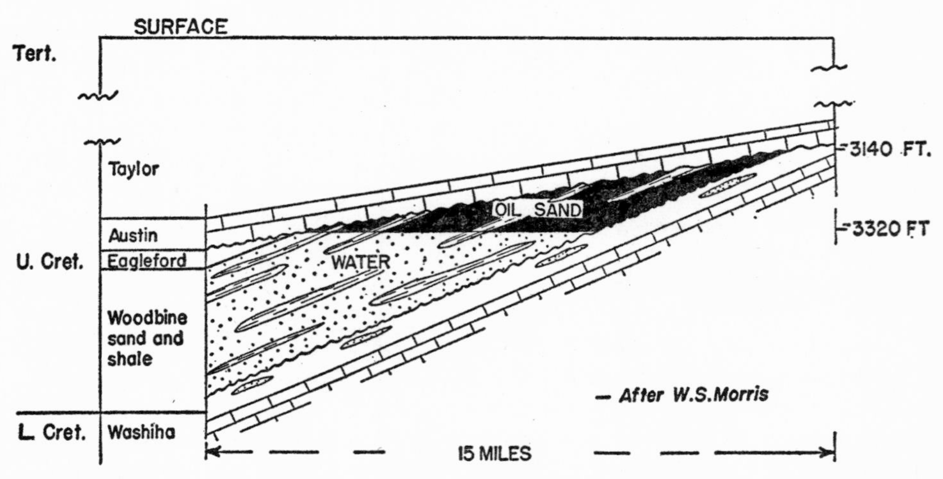

FIG. 30. Generalized west-east cross section of the East Texas field. The wedge shape of the producing formation is repeated in various ways in many oil and gas fields. (Sketch reproduced by permission of Geological Society of America.)

leum province is richly developed, should not be the site of hundreds upon hundreds of oil fields yet to be discovered. Deposits such as the Athabaska oil sands of Canada, or the Golden Lane and Poza Rica fields in Mexico, give a measure of the possible size of some of these undiscovered fields. Not until thousands of additional test wells have been drilled throughout both regions can we say that the exploration of these areas has been completed.

(3) *The Southeastern United States.* The dome structure in contours might be called the symbol of the petroleum geologist of the 1920's; it was practically the sole objective in his search for traps. While a dome structure is still as desirable a guide to an oil pool as ever, it is gradually being learned that stratigraphic variations are probably equally as important as folding and faulting in the formation of traps. One of the greatest oil fields in the world, the East Texas field, has a cross section which is symbolic of the changing emphasis (Fig. 30) and which is repeated in many forms in countless oil and gas fields. Whether it is an oil field in the small sand lens in Kansas or Alberta, the porous limestone reefs of West Texas or Mexico, the dolomite grading into limestone in Indiana and Ohio, the

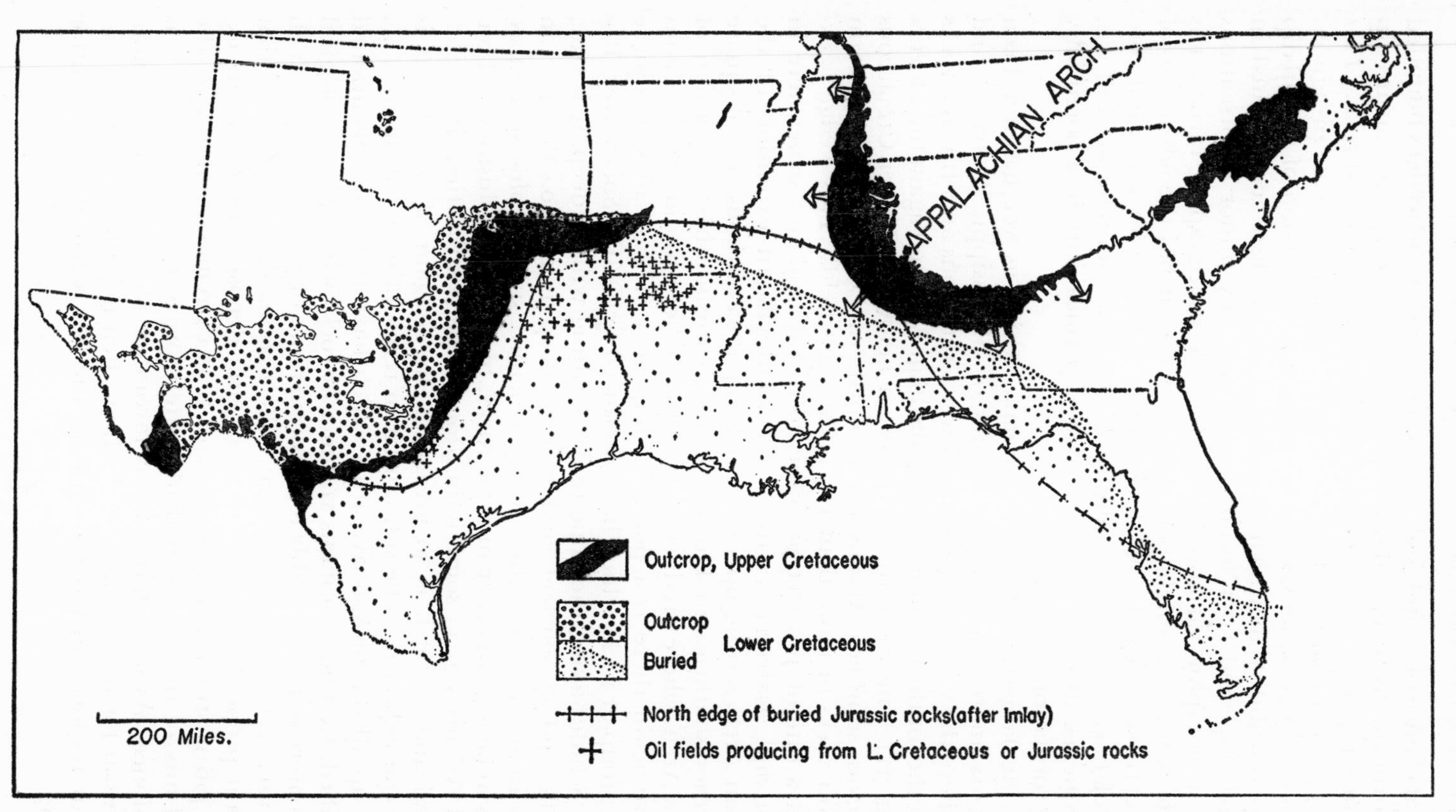

FIG. 31. Southeastern United States: Lower Cretaceous and Jurassic sediments wedging out up-dip across the southwestern-plunging Appalachian Arch and again across the Florida Uplift, thereby forming two large potential trap areas favorable to petroleum accumulation. (Sketch map reproduced by permission of Geological Society of America.)

rapid lateral change of a sand to shale in California, or the wedges bounded by unconformities as in East Texas, the principle is the same: an up-dip wedging-out of porosity and permeability in the reservoir rock. This principle applies not only to individual fields, but also to many provinces. Such a province may be looked upon as a large-scale version of the up-dip wedge of permeability in which the oil and gas are first accumulated in the regional trap, and later localized into pools by minor deformations and local stratigraphic variations. Whether the stratigraphic variation is on the scale of either a single pool or a province, it has the advantage of being the earliest trap in the geologic life of the reservoir rock, and consequently of having had more time in which to accumulate oil and gas. As a consequence, any up-dip wedge edge of permeability in a potentially producing formation is significant.

One of the largest of these phenomena, which as yet has not been tested, occurs in the southeastern States of Mississippi, Alabama, and Florida, where the wedge edge of the Lower Cretaceous and Jurassic rocks crosses first the southwest-plunging Appalachian Arch and again the Florida Uplift (Fig. 31). These two rock series occur below the Upper Cretaceous in East Texas, southern Arkansas, and northern Louisiana, and contain the reservoir rocks of more than 125 oil and gas fields. This wedge is expanding at a rate of 100 feet per mile down the dip, and somewhere in Mississippi and Alabama it crosses the southwestern extension of the Appalachian Arch, setting up a potential trap area of thousands of square miles. Farther southeast in Florida it crosses the southern extension of the Florida Arch, where production has already been found in Cretaceous rocks. Here again there is a potential trap area of thousands of square miles in a group of rocks which have produced fields like Smackover, Rodessa, and Schuler along the same trend. Drilling will be deep, it is true, probably beyond 15,000 feet, and even as deep as 20,000. No one can even guess in advance of drilling what structure, stratigraphic sequences, or changes in facies will occur in these rocks, but their past history in Louisiana and Arkansas gives some indication of the possibilities. The scale, both lateral and vertical, is so great that here, again, if production is found, we can expect it to be substantial in terms of national needs. . . .

Adding up all the prospective areas which have been both mentioned and described, we find enormous volumes of rocks which have geological conditions more or less favorable to the occurrence of petroleum, yet which have had relatively little exploration. The reasonable conclusion is that they contain proportionately large petroleum deposits, and that our exploration job is still far from complete. It is extremely doubtful if the geology of any of these areas will ultimately prove to be exactly as it is now envisioned. As long, however, as favorable geologic factors remain, there is reason for drilling and, if drilling, then hope for additional discovery. Our responsibility is to maintain the best scientific guidance for the drilling.

8. THE MIDDLE EAST

BY G. M. LEES *

THE use of petroleum in the Middle East [1] is older than the Book of Genesis. Noah, a practical man, used it for waterproofing the Ark, constructed of gopher wood, lined inside and out with pitch. The mother of Moses the Law-Giver put her baby in a bitumen-lined cradle for Pharaoh's daughter to find among the flags at the edge of the River Nile. Throughout the hidden ages and the bright, sun-baked countries that cradled civilization, peasant peoples, living humbly and in fear of an unknown god, worshipped the mysterious fire that burnt forever without cause, and they collected oil oozing from seepages for their lamps, for primitive domestic uses, and as a burning weapon against their enemies.

In Persia archaeological researches at Susa (modern Shush, between Dizful and Ahwaz) have revealed that bitumen was used there in Sumerian times, that is, some five or six thousand years ago, when it was employed not only as a bonding material in buildings, but also for fixing the blades of tools into their handles, as a setting for jewels, for water-proofing pottery, and for caulking craft.[2]

The earliest description of petroleum production in Persia is by Herodotus, who, writing in the fifth century B.C., gives the following account (Book VI, 119, A. D. Godley translation): "Ardericca [3] . . . is 210 furlongs [stadia] distant from Susa, and 40 from the well that is of three kinds whence men bring up asphalt [bitumen] and salt and oil. This is the manner of their doing it: a windlass is used in the drawing, with a skin made fast to it in place of a bucket; therewith he that dips into the well then pours into a tank, whence what is drawn is poured into another tank, and goes three ways; the asphalt [bitumen] and the salt go

* Chief Geologist, Anglo-Iranian Oil Company, Ltd.

Acknowledgements: I wish to express my indebtedness to Professor W. B. Fisher, University of Aberdeen, for assistance in the assembly of this paper. I am also indebted to Mr. B. K. N. Wyllie, Dr. L. Lockhart, and Miss P. B. Lapworth for many suggestions during the preparation of the manuscript.

[1] As a geographical concept the Middle East moved slightly to the west during the campaigns of the second World War. It is, therefore, not proposed to discuss where the old-fashioned boundaries of the Near, Middle, and Far East were once supposed to lie, but to accept, with the sanction of present-day usage, that the Middle East comprises Egypt, Palestine, Jordan, Syria, Lebanon, Turkey, Saudi Arabia, Iraq, Persia, and the countries of the Arabian coast of the Persian Gulf (Kuwait, Bahrein, Qatar, the Trucial Coast Shaikhdoms, Muscat, and Oman).

[2] Lockhart, L., "Iranian Petroleum in Ancient and Medieval Times," *Journ. Inst. Petrol.*, Vol. 25, 1939, p. 1.

[3] Perhaps identical with Masjid-i-Sulaiman.

forthwith solid; the oil, which the Persians call Rhadinace, is dark and evil smelling."

Oil and gas from seepages formed the fuel for the "eternal fires" which were the focal point of the fire-worshipping religion of the ancient Persians; and fire temples were constructed at Baku (belonging to Persia in those days) and at several places in north Persia. A ruined temple, possibly a fire temple, has given the name to the place and oil field of Masjid-i-Sulaiman (i.e. the Mosque of Solomon).

Petroleum warfare was first practiced in ancient Persia, and there are several recorded cases of the use of incendiary arrows. The poet Firdausi describes in his *Shah-Nama* how Alexander during his invasion of India scared and scattered the elephants of Fur by sending among them a thousand horsemen, each fitted with a blazing naphtha-container. A later variant of this ruse was employed by Nadir Shah during his invasion of India in 1739. He impregnated the humps of camels with oil, and drove them blazing among the war elephants of the Indian army, to the complete discomfiture of the latter. The writer has in his possession an engraved tray on which this entertaining scene is depicted (Plate 54). Today the preferred weapon is the flame-thrower (*flammenwerfer*).

The oil seepages, bitumen occurrences, and pitch-lakes of Iraq are so numerous, and are on such a scale, that it is not surprising that classical writings and inscriptions make numerous references to them and their uses.[4] Thus, King Sennacherib (704 to 682 B.C.) was so proud of an engineering achievement that he had it inscribed that he "covered the bed of the diverted river Telbiti with rush matting at the bottom and quarried stone on top, cemented together with natural pitch (asphalt). I thus had a stretch of land 454 ells long and 289 ells wide, raised out of the water and changed into dry land."

Herodotus (484 to 425 B.C.) describes the Hit seepages in these words: "There is a city called Is [Hit], eight days' journey from Babylon, where a little river flows, also named Is, a tributary stream of the river Euphrates. From the source of this river many 'gouts' of asphalt rise with the water; and from thence the asphalt is brought for the walls of Babylon."

Twenty-two centuries later a British traveler, George Rawlinson (1745), described the same locality as follows: "Having spent three days or better among the ruins of old Babylon, we came into a town called Hit, inhabited only by Arabians, but very ruinous. Near unto this town is a valley of pitch, very marvelous to behold and a thing almost incredible, wherein are many springs throwing out abundantly a kind of black substance, like unto tar and pitch, which serveth all the countries thereabouts to make staunch their barks and boats. Every one of which springs maketh a noise like a smith's forge in puffing out the matter, which never ceaseth night or day,

[4] A good historical review of references to asphalt and bitumen is contained in Abraham, H., *Asphalts and Allied Substances,* 5th ed., New York, 1945, pp. 1–43.

and the noise is heard a mile off, swallowing all the weighty things that come upon it. The Moors call it the Mouth of Hell." [5]

The Contemporary Importance of the Middle Eastern Oil Province

The immense historical antiquity of Middle Eastern oil in wars and religions and the everyday life of the world's oldest civilizations makes its meteoric exploitation in our day one of the great dramas of history. Through aeons of geological ages, the vast secret hoard of treasure lay silent and idle beneath the earth's crust, passively biding its time. Here and there an ignited flare lapped eternally when a gas seepage blew off in the arid foothills of Mesopotamia. Through such a fiery furnace passed, long ago, Shadrach, Meshach, and Abednego, remote young men in the recesses of Jewish history. In more recent times travelers' tales brought news to Europe of the ingenious marvels contrived from petroleum in the Middle East for man's use and even for his entertainment: the Duke of Holstein, passing by Arbadil on his way to Isfahan in 1637, was honored with a firework display of "little Castles, Towrs, Squibs, and Crackers," impregnated, according to the historian, "with a white Naphtha, which is a kind of petroleum." Fable and legend and miracle—the salt dome that was Lot's wife turned into a pillar of salt, the young men passing through the fire, the iron horse of Alexander—all these became the picturesque backcloth against which was outlined the drama of modern oil, opening in 1908 with the success in Persia of William Knox D'Arcy, reaching a brilliant dénouement in two swift decades, becoming the pivot of modern economic life.

Until quite recently, however, the Middle East as an oil-producing region has played a relatively minor part in the petroleum economy of the world: up to the end of 1948, only 4.9% (exclusive of Egyptian production) of the world's cumulative production of oil had been produced there (Table 14). Its development had been much retarded by its unfavorable geographical position relative to world markets, by the quality of its crude oils, and by other causes.

But these adverse factors are changing to the region's advantage, and recent discoveries have demonstrated that the potential oil reserves of the Middle East are on a scale far greater than had been thought some years ago. DeGolyer in 1944 estimated the proved and semi-proved reserves at about 18 billion barrels, and in 1948 he indicated that this figure should be scaled up to at least 32 billion. This figure is four billion barrels greater than that for the current proved reserves estimate for the United States, and recent developments have shown that a further and substantial upward revision is now required. In 1949 five important new discoveries

[5] Rawlinson, G., *Collection of Voyages and Travels,* Vol. II, London, 1745, p. 752, as cited in Abraham, *op. cit.,* p. 31.

Table 14. The Middle East's discovered oil: production and reserves of crude petroleum, by principal producing countries, January 1, 1949.

	Cumulative Production, from Beginning to January 1, 1949		Annual Production in 1948	Estimated Reserves on January 1, 1949
	Beginning Year	Crude Oil Produced		
		(In million barrels)		
The Persian Gulf Oil Province				
Iran	1913	1,938.2	190.395	7,000.0
Iraq	1927	411.6	26.466	5,000.0
Kuwait	1946	68.7	46.547	10,950.0
Qatar	1940	26.0	—	500.0
Saudi Arabia	1936	345.0	142.853	9,000.0
Bahrein	1933	98.9	10.915	170.0
		2,888.4	417.176	32,620.0
Egypt	1911	120.3	13.173	120.0
Turkey	1948	—	0.025	1.0
Total, Middle East [a]		3,008.7	430.374	32,741.0

[a] The "mediterranean" area of the Middle East, broadly conceived, includes also the producing areas of the Black and Caspian sea borders, in Rumania and the U.S.S.R.

Sources: Same as for Table 1.

were reported: at Nahr Umr, 19 miles northwest of Basra, and at Zubair, 12 miles southwest of Basra, both in Iraq; and at Fadhili, Ain Dar, and Haradh in Saudi Arabia. There is, of course, still great scope for new discovery, so much so that the Middle East may eventually be found to contain a substantial proportion of the world's total oil.

The importance of the Middle Eastern oil province depends partly on the immense size of its total reserves, partly on the high productivity both of individual oil fields and of individual wells. The Kirkuk and Kuwait oil fields are credited with reserves of about 5 and 11 billion barrels, respectively, and plans are now being developed to increase their offtake to around 500,000 b/d from each field. Other units in Persia and Arabia may have comparable production rates, all under controlled conditions which ensure uniform drainage from these great reservoirs. The capacity of the individual wells is, in some cases, 20,000 b/d, but the controlled rates mostly range from 2,000 to 15,000 b/d. The spacing interval between wells is as much as two miles, in the case of fields with thick and freely fissured limestone reservoirs.

The occurrence of so many giant oil fields in one area is unique in oilfield experience throughout the world. The only other single field which is comparable in total reserves is that of East Texas, and its current production rate of about 380,000 b/d will be overtaken within a few years by that of two or three Middle East fields.

The rapid expansion of production from the Middle East during the postwar years, and the still greater expansion now being planned, is the result partly of the realization of the immensity of the reserves, partly of the inability of the United States' fields to satisfy internal demand, and partly of changing economic factors to the advantage of the Middle East. Initial development costs in the Middle East are high, but, once new oil-field centers are established, production can be increased rapidly without corresponding increase in production costs. Transportation charges, hitherto under the heavy handicaps of a long sea haul around Arabia and of costly Suez Canal dues, will also be greatly reduced when the large-diameter trans-desert pipe lines now under construction are completed.

The first pipe lines to cross the Syrian desert were those of the Iraq Petroleum Company, connecting the Kirkuk oil field to two Mediterranean terminals, Haifa, in Palestine, and Tripoli, in Lebanon, by 12-inch lines about 650 miles in length. A 16-inch line to Haifa was under construction until interrupted by the Palestine troubles in 1948. Another 16-inch line to Tripoli was completed in 1949, while a larger diameter line is under consideration for a later stage. The Trans-Arabian Pipe Line Company commenced construction at the end of 1947 of a 1,000-mile 30-inch line, with a capacity of 300,000 b/d, to connect the Arabian American Oil Company's fields in the Persian Gulf area to a Mediterranean terminal near Sidon, in the state of Lebanon. This line will be followed by a still bigger diameter line, to be constructed by the Middle East Pipe Line Company, to carry oil from the south Persian fields and from Kuwait to the Mediterranean. The capacity of this line, a joint undertaking of the Anglo-Iranian Oil Company and the Standard Oil Company (New Jersey), may be as much as 500,000 b/d.

In addition to pipe-line projects directed primarily to serving terminals for crude-oil shipments, there will be substantial refinery developments both in the Persian Gulf area and on the Mediterranean coast. At present the Abadan refinery, with a capacity of 495,000 b/d, is the largest in the world, and a further increase in capacity is being effected. The refineries at Bahrein and Ras Tanura on the Persian Gulf coast, and at Haifa on the Mediterranean, are all in various stages of expansion. In past years the sulphurous and asphaltic crudes of Persia and Iraq were particularly intractable for yield of high-grade products, but improved refinery technique has made possible the production of the highest quality aviation spirit, and of quality products all down the scale.

These gigantic undertakings will have far-reaching consequences on the economy and social developments of all the Middle East countries. By 1955 the production may have reached a total of 1,500,000 to 2,000,000 b/d, and the increase in national income from the royalty payments to the various states will be a factor of great importance; a substantial improvement in the general standard of life will inevitably follow. The opportunities for lucrative employment will improve the present slow tempo of educational development, and even the harshness of the Middle Eastern summer can be softened by modern air-conditioning.

As a result of the changed conditions, the output of the Middle East (excluding Egypt and Turkey) in 1948 reached a total of 417,176,000 barrels, making 12% of the world's production for that year; and a further increase of both total and percentage figures was forecast for 1949. [Preliminary figures for the Middle East's production in 1949 are given by *World Oil* in its *Review-Forecast Issue* of February 15, 1950, as 509,544,000 barrels, or 15% of world production. *Editor's Note.*] The sudden emergence of this rich region into the forefront of oil politics and commerce calls for some discussion, from this particular point of view, of its salient geographical and economic features.

Structure and Topography

The oil province of the Middle East, as at present known, lies between the 24th and 37th parallels of north latitude (Fig. 32). Two principal elements can be distinguished in the structure of the area: the stable plateau block of Arabia, composed in the main of ancient and resistant rocks; and the intensely disturbed mountain ranges belonging to the Alpine-Himalayan system, a part of which, the Zagros Mountains, extends from eastern Turkey through northeastern Iraq and western and southern Persia to form the northern and eastern boundaries of the oil province. A broad belt of gentle folds separates the two regions. Close to the mountain front, rock outcrops form long ridges of low foothills; further southwest the solid geology is concealed by the alluvial deposits of the Tigris-Euphrates-Karun basin. To the west the land surface rises to form the plateaus of Syria, Jordan, and Palestine, diversified by occasional anticlinal fold ranges and by the rift-valley system that can be traced from the Red Sea and Gulf of Aqaba to the Dead Sea and Jordan Valley.

It is on the margins of the median belt, particularly towards the east, that the main oil fields of the Middle East occur. One group of fields is located in the foothill zone between the Zagros Mountains and the lowlands of Iraq and Persia; other fields occur on the southern margins, where the median belt abuts on the Arabian foreland. Oil possibilities, not yet tested, extend throughout central Persia, the Caspian coastal zone, and, perhaps, the desert areas of eastern Persia and Baluchistan. The possibilities of Syria and Palestine have not yet been adequately explored. The oil fields of Egypt, not unimportant, belong to a very different geological province.

The Zagros is the collective name given to the composite mountain chain which extends from the Turkish frontier north of Mosul to the entrance of the Persian Gulf, a distance of one thousand miles. Structurally, the mountains have in the main a simple fold pattern, though in some cases, particularly in the more northeasterly ranges, there is substantial overthrusting towards the southwest. The high mountains are, with few exceptions, composed of massive limestones, with a Middle and Lower Cretaceous limestone as the most prominent feature-forming member in the high ranges, and a Palaeogene limestone in the lower ranges.

Heights of over 10,000 feet above sea level are frequent, the highest point of the whole system being 14,500 feet.

The ranges are gashed at intervals by great river gorges, or "tangs," and there are many spectacular instances of superimposed drainage. The synclinal valleys between the ranges are, in many cases, sufficiently flat to allow local cultivation; and in many of the narrower mountain valleys terraced gardens of walnut, mulberry, poplar, and vine are irrigated from the mountain streams.

The foothill zone on the southwest flank of the Zagros ranges is in striking contrast to the simple grandeur of the mountains. Its height above sea-level is between 500 and 2,500 feet, and its character varies greatly according to the rock formation of which it is composed. In the north, from Mosul to Kirkuk and Kifri, the foothills are formed by narrow parallel ridges of sandstone or conglomerate, forming the axes or flanks of long anticlines, between which broad alluvial plains occupy the synclines. Locally, as at Kirkuk and Kifri, there are narrow exposures of gypseous strata, the white color of which makes a striking contrast to the buff and reddish appearance of the overlying sandstones and marls.

The general strike of the mountain ranges of the Zagros has a very constant northwest-southeast direction, the great anticlinal lines rising to local culminations in different sectors. The high mountains of Pusht-i-Kuh form one such fold-culmination, and from a point roughly northeast of Badra the ranges plunge slowly down in either direction, to the northwest and to the southeast. The effect is to produce two great embayments into the mountain front: the Sirwan River embayment and the Diz-Karkheh River embayment. Below the high Pusht-i-Kuh ranges the foothill zone is only about five miles in width, whereas in the Diz-Karkheh embayment the foothills are as much as eighty miles wide. The oil fields of Naft Khaneh, Naft-i-Shah, and Kirkuk are situated in the former embayment; and Masjid-i-Sulaiman, Haft Kel, and the other oil fields of southwest Persia in the latter.

The character of the foothills in the sector between the Diz River and the neighborhood of Kazerun, northeast of Bushire, is very different from that described above in the Kirkuk area. This difference is caused by a much thicker development of the Miocene Lower Fars Series, a formation composed of gypsum, and red and gray marls with great thicknesses of rock salt. The relative plasticity of this formation has resulted in intense minor folding and overthrust sheets, which in places are several miles in

Fig. 32 (*on following pages*). The Near and Middle East: location map. The three rectangles outlined by broken lines indicate oil-producing areas covered in more detail in Figs. 33, 35, and 36. Concessions, fields, and petroleum facilities in the region as a whole are indicated in Fig. 34. This location map is based on the Persian Gulf and Gulf of Aden sheets of the 1:4,000,000 series issued by the Geographical Section of the British General Staff. These, and the Bartholomew "Map of the Middle East and Near East" on the scale 1:4,000,000, show contours in color. In Fig. 32 the only contour shown is the 200-meter, indicated by a fine black line.

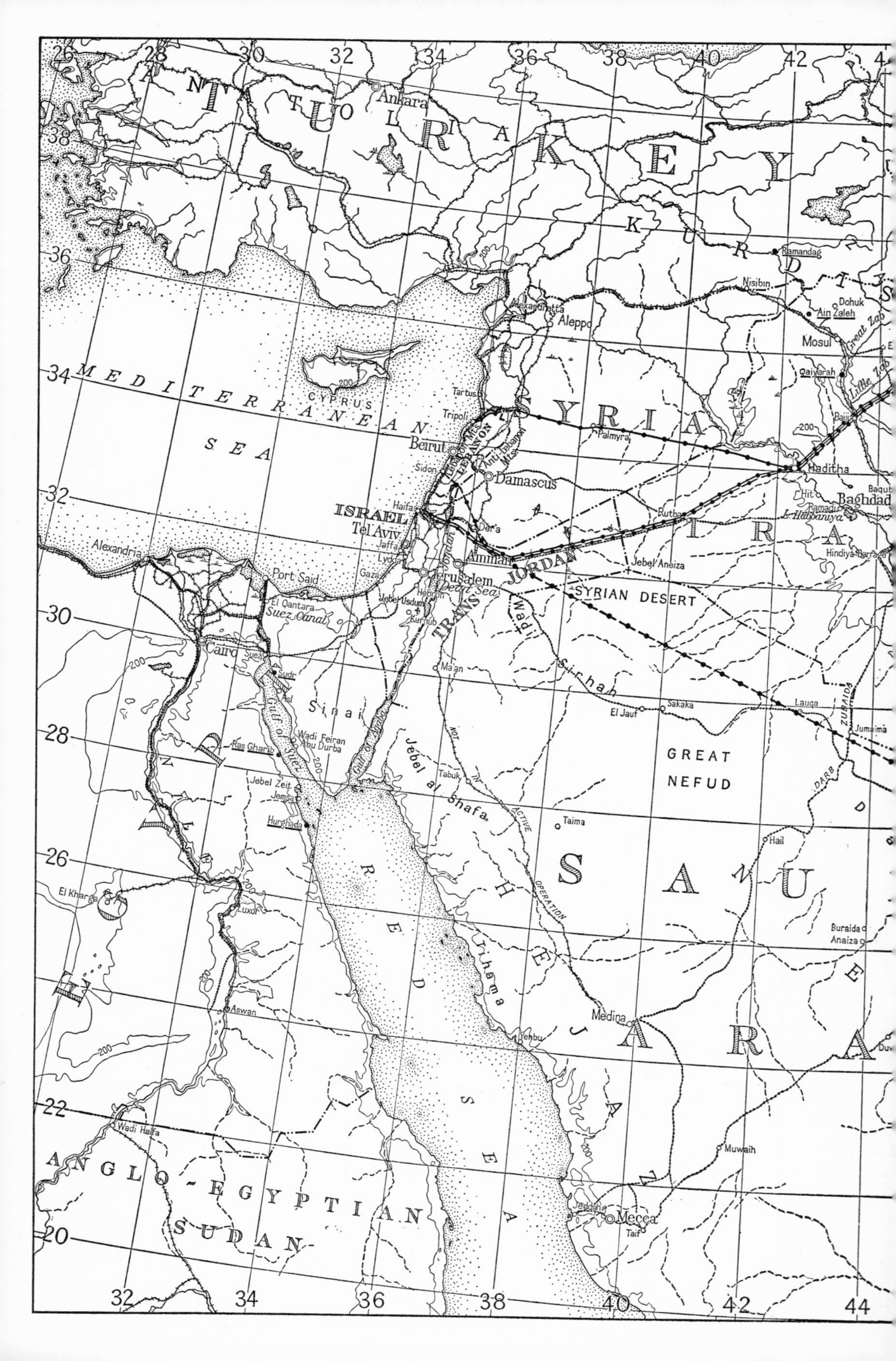

TURKEY
Ankara
KURDISTAN
Ramandag
Nisibin
Dohuk
Ain Zaleh
Mosul
Great Zab
Qaiyarah
Little Zab
Baiji
Haditha
Baghdad
Hit
Ramadi
L. Habbaniya
Hindiya Barrage
IRAQ
MEDITERRANEAN SEA
CYPRUS
Alexandretta
Aleppo
Tartus
Tripoli
SYRIA
Palmyra
Beirut
Sidon
LEBANON
Lebanon Mts.
Anti-Lebanon Mts.
Damascus
Haifa
ISRAEL
Tel Aviv
Jaffa
Lydda
Gaza
Dar'a
Amman
Jerusalem
Hebron
Dead Sea
Jebel Usdum
Kurnub
Jordan
TRANS JORDAN
Rutba
Jebel Aneiza
SYRIAN DESERT
Wadi Sirhan
Alexandria
Port Said
El Qantara
Suez Canal
Cairo
Suez
Sudr
Tor
Sinai
Gulf of Suez
Gulf of Aqaba
Ma'an
Wadi Feiran
Abu Durba
Ras Gharib
Jebel Zeit
Jemsa
Hurghada
NOT IN ACTIVE OPERATION
Jebel al Shafa
Tabuk
Taima
El Jauf
Sakaka
Lauqa
Jumaima
GREAT NEFUD
DARB ZURAIDA
Hail
SAUDI ARABIA
HEJAZ
Buraida
Anaiza
EGYPT
NILE
El Kharga
Luxor
Aswan
RED SEA
Ithama
Medina
Yenbu
Wadi Halfa
ANGLO-EGYPTIAN SUDAN
Muwaih
Jedda
Mecca
Taif
200
26
28
30
32
34
36
38
40
42
44

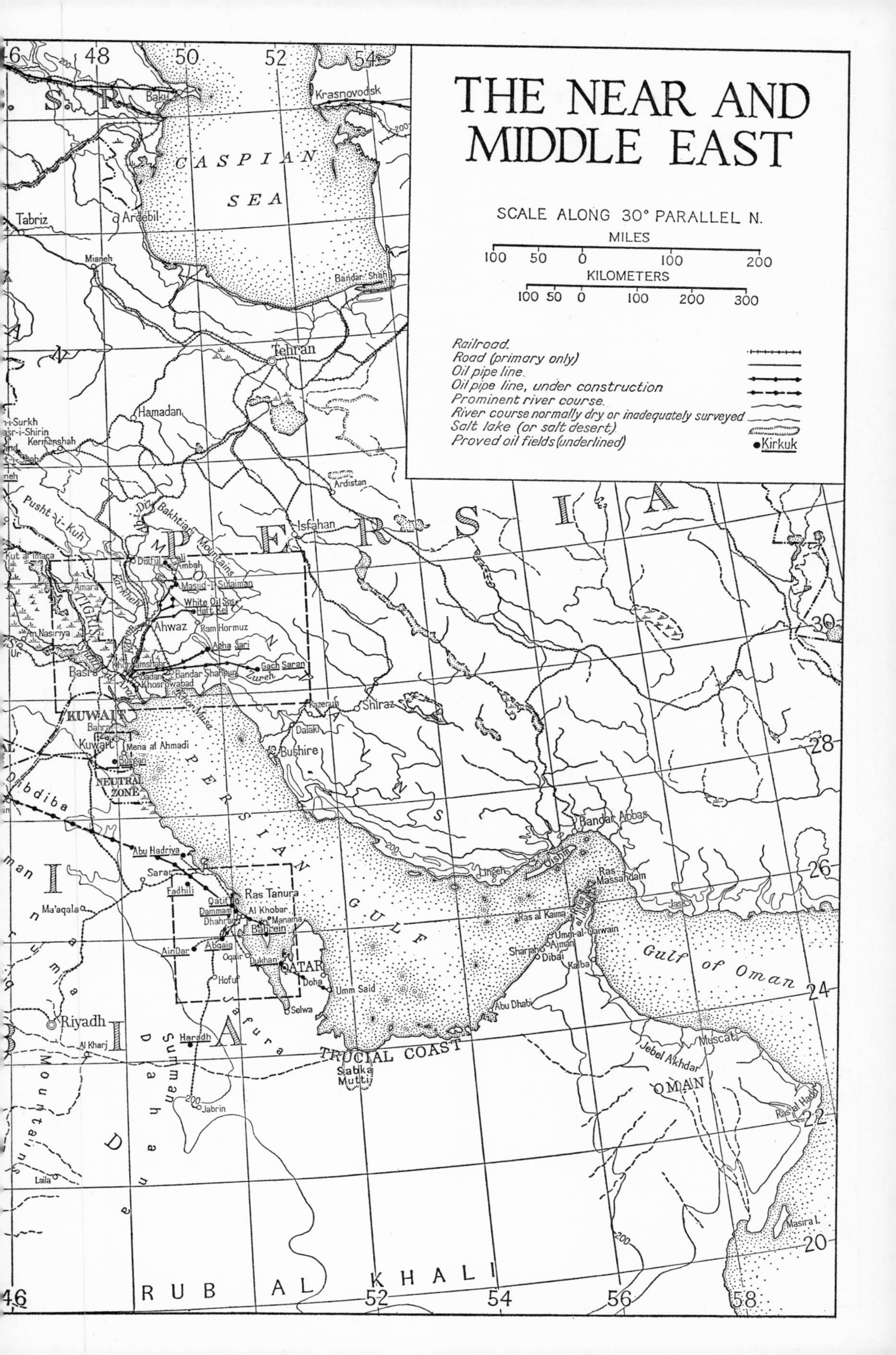
THE NEAR AND MIDDLE EAST
SCALE ALONG 30° PARALLEL N.
MILES
100 50 0 100 200
KILOMETERS
100 50 0 100 200 300
Railroad
Road (primary only)
Oil pipe line
Oil pipe line, under construction
Prominent river course
River course normally dry or inadequately surveyed
Salt lake (or salt desert)
Proved oil fields (underlined)
Kirkuk
CASPIAN SEA
Baku
Krasnovodsk
Tabriz
Ardebil
Mianeh
Bandar Shah
Tehran
Hamadan
Kermanshah
Ardistan
Isfahan
PERSIA
Pusht-i-Kuh
Bakhtiari Mountains
Kut al Imara
Dizful
Ambal
Masjid-i-Sulaiman
Amara
TIGRIS
Karkheh
White Oil Sps.
Haft Kel
Ahwaz
Ram Hormuz
An Nasiriya
Agha Jari
Ur
Basra
Khorramshahr
Bandar Shahpur
Gach Saran
Zureh
Abadan
Khosrowabad
Khor Musa
KUWAIT
Kuwait
Kazerun
Shiraz
Dalaki
Bahra
Mena al Ahmadi
Burgan
NEUTRAL ZONE
Bushire
Dibdiba
PERSIAN GULF
Bandar Abbas
Abu Hadriya
Lingeh
Kishm
Sarar
Ras Massandam
Fadhili
Ras Tanura
Qatif
Jask
Ma'aqala
Dammam
Al Khobar
Dhahran
Manama
Bahrein
Ras al Kaima
Umm-al-Qaiwain
Ain Dar
Abqaiq
Oqair
Sharjah
Ajman
Dibai
Dukhan
QATAR
Kalba
Gulf of Oman
Hofuf
Doha
Umm Said
Selwa
Abu Dhabi
Riyadh
Al Kharj
Haradh
Jafura
Summan
TRUCIAL COAST
Sabkha Mutti
Jebel Akhdar
Muscat
OMAN
Ras al Hadd
Jabrin
Dahana
Laila
Masira I.
RUB AL KHALI
46 48 50 52 54 56 58
30 28 26 24 22 20

width. These exposures of the gypsum series form a bizarre pattern of very broken white hills streaked with red and gray bands. The solubility of the gypsum leads to the development of innumerable swallow holes and underground drainage channels, upon which the removal of salt by solution must also have had some effect. The only known case of Lower Fars salt outcropping at the surface is at Ambal on the Karun River, northwest of Masjid-i-Sulaiman, where there is a small salt dome. Elsewhere the existence of salt in large quantities underground, locally many thousands of feet, has been demonstrated in a number of borings. The photographs of the Masjid-i-Sulaiman and Gach Saran country (Plates 55, 56, 57) illustrate this curious topography, a type quite unique on such a scale in the world, as far as the writer is aware.

At Masjid-i-Sulaiman and Haft Kel the lower flats and valleys are 600 to 800 feet above sea level, and the surrounding hills and high-level plateaus of Lower Fars, about 1,500 to 1,800 feet. In the Gach Saran area there is an extensive relic of a dissected old high-level terrace about 2,500 feet above sea level, whereas the main drainage of the area has been incised to 1,250 feet in the Zureh River. In this area the Lower Fars is overlain by strong limestones of Middle Fars age, which form some remarkable synclinal outliers. On one of these, the Kuh-i-Seh Qalehtun, the central production unit of the oil field is situated at 3,000 feet above sea level. The oil gravitates by pipe line to Abadan, 160 miles distant. (Fig. 33.)

South of Bushire there is another bulge forward of the higher mountain ranges, and the coast between Bushire and Lingeh forms the actual mountain front with no intervening foothill zone. Between Lingeh and Jask the mountains retreat from the coast, and there are several ridges of foothills separated by broad alluvial plains.

The numerous salt domes exposed at the surface in the sector between Bushire and Bandar Abbas are of great geological and geographical interest. These are extrusions of Cambrian age (i.e. very much older than the salt of the Lower Fars described above). The salt domes form circular inliers four or five miles in diameter; in some cases the domes are "dead," and erosion has reduced them to a low jumble of hills; in other cases the salt must still be in active movement, as it forms the highest point of the landscape, 4,000 to 5,000 feet above sea level, and higher than the surrounding limestone scarps—this in spite of a yearly rainfall of about five inches. These magnificent salt mountains are a most spectacular phenomenon (Plate 59).

The coastal zone between Jask, at the entrance of the Persian Gulf, and the Indian frontier is of quite a different character from anything elsewhere in Persia. Much of it could claim a world's record for badness of "bad lands," and the remainder is a drab area of low broken hills, interspersed with stony plains and broad, dry river-courses.

The foothills of the Zagros terminate abruptly along well-defined, almost straight lines, and give place towards the southwest to flat bare plains and, in the southern part of Iraq, to extensive marshes. The sunken land has been built by the silt carried down by the Tigris and the Euphrates, and by the left-bank tributaries of the Tigris—the Greater and Lesser Zab, the Sirwan (or Diyala in its lower reaches), and the Karkheh; but at present the extension of the Shatt-al-Arab delta forward into the Persian Gulf is the work of the Karun River, unaided by its larger brothers. The Karun, which has a shorter course from its mountain catchment area to the sea, is able to build forward at a rate greater than the building tempo of the Tigris and Euphrates. These two rivers must first fill up the extensive marsh areas between Amara or Nasiriya and Basra before their silt is available for onward carriage into the Persian Gulf. The enormous volume of the Tigris in flood causes a rise at Baghdad of about twenty feet above low-water level; but in the "narrows" below Amara there is very little seasonal variation, for the flood-water spreads out into the marshes, which function as vast settling tanks for its silt. Archaeological opinion is that the head of the Persian Gulf is in retreat, but there is geographical evidence for a reverse tendency; in fact extensive tracts below Basra, formerly cultivated, are now inundated from marine creeks.

Both Tigris and Euphrates have meandering courses typical of mature rivers. Thus the stretch from Baghdad to Kut-al-Imara is 103 miles by direct distance, but 213 miles by river. The Tigris at Baghdad at mean low-water level is only 94.7 feet above sea-level, although Baghdad is 358 miles in direct distance, and 549 miles by river, from its mouth. The effect of the tides from the Gulf extends upstream for 100 miles, that is, to the vicinity of Al Qurna, but the Shatt water is quite fresh at Basra and Abadan, and only slightly brackish at Fao.

The alluvial plains of Iraq are bounded on the west by the edge of the Arabian and Syrian deserts, a low scarp of rubbly limestone twenty feet or so in height, which runs parallel to the Euphrates along much of its course. This unpretentious scarp is such a striking contrast to the monotonous flatness of the plains that it has given the country its name, for "Iraq" in Arabic means "cliff." From the edge of this scarp the desert rises gradually, almost imperceptibly, to heights of over 3,000 feet (Jebel Aneiza, 3,084 feet; Rutba Fort, 2,018 feet). The desert surface within Iraq territory is for the most part formed of hardened clay, locally boulder strewn, and in some places of broken rubbly limestone.

The Arabian coasts of the Persian Gulf are for the most part bare, featureless, and semi-desert. Kuwait town has an Arab population of about 100,000 most of whose foodstuffs and even drinking water must be imported—the water in native sailing boats and water tankers from the Shatt-al-Arab estuary.

The only oases of consequence along or near the coast are Bahrein, Hasa, Doha, Abu Dhabi, Sharjah, and Dibai; these owe their existence to springs or wells of water, not too brackish to be usable, supplied by an Eocene limestone aquifer. The most spectacular of such springs are on Bahrein Island, where the water has a sufficient yield to irrigate extensive date-gardens. One spring, regarded as the best in quality of water, is in the sea just off the island. The Arabs used to dive to the sea-bottom to fill their goat-skins by holding open the neck in the stream of fresh water: recently, however, a prosaic engineer has driven a pipe and led the water to the surface. A calculation of the amount of rainfall falling on Bahrein and a comparison with the output of the springs show clearly that the water cannot be locally derived; it must indeed have its ultimate source in the highlands of Nejd, about 300 miles distant and about 3,000 feet above sea-level.

This water-bearing Eocene limestone is buried by later sediments along most of the coastal zone, and outcrops only on anticlinal summits, but water can be tapped by relatively shallow wells. The Iraq government has demonstrated this in their desert areas west of the Euphrates. In Kuwait, the Eocene limestone water is brackish in some areas, but in others potable for Arabs and their flocks.

The Cambrian salt domes in the Persian coastal sector between Bushire and Bandar Abbas have already been described. Other such domes form most of the islands in the southern part of the Gulf, and also form isolated hills along the Trucial Coast west of Abu Dhabi.

Further south, in Saudi Arabia, there are extensive tracts of sand dunes in a zone lying between the coastal strip and the highlands of Nejd. The great sand desert of southern Arabia, the Rub' al Khali, or Empty Quarter, is aptly named, although Bertram Thomas, who was the first Westerner to cross it, showed that it is not quite so bad as its reputation, though perhaps nearly so.

The mountains of Oman form a most striking contrast to the featureless desert areas of southeastern Arabia. Geologically they are part of an outer arc of the Zagros fold system, although it is uncertain in what way they connect with it toward the east. The fold system passes out to sea between Ras al Hadd and Masira Island, and may have a submarine connection with the ranges of Sind, which strike out to sea at Cape Monze. An alternative possibility is that both these ranges continue southward down the center of the Indian Ocean. The highest point of the Oman mountains is the great limestone scarp of Jebel Akhdar at 10,190 feet above sea level. The sector of the hills north of Jebel Akhdar and south of the Masandam Peninsula is formed of dark igneous rocks—serpentine, basalt, and diorite—which have weathered into a forbidding maze of bare, shining, black and greenish peaks and razor edges. Here and there among this wild tangle of hills there is sufficient perennial water to support small villages and gardens of date palms, the produce of which is highly valued for delicacy of flavor. The inlets into the rocky headland of Masandam Peninsula, or Ruus al Jibal, are drowned valleys. The valleys were excavated by erosion when the relative sea level was several hundred feet lower than at present.

Structurally and topographically Egypt falls into five main divisions: [6] (1) the belt of deep depressions in the extreme west, constituting the Oases and occupied by Cretaceous rocks; (2) the broad, waterless expanse of the Libyan Desert to the west of the Nile and the corresponding plateau region (Maaza limestone plateau) to the east of it, composed of Tertiary, especially Eocene, rocks; (3) the Nile valley and delta, formed of Pliocene and Pleistocene deposits; (4) the wilderness of the Red Sea Hills and southern Sinai, composed largely of ancient crystalline rocks; and (5) the Red Sea and the Gulfs of Suez and Aqaba, with the coastal plains.

While the first three regions belong geologically to the Mediterranean area, the fourth and fifth, in which the Egyptian oil fields are located, form part of Krenkel's "Syrabia," or the African-Arabian crystalline shield. Two principal periods of earth movements have affected this area, the first during the Upper Oligocene and the second during the Pliocene. During the first, faulting at right angles to the direction of folding caused the formation of the rift valley of the Gulf of Suez and the shallow border zone of the Red Sea. The deeper area of the rift in the central part of the Red Sea and Gulf of Aqaba was formed in several stages during the later movements. These faults are intimately connected with those causing the East African and Palestinian rift systems, and their upthrown sides form the central Red Sea Hills and the main range of Sinai. Step-faulting parallel to the Gulf has thrown down whole blocks of strata, which have been tilted so that the igneous rocks form ridges, such as Abu Durba and Jebel Zeit, facing the Gulf of Suez, whilst on the inland side the sediments dip away from the sea in succession. Where the rigid blocks have a cover of more plastic Miocene salt, gypsum, and marls, subsequent movements have caused squeezing and superficial folding, as at Hurghada and eastern Gemsa.

The Red Sea Hills are composed mainly of granite, gneiss, and schists, and reach heights ranging from 6,000 to 7,800 feet. They are of rugged character and varying forms according to the type of rock. The geological structure outlined above gives rise to long narrow ridges with precipitous slopes, and often with serrated, knife-like crests which rise suddenly from among a complex of narrow valleys and minor hills. Where the rock is granite or gneiss, the gray slopes are steep and smooth, and unrelieved by any vegetation except where storm waters produce sudden small patches of brilliant verdure. Where the rocks change to schists, the ridges are less bold in appearance, the summits conical in shape, dull green in color, and covered by broken rock-fragments; the valleys are arid and stony.[7]

[6] W. F. Hume, quoted by F. R. C. Reed, in *Geology of the British Empire,* London, 1921, Chap. II.

[7] Hume, W. F., *Geology of Egypt,* Vol. I, Cairo, 1925.

Oil-field discovery in Egypt has, so far, been confined to the Gulf of Suez, where the fields of Hurghada, Ras Gharib, and Sudr are situated, but exploration work is active in central and northern Sinai and in the Western Desert.

Oil-Field Development

PERSIA

The first scientific record of the oil seepages of southwest Persia was made by Loftus in 1855.[8] In 1872, Baron Julius De Reuter, the founder of Reuters' News Agency, obtained an exclusive concession from the Persian government to prospect for certain minerals, including petroleum, but a year later the concession was canceled under pressure from the Russian government, which resented such British infiltration. Undeterred, De Reuter tried again, some years later, and in 1889 received a second concession, including the right to found a bank. The oil exploration was undertaken by the Persian Bank Mining Rights Corporation (with British-German capital), and in the three years 1891 to 1893 two wells were drilled at Dalaki, northeast of Bushire, to something over 800 feet, and one was drilled on Qishm Island to 700 feet. All three wells were unsuccessful, and in 1894 the company went into liquidation.

In 1901 an oil concession for the whole of Persia, except the five northern provinces, was granted to William Knox D'Arcy, an Englishman who had made a considerable fortune in gold mines in Australia; and in 1902 drilling operations commenced at Chia-Surkh (Chah-i-Surkh on map).[9] Some encouraging showings of oil were struck, but no sustained production resulted. In 1906 two test wells were drilled at Mamatain, near Ram Hormuz, but again without success, and in 1907 operations were transferred to Masjid-i-Sulaiman. On May 26, 1908, oil was struck in No. 1 well at a depth of 1,180 feet, and ten days later No. 2 well reached oil and gas at 1,010 feet. Further development followed rapidly. In 1909–1910 construction commenced of a pipe line to Abadan and of a refinery there, and in 1913 the plant was finally commissioned.

The original capital of Mr. D'Arcy had proved insufficient to meet the cost of subsequent requirements, and additional capital was subscribed by the Burmah Oil Company and certain individuals forming the Anglo-Persian Oil Company, Ltd. In May 1914, the British government acquired a substantial financial interest in the company, largely at the instigation of Winston Churchill, then First Lord of the Admiralty, in order that the fuel-oil supplies for the Navy might be sufficiently safeguarded. The outbreak of war with Turkey in November 1914 threatened the security of the Persian production, to protect which a British force was landed.

[8] Loftus, W. K., "On the Geology of Portions of the Turco-Persian Frontier and of the Districts Adjoining," *Quart. Journ. Geol. Soc. London,* Vol. XI, 1855, pp. 247–344.

[9] The Chia-Surkh area was transferred from Persia to Turkey by a frontier adjustment in 1914.

The Mesopotamian campaign, destined to last till 1918 with a checkered history of successes and disasters, was a consequence.

After 1918, the production of Masjid-i-Sulaiman increased steadily, and the refining capacity of Abadan was expanded. In 1928 the Haft Kel field commenced production, then in 1941 Gach Saran, in 1944 Agha Jari, in 1945 White Oil Springs, and in 1948 Lali. The average production in 1948 was at the rate of about 518,000 b/d, and up to the end of 1948 Persia had produced a total of 1,938 million barrels. There is a small topping plant at Masjid-i-Sulaiman, which supplies the transport of the oil field with gasoline.

The Persian oil fields are large simple anticlines, with the 1,000-foot

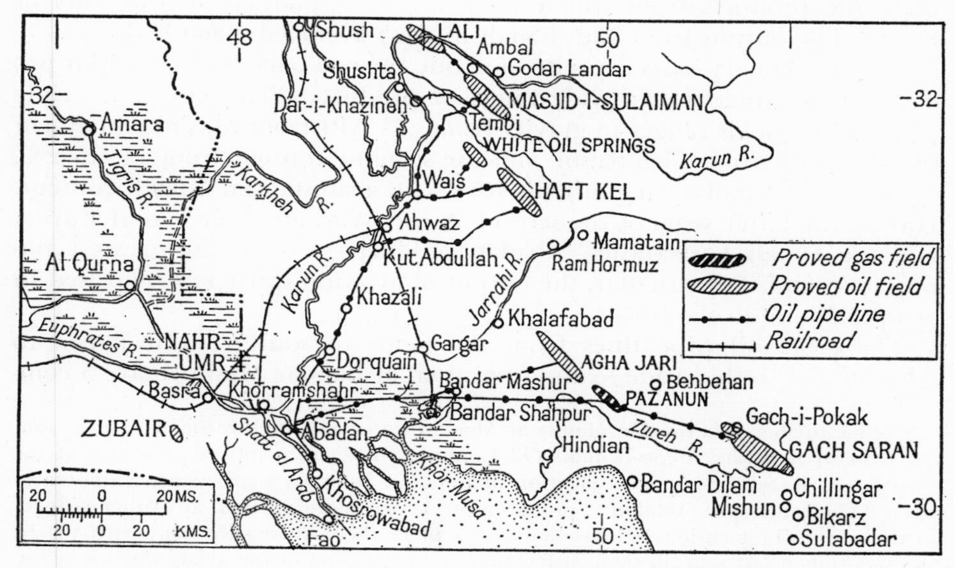

FIG. 33. Southwestern Persia and southeastern Iraq: oil-field development and transport. The facilities for pipe-line transport and for shipping afforded by Persia and Iraq in this area are described by Dr. Lees in his section "Communications," and the facilities at Abadan are described in his footnote #10.

thick Asmari Limestone, of Lower Miocene to Oligocene age, as reservoir rock. The folding is acute; the southwest flanks are steep and in some cases nearly vertical; and the limestone is in consequence freely fissured. The limestone is of low average porosity, and a well must encounter a fissure to achieve worthwhile production. The reservoir connection is so free throughout most of the fields that a uniform pressure-fall, consequent on production, can be detected throughout the fifteen to twenty miles length of the individual fields. Well spacing is one to two miles, and movements of gas-oil and oil-water levels are closely watched by observation wells.

The Asmari Limestone is overlain by an anhydrite-salt-shale series called the Lower Fars, which forms the cap-rocks. Seepages of oil or gas mark the position of the shallower oil fields, where the relation of weight of cover to fluid pressure in the underground reservoir was insufficient. The plasticity of the salt-bearing series has allowed an unusual degree of disharmony between the surface formations and the massive Asmari Limestone below, and in some cases surface synclines overlie buried anticlines. Seismic refraction-arc surveys are employed for structural definition.

Naft-i-Shah is a small oil field detached from the main group of the Persian fields. It is situated on the Iraq-Persian border northeast of Baghdad (the sector of the anticline in Iraq is known as the Naft Khaneh field). Naft-i-Shah field is connected by a 3-inch pipe line to a refinery at Kermanshah, the products from which serve the local markets in this part of Persia. The output from Naft-i-Shah in 1947 averaged 2,800 b/d.

Abadan refinery has a capacity of about 495,000 barrels crude input per day, and manufactures a wide range of products (Plate 16). The crude oil from the fields ranges in gravity from 38° API from Masjid-i-Sulaiman to 32.5° API from Gach Saran, and the sulphur content from 1% to 2%. The oil is of paraffinic-naphthenic base with a substantial asphalt content. During the latter years of the second World War, after the loss of Burma and the Dutch East Indies, Abadan products were of the greatest importance, and, in particular, the output of aviation spirit was increased in 1945 to about 20,000 b/d.[10]

The crude-oil pipe lines from the fields of southwestern Persia to Abadan and to the crude loading terminal at Bandar Mashur have a com-

[10] Some further data on the refinery at Abadan were presented by Sir William Fraser, the Chairman of the Anglo-Iranian Oil Company, Ltd., in a report presented at the thirty-seventh ordinary general meeting in London, June 30, 1946, which is quoted in part: "The refinery at Abadan is probably processing more oil than any other refinery in the world. Tremendous achievements took place there during the war, particularly, as I mentioned last year, in the construction and expansion of the aviation spirit plant. This plant is a major works in itself, but the refinery has also been expanded by additional distillation equipment and other plant of all kinds, and by the complementary provision of workshops, power and water supplies, oil-loading facilities, storage, wharfage, transport, stores, and many other ancillary services. The result of these efforts, many of them improvised and consummated under extreme difficulties, is that the capacity of Abadan refinery is now almost double what it was before the war. . . .

"The regions in the south in which the company operates are mainly arid, and were formerly virtually unpopulated. Before the period of its use by the company, Abadan Island was desert, fringed by a riverside area of date gardens and adjoined by a small village. It is now the site of this great refinery and of one of Iran's largest townships, and most of its inhabitants are company employees or their dependents. In the oil fields and at Abadan the company has built over 14,000 houses and quarters for its employees—and it must be remembered that it is the company that has to provide roads, sewage disposal, electricity, water supply and other facilities for each of these houses. Every house, moreover, needs imported cement, woodwork, roofing, kitchen and sanitary arrangements and household fittings." A description of the Abadan refinery, illustrated by diagrams and photographs, appeared in the *Petroleum Times Review of Middle East Oil,* London, June 1948, pp. 24–32.

bined capacity of about 650,000 b/d. Masjid-i-Sulaiman, Lali, Haft Kel, and Naft Safid are connected to Abadan by six 10-inch lines; Gach Saran to Abadan by one 12-inch line; Agha Jari to Abadan by one 12-inch line, and to Bandar Mashur by one 12-inch and one 22-inch line. Owing to the elevation of the production units at Gach Saran and Agha Jari the crude from these fields is delivered to the terminals by gravity flow, but pumping stations are required for the other fields. There are topping units at Masjid-i-Sulaiman and small distillation plants at the other fields for local fuel needs. At Abadan the storage-tank capacity of some 800,000 barrels of crude is normally kept about half full.

The original concession acquired by D'Arcy covered 480,000 square miles, excluding from the whole of Persia only the five northern provinces of Azerbaijan, Gilan, Mazanderan, Asterabad, and Khurasan. It was for sixty years dating from 1901, and the royalty was fixed as 16% of the net profits. During the subsequent operation of the concession, after production had commenced, a number of disagreements arose over the definition of "net profits," and the Persian government found that the oscillation in the size of the annual payments was a cause of vexation: the payments depended on trading profits, and thus on the state of world markets. Negotiations for a fresh agreement were actually well advanced when, in 1932, the government unilaterally canceled the concession, the immediate cause being the sudden drop in profits in 1931, due to the world depression and the great fall in prices. The concession dispute was brought before the League of Nations at Geneva, and eventually an acceptable settlement was achieved between the company and the Persian government. A new concession was agreed, whereby the government received a fixed royalty (4 shillings per ton, the value of the shilling to bear an agreed relation to gold) plus a share in the company's profits when these exceeded 5%. The life of the new concession was to be sixty years from its commencement (1933), but its extent was reduced to 100,000 square miles (Fig. 34).

In 1937 a concession in northeast and east Persia was granted to the Amiranian Oil Company, a subsidiary of the Seaboard Oil Company of Delaware, and extensive exploration-surveys were carried out during the two following years. Eventually the concession was relinquished, as the company, in which Caltex had in the meantime acquired an interest, failed to find prospects of sufficient promise to justify development in such an unfavorable geographical situation. The geological results in eastern Persia have been admirably described by F. G. Clapp.[11]

In 1943 and 1944 British and American interests attempted to acquire concessions in central, eastern, and southeastern Persia, and the Soviet Union insisted on concessionary rights in the north Persian zone. The government, to avoid political embarrassment, decided to defer all further

[11] Clapp, F. G., "The Geology of Eastern Iran," *Bull. Geol. Soc. Amer.*, Vol. 51, 1940, pp. 1–102.

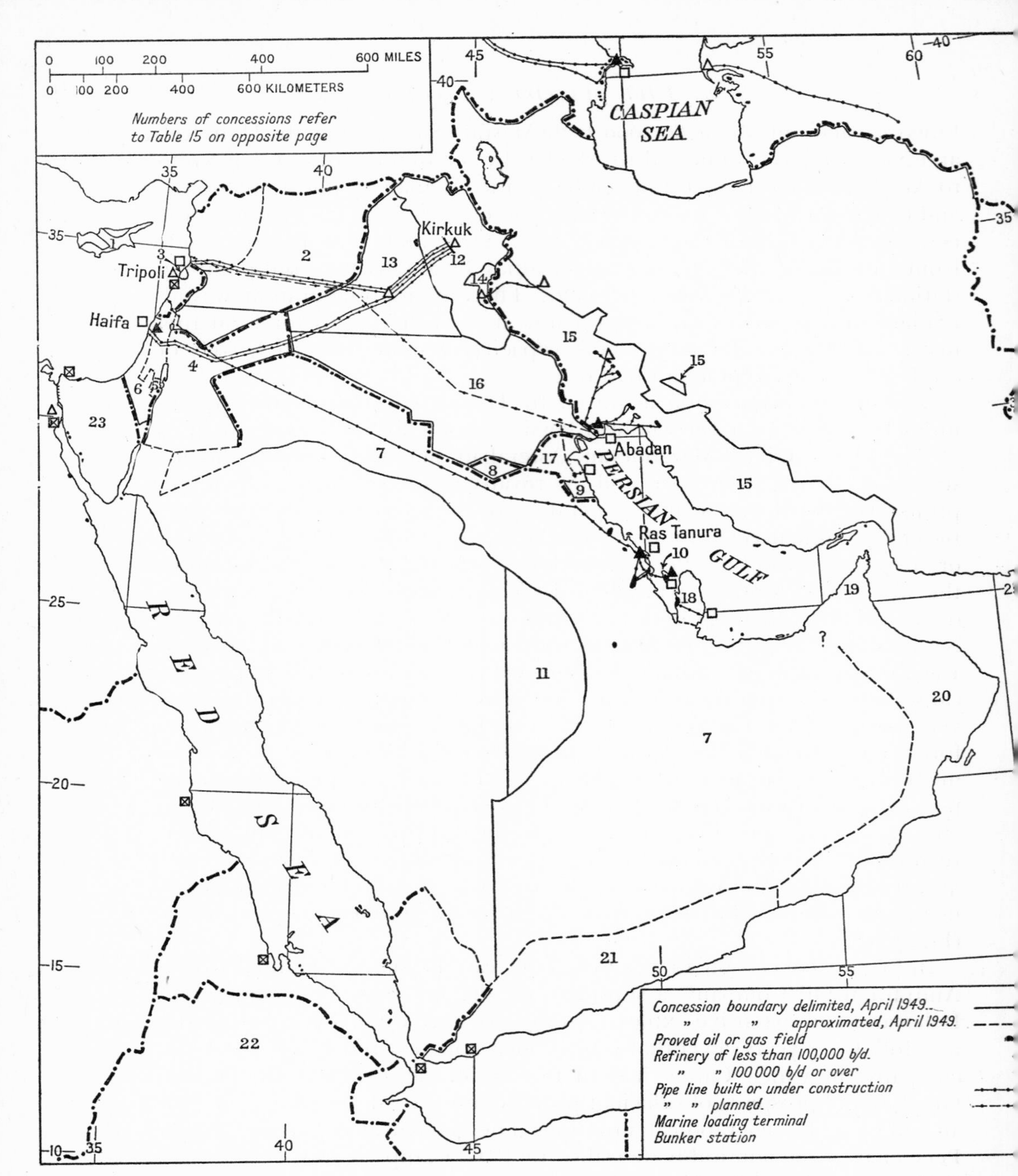

Fig. 34. The Middle East: petroleum concessions, oil and gas fields, refineries, and transport facilities a bunker stations, April 1949. The pipe lines shown by solid lines were in the following stages of completi (mid-1949): from Kirkuk to Tripoli, one line 12¾ inches in diameter in use and one line 16 inches in ameter still under construction; from Kirkuk to Haifa, one 12¾-inch line temporarily out of use and o 16-inch line in a late stage of construction; from Abqaiq to Sidon, the 30/31-inch line completed fr Abqaiq to Qatif Junction (44.7 miles) and thence to Abu Hadriya (80 miles). The pipe line shown by brok line is the planned Middle East pipe line of 34-inch diameter from Abadan to Tartus, north of Trip Additional pipe lines of 30-inch diameter have been contemplated, to run from Kuwait to Tartus and fr Kirkuk to Baniyas, north of Tartus. The capacities of the principal refineries shown, omitting those in t U.S.S.R., are: Abadan, 495,000 b/d crude and 125,000 b/d cracking; Haifa, 90,000 b/d crude, 17,300 b cracking; Bahrein, 155,000 b/d crude, 15,000 b/d cracking; Ras Tanura, 140,000 b/d crude only.

8. THE MIDDLE EAST

Table 15. Petroleum concessions in the Middle East, April 1949.

Reference No. on Map	Concession Area	Concessionaire	Length of Term: Date	Effective Years	Ownership of Concessionaire Company
1.	2,000 sq. mi. in Cyprus	Petroleum Development (Cyprus), Ltd.	Relinquished, Jan. 1949		Petroleum Concessions, Ltd.
2.	41,700 sq. mi. in Syria	Syria Petroleum Co., Ltd.	Mar. 26, '40	75	Same group as Iraq Petroleum Co., Ltd. (#12)
3.	193 sq. mi. in Lebanon	Lebanon Petroleum Co., Ltd.	...[a]	...[a]	Petroleum Concessions, Ltd.
4.	All of Transjordan	Trans-Jordan Petroleum Co., Ltd.	May 10, '47	75	Petroleum Concessions, Ltd.
5.	386 sq. mi. in Palestine	Jordan Exploration Co., Ltd.	...[a]	...[a]	Palestine Potash Syndicate
6.	5,270 sq. mi. in Palestine	Petroleum Development (Palestine), Ltd.	...[a]	...[a]	Petroleum Concessions, Ltd.
7.	Original area, 360,000 sq. mi. in Saudi Arabia	Arabian American Oil Company	Jul. 14, '33	66	Standard Oil Co. of Calif., 30% The Texas Co., 30% Standard Oil Co. (N.J.), 30% Socony-Vacuum Oil Co., 10%
	Supplemental area, 80,000 sq. mi. at effective date [b]	"	Jul. 21, '39	66	"
8.	All of Iraq-Saudi Arabian Neutral Zone [b]	Arabian American Oil Co., Saudi-Arabian half-interest and	Jul. 21, '39	66	As listed under #7
		Basrah Petroleum Co., Ltd., Iraq half-interest	...	...	As listed under #16
9.	All of Kuwait-Saudi Arabian Neutral Zone (about 2,000 sq. mi.)	American Independent Oil Co., Kuwait half-interest, and	June 28, '48	60	...[c]
		Pacific Western Oil Corp., Saudi Arabian half-interest	Feb. 20, '49	...[c]	...[c]
10.	All of Bahrein Is. and territorial waters	Bahrein Petroleum Co., Ltd.	June 19, '40	55	Standard Oil Co. of Calif., 50% The Texas Co., 50%
11.	Area of preferential rights in Saudi Arabia [b]	Arabian American Oil Co.	Jul. 21, '39	60	As listed under #7.

Petroleum Concessions in the Middle East, April 1949 (Continued)

Reference No. on Map	Concession Area	Concessionaire	Length of Term: Date	Effective Years	Ownership of Concessionaire Company
12.	32,000 sq. mi. in Iraq (Mosul and Baghdad vilayet east of Tigris R.)	Iraq Petroleum Co., Ltd.	Mar. 14, '25 [d]	75	Anglo-Iranian Oil Co., Ltd., 23.75% Royal Dutch Shell group, 23.75% Cie. Francaise des Petroles, 23.75% Standard Oil Co. (N.J.), 11.875% Socony-Vacuum Oil Co., 11.875% C. S. Gulbenkian, 5%
13.	46,000 sq. mi. All of Iraq west of Tigris R. and north of 33°N.	Mosul Petroleum Co., Ltd. (formerly British Oil Development, Ltd.)	May 25, '32	75	Same group as Iraq Petroleum Co., Ltd. (#12)
14.	684 sq. mi. in Iraq	Khanaqin Oil Co., Ltd.	Aug. 30, '25	70	Anglo-Iranian Oil Co., Ltd.
15.	100,000 sq. mi. in Persia	Anglo-Iranian Oil Co., Ltd.	1933	60 [e]	British Govt., 52.55% [f] Burmah Oil Co., 24.95% Individuals, 22.50%
16.	93,000 sq. mi. All of Iraq not covered by I.P.C., M.P.C., and A.I.O.C. concessions	Basrah Petroleum Co., Ltd.	Nov. 30, '38	75	Same group as Iraq Petroleum Co., Ltd. (#12)
17.	All of Kuwait	Kuwait Oil Co., Ltd.	Dec. 23, '34	75	Anglo-Iranian Oil Co., Ltd., 50% Gulf Exploration Co., 50%
18.	4,500 sq. mi., all of Qatar	Petroleum Development (Qatar), Ltd.	May 17, '35	75	Petroleum Concessions, Ltd.
19.	All areas of Abu Dhabi, Dubai, Sharjah, Ras al Khaima, Ajman, Um al Qawain, Kalbah	Petroleum Development (Trucial Coast), Ltd.	...	...	Petroleum Concessions, Ltd.
20.	All of Oman and Dhofar	Petroleum Development (Oman and Dhofar), Ltd.	June 24, '37 [a]	75 [a]	Petroleum Concessions, Ltd.
21.	Hadhramaut	Petroleum Concessions, Ltd.	... [a]	... [a]	Same group as the Iraq Petroleum Co., Ltd. (#12)
22.	All of Ethiopia	Sinclair Petroleum Co.	...	50	Sinclair Oil Corp., 100%

Petroleum Concessions in the Middle East, April 1949 (Continued)

Reference No. on Map	Concession Area	Concessionaire	Length of Term: Date	Effective Years	Ownership of Concessionaire Company
23.	North and west parts of Sinai Peninsula	Scattered parcels to Socony-Vacuum Co., Standard Oil Co. of Egypt, Anglo-Egyptian Oilfields, Ltd., Beckwith and Co., Ltd., respectively [g]	...	...	...

Note: This table does not show the many concessions negotiated but not finalized in earlier years, nor concessions that had lapsed prior to 1949.

[a] Exploration permit only.

[b] The supplemental area on the effective date included those areas held under preferential rights in Saudi Arabia and the Saudi-Arabian undivided half-interest in the Kuwait and Iraq neutral zones. In 1948 Aramco relinquished preferential rights over a part of Saudi Arabia west of 46°E and over the Saudi-Arabian interest in the Kuwait Neutral Zone.

[c] The conclusion of concession arrangements between the Shaikhdom of Kuwait and the American Independent Oil Company covering the Kuwait half-interest was announced on July 5, 1948, by Mr. Ralph K. Davies, president of the company. The American Independent Oil Company was incorporated in August 1947 by a group of two individuals and eight independent companies engaged in the production, refining, and distribution of petroleum in the United States. In alphabetical order the incorporators are: J. S. Abercrombie, Houston, Texas; Allied Oil Company, Inc., of Cleveland, Ohio; Ashland Oil and Refining Company of Ashland, Kentucky; Ralph K. Davies of San Francisco, California; Deep Rock Oil Corporation of Chicago, Illinois; Globe Oil and Refining Company of Wichita, Kansas; Hancock Oil Company of Long Beach, California; Phillips Petroleum Company of Bartlesville, Oklahoma; Signal Oil and Gas Company of Los Angeles, California; and Sunray Oil Company of Tulsa, Oklahoma. On May 10, 1949, according to *Standard Corporation Records,* J. Paul Getty, individually and as trustee, held 84% of the stock of the Pacific Western Oil Corporation. Under an agreement between the two concessionaires, the American Independent Oil Company is carrying forward the development work and drilling for the joint account.

[d] Modified March 29, 1931.

[e] Expiring December 31, 1993.

[f] The present issued capital stock of the Anglo-Iranian Oil Company is £32,843,752, made up as follows (all stock in £1 units):

Ordinary stock (carrying two voting rights per share)	£20,137,500
Preference stock (carrying one-fifth voting right per share)	
8% cumulative first preference stock	£ 7,232,838
9% cumulative second preference stock	5,473,414
The British Government's holding is as follows:	
Ordinary stock	£11,250,000
Preference stock	1,000
The Burmah Oil Company's holding is:	
Ordinary stock	£ 5,342,985

The balance is held by individuals.

[g] *World Oil Atlas 1948,* map on p. 281.

Sources: Information obtained from representatives of the companies, except as noted otherwise in footnotes.

grants of concessions until foreign troops had been withdrawn. In April 1946, a concession for a substantial area in north Persia was provisionally granted to a Russo-Persian company, a circumstance which occasioned much debate in the meetings of the United Nations Organization. This concession was subject to ratification by the Persian Majlis (Parliament), but when the Majlis met in November 1947 it refused to ratify the concession, and proposals were made that only Persian nationals might in future acquire oil-exploration rights. The prospects of these areas are quite uncertain; numerous oil indications are known, but geologically they have little in common with the southwest Persian oil zone.

IRAQ

The discovery of oil at Masjid-i-Sulaiman in Persia in 1908 drew attention to the possibilities of Iraq, although Mr. D'Arcy had been negotiating for a concession as early as 1901–1902. A concession was finally granted in 1914 by Turkey to British and German interests which formed the Turkish Petroleum Company, Ltd. The newly created state of Iraq fell heir to the oil rights after the defeat of Turkey in 1918, and a new concession was negotiated by a re-formed Turkish Petroleum Company, subsequently named the Iraq Petroleum Company. A French group was allotted the earlier German share in 1920 by the terms of the so-called San Remo Agreement, and in 1922, as a result of representations made by the State Department, a group of American oil companies were conceded one-half of the share of the Anglo-Iranian Oil Company, the latter receiving a royalty interest as compensation. Originally fourteen American companies were interested in this Iraq opportunity, but only five accepted participation in the American group forming the Near East Development Company, and only two finally remained. These arrangements resulted in the following shareholding of the present company:

Anglo-Iranian Oil Company	23¾%
Royal Dutch-Shell Group	23¾%
Near East Development Company (Standard Oil Company, N.J., and Standard Vacuum)	23¾%
Compagnie Française des Pétroles	23¾%
Mr. C. S. Gulbenkian (the Armenian intermediary of the original concession)	5 %

The concession of the Iraq Petroleum Company, as revised in 1931, was limited to that part of Iraq north of the 33° parallel of latitude on the east side of the Tigris. A concession over the remainder of Iraq north of the same parallel was granted in 1932 to a British syndicate termed "B.O.D., Ltd.," in which by 1935 Italian and German interests had acquired control. These, and other interests, were later bought out by the groups forming the Iraq Petroleum Company, who constituted the Mosul Petroleum Company to hold and operate the concession. In 1938 a concession over southern Iraq was granted to the Basrah Petroleum Company, also of similar composition to the Iraq Petroleum. Thus the whole of Iraq is now held in three concessions by the same groups, with the exception of a small area near Khanaqin, northeast of Baghdad. This area was transferred from Persia to Turkey in 1914 as one of several minor frontier adjustments; and, as the area formed part of the D'Arcy concession held by the Anglo-Iranian Oil Company, their rights were recognized by the Turkish and later the Iraqi governments. The concession has been developed by an A.I.O.C. subsidiary, the Khanaqin Oil Company, Ltd., which operates an oil field at Naft Khaneh and a small refinery thirty miles distant at Alwand, close to Khanaqin. The products from this refinery serve the local Iraq market.

The exploration of the I.P.C. concession resulted in the discovery of the Kirkuk oil field in 1927; and after very few wells had been drilled and tests completed, it was clear that the field was of giant size. It is a simple narrow anticline, productive throughout a length of nearly sixty miles from a porous limestone reservoir rock of Miocene to Middle Eocene age. Its crest-maximum is marked by copious oil and gas seepages; the latter, at Baba Gurgur, are linked in local tradition with Shadrach, Meshach, and Abednego (*Daniel,* iii: 8–30). The permeability of the reservoir is so high that the structure was effectively drained, at a rate of 80,000 b/d, for some years by six wells grouped around the central crest-maximum of the anticline. Plans were developed in 1947 for an increase in offtake to 400,000 b/d by about 1955.

Pipe lines to carry the output from Kirkuk to the Mediterranean were completed in 1934, one line going from Iraq through Syria to a terminal at Tripoli in the Lebanon Republic. At present the crude is shipped from Tripoli to various refineries in Europe. A southern line takes a route through Jordan and Palestine to a terminal at Haifa. These lines are partly 10- and partly 12-inch in diameter, and have a combined capacity of 84,000 b/d (4,000,000 tons per year). A refinery was constructed at Haifa by the A.I.O.C. and Shell groups to process their share of the Kirkuk output; it was completed in the early days of the war and proved a great asset to the Allied cause. The refinery was later extended to a capacity of 90,000 b/d. A large refinery on the Mediterranean coast is being planned by the Standard Vacuum and New Jersey companies. Kirkuk crude is 35.6° API gravity and has 2% sulphur content.

In 1945 plans were developed for a 16-inch diameter pipe line from Kirkuk to Haifa, and by 1948 it had been completed to within forty miles of Haifa when construction was halted by the Israeli-Arab war. A second 16-inch pipe line from Kirkuk to Tripoli was under construction in 1949, and is planned for completion in 1950. The expectation, of which these plans are evidence, that a production of about 400,000 b/d can be drawn from the field, gives some measure of the size of this immense oil reserve. In 1944 DeGolyer published an estimate of a proven reserve of 4,000 million barrels for Kirkuk, and in 1947 L. F. McCollum assessed these reserves at 7,500 million barrels (*Oil Forum,* July 1947).

Exploration of the B.O.D. concession has been less fortunate. Intensive drilling in the Qaiyarah area on the Tigris and between Hit and Ramadi on the Euphrates has revealed large oil accumulations, but the crude oil is of such heavy gravity, 12° to 16° API, and with such a high sulphur-content, 5% to 10%, that it has not been found practicable to produce it profitably. Just prior to the outbreak of war in 1939 a small field yielding oil of similar quality to that of Kirkuk was found at Ain Zaleh, north of Mosul; its development was retarded by the circumstances of the war, but drilling was resumed in October 1947. In the concession of the Basrah Petroleum Company, exploratory drilling is in progress. Two wells were commenced in early 1947 at Nahr Umr and Zubair, northwest and south of Basra, respectively, and in early 1949 both these wells proved large-scale production. The crude of the former is of 42° API gravity, being thus the best quality oil so far found in the Middle East.

The transdesert pipe line from Kirkuk to the Mediterranean was thought to be an outstanding engineering achievement in its day, but it may be of interest to recall that it is not the first recorded transdesert pipe line in history. In 525 B.C., Cambyses, King of Persia, decided to invade Egypt for certain domestic reasons. The Arabian desert was a formidable barrier to the passage of his army, but, fortunately for him, the King of the Arabians accepted a contract for his water supply. There is a tale, according to Herodotus, that it was accomplished thus:

"There is a great river in Arabia called Corys. From this river (it is said) the king of the Arabians carried water by a duct of sewn ox-hides and other hides of a length sufficient to reach to the dry country; and he had great tanks dug in that country to receive and keep the water. It is a twelve days' journey from the river to that desert. By three ducts (they say) he led the water to three separate places." (Herodotus, Book III, 9, A. D. Godley translation.)

Unfortunately no further details are known. Engineers would have been interested in pumping pressures and hydraulic gradients, but, alas, history is silent!

KUWAIT

A concession for the oil rights of Kuwait was granted in 1934 to the Kuwait Oil Company, Ltd., jointly owned and managed by the Anglo-Iranian Oil Company, Ltd., and the Gulf Exploration Company (Gulf Oil Corporation). The first test well drilled at Bahra, north of Kuwait Bay, was unsuccessful, but the second well, drilled in 1937 at Burgan, thirty miles south of Kuwait town, struck oil in quantity in a Middle Cretaceous sandstone at 3,692 feet. Development drilling followed, though interrupted by wartime circumstances. Production at the rate of about 40,000 b/d commenced in mid-1946, the average rate in 1948 was 126,000 b/d, and in early 1949 it reached 240,000 b/d.

The structure of the field (Fig. 35) is a broad dome with low dips, and the reservoir rocks are thick sandstones of Middle Cretaceous age and a thin Orbitolina limestone. The strong development of sandstones at Kuwait makes a striking contrast to the oil fields of Persia, Iraq, and elsewhere in the Persian Gulf, in all of which the reservoir rocks are limestones. The reserves of the Burgan field have been estimated at about 11,000 million barrels, making it one of the largest single oil fields in the world. The Kuwait crude is of asphaltic type, has an average specific gravity of 33° API, and contains 2% to 2½% sulphur.

QATAR, TRUCIAL COAST, AND OMAN

The oil concession for the Shaikhdom of Qatar was granted by the ruling Shaikh in 1935 to Petroleum Development (Qatar), Ltd., a company of the same composition as the Iraq Petroleum Company. Drilling commenced at Jebel Dukhan in 1938, and in 1939 the first well struck oil in quantity of the order of 3,000 b/d. No. 2 well was also a producer, but No. 3 was off structure, and at this stage development was interrupted by the war. The oil occurs in a long narrow anticline in a limestone reservoir rock of Jurassic age. The depth to commercial production is around 6,000 feet. A program of drilling and engineering development of the Dukhan field commenced in 1947, and by 1949 nine wells had been drilled. A pipe line was laid across the peninsula to a loading terminal at Umm Said on the east coast, and production was planned to commence at the end of 1949 at the rate of 40,000 b/d.

Oil concessions covering the Shaikhdoms of the Trucial Coast and the State of Muscat are held by Petroleum Development companies associated with the Iraq Petroleum group. Up to 1949 exploration in the Trucial Coast and Muscat had not progressed beyond the geological and geophysical survey stage, and the great extent of blown sand had proved a substantial handicap to exploration of much of the area.

TURKEY

First production was in 1940, from a well in the Ramandag area, in the Tigris valley in southeastern Turkey, some fifty-eight miles north of the Syrian frontier and a little less than one hundred miles northeast of the Mosul Petroleum Company's field at Ain Zaleh in northern Iraq. This well produced some 70 b/d for a short period only, but a number of exploratory wells were sunk in the same area, and in 1947 Raman 9 came in with an estimated 300 b/d. At the end of 1948, three wells were producing and three further tests were being drilled. Production was from the top of massive Cretaceous limestone, and the average depth of the producing zone was about 4,500 feet. Pumping of the producing wells was discontinuous in 1948, and total production was held at some 25,000 barrels for the year, because of a lack of storage facilities. A topping plant in the field, with a nominal capacity of 200 b/d, handled this output. A second plant, with capacity of 350 b/d, was being built, and the

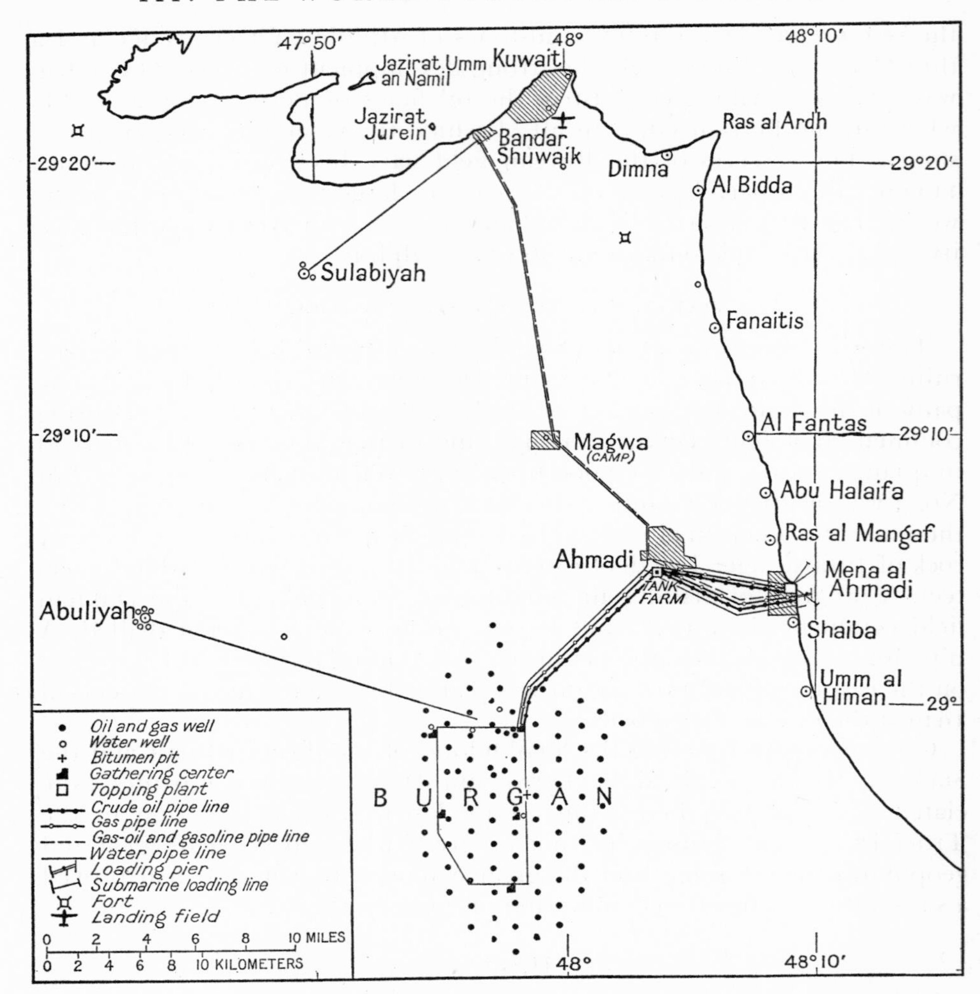

FIG. 35. Kuwait: oil-field development and facilities. The productive area of the Burgan [Burghan] field, a broad anticline with a north-south axis, comprised about 75 square miles at the beginning of 1950. The development plan allows a 600-acre space surrounding each well, and at the beginning of 1950 there were 87 completed wells. The additional wells shown on the map indicate the pattern to be followed.

This diagrammatic map, simplified as it is, may serve as an example of the layout of a modern oil field. The depths from which oil is obtained vary between 3,300 and 4,800 feet. The oil is of asphaltic type, and some 500 cubic feet of non-toxic gas is produced with each barrel of crude.

The wells are connected by flow lines, varying from 4 to 6 inches in diameter, to several 7-stage separator units at main gathering centers. From there the separated oil and gas is transported 10 miles to the tank farm at Ahmadi, by two 20-inch crude-oil pipe lines and by two gas lines, one of 12-inch and one of 3-inch diameter. As there is a rise of 140 feet en route, the flow in the pipe lines is maintained by centrifugal pumps. The tank farm when completed will have storage capacity of about 3.5 million barrels.

From the tank farm the greater part of the crude oil flows by gravity the 6 miles to intention was to expand production from the Ramandag area to the point where it could supply a large part of Turkey's domestic requirements. The crude is of 20° API gravity, and contains 4% sulphur. An important Ramandag product is asphalt for road-making. Turkey's oil industry is state-owned, and under the direction of the Mining Research and Exploration Institute, but the actual drilling in the Ramandag area since 1947 has been undertaken under contract by the Drilling and Exploration Company of Dallas, Texas.

The second main area of current oil interest is the Adana basin in south-central Turkey, named for the city of Adana on the Seihun or Seyhan river, some twenty-six miles north of the Mediterranean coast at Karatash. Geological parties and seismograph geophysical crews have been active here, part of the geophysical work being undertaken under contract by the United Geophysical Company of Houston, Texas. There has also been geological work in a third area, along the Black Sea coast in the neighborhood of Mapavri, about half-way between Trebizond and Batum.

Oil shales are known to exist in important amounts near Ankara, the capital, and in several areas of northwestern and western Turkey in Asia.

---

the loading terminal at Mena Al Ahmadi, through three pipe lines of 24-inch diameter and one line of 22-inch; the remainder flowing through a 22-inch line the 5 miles direct to the topping plant in the north section of Mena Al Ahmadi. This refinery, with a capacity of 25,000 b/d, produces gas oil, kerosene, and gasoline for local use, and these products are circulated back to Ahmadi by one 4-inch line for each, parallel to the larger crude lines in the south. One 3-inch line, used alternatively for gas oil or gasoline, traverses the 20 miles from Ahmadi to Bandar Shuwaikh, the port for Kuwait. Gas for the power plant and the topping plant in the industrial port area at Mena Al Ahmadi is brought from Ahmadi by a 10-inch line along the southern route.

The loading terminal at Mena Al Ahmadi is equipped with a 6-berth loading jetty for 25,000-ton tankers, and with a 2-berth cargo jetty for receiving oil-field equipment and other freight, as well as the fresh water brought in by tanker. Special fenders permit ships to remain at the berths under normal sea conditions.

In addition to the loading terminal itself, a small harbor has had to be constructed for the tugs, barges, launches, and other craft essential to the port.

During 1949 the Kuwait Oil Company personnel reached a total of 18,000, and 450,000 gallons of fresh water were required daily, all of it imported from the Shatt-Al-Arab; but evaporation plants to provide 600,000 gallons of fresh water daily are under construction at Ahmadi.

Both fresh water (imported) and salt water are conducted by pipe line from the port to Ahmadi. The water wells in and about the oil field supply brackish water, which is circulated through the field by the pipe circuit shown on the map, and fresh water is transported from Ahmadi by four pipe lines varying from 2 to 8 inches in diameter. One 6-inch fresh-water line from Ahmadi continues north to Bandar Shuwaikh.

The above data have been derived from the *Petroleum Times* (London) "Review of Middle East Oil," June 1948; from information kindly supplied by the Gulf Exploration Corporation; and from other sources.

SYRIA AND LEBANON

The northern part of Syria contains a number of large anticlines which form a westerly continuation of the fold-belt of Iraq, and it is possible that oil fields may be discovered somewhere throughout this extensive area. The Syria Petroleum Company, Ltd., another company of the Petroleum Concessions, Ltd., group, of similar composition to the Iraq Petroleum Company, held concessionary rights over most of northern Syria until 1949, and completed a number of deep borings west and east of Palmyra, north of Aleppo, and east of the Euphrates, but without results. In 1949 the western part of this concession was relinquished, and further drilling was concentrated in the eastern section.

The same company also holds concessions in the Lebanon Republic, and a first exploration boring close to Tripoli, begun in May 1947, was abandoned at the end of 1948.

PALESTINE AND JORDAN

The Dead Sea was called by the Romans *lacus asphaltites* on account of large masses of asphalt which periodically floated to the surface of the lake, particularly after earthquakes, and were known as "bitumen Judaicum." The Greek geographer Strabo (63 B.C. to 24 A.D.) describes the Dead Sea occurrences thus:

"The Dead Sea is full of asphalt. It comes to the surface irregularly at the center, with noisy disturbance of the water, which appears as though it were boiling. The visible portion of the asphalt lumps is round and has the form of a hillock. Much sooty matter accompanies the gaseous emanation, which tarnishes copper, silver and all bright metals, and even causes gold to rust. . . . The asphalt remains on the surface of the water, owing to its salty nature. People go in rafts to the asphalt, hack it to pieces, and take as much of it away with them as they can."

Hence Palestine has the sanction of classical authority for claim to be potential oil-territory.

Exploration for oil has extended over many years in recent history, but the prospects have been regarded as borderline in attractiveness. In 1914 the Standard Oil Company of New Jersey decided to bore at Kurnub, south of Hebron, but World War I overtook their intentions, and as a result of surveys after 1918 the company reconsidered the decision to drill. In 1939 the Petroleum Development (Palestine), Ltd., a subsidiary of Petroleum Concessions, Ltd., a company of similar composition to the Iraq Petroleum Company, Ltd., was attracted to the region by a large gas escape and decided to drill a test well near Gaza, but again an intention to drill was overtaken by the outbreak of war. However, drilling eventually commenced in October 1947, but it was suspended in 1948 by the Arab-Jewish troubles. Concessions held by the Petroleum Development (Palestine), Ltd., are mostly in the coastal area and in southern Palestine.

Other concessions in the Dead Sea area are held by the Palestine Potash Syndicate and the Jordan Exploration Co., Ltd., but no deep tests have as yet been drilled. Jebel Usdum (Arabic name for Sodom), a salt dome at the southern end of the Dead Sea, has attracted attention by reason of its structural nature. An active oil seepage occurs on the Dead Sea coast north of the Lisan Peninsula, and gas escapes strongly in the sea nearby. East of Lisan the Nubean sandstone is strongly impregnated with oil residue.

A concession for the oil rights of all Jordan is held by the Trans-Jordan Petroleum Company, Ltd., an associated company of the Iraq Petroleum Company, but exploration has not progressed beyond the geological and geophysical phase.

EGYPT

The impressive oil indications in the Gulf of Suez area drew the attention of oil companies at an early date, but the unusual complexity of structural conditions has resulted in a very high ratio of failures in exploration wells. Only four oil fields of importance have been discovered: Hurghada, Ras Gharib, Sudr, and Asl. Two others, Gemsa and Abu Durba, now abandoned, had a small production.

The early oil exploration in Egypt was almost exclusively a British enterprise, but during recent years American companies have taken an increasing share. Wildcat drilling and exploration surveys were being actively carried out until the outbreak of the second World War, but owing to wartime conditions the Egyptian government granted a moratorium on concession obligations. The exploration programs were recommenced in 1945, and exploration permits have been taken out in extensive areas in the Gulf of Suez region, in central and northern Sinai, and in the Western Desert between the delta and the Libyan frontier.

The Hurghada field, discovered in 1913, is now in an advanced state of decline, and the average output in 1948 was only about 900 b/d. The cumulative production to the end of 1948 was 37½ million. The crude has an average gravity of 22° API and has a high asphalt and sulphur content.

The development of the Ras Gharib field, discovered in 1938, was accelerated by the wartime necessity of oil products for the Middle East campaign. The production in 1948 averaged 25,400 b/d, and the cumulative total at the end of 1948 was about 78 million barrels. The crude is of 26° API gravity. A new discovery was made in 1946 at Sudr, twenty miles southeast of Suez. Commercial production commenced in 1948, and about 3.5 million barrels were produced in that year. Additional discoveries made in 1949, notably at Asl, promise greatly augmented production.

The crude oil from the Hurghada and Ras Gharib fields is shipped by tanker to two refineries at Suez, and the products are distributed from there to the Egyptian market. The principal refinery is operated by the

Anglo-Egyptian Oilfields, Ltd.; the second is a small government-owned refinery in which the government-royalty oil is treated.

The Hurghada and Ras Gharib oil fields are operated by Anglo-Egyptian Oilfields, Ltd., a company in which the Shell and Anglo-Iranian groups have a large holding. The Sudr field is on acreage held jointly by Anglo-Egyptian Oilfields, Ltd., and Socony-Vacuum Oil Company.

Communications

BY SEA

1. Mediterranean. Alexandria and Port Said are first-class well-equipped harbors capable of handling ships of all sizes and discharging heavy lifts. Suez is an anchorage only, and ships must be lightered. Oil tankers carrying crude oil from the Ras Gharib field to Suez refinery discharge through a sea line.

Palestine has only one protected port, Haifa, completed in 1933. It has good discharging facilities and an oil dock, though the tankers loading from the Iraq Petroleum Company terminal do so by sea lines outside the harbor. Interruptions to loading by unfavorable weather average only a few days in the year. Notwithstanding the development of Haifa, some traffic is still handled through Jaffa, but ships must anchor off the port and be lightered.

The chief harbors of Syria and Lebanon are Tripoli and Beirut, the latter being the main port for general cargo and passenger ships, the former the loading point for crude oil from the Syrian terminal of the pipe line from Kirkuk. The ancient harbor town of Sidon (population about 10,000), in the Lebanon Republic, has been chosen as the terminal of the Trans-Arabian pipe line. The terminal of the new Middle East pipe line will be at Tartus, in Syria.

Alexandretta, a small port, is now in Turkish territory.

2. Persian Gulf. The entry to the ports of Abadan and Khorramshahr, in Persia, and of Basra, in Iraq, is via the Shatt-al-Arab, lying between Iraq and Persian territory. The Port of Basra Authority controls the dredging of the bar and pilotage.

The dredged channel limits ships to a draught of 30 feet and a length of 600 feet, but for the best working results the draught should not exceed 25 feet and the length 400 to 500 feet. The port of Basra has ten berths and numerous moorings in midstream. The port of Khorramshahr, in Persia at the mouth of the Karun, was constructed in 1943 by U.S. engineers in order to accelerate the discharge of lend-lease goods to Russia, and a branch railway was built to connect at Ahwaz with the existing Persian main line. Khorramshahr has six berths.

The port of Abadan serves the Anglo-Iranian Company's refinery, and does not handle general cargo other than that serving the requirements of the company. The berths for large ships are all at quays and jetties built out into the river, and there are other tanker jetties at Khosrowabad, thirteen miles downstream.

The port of Bandar Shahpur, constructed as the cargo terminal for the trans-Persian railway, and opened in 1932, was chosen in order that Persia might have a port independent of Iraqi control. It is approached by the deep-water creek of Khor Musa, which has 29.5 feet of water over its bar at high tide. The channel is easy, and the Khor can accommodate many large ships at anchor. The surroundings of the Khor are dreary mud flats, and the Bandar Shahpur berths are at the ends of a 1,000-foot long jetty built on piles. The number of berths was doubled in 1943, to six, to expand the discharging capacity of the port for handling supplies to Russia. The Anglo-Iranian Oil Company has crude-oil loading berths at Bandar Mashur, also in the Khor Musa.

Kuwait Bay, on the Arabian side of the Persian Gulf, has a good sheltered anchorage, and is a possible site for an enclosed harbor. It was, in fact, projected for such development as the terminal of the Berlin-Baghdad railway prior to 1914; but during the 1914–1918 war Basra developed as a modern port, and the Kuwait scheme was not revived subsequently. The British-India Steam Navigation Company ships call at Kuwait weekly, but they must anchor far offshore, and be lightered. The development of the oil field at Burgan in the State of Kuwait has led to the construction of extensive loading facilities at Mena al Ahmadi (formerly known as Fahaheel), twenty-five miles south of Kuwait town and on the Persian Gulf (see Fig. 35).

Bahrein and Ras Tanura refineries, also on the Arabian side, have both moorings and jetties for tanker loading. Elsewhere around the Gulf the small towns are all served from ships lying at open anchorages and discharging into lighters, in some cases the fall of the coast being so small that steamers must lie about two miles offshore, as, for example, at Bandar Abbas. The British-India Company ships make a weekly tour of the Gulf towns on a journey between Basra and Karachi, an institution aptly named the "slow mail," in contrast to the "fast mail" steamers which make a more direct sailing.

BY RIVER

Navigable rivers throughout the Middle East are the Nile, the Tigris, the Euphrates, and the Karun, but these can be used only by small river craft. The Tigris is navigable as far as Baghdad, 549 miles from its mouth, by river steamers with draughts of 4½ feet in high-river season and 4 feet in low. The steamers usually tow two barges alongside, the journey upstream taking from four to fourteen days according to the state of the river. Small steamers of 3-foot draught can reach Mosul, 294 miles above Baghdad, in high-water season, though with difficulty. The Euphrates is too shallow for steamers, but native sailing craft carry a fair volume of traffic.

The Karun is navigable by river steamers from its confluence with the Shatt-al-Arab up to Ahwaz, where there are rapids. The Anglo-Iranian Oil Company moved much of its material for the oil field of Masjid-i-Sulaiman by such steamers and by towed barges. The material had to be discharged at Ahwaz and moved by road transport to a point above the rapids, where it was loaded again into barges and towed upstream to Dar-i-Khazineh, there to be transferred to a narrow-gauge railway for the oil-field center. This system has now been abandoned in favor of road transport.

BY RAILWAY

Egypt has a railway network serving most of the towns in the Delta and the Suez Canal area, with one line extending upstream to Luxor. The Egyptian system connects with the Palestine railway at El Qantara on the Suez Canal, whence the line crosses northern Sinai to Gaza, Lydda, and Haifa. A line to Jerusalem branches off at Lydda. In 1944 the railway was extended along the coast from Haifa to connect with the Syrian and Turkish railway system, and it will thus be possible eventually to travel by rail from Europe to Egypt by the Simplon-Orient route via Syria. A narrow-gauge cog railway connects Beirut with Damascus, crossing the Lebanon and Anti-Lebanon mountain ranges.

In the years before 1914 Damascus was connected with Medina by a railway built specially as the Moslem pilgrim route, but as a result of the activities of T. E. Lawrence most of the bridges were spoilt during World War I and have not since been repaired. The line now operates as far as Ma'an, but a very occasional train satisfies the traffic demand. This railway line is connected with Haifa by a branch line with a junction at Der'a.

Iraq has railways of two gauges, a standard European gauge from Baghdad to the Turkish frontier near Nisibin, and so connecting via Aleppo with the Anatolian railway; and a meter-gauge railway from Baghdad to Basra and from Baghdad to Kirkuk, with an extension to Erbil now under construction. The meter-gauge line was built by British Army engineers during World War I, utilizing track and rolling stock from India; in the circumstances of the time the problem of uniformity of gauge with a Turkish railway did not arise. The German plan of a railway connection between Berlin and Baghdad was in its early stages of development, and in 1914 the railway had only been completed between Baiji in the north and Baghdad.

The Trans-Persian railway was planned by Shah Reza Pahlavi as a means of making Persia economically independent of Russia, by giving an alternative export route for the products of the rich Caspian provinces. But the railway, built between the years 1927 and 1938, was an invaluable communication between the Anglo-American forces and Russia, and the combined invasion of Persia by these powers to counter German schemes led to the Shah's downfall.

8. THE MIDDLE EAST

Climate and Vegetation

The climate is a substantial handicap to the petroleum industry in that British and American technical personnel require special inducements and more elaborate amenities to compensate for the hardship of the extreme summer heat.

The climate of the Middle East is influenced by its situation as a marginal but integral part of the greatest land mass of the world. In spite of the proximity of two large seas, climatic conditions are markedly continental and show extreme seasonal variation, which is modified only in a restricted coastal zone. However, although Asiatic influences predominate, in winter and spring air currents originating in Europe give spells of characteristically changeable weather, and from time to time, also, a "breath of desert" spreads across the region from northeast Africa.

Summer conditions are controlled by the development of an area of intense low pressure over the southern Persian Gulf and northwest India. A minor but persistent center of low pressure also occurs over Cyprus, and serves to prolong the zone of low pressure as far as the Aegean Sea and Cyrenaica. The air currents produced by the low pressure move from northern India over southwest Persia, thence northwestward over northwest Persia to Turkey, turning southward over the Aegean, then eastward over the Levant, returning to the center of low pressure as westerly winds in Syria, and northwesterly winds in Iraq. The moisture held by these winds is deposited in India, and for the remainder of the track there is no rainfall of any description, beyond extremely occasional thunderstorms in mountainous localities close to the sea.

Winter conditions are more variable. From time to time active depressions move eastward along the Mediterranean, and continue southeastward, crossing Syria or Palestine and Iraq to reach the Persian Gulf; or they may take a more northerly track over Turkey and the Black Sea. Such depressions give by far the greater part of Middle Eastern rainfall, which, however, is unevenly distributed, owing to the topography.

The influence of topography is well seen in the case of Egypt, which, although from time to time in the track of depressions, may not receive any rain inland for several years, and normally has less than 10 inches on the coast. The westward-facing mountains and plateaus of the eastern Mediterranean littoral receive an unduly high proportion of rain, at the expense of the areas to the east, which lie in a rain shadow and are thus barren steppe land merging rapidly into desert. Similarly, the Zagros Mountains, which, although further from the sea, are higher and more continuous, receive a heavier rainfall and, in turn, create a second rain shadow area in central Persia. As much as 50 inches of rain may fall in the mountain areas, but no more than 5 or 10 inches in the plains and lower plateaus of the rain-shadow areas.

Rainfall may be heavy, but is usually of short duration, with rapid onset and equally rapid clearance; and the amount received varies considerably from one year to another. For example, in Beirut, which has an annual average of 36 inches, only 13 inches fell during 1943. The average rainfall in the coastal zone between Kuwait and the entrance to the Persian Gulf is under 5 inches per year, and some years may be completely rainless.

Because of the absence of cloud in summer, temperatures reach greater values than are normal in the wet tropics. Although the lands bordering the Mediterranean are influenced by the cooling effect of breezes from the sea, which are more strongly developed than those of temperate latitudes, further inland day temperatures rise to 110° to 120°F and exceptionally to 128°F in the shade, with sun temperatures of the order of 180°F. Fortunately, there is a substantial fall in the hours of darkness, at least in the more open inland country. In coastal regions very high humidity, in some places at a maximum in summer, adds greatly to the discomfort of summer conditions, but humidity is low in areas remote from the sea. In many areas "dust devils"—swirling columns of dust due to intense local heating, sometimes reaching 6,000 feet in height—are a striking but minor feature. In Iraq blowing dust may reduce visibility for days at a time.

Winter temperatures remain fairly high near the coast, as the result of the proximity of warm sea water; but inland, increasing continentality, reinforced in many cases by the effects of altitude, gives low temperatures, with many periods of frost. Damascus, at an altitude of 3,000 feet, has an average night minimum temperature of 30°F during January, and in most areas of the Zagros Mountains and the Persian plateau the average minimum is even lower.

In winter and early spring, depressions are often followed by waves of cold air from eastern Europe. Snow may fall, succeeded by cold clear conditions; thunderstorms are a common accompaniment. The mountain zones of Syria, Kurdistan, and Persia receive heavy snowfalls which disrupt communications; but snowfall is exceptional in the plains of Iraq and central Persia. However, no part of the Middle East is entirely free from it. Summer heat melts most of the snow on the mountains, with the exception of a few drifts on the northern slopes of the higher ranges. A permanent glacier was recorded [12] in the Bakhtiari mountains of southwest Persia, but a subsequent observer, Mr. N. L. Falcon, found only a snow field and extensive scree deposits.

The mountain ranges of the Zagros carry sparse vegetation. The highest zones are above the tree level, but at intermediate heights scrub oak can grow where spared by man and his flocks. Conifers are limited to isolated localities, as in the vicinity of Dohuk in Iraq Kurdistan, where the Aleppo pine (*Pinus halepensis*) forms local forests. A feature of much of the

[12] Desio, A., "Appunti geografici e geologici sulla catena dello Zardeh Kuh in Persia," *Mem. Geol. e Geografiche di Giotti Dainelli,* Vol. IV, 1934.

mountain country is the number of plants adapted to dry conditions (*xerophytes*), including members of the sage group and numerous kinds of thistles. In early summer the mountain sides are bright with wild flowers, narcissus, tulip, anemone, wild onion, fritillary, grape-hyacinth, iris, crocus, and many others. Hollyhocks and rhododendrons are a feature of many mountain valleys.

In the Iraq and Persian foothills vegetation is much sparser than in the mountain country, by reason of lower rainfall and hotter summers. In years of early rainfall the hill slopes and valley bottoms may acquire a thin grassy verdure with an attractive color pattern of wild flowers, among which the most common are red, white, and mauve anemones and red tulips; but by April or May increasing heat and drought dry out both soil and vegetation, the last survivors being various thistles and thorny plants. The spoliation of tribesmen and villagers has denuded the foothill region of almost all trees, and even of most bushes; but the occasional preservation of a clump of trees sanctified by some holy shrine demonstrates that tree life is possible, given sufficient encouragement and protection. Scrub oak, myrtle, juniper, maple, almond, an acacia with the Persian name of *Kunar,* tamarisk, and several other trees can achieve an existence, but the very small number of these serves only to emphasize more strongly the general barrenness of the hills.

The vegetation of the unirrigated plains of Iraq is uninspiring. They are completely treeless. In places a gorse-like camel thorn grows freely, also a small much-branched prickly shrub with reddish pea-shaped flowers. Licorice grows in certain areas, and the roots are collected in the Basra vicinity. In saline areas halophytic plants are common, but there are many patches in the plains where a salty crust prohibits the growth even of such plants.

The extensive marshes of lower Iraq are a world of their own. They are occupied by dense communities of a perennial grass, *Phragnitis communis,* and a bulrush, *Typha angustala,* growing up to ten feet in height and forming dense thickets. Willows, the Euphrates poplar, and tamarisk are common, though almost everywhere stunted in growth through cutting for fuel. The grasses and reeds of the marshes are used for matting, for hutmaking, and in many other ways, so that their manufacture forms an important part in the economy of the amphibious marsh Arabs.

On the Arabian coasts of the Persian Gulf vegetation is so sparse that only the Arab camel can make a living. Palgrave [13] painted a vivid pen picture of Qatar peninsula during a visit in 1862–1863 as follows:

"To have an idea of Katar, my readers must figure to themselves miles on miles of low barren hills, bleak and sun-scorched, with hardly a single tree to vary their dry monotonous outline: below these a muddy beach extends for a quarter of a mile seaward in slimy quicksands, bordered by a

[13] Palgrave, W. H., *One Year's Journey Through Central & Eastern Arabia,* 2 vols., London, 1865.

rim of sludge and seaweed. If we look landwards beyond the hills, we see what by extreme courtesy may be called pasture land, dreary downs with twenty pebbles for every blade of grass."

Agriculture and Industry

It follows from the description of climate and topography that the main density of population is in areas capable of irrigation. In consequence of the low winter rainfall and absence of summer rains, the greater part of the Middle East has a short growing season. Agriculture, especially in the drier areas, therefore tends to become precarious, a fact of great importance in the human geography of the region. Wheat and barley cultivation in unirrigated land depends for a successful crop on a good season of rain, and complete or partial crop failures may be expected perhaps one year in five.

Cultivation by irrigation provides the real prosperity of the Middle East, and the lands of the great rivers are the most favored. In Egypt the flood waters of the Nile, due to summer rains in Abyssinia and central Africa, give a yearly inundation and enrichment to the fields of the delta. The heavy crops of wheat, rice, sugar cane, and cotton form the basis of the country's wealth. In Iraq—or Mesopotamia, "the Land of the Rivers" —however, the Tigris' and Euphrates' yearly floods arise from heavy winter rain and from the spring melting of snow in the Anatolian highlands, and the flood waters pour into the extensive marsh areas in lower Iraq between Basra and Amarah, largely useless for irrigation. In central and upper Iraq the rivers are deeply entrenched, and the water in the summer time must be lifted ten or even twenty feet to irrigate the adjacent areas.

At the period of Iraq's greatest prosperity, from the times of the Babylonians to the twelfth century, a highly developed system of canals traversed the country, but the headwaters of these canals were far upstream of the area of their deployment, and their maintenance required constant attention. A traveler by air across Iraq cannot fail to observe the pattern of these canals, with their elevated banks resulting from centuries of silt clearing. Modern Iraq has few and relatively unimportant canals of the ancient type, one such being on the Euphrates, fed from the Hindiya Barrage (built in 1912). Recently a barrage at Kut has diverted part of the Tigris into a branch course, and thence to some canal systems. The modern trend in Iraq and in the Persian plains of Khuzistan is towards pumping installations, using cheap oil as fuel; a large number of small canals fed in this way are taking the place of the major units of the old regime.

In lower Iraq crops of rice are either irrigated by canals or planted around the margins of the extensive marsh areas from which the water recedes in late summer as the ripening season commences. In this way excellent crops are obtained by a notable economy of effort, which from an Arab point of view is wholly estimable. In the lowest reach of the river below Qurna, where the Euphrates and Tigris unite to form the Shatt-al-Arab, the banks of the Shatt between Qurna and Fao carry extensive date gardens, the produce of which is world famous. The date groves are watered by a system of small canals leading off from the Shatt. The river in this section is tidal, with a rise and fall of as much as four feet, and as the river banks are only slightly above the high-water level, the twice daily rise of the tide gives the palms their daily water.

In irrigated gardens date palms are dominant. In Baquba and Khanaqin oranges are grown extensively, also apricots, grapes, and other fruits. Bananas are grown on a small scale in Basra, where the shelter of the date gardens protects them from the worst of the cold winds of winter.

The economic level of the inhabitants in areas remote from perennial water is marginal. The nomadic tribal system has evolved from the natural circumstances, and flocks of sheep, goats, or camels are driven over great areas in search of sufficient pasture. The black goats' hair tents of the tribesmen and all their apparatus of life are packed on camels or donkeys during the frequent moves. In Iraq and Persia the nomadic tribes among Kurds and Lurs spend the winter months in the low-lying plains and foothills; then, as the onset of early summer dries up the pastures, they drive their flocks and herds into the higher mountain country, where cooler temperatures and later rainfall give them richer grazing through the summer months. The Arab nomadic tribes also roam over great areas in search of grazing, though they lack the strong contrast between low country and highlands. The camel carries in its hump and the sheep in its tail an emergency ration of fat as an insurance against occasional lean periods.

The desert population is scanty in the extreme, but money to be earned will always attract sufficient labor. The great shortage for the petroleum industry is in the skilled and semi-skilled categories: clerks, fitters, turners, drivers, etc. These must be specially trained by the oil companies, and it takes years before sufficient numbers are available in any new area. The Anglo-Iranian Oil Company runs a Technical College, Apprentice Schools, and artisan training schemes to ensure its requirements of skilled employees.

Until the discovery of oil, the only important industry of the Persian Gulf was pearling. The best pearl banks are in the shallow waters off the Trucial Coast, east of Qatar Peninsula. Dhows are sent during the season from nearly all towns on both sides of the Gulf to join in the pearling, but the great pearl market is Bahrein. The Persian Gulf pearls are world famous for their size and quality, and in good years the value of the annual catch has been as much as £220,000. The oyster catching is entirely manual, without the aid of any diving apparatus, other than a nose clip and a sinker stone; and the conditions are very exacting to the divers. Sharks are common, but they are less feared than the evil saw-fish, though actually casualties from such causes are surprisingly infrequent.

Mining and manufacturing throughout the Middle East are at present negligible in importance. Kuwait town, developed as a trading port serving the Arabian hinterland, is also famous for its shipbuilding, the teak for which is imported from India; the dhows of Kuwait are used for trade and for pearling. Palestine is, of course, the most industrialized country of the Middle East, but even here the only important industries up to the present have been potash extraction from the Dead Sea and cement manufacture at Haifa, though the oil refining at Haifa is also a locally important industry. Iraq has no mining or industry worthy of the name. Persia has during recent years attempted to stimulate local manufactures, and there are in existence a few sugar refineries, cotton and woolen mills, a tobacco factory, and a cement works, but all in all they are not on a scale to give it any great status as an industrial country.

Table 16. The Middle Eastern countries: estimated areas and populations at midyear, 1948.

	Area	Population
	(In square miles)	(In thousands)
Persian Gulf and Arabian Peninsula		
Persia	630,000	17,000 [ab]
Iraq	116,000–175,000	4,950 [b]
Saudi Arabia (Nejd, Hejaz, Asir)	913,000–927,000	6,000 [b]
Kuwait	6,000	120 [a]
Bahrein	200	125 [b]
Qatar	8,000	16 [a]
Trucial Coast Shaikhdoms	6,000	105 [a]
Muscat & Oman	82,000	830 [b]
Yemen	. . .	7,000 [b]
Aden Colony	115	81 [a]
Aden Settlement	80	
Aden Islands (Kuria Muria & Perim)	35	
Aden Protectorate	112,000	650 [a]
Mediterranean		
Turkey in Asia	285,250	17,930 [ab]
Syria	54,000	3,750 [b]
Lebanon	3,600	1,208 [ab]
Palestine	10,400	2,000
Israel	. . .	713 [a]
Palestine, Other	. . .	1,287 [b]
Jordan	30,000	400 [b]
Egypt	363,200–386,100	19,528 [a]
Cyprus	3,600	460 [a]

[a] Figures from official statistical sources of each country.
[ab] Figures estimated by United Nations and approved by country.
[b] Figures estimated by United Nations, or taken from unofficial sources.
Note: In certain cases where boundary claims are in conflict, alternative estimates of area are given.
Sources: Areas from *Whitaker's Almanack, 1949,* pp. 187–190, and supplied by the author from private sources; populations from "Population and Vital Statistics Reports," United Nations Statistical Office, *Statistical Papers Ser. A 7–8,* Aug. 1, 1949.

The average level of prosperity throughout the Middle East is very low, and in no section is the local market for petroleum products of real importance. Kerosene sales rank ahead of gasoline. Because in most of the oil-field areas communications were non-existent, the operating companies have had to develop their own system of roads. An exception to this generalization is the Kirkuk oil field in Iraq, which is served by the Iraq railway system, and where the only company-built roads are within the confines of the oil field itself. However, the war years have left a greatly improved inheritance of motor roads, and doubtless during the years to come there will be considerable increase in the number of automobiles and trucks in use.

The Political Background

PERSIA

For 2,500 years Persia has possessed an educated ruling class and a tradition of civilization which has survived the most barbarous invasions and the most alien conquests. Time after time it has been invaded and conquered, but each time it has managed either to eject or to assimilate its conquerors, and to emerge independent. The country is in reality a federation of provinces differing in speech and custom and lacking a traditional capital, the unifying principle being a monarchic system. The present capital, Tehran, became so only under the Qajars, the dynasty which preceded the reigning Pahlavi dynasty. Throughout the centuries Persia has demonstrated a capacity for balancing on the brink of administrative chaos and yet being steadied time after time by a new strong hand.

Modern Western thought has had its effect on Persia, and in the early years of the present century there was agitation for a constitution and for government by an elected Parliament, or Majlis. These reforms were granted nominally in 1906, but the early developments were confused by provincial rebellions and by foreign influence.

In the 1914–1918 war Persia remained neutral, though the armies of Russia, Turkey, and Great Britain campaigned across her territory. After the war an outstanding personality, Riza Khan, rose to power through control of the army. In 1923 he became Premier, and in 1925 Dictator; later, following the deposition of Ahmad Shah, he became Shah under the title Riza Shah Pahlavi.

The despotic and oppressive rule of Riza Shah has to its credit many material achievements: the construction of the Trans-Iranian railway without foreign financial aid, the building of many industrial factories for cotton and woolen goods, sugar, tobacco, etc. and of many modern administrative offices, hospitals, and museums. However, it would seem that the tempo was too fast. Education and social reform were energetically pressed on, but the taxes necessary to maintain them kept the mass of the people in a state of great poverty.

As has been seen, the Trans-Iranian railway, which was conceived from a desire to make Persia economically independent of Russia, proved to be Riza Shah's undoing. After the invasion of Russia by Germany in 1941,

Persia became an essential line of communication for British and American supplies to the Soviet Union, and a military invasion and occupation followed. Riza Shah abdicated, and his son, Muhammad Riza, became Shah. Persia later declared war on the Axis Powers, and thus qualified for membership in the United Nations.

The Majlis has gained in authority since the disappearance from the scene of Riza Shah, but its willingness to change with changing times has been slight. The tribal system was greatly weakened by Riza Shah, but has subsequently made some recovery. The great mass of the people still live under feudal conditions, and the many minority constituents of the Persian state receive scant sympathy from the central government, which has attempted to standardize the governmental machine into a Persian pattern regardless of whether the population of any particular province is of Arab, Turk, or Kurd stock.

IRAQ

After its liberation from the Turks by British Forces in 1918, Iraq was administered directly by British officials, pending a decision on its final state. The League of Nations placed the country under British mandate in 1919, and Great Britain undertook to allow Iraq self-government by degrees. The process was accelerated by an Arab rebellion in 1920, following which Emir Faisal, eldest son of King Husain of the Hejaz, was created King of Iraq in 1921, and the independence of the State recognized. Iraq is today a constitutional monarchy under King Faisal II (born 1935), with his maternal uncle, Amir Abdul-Illah, as regent. The government consists of a Senate and a Chamber of Deputies, the members of the former being nominated by the King and the latter being elected. The administration rests in the hands of a cabinet of eight or nine members under a Prime Minister.

The population of Iraq is predominantly Arab, but its territory includes several "minority" areas. These have been a source of trouble to the State throughout its short existence, owing to the inability of the Iraqi officials to grant a sufficient degree of local self-government to satisfy local feeling. Kurdistan lies in the mountain belt along the northern border of Iraq, but it extends into both Persia and Turkey, and none of these three powers has been able to demonstrate sufficiently generous statesmanship towards the Kurds to avoid periodic insurrections occasioned by reasonable grievances. At times there has been a movement towards Kurdish unity and autonomy, but geographical circumstances would render this difficult or impossible of realization. The oil field of Kirkuk lies along the southwestern margin of Kurdish country, although the people of Kirkuk town are of Turkish stock.

After the outbreak of war in 1939, Iraq broke off relations with Germany, but Axis influence gained importance after the fall of France with, in general belief, the certain defeat of Great Britain to follow. Great Britain

54. Persia. An engraved brass tray depicting petroleum warfare in the mid-eighteenth century. Nadir Shah f Persia is attacking an Indian army by driving "fire camels" against the elephants. (Anglo-Iranian Oil Comany, Ltd.)

5. Persia. Aerial view of Masjid-i-Sulaiman oil field, showing administrative center and houses. (Anglo-Iranian Oil Company, Ltd., photo by Hunting Aerosurveys, Ltd.)

56. Persia. Aerial view of part of Masjid-i-Sulaiman oil field showing Gas Absorption and Topping plant
(Anglo-Iranian Oil Company, Ltd., photo by Hunting Aerosurveys, Ltd.)

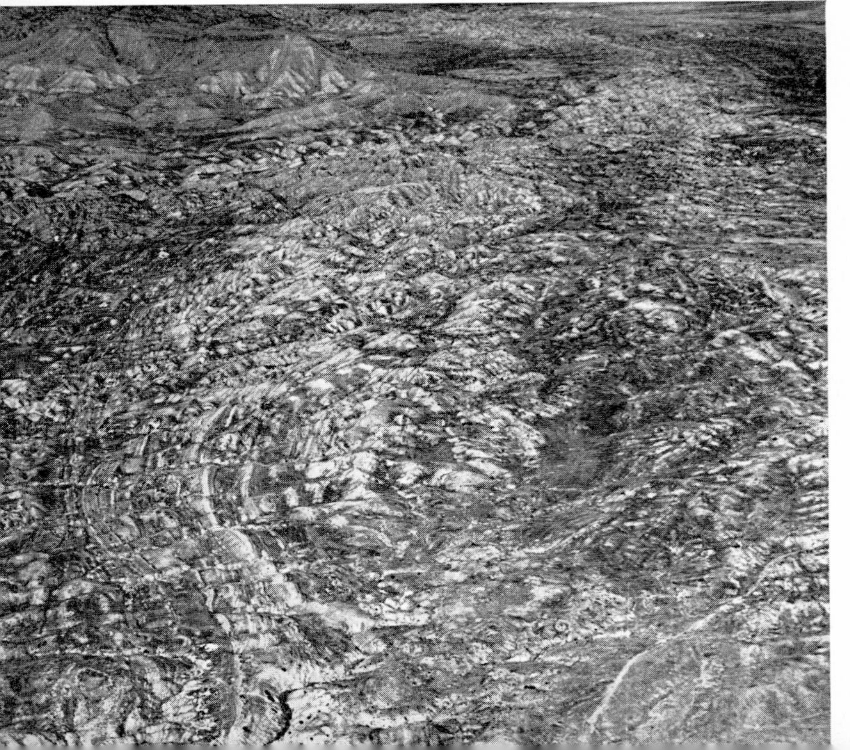

57. Persia. Aerial view of G
Saran area, showing the confu
gypsum strata of Lower Fars a
an outlier of Middle and Up
Fars in top center of photo. (Ang
Iranian Oil Company, Ltd., ph
by Hunting Aerosurveys, Ltd.)

58. Iraq. Baba Gurgur, Kirkuk, stabilization plant and residential area. (Iraq Petroleum Company.)

59. Persia. Kuh-i-Namak salt plug, east of Bushire. The salt extrudes through Cretaceous limestone, and the flow of salt downhill at either side of the plug can be seen. The height of the top of the salt is about 4,500 feet above sea level, and the relative height above the plain in the foreground is about 3,500 feet. (Anglo-Iranian Oil Company, Ltd., photo by Hunting Aerosurveys, Ltd.)

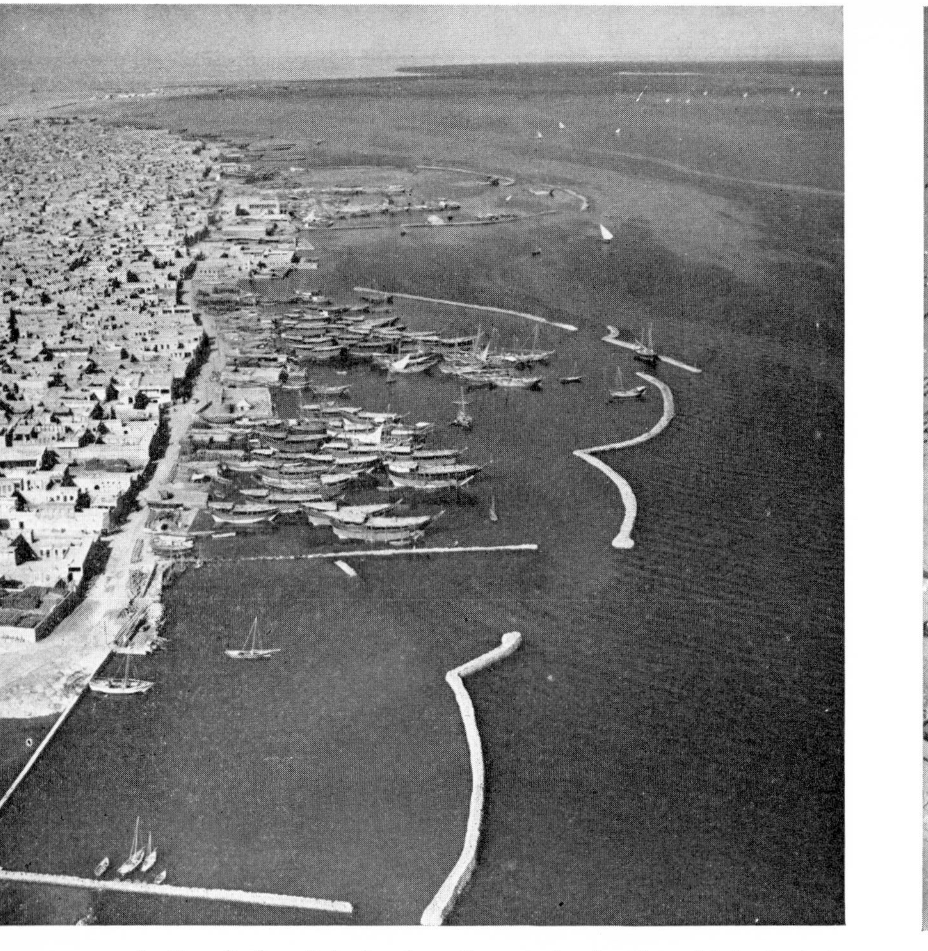

60. Kuwait. Kuwait harbor from the east, showing Shuwaikh in the background. (Hunting Aerosurveys, Ltd.)

61. Kuwait. Construction camp at Ahmadi, from the south. (Hunting Aerosurveys, Ltd.)

had, by earlier treaty with Iraq, the right to pass troops through the country in time of war, but when she attempted to do so in April 1941, in order to counter German infiltration, her troops and the air base at Habbaniya were attacked by Iraqi forces. But German air-borne assistance was too little and too late. Iraq was occupied by British troops, and extensive bases were formed against a possible German break-through, whether from North Africa to Egypt or through the Caucasus, but neither eventuality materialized. Iraq later declared war on Germany and became a member of the United Nations.

SAUDI ARABIA

Turkish authority in Arabia before 1914 had been more nominal than real, and after the defeat of Turkey the Ottoman Empire dissolved into its component parts. Arabia being in the main divided into two kingdoms, Nejd and the Hejaz. The Wahhabi power, which first arose in Nejd under the leadership of the Saudi family in the late eighteenth century, gained fresh impetus in the twentieth from the appearance of a great personality, Abdul Aziz Ibn Saud. By defeating his central Arabian rival, Ibn Rashid, in 1918–1921, and King Husain of the Hejaz in 1923–1924, Ibn Saud created the modern state of Saudi Arabia. His government is patriarchal in character, and well suited to the tribal conditions prevailing throughout the greater part of the country. The center of government is at Mecca, but as no unbeliever may enter this holy city, all contacts with the outside world are through foreign embassies at Jidda. Riyadh, the capital of Nejd, from which Ibn Saud extended his power, is also an administrative center of importance.

Saudi Arabia has concluded treaties of friendship with the neighboring Arab states—Iraq, Yemen, Egypt, Syria, and Jordan—and during recent years there has been a movement towards some degree of Arab federation. So far, the only progress has been the formation of an Arab League, and further development on these lines has been hampered by conflicting views among these friendly and racially allied states as to which should take the leadership.

Saudi Arabia remained neutral throughout the second World War, though Ibn Saud's sympathies were undoubtedly with the Allied powers. He finally declared war on Germany in 1945, in time to qualify for inclusion among the United Nations.

KUWAIT

The Shaikhdom of Kuwait formed part of the Turkish Empire until 1899, when it gained its independence and entered into treaty relationship with the British government. International competition for its harbor rights was keen around the turn of the century, when Russia was interested in establishing a port or coaling station. In 1900 Kuwait was visited by a German Railway Commission in quest of a suitable terminus for the projected Berlin-Baghdad Railway, and in 1902–1903 it was visited by French and Russian cruisers. Thus this small desert corner of the Persian Gulf became the scene of considerable international rivalry.[14] The ruling Shaikh, Shaikh Mubarak, accepted British protection, and the British were careful not to violate his independence. During the war with Turkey in 1914–1918 the British Forces developed Basra as their port of entry into Iraq, and with the construction of the Basra-Baghdad railway the possible importance of Kuwait as a railway terminal ceased. But in 1946 it again achieved importance, this time as an oil-producing state. As such it ranked seventh among the countries of the world in 1948.

BAHREIN

From the sixteenth to the early part of the nineteenth century Bahrein was occupied or claimed successively by Portuguese, Persians, Muscat Arabs, Wahhabi Arabs, and Turks. Bahrein first entered into treaty relationship with Great Britain in 1820, and in 1861 a convention was signed whereby the Shaikh promised to abstain from "the prosecution of war, piracy and slavery by sea" in return for the support of the British government against external aggression. When the Turks annexed Hasa in 1871 they were prevented by this treaty from including Bahrein among their conquests. Since then the history of the island has been relatively uneventful, excepting perhaps some domestic quarrels and, in 1895, an invasion from Qatar. In 1913 Turkey formally renounced her claim on Bahrein when she signed the Anglo-Turkish Convention of that date.

QATAR AND THE TRUCIAL COAST

The Shaikhdoms of Qatar, Abu Dhabi, Dibai, Sharjah, Ras al Kaima, Ajman, Kalba, and Umm-al-Qaiwain, form the Trucial Coast of the Persian Gulf—so called when they entered into treaty relationships with Britain in the middle of the last century, under the terms of which they undertook to abstain from slave-trading and other unlawful practices at sea. In return the British government undertook to recognize and protect their independence.

OMAN

The sultanate of Oman was a source of much international friction during the nineteenth century. Britain endeavored, in spite of great difficulty, to suppress slave-trading and gun-running, but in doing so narrowly escaped complications at various times with France, Portugal, and Turkey. Disagreement between France and Britain over Muscat was finally referred to the arbitration of the Hague Tribunal, which declared in 1905 in favor of the British case.

[14] For further detail on Kuwait and other Gulf Shaikhdoms, see *The Persian Gulf* by Sir A. T. Wilson, Oxford, 1928, pp. 151–152.

8. THE MIDDLE EAST

SYRIA AND LEBANON

France received a mandate for Syria from the League of Nations in 1919, and at first attempted to rule the country by direct French administration. Arab aspirations for independence gave rise to many disturbances, and France endeavored to compromise and allow a considerable degree of autonomy. In 1941 British and Free French forces invaded Syria to prevent German use, with the connivance of the Vichy French, of air bases for operations in Iraq. Shortly after, the Free French movement undertook to allow Syria complete independence, and in 1945 this independence as a republic became a fact.

Under French administration the Lebanon was treated as a province and separated from Syria, to which it had been attached under the Ottoman Turks. It was granted a certain measure of local autonomy, and in 1945, parallel with developments in Syria, Lebanon declared itself an independent republic.

PALESTINE

Palestine was placed under British mandate by the League of Nations in 1919, but its subsequent history was grievously disturbed because of the conflicting interests of Arabs and Jews. The government was a direct British administration under a governor nominated by the British cabinet for a period of five years. Attempts to broaden the basis of the administration to include Arab and Jewish members failed, because of the inability of the two races to cooperate, until in 1947 the British government announced its intentions of relinquishing its mandate to the United Nations. The United Nations Assembly decided to partition the country between the Arabs and Jews, creating for each an independent state.

JORDAN

Trans-Jordan was placed under British mandate after the breakup of the Turkish Empire; and Abdulla, son of King Husain of the Hejaz, was appointed Emir in 1921. Its independence was recognized by the British government in 1927, though by treaty Britain retained certain rights. Its complete independence was attained in 1946, and in 1949 it assumed the name of Jordan (Hashemite Jordan Kingdom).

EGYPT

Egypt was until 1914 nominally a tributary state of the Turkish Empire, ruled by a Khedive appointed by the Sultan, but on the outbreak of World War I Great Britain declared a protectorate over Egypt and defended it from Turkish attacks. After World War I Egypt became an independent sovereign state under King Fuad, though Great Britain retained special rights, later regulated by treaty, for the defence of the Suez Canal Zone and other matters. The government is by a cabinet of ministers appointed by the King, in concurrence with a parliament. The parliament consists of two houses: a senate, of which two-fifths of the members are nominated, and a chamber of elected deputies. The present ruler is King Farouk, son of King Fuad.

9. SAUDI ARABIA AND BAHREIN

BY MAX STEINEKE * AND M. P. YACKEL **

THE oil province of the Persian Gulf Basin, as referred to in this article, includes the Tigris-Euphrates Valley, the Persian Gulf, and the Rub' al Khali desert. In view of the widespread favorable signs of oil in a region where the sedimentary section attains a thickness of 50,000 feet, this basin is one of the greatest oil basins in the world, if not the greatest. It comprises an area of roughly 600,000 square miles: its length, from the foothills of northern Iraq to Wadi Hadhramaut in southern Arabia, is approximately 1,800 miles; and its width varies from 200 miles in the Tigris-Euphrates valley to 450 miles in the Rub' al Khali.

In the north are the oil fields and oil seepages of Iran and Iraq. In the central portion are the oil fields in the coastal area of the western Persian Gulf, extending from Kuwait to the Qatar Peninsula. The great Rub' al Khali of southeastern Arabia is still unexplored, but this part of the Persian Gulf Basin may contain a favorable stratigraphic section for the accumulation of oil, and oil shales are reported along its southwestern extremity in the Wadi Hadhramaut area.

Prior to 1931 all the proved oil fields in the Persian Gulf Basin were in Iran and Iraq, on the highly folded foothill belt between the main portion of the Iranian ranges and the great depression comprising the Tigris-Euphrates valley and the Persian Gulf. This area was then the focal point of Middle East oil; the numerous oil and gas seepages of Iran and Iraq evidenced the huge stores of petroleum trapped beneath the surface. The abundant production from drilled wells came from Lower Tertiary strata; the underlying Mesozoic strata were not productive.

Farther south, on the western side of the Persian Gulf a few small oil seepages were found, but the Tertiary sediments, productive in Iran and Iraq, occurred in strata of minor thicknesses. For this reason the eastern side of the Arabian Peninsula was long regarded as of minor importance as an oil-bearing region. Bahrein Island was known to contain an ideally shaped structural feature for the accumulation of oil, but the Middle Eocene sediments exposed were lower stratigraphically than those from which production had been obtained on the eastern side of the Persian Gulf. Despite this fact, the Bahrein Petroleum Company, Ltd., a

* Consulting Geologist, The Arabian American Oil Company.
** Geologist, The Arabian American Oil Company.

Acknowledgements: It is impossible for the authors to enumerate all the Aramco and Bapco employees to whom they are indebted for the information used in this article. Mr. Roy Lebkicher, Mr. G. W. Rentz, and Mr. T. C. Barger have made outstanding contributions, however, and their aid is gratefully acknowledged.

subsidiary of Standard Oil of California, drilled a well in the central portion of the island, and commercial production from Middle Cretaceous sediments was realized on May 31, 1932.

In 1933 Standard Oil of California obtained an oil concession in eastern Saudi Arabia from King Ibn Saud, and exploration started the same year. The first oil in Arabia was discovered in August 1936 at Dhahran (Damman field), which is in sight and only 25 miles west of Bahrein Island. The zone encountered was at the same stratigraphic position as at Bahrein, but oil was not found in paying quantities. The results of drilling nine additional wells elsewhere on the structure were very discouraging.

Later on, however, by drilling 2,000 feet deeper, more prolific zones were discovered in Upper Jurassic limestones, and commercial production was a certainty when a well was brought in on August 30, 1938. Since that time, six other fields have been discovered: Abu Hadriya, 100 miles northwest of Dhahran, in 1939; Abqaiq,[1] 40 miles southwest of Dhahran, in 1940; Qatif, 14 miles northwest of Dhahran, in 1945; and Fadhili, 71 miles northwest of Dhahran, Ain Dar, 63 miles southwest of Dhahran, and Haradh, 156 miles southwest of Dhahran, in 1948 and 1949. Each of these fields will be discussed later.

In addition to the developments in Saudi Arabia the very prolific Burghan field was discovered in Kuwait in 1938, and the Dukhan field in Qatar Peninsula was discovered in 1940. The Kuwait concession is operated by the Kuwait Oil Company, Ltd., which is owned jointly by the Anglo-Iranian Oil Company and the Gulf Exploration Company; and the Dukhan field by Petroleum Development (Qatar), Ltd.

Before considering in detail this oil-rich region bordering the western side of the Persian Gulf, a brief review of the topography and structure of the Arabian Peninsula as a whole may be useful.[2]

[1] Three wells were drilled in the Buqqa sub-field, which is a northern extension of the Abqaiq field, in 1947. They are not being produced at present because there are no production facilities installed in the area.

[2] The authors have not attempted to discuss the climate of the Arabian Peninsula. Accurate weather observations over a considerable period are available for only a few localities. The climate of the Middle East as a whole has been described by Dr. Lees (Part III, Chapter 8), and the weather of certain sectors has been recorded by Musil, Doughty, and Philby, in the works listed in the bibliography. Anyone interested in the effect of desert conditions on human efficiency should read the excellent physiological analysis by Adolf.

Generally the weather of the Persian Gulf is relatively cool from the end of October until the end of April. In January and February, the coolest months, there is a slight amount of rainfall. July is the hottest month: in 1946 the averaged daily maximum at Ras Tanura was 101°F. The Arabs have successfully adjusted to this climate for thousands of years; Americans and Europeans adjust to it quite well, with the help of air-conditioned homes and dormitories. The climate of the interior is, of course, very different, with freezing temperatures in the north in winter.

The Arabian Peninsula as a Whole

The Arabian Peninsula is a rough trapezoid, with its longer axis oriented in a northwest-southeast direction, bounded on the northeast by the Gulf of Oman and the Persian Gulf, on the north by the Syrian Desert, on the southwest by the Red Sea, and on the south by the Gulf of Aden and the Arabian Sea. The total area of the peninsula is slightly over one million square miles, or about one-third the size of the United States.

TOPOGRAPHY AND STRUCTURE

There are two marked characteristics of the geomorphology of the Arabian Peninsula: all the mountainous areas occur on its periphery, and more than 90% of the drainage is toward the Persian Gulf or the Rub' al Khali basin.

The western half of the peninsula is made up largely of exposed igneous and metamorphic rocks, but in the eastern half these are covered by a tremendous thickness of sedimentary rocks, ranging in age from early Paleozoic to Recent. While the western half has good prospects for minerals [3] associated with plutonic or volcanic processes, it has little potentiality as an oil-bearing or coal-bearing area. The Red Sea coastal region, however, may have some oil-bearing strata.

The lowland bordering the Red Sea is known as the Tihama, although sometimes distinction is made between the Tihama of the Hejaz, that of Asir, and that of Yemen. Beyond the Tihama, at varying distances from the coast, are desolate, bold, fault-block mountains, running an almost uninterrupted course from north to south. In the Jebel as Shafa range of northern Hejaz, the highest point encountered is Jebel al Loz (8,461 feet), and neighboring peaks have elevations from six to seven thousand. East of the Holy City of Mecca, Jebel Daka has an elevation of 8,346 feet, and the elevations continue to rise southward, reaching their highest point at Jebel Maeth (12,336 feet), west of San'a, the capital of Yemen. The height of the mountains in Yemen and their intercepting of the monsoons of the Indian Ocean secures to that country the most abundant rainfall of any section of the Arabian Peninsula, and enables Yemen to support the densest population.

The drainage divide along the crest of the mountains of Hejaz, Asir, and Yemen varies from 50 to 75 miles from the coast. Short streams of high gradient have carved deep gorges in the steep-walled western flank,

[3] At Mahad, about 200 miles northwest of Jeddah, there are about 50 gold mines, the largest of which, Mahad Dhahab, is presently being worked by the Saudi Arabian Mining Syndicate, Ltd., a corporation in which Saudi Arabian nationals, the Saudi Arabian government, and American, British, and Canadian interests participate. Mahad Dhahab means "cradle of gold," and, according to Arab accounts, either King David's miners may originally have worked it, or it may have been one of King Solomon's mines. See Twitchell, K. S., *Saudi Arabia,* Princeton, 1947, pp. 123, 146, 157.

and longer patterns of drainage have been developed by consequent streams on the more gently inclined eastern slopes. While granitic and metamorphic rocks form the core of the western ranges, volcanic rocks cover most of their extension into southern Yemen.

The high mountains fringing the southeastern margin of the Arabian Peninsula extend from Aden to eastern Dhufar. In the western Hadhramaut bordering the Gulf of Aden the average elevation is from 6,000 to 8,000 feet, but one peak north of the town of Mukalla has an elevation of 11,124. Eastward the elevations decrease to 3,000 feet in Dhufar, except for a 5,506-foot peak north of Murbat.

The mountains of the Hadhramaut and Dhufar are composed essentially of sedimentary rocks dipping gently northward into the Rub' al Khali Basin, but their southern slopes are irregular and precipitous. It is believed by the writers that the structure of these mountains is essentially a great homoclinal block tilted gently northward into the Rub' al Khali Basin and abruptly terminated on the south by faulting, closely related to the *graben* forming the Gulf of Aden depression.

A lowland plateau about 150 miles long occurs between the eastern extension of the mountains of Dhufar and the southern part of the Oman ranges. In this area the average elevation is about 500 feet.

The Oman Mountains have a conspicuously rugged topography, with the higher elevations ranging from 7,000 to 9,900 feet. They are characterized by steep slopes, both eastward and westward, and have the general configuration of a hogback range, the orogeny of which was dominated by folding. These mountains are similar to the Iranian ranges in structure, topography, and geology, but are very dissimilar to the mountains on the southwestern and southeastern sides of the Arabian Peninsula.

DESCRIPTIVE CROSS SECTION OF THE ARABIAN PENINSULA, FROM THE RED SEA TO THE PERSIAN GULF

1. The Tihama and the Hejaz Mountains. The rise in elevation eastward from the narrow coastal strip (Tihama) bordering the Red Sea is abrupt as soon as the highly dissected, rugged Hejaz Mountains are reached. As the mountains of the western ranges are composed largely of igneous and metamorphic rocks, they are extremely resistant to erosion. While this rock composition is a factor in the maintenance of elevation, the initial high elevations were the result of intermittent upward movements along the western fault zone, perhaps over a considerable period of geological time. The western slope of the Hejaz Mountains is a gigantic fault scarp. The opposite slope is, however, the result of a gentle eastward tilting of the fault block. In the northeastern foothill belt, gently eastward dipping Paleozoic sandstones are found resting on the rocks of the pre-Cambrian basement complex, and south of Tabuk extensive basaltic lava flows, called "harra," cover the early Paleozoic sandstones.

The eastern slope of the Hejaz Mountains gradually merges with the central plateau region.

2. Central Plateau Region. East of the mountainous area described above, there is a broad, desertic central plateau, 300 miles wide. The greater portion of this area is covered by pre-Cambrian igneous and metamorphic rocks; however, several extensive lava flows of comparatively recent age occur in the western half, while almost horizontal beds of massive sandstone (Cambro-Ordovician) are found overlying the basement rocks between Medina and the southwestern edge of the Great Nefud. The great explorer, Doughty,[4] gives his readers an excellent account of this and other sectors in northern Nejd.

The elevations of the central plateau region vary from 3,500 to 4,500 feet; however, there are local depressions of 3,000 feet altitude, and ridges as high as 6,000 feet. As seen from the air, this land mass shows mature dissection by eastward, now intermittently flowing, consequent streams which had developed a well-integrated dendritic drainage pattern. Irregular hills of moderate relief, separated by wide wadis filled with alluvium, are the dominant features of the landscape. This region is barren, except for sparse vegetation in the form of thorn bushes and coarse desert grass found growing in isolated depressed areas.

3. The Escarpment Region. Between the central plateau region and the Dahana, there is a region about 200 miles wide with eastward homoclinal dip. The topography of this area is dominated by six to eight distinctive north-south ridges, typical cuestas with steep west-facing escarpments and very gentle eastern slopes. The Tuwaiq Mountains (Upper Jurassic limestone), 500 miles long, and the Aruma escarpment (Upper Cretaceous limestone), which with its extensions is of the same length, are the most prominent topographic features; both are representative of the general trend of the topography as well as of the geological structure. The elevation of the Tuwaiq Mountains is about 2,800 feet above sea level, and the top of the escarpment is around 800 feet above the level of the adjoining plains and minor escarpments to the west. The Aruma escarpment, about 1,800 feet above sea level, is less conspicuous, because it is only about 400 feet higher than the plain to the west of it. The remaining escarpments, which are roughly parallel to the two dominant ones, have lower elevations and are not so long. However, some of the minor escarpments appear to be major topographic features when contrasted with the adjacent featureless plains, which are covered by *nefuds* near many, but not all, of the escarpments' western margins. (A nefud—or *nufud, nafud*—is an extensive area covered with dunes, sand ridges, and sand mounds.)

The prevailing northwesterly winds of this region have a westerly component which moves the sands of the nefuds slightly eastward. The

[4] Doughty, C. M., *Travels in Arabia Deserta.* London, 1888; and subsequent editions.

escarpments, or outlying mesas and buttes, are natural barriers beyond which the sands cannot continue their eastward course; consequently, the sands are slowly but continuously moving southward, parallel to the general trend of the neighboring escarpments.

The cuestas and plains of the escarpment region were formed by differential weathering of resistant limestones, friable sandstones, and soft shales—all members of a uniformly gently-dipping sedimentary series. These rocks, named in the order of their abundance, are well exposed here, but eastward to the Persian Gulf they are covered by more recent strata, which are similar in composition and abundance. Sedimentary rocks such as these are favorable source rocks for oil and coal, but are rarely of consequence for metallic mineral deposits.

4. The Dahana. The Dahana, which closely parallels the Aruma escarpment and the Tuwaiq Mountains, is one of the most distinctive terrains in eastern Saudi Arabia. Its length, from the Great Nefud to the Rub' al Khali, is approximately 800 miles, and its width ranges from 15 to 50 miles, the elevation of the terrain averaging about 1,500 feet above sea level.

The sands of the northern half of the Dahana are mainly of the stabilized type, supporting a fair vegetation of small bushes, shrubs, and grass. Dunes and sand peaks which indicate migratory sand are rare. However, just south of the Ma'aqala-Riyadh road, dunes appear on the west side of the Dahana, and from there southward more and more dunes are found. Our data indicate that the southern end of the Dahana is composed mainly of migrating dune sands. The sands of the Dahana are medium- to fine-grained, and the color due to hematite stains approaches a deep orange red, especially in the morning and later afternoon.

The Dahana is a favorite grazing ground in the winter and early spring. The Bedouins today derive Dahana from the word *dihn,* meaning "fat," evidently with reference to the excellent grazing there, although it is more likely that Dahana means "red," a name that could have been taken from the sands' characteristic color. Water in the Dahana area is scarce, and the Bedouins who herd camels there go with only a small amount for weeks. The camels do not require water while grazing on green grass and bushes, and the Bedouins live mainly on dates and camel's milk.

5. The Summan Plateau. A plateau often called the Summan, 50 to 150 miles wide, covered with either gravel or rock outcrops, lies to the east of the Dahana. In most places the Summan near the Dahana consists of fairly flat rocky terrain, but farther east stream channels have cut into it, forming an irregular topography. The Hofuf-Sarar escarpment on the east margin of the Summan terrain is a highly irregular front, consisting of isolated mesas and extensive table lands projecting prominently into the coastal lowlands. The elevation of the Summan near this escarpment averages about 800 feet, and that of the western margin adjoining the Dahana about 1,300 feet, giving an eastward slope of from three to four feet per mile. The surface of the Summan plateau is generally featureless and barren, but after a rare season of appreciable rainfall, a fair growth of grass and numerous flowering plants spring up.

Northeast and adjacent to the northern portion of the Summan lie the *Dibdiba,* gravel plains about 150 miles long in a northwest-southeast direction and over 100 miles wide. "Dibdiba" means "gravel plain," and is said to be derived from an Arabic root word descriptive of the sound of camels' hoofs striking pebbles and cobbles as they cross it. The Dibdiba are slightly rolling, with some fairly extensive flat areas. The vegetation is slight, confined mainly to depressions where water accumulates during rainy seasons. These plains form a good surface for speedy travel by automobiles, and airplanes of medium type could land almost anywhere on them.

6. The Persian Gulf Coastal Region. This area includes a strip of the Arabian Peninsula, bordering the western Persian Gulf coast between the base of the Qatar Peninsula on the south and Kuwait on the north. The large oil fields developed in this region during recent years make this province one of the most important oil-producing centers in the Middle East. In the coastal region, lying in the zone of gentle folding between the homoclinal dip of the Arabian foreland and the highly disturbed structures in the mountains of Iran and Oman, seven oil fields have already been discovered, and other promising structures await detailed geological investigation or future tests by wildcat drilling. Still other large sectors, completely covered by sand or gravel plains, where limited subsurface work has been done, await exploration.

Recent sands and Miocene rocks cover most of the coastal district. Fairly large areas of Eocene limestones project through the Miocene cover at Dhahran and on the northern portion of the Abqaiq structure. Other small Eocene limestone outcrops are exposed, and these rocks are usually found at a depth of less than 500 feet over most of the coastal district. Except for Dammam dome, surface evidence of structure is slight. However, geophysical surveys and structure-drilling explorations have revealed that the entire coastal district has been folded in varying degree, and that Cretaceous strata have been folded to a far greater extent than the exposed or near-surface Eocene beds; but generally the dips do not exceed seven degrees, even at depth. How far this belt of folding extends westward is not known.

Although in general the eastern coastal area of Saudi Arabia is barren and featureless, flowing artesian wells are present and supply water for date gardens, of which Hofuf and Qatif are the most outstanding. Artesian water is found in the coastal area mainly because of the gentle eastward dip of strata between east-central Saudi Arabia and the Persian Gulf. The rainfall entering the rocks at a higher elevation in the interior of Arabia slowly seeps downward, and continues eastward through porous strata, which are sealed above and below by impervious layers. When these are punctured by a drilled or hand-dug well, the water in the porous strata rises to the surface, producing the artesian wells that are common in the Gulf area.

From Kuwait south to approximately Jubail, at 27°N, the area consists primarily of low rolling sand ridges covered by a sparse to fairly dense growth of small clumps of bushes and a coarse wiry bunch grass—a terrain called *dikaka* by the Arabs. The sand is generally firm because of the root systems of the grass and bushes. The hummocks and small sand ridges caused by the wind and vegetation form an extremely rough driving surface for automobiles, but a good road can be made by removing the irregularities with a heavy drag or road scraper.

The western margin of the southern portion of this dikaka area terminates rather abruptly at the east-facing Qidam-Sarar escarpment. Farther north, however, the escarpment terrain breaks into isolated mesas and buttes. West of Kuwait and the Neutral Zone, the coastal area merges with the dibdiba gravel plains.

Near Jubail sand dunes appear, and this belt widens southward, merging with the well known Jafura dune area, which in turn merges on its southern margin with the Rub' al Khali. The dunes of the area range from 75 to 150 feet in height and over 1,200 feet in width.

A prominent east-facing escarpment, called the Oqair-Selwa escarpment, flanks the Gulf of Selwa, but is lost under the Jafura sand as it turns westward near Oqair. Under the Jafura sands lie flat gravel plains, which appear again to the west between the sands and another escarpment, about 230 miles long, facing east in a nearly straight line from about 30 miles north of Hofuf to 40 miles south of Jabrin.

The general trend of the coast line between the base of the Qatar Peninsula and Kuwait is fairly straight, but because of the saline flats, sand spits, and shoals along the coast, the detailed configuration of this water front is extremely irregular. The topographic rise inland from the coast is at the rate of about five feet per mile.

The saline flats, or *sabkhas,* that border most of the coast have been developed by the filling of embayments by sand along with the evaporation of sea water. Water is always found within a few feet of the surface, and apparently the sabkhas are lakes or other water bodies filled by drifting sand, silt, and dust from the desert. Because of the evaporation, a varying amount of salt is deposited, and an impure mixture of salt and sand forms a crust a few inches deep. The flat surface is maintained at the level to which moisture rises above the static water level. Below the moisture level the sand, silt, and dust are damp and fixed in place, but above this level the sediments dry out and blow away, leaving a flat surface. These flats are, generally, fairly firm, and good roads requiring little upkeep can readily be constructed over them.

REMOTE, PARTIALLY EXPLORED, AND UNEXPLORED REGIONS

1. Northern Saudi Arabia. Northern Saudi Arabia, a portion of which is the southern extension of the great Syrian Desert, lies between the Great Nefud and the northern portion of the Dahana on the south, and the Iraq and Jordan borders on the north. Most of this region consists of slightly dissected rocky plains, interspersed with gravel and with comparatively smooth plains of sandy soil.

The *widyan* region along the border between Saudi Arabia and Iraq, however, contains some rather highly dissected terrain, due to the erosion of mature intermittent streams that flow northeast toward the Euphrates Valley. ("Widyan" is the Arabic plural of *wadi,* sometimes less correctly given as *wadiyan.* In the strict sense a wadi is a stream channel which is dry through much of the year. In many places, however, where the drainage system is poorly developed and obscure, long shallow depressions are also given this name.) Other pronounced topographic features include the Wadi Sirhan structural depression—approximately 200 miles long, 20 to 30 miles wide, and about 1,000 feet below the elevation of the adjoining plateau region—and the irregular terrain of the narrow belt of lava beds lying immediately to the east.

The maximum elevations in northern Arabia are found on the plateau between the northern extremity of the Great Nefud, near the town of El Jauf, and Jebel Aneiza, a common boundary point of Saudi Arabia, Iraq, and Jordan. This plateau attains an elevation of 2,900 feet, and forms a divide between the Euphrates drainage on the east and the inland drainage to Wadi Sirhan on the west.

The terrain of northern Arabia is of considerable interest as a natural corridor for roads, railroads, and oil pipe-line routes between the Mediterranean Sea and the Persian Gulf. For thousands of years Wadi Sirhan has been an historic trade and pilgrim route, and the geographical expedition of Musil [5] used it as an approach to Arabia Deserta. Aramco geologists and water-well drillers have shown that water is available in the "north country." However, wells must be drilled 400 to 700 feet deep to reach the water table in the plains areas; and in the plateau area, to a depth of 1,100 feet or more. In either case, an abundant supply can be obtained only by pumping. The Trans-Arabian pipe line, which will transport oil from the Aramco fields near the Persian Gulf to the Mediterranean sea coast at Sidon, Lebanon, will pass through this region.

2. *The Great Nefud.* This "sea of sand," irregularly elliptical in shape, covers approximately 27,000 square miles in northern Saudi Arabia. Most of the area covered by it consists of rolling sand ridges, supporting sparse wing-bunch grass and other vegetation. Patches of migrating dunes [6] and

[5] Musil, Alois, *Arabia Deserta,* New York, 1927.

[6] *Ibid.,* p. 132. The author describes here the Arab *fluk,* which is the barchan type of dune.

sand peaks are developed locally. The Great Nefud contains a few watering places, and it furnishes excellent grazing for camels, goats, and sheep during the winter and spring seasons.

The rocks underlying the central portion of the Great Nefud are unknown, but on the western and northern sides the sands rest on almost horizontal beds of Cambro-Ordovician sandstones. Locally, mesas of sandstone strata project through the sand dunes on the Great Nefud's periphery.

It is believed that the sands of the Great Nefud were largely derived locally, from the erosion of limonite and hematite-stained, friable sandstones, but some mineral grains have undoubtedly come from the rocks of the basement complex on its western and southwestern margins. The sand is predominantly fine- to medium-grained quartz, stained by hematite which gives it a reddish hue. On the eastern and northeastern edges of the Nefud the sands rest on gently northeast-dipping strata.

3. The Rub' al Khali. This desert occupies what is essentially a great basin, bordered by the Oman Mountains on the east, the mountains of Dhufar and Hadhramaut on the southeast, the foothills of the mountains of Yemen and Saudi Arabia on the west, and the plains with scattered sand areas of Trucial Oman and the Jafura on the north. The basin has a length of approximately 750 miles in a N 60° E direction, and a maximum width of nearly 400 miles. It covers an area of about 250,000 square miles, nearly the size of Texas.

This region is relatively unexplored. Bertram Thomas, the only explorer who ever crossed it, made the trip on a camel from Salala on the Arabian Sea to Doha on the Qatar Peninsula in 1931.[7] In 1932 Mr. H. St. John B. Philby explored the central and northwestern portions, which he so admirably described in his book.[8] Aramco geologists, in 1937 and 1940, mapped a large section of the northern half.

It is estimated that 80% or more of the Rub' al Khali Basin is covered by sand; without question, this is the largest continuous body of sand in the world. Company data indicate that about half of the area is covered by dune sand, and the remainder by rolling sand ridges and sand hills that support a sparse vegetation of bushes and some coarse grass. With the possible exception of some of the sand peaks of the northern Dahana and the Great Nefud, the highest and largest sand dunes in the Arabian Peninsula are found in the Rub' al Khali. These dunes attain heights of 500 feet or more.

Although the Rub' al Khali is a bleak and dreary waste, there are a few Bedouins who manage to live there. Except along its eastern margin, the water of the basin is not good, and the Bedouins depend largely on camel's milk instead.

Little is known about the geological structure and stratigraphy of the

[7] Thomas, Bertram, *Arabia Felix,* New York, 1932.

[8] Philby, H. St. John B., *The Empty Quarter,* New York, 1933.

rocks underlying the sands of the Rub' al Khali. But as it appears to be a great structural basin formed by gently dipping strata on the west, south, and east peripheries, their coalescence with strata on the Persian Gulf geosyncline on the north is highly probable. The Rub' al Khali basin may contain oil-bearing strata, but development of the area will be slow and arduous because of the terrain, the climate, and the lack of water. Future improvements of special equipment will undoubtedly overcome these obstacles, and make it possible to explore thoroughly this forbidding area.

The Development of the Oil-Field Areas in Saudi Arabia

The initial efforts of the Standard Oil Company of California to obtain permission from King Ibn Saud to send a party of geologists to Saudi Arabia for reconnaissance were unsuccessful. However, by mid-year of 1932 the King had received favorable reports from the Shaikh of Bahrein about the Bahrein Petroleum Company (Bapco), and when the company sent representatives to Jeddah in the spring of 1933 they received a friendly reception. The terms under which the association between the government and the company was to operate the concession were negotiated point by point, and an agreement signed on May 29, 1933.

Shortly after this, the Standard Oil Company of California organized the California Arabian Standard Oil Company. The Texas Company acquired a half-interest in both the Bahrein and Arabian concessions in 1936, and in 1944 the name, California Arabian Standard Oil Company, was changed to Arabian American Oil Company (Aramco). The original concession was supplemented on May 31, 1939, to include the area of about 440,000 square miles now embraced by the Aramco concession. The supplemental concession will expire in 2005, whereas the original terminates on May 29, 1999. The ownership of Aramco now is as follows: Standard of California, 30%; The Texas Company, 30%; Standard Oil Company (New Jersey), 30%; and Socony-Vacuum Oil Company, 10%.

OIL-FIELD DEVELOPMENT

1. Dammam Field. Dammam dome at Dhahran is a conspicuous structure, the major axis of which extends approximately 25 miles in a north-south direction. Its minor axis is about 20 miles long, thus making the structure nearly circular. The productive area on the crest is an oval with axes of 4½ and 4 miles, and an area of about 9,000 acres. A distinctive Eocene limestone "rim rock" occurs 3 to 4 miles out from the center of the dome, and the strata dip outward from this, with slopes of 3 to 5 degrees. Miocene and later sediments are found on the periphery.

The producing horizon is called the Arab Zone. This is divided into "A," "B," "C," and "D" members, and consists primarily of oolitic and dolomitic limestones. The Arab Zone is capped by a thick bed of finely crystalline anhydrite, and each member is separated from the next

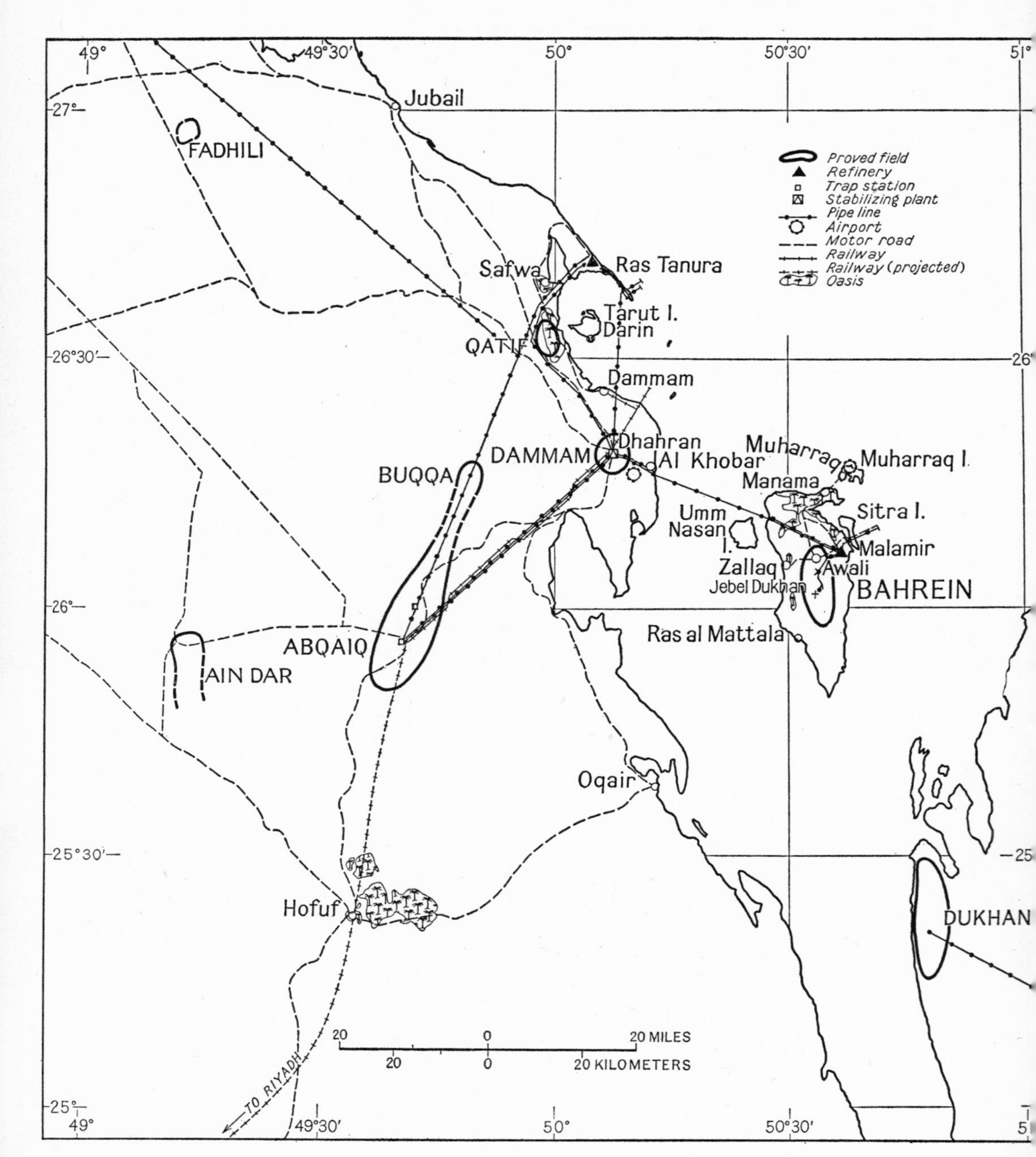

FIG. 36. Persian Gulf coast of Saudi Arabia, Bahrein, and Qatar: oil-field development and facilities. T fields with commercial production in mid-1949 (Abqaiq, Dammam, Qatif, and Bahrein) are linked by p lines, and share refinery and shipping facilities; they are described in Table 17 and in Chapter 9, in the t sections on oil-field development in Saudi Arabia and in Bahrein, respectively. The pipe line from Qatif the northwest, however, is not to be used until the later stages of the Trans-Arabian pipe line are complet

In the Abqaiq field the trap stations, to which the crude is brought from the wells by flow lines at w head pressure, are equipped with gas-oil separators. The crude is carried by pipe line either directly to

62. Saudi Arabia. Aerial view of marine terminal and storage tanks (in foreground) and refinery (in background) at Ras Tanura, a huge sandspit formed by longshore currents of the Persian Gulf. Note pipe lines for refined products connecting refinery and terminal. (Arabian American Oil Company photo by Robert Yarnall Ritchie.)

63. Persian Gulf. Pearl divers bringing up camel skin of fresh water from subterranean spring near Jubail. (Photo by Max Steineke.)

66. Saudi Arabia. The Tuwaiq mountain escarpment, northeast of Dhruma. The huge pinnacle, "Zig Hamud," rises 600 feet above the surrounding plain. It is an erosion remnant along the escarpment front formed by differential weathering of weak shales below and resistant limestones above. (Photo by Max Steineke.)

65. Saudi Arabia. Kenworth truck stringing 30-inch pipe for Qatif-Abqaiq pipe line. Each section of pipe is 90 feet long and weighs 8,100 pounds; the load on each fully laden truck is about 36 tons. (Trans-Arabian Pipeline Company.)

64. Saudi Arabia. A geological party in a "dikaka" area, 20 miles southwest of Abu Hadriya. (Photo by Max Steineke.)

67. Saudi Arabia. Aerial view of part of the Abqaiq field. (Arabian American Oil Company photo by Robert Yarnall Ritchie.)

68. Saudi Arabia. Tractor clearing the field for Al Kharj agricultural project. (Arabian American Oil Company photo by Robert Yarnall Ritchie.)

69. Saudi Arabia. Melons are ripe at Al Kharj. (Arabian American Oil Company.)

70. U.S.S.R. Kuybyshev. View of a large oil refinery built during the war. (Sovfoto.)

71. U.S.S.R. In the Zhiguli Hills. The you woman is engaged in de-emulsion work. (Sovf photo by N. Finikov.)

72. U.S.S.R. Oil fields in area of the Zhiguli Hills during the spring floods on the Volga, when row-boats are the only means of communication. (Sovfoto photo by N. Finikov.)

73. U.S.S.R. 29th anniversary of the Revoluti Wooden derricks have disappeared in the Baku field; modern equipment has been installed. (Sovf photo by G. Sedykh.)

by a stratum of anhydrite. The age of this zone is thought to be Upper Jurassic, although no diagnostic fossils have as yet been found in it. However, fossils of Upper Jurassic age are found immediately below the zone, and Lower Cretaceous fossils occur just above it.

The top of the producing horizon in the Arab Zone is reached at an average depth of 4,500 feet. The present production rate (April 1949) from 34 wells is 83,000 b/d. The oil has a gravity of 34° API, and because of its hydrogen-sulphide content, the crude is stablized for tanker shipment.

2. *Abu Hadriya Field.* In 1939 a wildcat well was drilled on the Abu Hadriya structure, about 100 miles northwest of Dhahran, after it had been mapped by surface and seismic methods. Oil in the Hadriya Zone, which is 500 feet below the Arab Zone, was discovered in commercial quantities at 10,110 feet. Production from this well, however, and further development of the field will not be undertaken until additional production is needed to supplement that from more prolific and shallower fields. In the meantime, subsequent structure drilling in the area has revealed that the Abu Hadriya structure is nearly circular and that the apex of the dome is almost four miles south of the discovery well. For this reason production from the Bahrein Zone (Middle Cretaceous) and the Arab Zone (Upper Jurassic) appears to be a possibility when wells are drilled on the apex in the future.

3. *Abqaiq Field.* Abqaiq is the largest and most prolific oil field in Saudi Arabia today. The Abqaiq structure is a huge anticline, with an indicated length of 50 miles and a maximum width of 25 miles. It was most difficult to find because of the sand-dune cover, the superficial solution features superimposed on it, and the very poor outcrops. The southern half of the Abqaiq anticline is a well-developed, elliptically shaped dome, whereas its northern half is a narrow tongue-shaped extension of the southern part.

The productive crest of this structure has, up to now, a proved length of 32 miles along the N 20° E trend of its axis, and a maximum width of 5.7 miles. The crestal area of the structure is quite broad, and outward

Tanura refinery or to the main gathering center at Dhahran, where crude from the Dammam field is col-ed also. Part of the collected crude is rerouted to the Ras Tanura and Bahrein refineries, and that to be orted is treated at the stabilizing plant to eliminate toxic hydrogen sulphide. Crude from the Qatif field to Ras Tanura by using the pipe line from Dhahran. Storage tank farms are maintained at the transfer ts from one pipe line system to another, at the processing plants, and at the shipping terminals. The cipal residential areas for Aramco personnel are at Dhahran, Ras Tanura, and Abqaiq, and that for co personnel is at Awali.

Qatar the Dukhan field was under development in 1949 in preparation for commercial production in y 1950. Four wells were producing, three at the northern end of the field and one ten miles to the south, exploratory drilling was proceeding. Development plans call for well-head separators and a central ti-stage separation unit, with three storage tanks of 90,000 barrels combined capacity and a 12-inch pipe to Umm Said on the east coast. The problem of supplying fresh water is to be met in part by salt-water oration plants. (Data on Qatar from *Petroleum Times Review of Middle East Oil,* London, June 1948.)

from this the dip of the beds increases gradually, reaching a known maximum of 8 degrees on the flanks.

Two producing members of the Arab Zone and one member of the Hadriya Zone have been discovered to date in the Abqaiq field. The upper horizon ("C" member of the Arab Zone) is of comparatively limited extent: about 16,000 acres. The "D" member is the main producing horizon and has an extent of 70,000 acres. The extent of the Hadriya Zone, recently discovered by deeper drilling, is unknown. The depth to the top of the "D" member varies from 5,878 to 7,300 feet, and the oil from it has a gravity of 39° API.

Daily production at present (April 1949) is limited to about 425,000 b/d from 36 wells, which fill to capacity the three pipe lines from the Abqaiq Field to the coast. The known reserves of this field alone are estimated at 5.4 billion barrels. The eastern terminal of the Trans-Arabian pipe line will be at Abqaiq.

4. Qatif Field. Near Qatif, an old Arabian sea-coast town famous for its date gardens, and about 14 miles northwest of Dhahran, an anticlinal structure was detected by surface mapping, and later checked by structure drilling. A wildcat was drilled in 1945, and to date six producing wells have been completed. The structural axis has an indicated length of 15 miles, and the width of the anticline is about 8 miles. Up to the present production has been proved for about 8 miles along the axis; the width of the producing area will probably not exceed 4 miles.

5. Ain Dar Field. This field was discovered during 1948 some 63 miles southwest of Dhahran. Four wells have been completed to date, but the extent of the field cannot be estimated until more wells are drilled. The producing zone is the "D" member of the Arab Zone with potential productivity in the vicinity of 6,000 b/d. Pipe-line facilities are not currently available to transport this crude from the wells.

6. Fadhili Field. Located 71 miles northwest of Dhahran, this wildcat was brought in early in 1949 when oil was discovered in the Fadhili Zone. The drilling of additional wells is necessary before the extent of the producing horizon can be determined.

7. Haradh Field. This wildcat was brought in early in 1949. It is located 156 miles southwest of Dhahran. Production is from the "D" member of the Arab Zone, but additional wells must be completed in this area before the extent of the field can be determined.

Production during the month of April 1949 averaged 524,000 b/d from the fields at Abqaiq, Dammam, and Qatif. Present plans call for the maintenance of a production of about 500,000 b/d from those fields through 1949 and 1950. Currently the estimated reserves of Aramco oil fields in Saudi Arabia are estimated at 6.7 billion barrels. The potential reserves are probably many times greater.

The anticipated increase in Arabian oil production a few years hence will be largely delivered overseas in the form of crude oil, for refining in Europe and other marketing areas. The Trans-Arabian Pipe Line Company (Tapline), an Aramco affiliate, is constructing a combination 30/31 inch pipe line from Abqaiq to the Mediterranean terminus at Sidon, Lebanon. This line, which is to be 1,070 miles long, has a rated capacity of 300,000 b/d and is scheduled to start operation late in 1951. At the present time about one-half of Saudi Arabian crude oil is refined at the Ras Tanura refinery and at the Bapco refinery on Bahrein Island. Crude is primarily moved by tankers headed for European ports; occasionally it is shipped to the Western Hemisphere and to Africa.

The Ras Tanura refinery, designed in 1943 as a war project, went into production in 1945, and since then has played an important role in maintaining the fuel supply of the U.S. Navy in the Far East. The marine terminal and refinery are indicated in Fig. 36. This refinery, which when first completed had a rated capacity of 50,000 b/d, averaged 140,000 barrels of crude charge per day early in 1949. The following is an approximate percentage output of the various products refined on an average day's run: gasoline (70–72 octane), 24%; kerosene, 6%; diesel oil, 24%; and fuel oil, 41%. During early 1949 the U.S. armed forces purchased about 25% of the products from Ras Tanura.

PROBLEMS OF GEOGRAPHICAL ADAPTATION

Both outgoing cargoes of crude oil and products and incoming freight for the oil-field areas are dependent on sea transport, which is made difficult in eastern Arabia by the shallow depths along the shore of the Persian Gulf. As there is no natural harbor with sufficient depth for cargo ships to anchor at the water's edge, the port of Ras Tanura has been constructed by Aramco for its own use. In March 1948 the facilities included a floating dock with four submarine loading lines and a T-shaped pier accommodating four ships and connected with the shore by trestle and ten loading lines. These facilities are on the eastern side of the cape and are used almost entirely for oil shipments. Incoming supplies are transferred by barge to a pier on the west shore of the cape. Additional tanks for oil storage and a new shipping wharf were scheduled for completion in July 1949. General cargo is also lightered ashore at Al Khobar, a small shallow harbor five miles east of Dhahran.

Inland transport in Arabia has traditionally been dependent on the camel and the donkey, but motor transport has supplanted them to a considerable extent in industrial operations, in spite of a lack of roads. Aramco's fleet of automobiles, trucks, and tractors are equipped with large low-pressure tires for use on soft sand. By this means, and the use of specially designed construction, trucks up to 12-tons capacity and tractor-drawn trailers moving units of drilling rigs of as much as 125 tons successfully navigate the sand.

Table 17. Saudi Arabia: Pipe-line status at end of 1948.

From	To	Year Completed	Diameter (Inches)	Capacity (b/d)	Length (Miles)
Dhahran	Ras Tanura Refinery [a]	1939	10	63,000	36
Dhahran	Ras Tanura Terminal [b]	1946	12	123,000	23
Dhahran	Bahrein [c]	1945	12	123,000	50
Abqaiq	Dhahran	1946	12–14	105,000	38
Abqaiq	Dhahran	1947	14	110,000	39
Abqaiq	Ras Tanura Refinery	1948	30–20–22	326,000	63

[a] Receives feed line from Qatif field.
[b] Submarine line across Tarut Bay.
[c] Submarine line across Gulf, plus land loops.
Source: Arabian American Oil Company.

Although almost all of the great sand areas can be traversed without roads by using proper equipment, tracks through any large sand area like the Dahana or Jafura become soft with heavy truck travel, and cause endless trouble. Good roads can be built through the dikaka areas, which generally have enough roots and other vegetable matter to keep the sand firm, and roads are needed; driving on this type of terrain without a road is slow, and it is practically impossible to exceed ten miles an hour across the grain of the sand hummocks. By scraping off the superficial vegetation and smoothing off the small sand ridges with a heavy drag or road grader, a good road can be constructed cheaply, suited to speeds of 30 to 40 miles an hour.

The greater portion of the northern two-thirds of the Arabian Peninsula, which consists essentially of rock and gravel plains, except for the Hejaz Mountains bordering the Red Sea, can be traversed by automobile almost anywhere without roads. One stretch of the proposed Persian Gulf-Mediterranean pipe-line route follows an almost straight line 600 miles long, mainly over gravel and rock plains.

The Great Nefud and the Rub' al Khali, however, have never been crossed by automobile, although the company's cars have gone far enough to know that properly equipped cars can be driven over them. Road construction and maintenance for development purposes will here be extremely difficult and expensive.

The principal automobile routes in the Arabian Peninsula are described in the final section of this chapter.

Automobile transport is supplemented by airplanes. From Dhahran airport, which is operated by the Saudi Arab government in conjunction with U.S. Army, regular flights are made by Saudi Arab government and Aramco planes to Al Kharj, Riyadh, and Jeddah; and four flights a week to the United States are available via Trans-World Airlines. Aramco operates a fleet of fifteen planes, ranging from DC-4's for trips to the United States to two-passenger planes for local use.

Dhahran is now connected by a railroad 40 miles long to Abqaiq and a line 10 miles long to Dammam town, where Aramco is constructing deep-water port facilities for the Arabian government. This undertaking was planned with the two-fold purpose of alleviating the freight-handling situation and of developing an improved Arab community, where employees could live with their families and commute to their work. His Majesty King Ibn Saud had had plans for a government railroad from the east coast to Riyadh, to solve the government's own transportation problem and to develop settled communities in the interior wherever water is sufficient to support agriculture. When the King heard of Aramco's plan, he requested that the company's harbor facilities and railroad be made a part of the government project, and the construction is proceeding on this basis. Aramco is doing the work and financing it, but will eventually be reimbursed. It involves a 7-mile pier to reach deep water, at the end of which a large unloading wharf is being constructed, with customs warehouses and other facilities at Dammam. The railroad runs directly to Abqaiq, with a spur to Dhahran. The town site of Dammam has been laid out, additional water wells have been drilled, and Aramco has assisted many employees to build homes.

Aramco has also provided assistance to the Saudi Arab government by operating truck transport over some routes, supervising it over others, and servicing government motor-transport equipment in the Dhahran garages. It has also acted as purchasing agent for a great variety of industrial and domestic appliances, and procured expert advisers and technicians. The development of the oil fields has brought the Industrial Revolution to Arabia, and there will be an abundant supply of oil to run the imported machinery. Arabs from the desert and the towns are manipulating tools, repairing or operating engines, running tests in laboratories, constructing modern houses, drilling oil wells. There are unlimited opportunities for advancement, unknown a decade ago, for those Arabs who have the needed ability and skill.

In the development of the oil fields in the desert it has been necessary for the company itself to provide most of the supporting industrial and personnel facilities.[9] These include public utilities (electric-power plants, transmission lines, and transformer stations; pipe-line pumping stations; reclamation plant; water-distributing systems, including water wells and distillation plants; sewage systems and garbage collection and disposal; natural-gas distributing systems; telephone systems); housing (dormitories and homes; central air-cooling plants and distribution systems; steam laundries and dry-cleaning plants); distribution (huge storehouses to substitute for the wholesale supply centers and the retail and oil-field supply houses common in the United States; large commissaries for the dis-

[9] In addition to 14,000 employees of Saudi Arabian and neighboring nationalities, there were in 1948 almost 4,500 American employees in residence, including about 1,900 employees of Tapline and of American contractors engaged in work for Aramco. There were also several hundred American wives and children.

tribution and storage of food, including refrigeration plants); service shops (machine, carpentry, foundry, welding, airplane repair, and paint shops; garages; etc., as well as laboratories and offices). Radio connects the major oil fields with field parties and camps in the interior, and a line connects Aramco's Jeddah office with headquarters at Dhahran. The Saudi Arab government has with Aramco's aid established a radio network linking the principal headquarters of the government with the capital at Riyadh.

Particular attention has been paid to sanitation and the control of disease. The irrigated farm lands at Dammam and the date gardens at Qatif are dusted and oiled in order to control the malarial mosquito, and every precaution is taken to eliminate other potential breeding places; an entomologist has been engaged to study the species of disease-carrying mosquitos in the Persian Gulf region. Trachoma is common, and company doctors are attacking it among the workers. A special effort has been made to eliminate amoebic dysentery; fresh pure water is piped to the camps, and sewage is disposed of by sanitary methods.

A large Arab hospital was completed in 1945 at Ras Tanura, and another in 1949 at Dhahran. American hospitals are found at both. All four units are equipped and operated in accordance with the best standards, and though medical and surgical care is primarily for employees, it is shared as far as possible by their families.

THE OIL INDUSTRY AND THE ARABIAN PEOPLE

The oil-field developments have given rise to a variety of new employment, which has attracted both urban folk and nomads. In addition to about 11,300 Saudi Arab employees, there are about 2,600 skilled and semi-skilled Indians and Pakistanis, Adenese, Sudanese, and others from neighboring countries. The wages paid, providing new purchasing power for the workers and their families, have brought a greater demand for goods; and many an Arab has gone into business, supplying the population around the oil towns.

The company has organized schools, the first modern primary education in the country. Two types have been established, one devoted to the reading and writing of English and Arabic, plus arithmetic, and the second, known as "Trade Schools," devoted to the teaching of manual crafts and an understanding of mechanical equipment. Boys and young men over sixteen attend these schools, at full pay in the majority of cases, dividing their time between school work and work on actual jobs. There are also evening "opportunity schools" open to adult Arabs on a voluntary basis. In 1948 over 800 men and 300 boys attended classes in the preparatory section, and over 2,600 men and 80 boys were enrolled in the trade and industrial section. The company also has put into operation an expanded educational program for about 500 young Arabs, who will be fully maintained while attending a six-year course of basic education. Upon completing it the Arab boys will attend a company trade school, and specialize in the vocation of their choice.

Apart from the oil industry, the people of the coastal provinces are engaged largely in fishing, pearling, date growing, and truck gardening. Tarut Island, a few miles east of Qatif, and the sea-coast town of Jubail are important pearl-fishing centers, and good pearling reefs occur a few miles southeast of Ras Tanura. The town of Darin on Tarut Island, Qatif, and many other cities along the eastern seaboard are of considerable interest to archeologists and historians.[10]

In the nearby desert, brush and short grass grow after the infrequent rains, and the domestic animals native to Arabia—camels, donkeys, goats, and sheep—subsist on this sparse forage.

Except for some of the coastal ports and pearl-fishing centers, nearly all towns of Saudi Arabia are built around oases, where the most important agricultural product is dates. The Arabs living in the cities [11] and in small towns or villages in the date gardens constitute the settled portion of the population; the remaining inhabitants of Saudi Arabia are Bedouins, who trek across the deserts and mountains with their families, searching for food for their flocks of goats and sheep and for their herds of camels. Except for its northeastern corner, the great Rub' al Khali desert has a population of only a few thousand wandering Bedouins.

In the interior, about 50 miles southeast of Riyadh, a notable agricultural project has been developed by the cooperative efforts of the Saudi Arab government, the United States government, and the Arabian American Oil Company. The source of water is a huge natural pit (one of three in the area), which reaches down through the limestone strata to a deep-seated ground-water source. The Saudi Arab government started this project during the 1930's, with the aid of an irrigation engineer from Iraq. During the 1940's the project was extended by the installation of large pumps and the construction of an 11-mile canal under the direction of engineers and construction supervisors supplied by Aramco. Additional land was brought under cultivation under supervision of an agricultural mission supplied by the U.S. government under the Federal Economic Administration. This mission returned to the United States with the winding up of the

[10] Forster, C., *The Historical Geography of Arabia,* London, 1844, p. 209; and Cornwall, P. B., "In Search of Saudi Arabia's Past," *Nat. Geogr. Mag.,* April 1948, pp. 493–522.

[11] Estimated populations of the ten larger cities of Saudi Arabia: Hofuf (Al Hasa Oasis), 80,000–100,000; Mecca, 80,000; Jeddah, 50,000; Medina, 20,000–40,000; Riyadh, 20,000–35,000; Qatif, 15,000–20,000; Buraida, 20,000; Hail, 15,000; Anaiza, 15,000; Bisha, 15,000. Other important cities, having estimated populations of 7,000 to 15,000, are Taima, Sakaka, Yenbu', Taif, Nejran, Laila, Abha, and Shaqra. There are also dozens of smaller towns, with popultions of a few score to 5,000 people, built around small oases with limited water supply.

F.E.A. at the end of June 1946, and the project has since been supervised by American agriculturalists engaged for the Saudi Arab government by Aramco.

The project now covers some 3,000 acres. Principal crops are dates, wheat, barley, alfalfa, melons, and various vegetables; and beans, grapes, potatoes, and other crops are being tested. This first serious attempt to apply modern agricultural methods in Saudi Arabia has been generally successful, and plans are under way for the establishment, under the same group of agriculturalists, of four demonstration farms in different parts of the Kingdom, where water is available.

While Aramco's mission in Saudi Arabia is essentially commercial—the exploration and development of oil fields—the activities of the company, it has been seen, are widespread, and include various long-range programs that will enable the Arabs to help themselves, and raise their standard of living. The responsibility placed upon Aramco to deserve the trust with which Americans were welcomed into Saudi Arabia—to live and work with the Arabs harmoniously, to understand and respect their traditions, ideas, and feelings—is a vital element in the planning of its enterprises. Human relations must be handled with foresight, sympathy, and wisdom. From the very start the company has conducted its operations on the basis of partnership with the King and his subjects, and great care is exercised to formulate policies that will maintain mutual respect and fair dealing.

Before oil was discovered, the Saudi Arab government's income was derived largely from the annual pilgrimage (Hadj) to the Holy Cities of Mecca and Medina. In the years following World War II this revenue has risen to the unprecedented high of seven million dollars, as against two to four million dollars in prewar years. During the war, when the great reduction in revenues from the pilgrim traffic and other sources placed the government in financial difficulty, the company made advances totaling some million dollars.

Today by far the larger part of the governmental revenue is derived from the royalty of 22 United States cents per barrel paid to King Ibn Saud by the company. Currently, Aramco's daily production is averaging around 500,000 barrels, and on this basis, His Majesty receives about $40 million per year.

This new source of revenue has enabled the King to import increasing quantities of food for his people, to extend the agricultural project at Al Kharj, and to develop the water resources of the country. Recently he has initiated plans for the establishment of a Ministry of Agriculture, and his dreams of large farms in other sections of the Kingdom are slowly but surely being realized. King Ibn Saud is a wise and just King, who sees clearly the economic development that is necessary. In a recent statement to Aramco officials, His Majesty said: "We want to teach our people to help themselves to become better and more useful citizens of the modern world. We are very happy that in this enterprise we have as our good partners the representatives of the United States of America."

Bahrein Island

The Bahrein Islands are situated in an arm of the Persian Gulf, bounded on the west by Al Hasa Province of Saudi Arabia, and on the east by the Qatar Peninsula (Fig. 36). The most noteworthy of the group are Bahrein, Muharraq, Sitra, and Umm Nasan. The entire archipelago takes its name from Bahrein Island, which is by far the largest and most populous and has given its name to the Bahrein oil field in the center of the island.

TOPOGRAPHY AND STRUCTURE

Regionally, Bahrein Island lies near the flank of a large ancient geosynclinal basin, of which the Persian Gulf is a present expression. The island is within the wide band of folding that exists along the eastern coast of Arabia.

Bahrein Island itself is the topographic expression of a large simple anticline. A recent study of water-well and structure-hole logs has shown that Eocene strata dip gently (½° to 1½°), in all directions from the crest, down under Miocene and more recent formations around the periphery of the island and out under the Gulf for several miles. The fold, therefore, is of considerably greater extent and closure than the surface above sea level indicates, the oil field occupying a relatively small area on top of the structure.

The island, a rough oval, is approximately 28 miles long and 10 miles wide. Rocky slopes dip down and outward from the rim of a central erosional basin and disappear into the Gulf or pass under sandy beaches. The central basin is about 12 miles long and 4 miles wide, and is clearly outlined by cliffs—the "Rim Rock"—which form a nearly continuous inward-facing wall around it. This Rim-Rock wall varies from 20 to 100 feet in height and is deeply serrated by steep and narrow valleys, which make the building of roads over the rim somewhat difficult.

Within the central basin and near its center is a cluster of relatively high peaks, rising about 450 feet above sea level, known collectively as Jebel Dukhan (Hill of Smoke). Excluding these hills, the highest area within the basin is approximately 220 feet above sea level. The major part of Bahrein Island consists of exposures of Eocene limestone, but fossils found on Jebel Dukhan indicate that Miocene marine deposits once covered it completely.

A belt of wind-borne and alluvial material has accumulated within the central basin close to the Rim Rock, and these sediments serve as a reservoir for surface waters. Shallow wells sunk in this belt produce fresh water, the level of which varies directly with the rainfall. During relatively wet seasons wheat and other grains are grown at favorable spots in the alluvial belt. Shallow lakes are formed in this belt during periods of abnormal rainfall, but they soon disappear, through drainage outlets to the Gulf, by evaporation, or by seepage into the ground.

While the major portion of the island consists of barren limestone exposures, a broad belt of relatively fertile soil, devoted to the growing of alfalfa and dates, forms a horseshoe around the northern end, immediately above the shoreline, and extends along both eastern and western shores for approximately a third of the island's length: this belt corresponds to the area within which artesian wells and natural springs are found. Water resources have been developed by the Bahrein Petroleum Company, Ltd. (Bapco), and many artesian wells drilled by the company are located in the agricultural belt mentioned above.

The shorelines of the island are characterized by wide, flat, sandy, coquina-like beaches that slope gently beneath the waters of the Gulf; on the southern end of the island a roughly triangular sandy beach terminates in an elongated spit. There are a number of salt swamps on the western side of Bahrein, the most prominent of which lies between Zallaq and Ras al Mattala. Near the town of Malamir, on the eastern side, there is a large area often covered by ordinary high tides and, consequently, not adapted to cultivation.

Surrounding the island are a number of reefs, varying in size from small coral patches a few square yards in extent to areas of several thousand acres. The larger reefs, in many cases, are exposed at low tide, and at high tide covered by 3 to 5 feet of water.

THE DEVELOPMENT OF THE OIL INDUSTRY

The Bahrein oil concession, now owned exclusively and equally by the Standard Oil Company of California and the Texas Company, was originally granted in 1925 to the Eastern and General Syndicate, an independent British group. In 1927 this group granted two option contracts to the Eastern Gulf Oil Company, an American corporation. One of these contracts covered Kuwait, and the other covered the Bahrein concession, which in 1928 was transferred to the Standard Oil Company of California. This company organized the Bahrein Petroleum Company, Ltd. (the Texas Company became a partner in 1936), and by 1930 it was registered under the laws of Canada as a British corporation and approved by the British government. The original concession contemplated a mining lease of a hundred thousand acres, but it has since been extended to cover the entire Bahrein archipelago. These negotiations are adequately described by Fanning,[12] where references are given to memoranda of the U.S. Department of State. The concession expires in June 1999.

Since the beginning of commercial production in 1934, a total of 98.8 million barrels of crude had been produced in Bahrein by the end of 1948. On January 1, 1949, there were 66 producing wells on Bahrein, and the

[12] Fanning, L. M., *American Oil Operations Abroad,* New York, 1947.

average daily production for 1948 was 30,000 barrels. The average depth of the producing zones is around 2,300 feet in Middle Cretaceous limestone.

The Bahrein Petroleum Company, Ltd., operates a refinery of about 155,000 b/d capacity on the northeastern portion of the island only a few miles from the oil field. The refinery facilities include two thermal-cracking units, a fluid catalytic-cracking unit, two thermal-reforming units, asphalt-manufacturing facilities, sulfur-dioxide treatment plant, crude stills and vacuum units, 100-octane gasoline-manufacturing facilities, acid-manufacturing and recovery plants, and various treating and rerunning facilities. In addition to more than 30,000 b/d of Bahrein crude, it handles 125,000 b/d of crude transported by a 12-inch, 34-mile submarine pipe line from Dhahran. The refined products are delivered by pipe line to a shipping tank farm on Sitra Island, whence thirteen lines of varying diameter (6–18 inches) and three miles long conduct the products to a deep-water pier, connected to the shore by a causeway and trestle, where tankers may load, two at a time. Incoming cargo is lightered ashore to a company pier on Sitra Island.

The reefs that surround the islands are a decided hindrance to navigation, and, combined with gently dipping land surfaces near the shoreline, make it necessary for any marine craft drawing more than three or four feet of water to anchor some distance off shore. American and British cargo ships are usually to be seen, unloading into lighters or barges. Manama, the principal port (Fig. 36), is always filled with native dhows of the pearling fleet or the regular fishing fleet, the latter engaged in catching barracuda and shrimp. Many small craft are customarily tied up, loading, or unloading cargo from Basrah or India. Bahrein is easily approached by sea and air. A modern airfield and a seaplane landing are located on Muharraq Island; both are operated by British Overseas Airways Corporation. On the island itself there is ample automobile transport, and causeways connect Bahrein Island with Muharraq and Sitra.

The political history of Bahrein has been dealt with by Dr. Lees (Part III, Chapter 8). The British government, by means of a series of commercial treaties with the present ruler of Bahrein, His Highness Shaikh Sir Salman bin Hamed al Khalifa, controls the monetary system of the government and acts in an advisory capacity on all political and economic matters. The royalties from the oil are distributed equally to His Highness, for disbursements to members of his feudal society, to the State of Bahrein for investment, and to public works, education, police, and general administrative expenses.

Soon after the Bahrein Petroleum Company, Ltd., initiated exploration and development of the oil resources of the island, it gave considerable financial aid to the American Mission Hospital at Manama in order to provide for the medical needs of Arab employees and their families. Recently Bapco has completed a 50-room Arab hospital at its main camp in Awali. This hospital is fitted with the best modern equipment, and staffed by Bapco's medical department.

Bapco has cooperated with the local governmental agencies and His Highness the Shaikh to improve all housing facilities, for its employees and for the Bahreinis as well. The native *barrasti* of palm leaves and reeds is being replaced by permanent dwellings provided with piped water, electric light, and cooking and sanitary facilities.

The government's technical school for Arab boys desiring to pursue various mechanical trades has since the war been assisted by the oil company. Graduates of this school are employed to a great extent by Bapco, where their training is continued on the job. The young men who show real ability are soon advanced to positions of responsibility, and it has not taken the Arabs long to discover that pay increases are the fruit of skilled work.

Other phases of Bapco-sponsored programs include safety education, and cooperation with the government in measures for better sanitation, the limitation and elimination of disease,[13] the development of irrigation projects and water resources, the conservation of fresh-water supplies, and the improvement of agricultural produce by the use of selected seeds imported from the United States.

The population of the Bahrein archipelago is estimated at about 125,000 inhabitants, most of whom are Arabs of the Moslem faith. Manama, the capital, has an estimated population of 40,000, while Muharraq has half that number. At the beginning of 1948, the number of Bapco employees in Bahrein exceeded 6,000, and more than three-fourths of them were Bahreinis.

The one great industry in this whole region before the discoveries of petroleum was pearl fishing (for more than 2,000 years, according to Kunz and Stevenson [14]). Pearls were the chief source of Bahrein's income, which in very good years was as high as $9 million, according to Williams.[15] Kunz has some interesting estimates on the income derived from pearls,[16] but he shows what a small share of the profits are passed on to the hard-working divers and others engaged in the industry.

The pearl-fishing season gets under way in early June and continues until late September. During the season nearly everybody on Bahrein Island used to be involved, one way or another, with the pearl fisheries. In April and May, when the water on the deep banks is too cold for the pearl divers, many fishermen explore the shallower warmer areas of the Gulf. In the winter, from October to May, the strong northwest winds

[13] Harrison, P. W., *Doctor in Arabia,* New York, 1940. This book describes the prevalent diseases of the Middle East.

[14] Kunz, G. W., and Stevenson, C. H., *The Book of the Pearl,* New York, 1908, p. 85.

[15] Williams, M. O., "Bahrein: Port of Pearls and Petroleum," *Nat. Geogr. Mag.,* February 1946, p. 195.

[16] Kunz and Stevenson, *op. cit.,* pp. 80, 88.

interfere with fishing, but some oysters are obtainable in the smaller protected bays and inlets throughout the year.

The opportunity for steady employment and a higher annual wage in the oil industry has lured many pearl divers. The indefinite income from pearl fishing, and the period of unemployment from October to May, during which the average fisherman would run out of money and "go on the books" with a pearl trader for food, had always been serious disadvantages of this occupation, not to mention the physical hazards—and the time-worn custom of debt inheritance, which has forced many a man into the pearl fisheries against his will. Generally speaking, former pearl fishermen have been the most readily trained,[17] and ultimately the best, workers in the oil fields, especially in well-rig construction and other pursuits requiring physical aptitude.

Bahrein's principal exports at present are oil, pearls, dried fish, and dates. The fabulous pearl wealth goes largely to India, from which wood, sacked coal, brass trays, Kashmir shawls, sandals, and many other necessities and luxuries (jewelry and gemstones, especially star rubies and sapphires) are imported. However, in Bahrein's native bazaars there are many standard American and British brands of canned food and other articles, and automobiles from both these countries are a common sight on the narrow streets of downtown Manama.

A Note on Automobile Routes in Saudi Arabia and Neighboring Areas

The roads connecting the present centers of Aramco operations consist of rough graded, and in some cases oiled, roads across sand areas. A fairly good road follows the coastal area from Dhahran to Kuwait and Iraq. The southern half of the route is mostly on well-packed sabkhas, and the northern portion over gravel plains and dikaka.

The road to Transjordan branches off from the Dhahran-Kuwait road slightly beyond El Jauf wildcat camp. From this point northwest to the intersection with the Darb Zubaida at Jumaima, a distance of 350 miles, the road is plainly marked, and passes over nearly flat gravel and rock plains on which very fast driving can generally be maintained.

Westward from Jumaima there is less travel, but a rough road, stretches of which are poorly defined, connects Lauqa, Duwaid, Sakaka, El Jauf, and Qariyat al Milh. Much of the present automobile road follows a camel route which takes a more or less straight path from one watering place to the next, and parts are bad for automobile travel. However, after an adequate survey, a good fast road could be constructed through northwestern Arabia at comparatively small cost.

The Darb Zubaida is an ancient road, one of the main pilgrim routes, from Iraq and Iran to Mecca and Medina. The road was named after

[17] The writers are indebted to Mr. Dale Nix, who was in charge of production at Bapco for many years, for this personal observation.

Harun al Rashid's pious wife, Zubaida, who furnished the money to build it and a series of birkas (cisterns) for catching water. These birkas are spaced at intervals of a day's camel ride (15 to 20 miles) along the route. They are 50 to 75 feet square, built of well-shaped blocks hewn from nearby limestone outcrops. Because of the scant rainfall in recent years, these cisterns are filled only occasionally, and are of little use at present. However, they are less necessary now, because of the faster transportation by automobile. The Darb Zubaida is a fair automobile road, and apparently cars with ordinary narrow tires can traverse the sands without special difficulty. In the bad sandy stretches the road is paved with a thick layer of brush and grass, which is satisfactory for passenger cars and light trucks, but would not stand up under heavy traffic.

Although there are several routes across Arabia, the Dhahran-Jeddah road via Riyadh, Duwadami, and Muwaih is most commonly used for travel between the Persian Gulf and the Red Sea. The airline distance between Dhahran and Jeddah is about 800 miles, but the road distance is about 1,050. Much of the road follows old established camel routes, as most Arabian auto roads do, and the distance could be shortened by making proper surveys. Except for crossing the sands of the Dahana and the nefuds, the Dhahran-Jeddah road may be compared to unpaved country roads through the featureless desert areas of New Mexico and West Texas. There are long stretches over smooth plains on which good time can be made. The trip across the Arabian Peninsula can be driven in 50 to 60 hours, or five fairly long days.

The road from Dhahran to Hofuf is maintained because it is on the regular route to the Al Kharj irrigation project and to Riyadh. The dragged road from Hofuf to Jabrin is now rather obscure, because there has been little traffic over it during the last four or five years. Most of it is over gravel plains and slightly dissected plains lying just east of the Hofuf-Jabrin escarpment. From Jabrin southwestward there is no established road, but nearly flat rock and gravel plains extend for over 300 miles to the southwest.

A coast road, following a chain of sabkhas, extends from Dhahran to some distance south of Oqair, but from the latter point on to Sharja there is no established road. Automobiles have made the trip, and the terrain is good for travel except for occasional sand areas. Gravel plains extend from about thirty miles south of Oqair to Sabkha Mutti at the western end of the Trucial Oman coast; by taking a selected course, one encounters very little sand. Our knowledge of the Trucial Oman coast is limited, but we know that sabkhas and plains border the Gulf, and the high sand dunes of the Rub' al Khali begin more than fifty miles inland. There are no regular automobile routes through the Oman Mountains, and because of the extremely rugged topography, road construction would be very difficult. However, passes in the mountains do exist, and we understand that in 1943 the British anti-locust mission made a trip by car from Sharja to Muscat. The natives of these mountains are the most unfriendly people of the Arabian Peninsula, and they would undoubtedly object to construction of automobile roads in their territory.

We have only slight information on the roads of western Arabia, although it is known that roads extend all the way from the Hadhramaut to Palestine and Jordan. Apparently a well traveled road trends from Mecca southeast to Nejran, and branch roads link it with Asir and the northern Yemen. It is also possible to travel from Nejran to Hadhramaut, although there are minor sand barriers.

A fair road links Mecca and Medina, the two Holy Cities, and a pilgrim road extends north to Jordan, closely following the route of the abandoned Turkish railroad which formerly connected Damascus and Medina.

The only paved road of any length in Saudi Arabia, outside of the roads built by the company in the oil-concession area, is the 45 miles of road between Jeddah and Mecca, macadamized a few years ago by Egyptian engineers. There is an extension of this road from Mecca to nearby Arafat, the site of the sacred mountain visited by the pilgrims.

10. THE UNION OF SOVIET SOCIALIST REPUBLICS

BY EUGENE STEBINGER

THE territories of the U.S.S.R. include more than one-seventh of the world's land area; out of a world land area of 57,500,000 square miles, the pre-war Soviet area amounted to about 8,200,000. The greater part is a relatively flat plain stretching from the Baltic to the Pacific Ocean with only comparatively minor interruptions, such as the Ural Mountains.

Climatically, the great plain of the Soviet Union comprises three east-west zones: a frozen and marshy tundra belt on the north, a broad forest belt, also with much marsh, and the steppe.

The first two of these have extreme winter cold and long winters. Verkhoyansk, a town in the Yakhutsk A.S.S.R., marks the center of the cold pole of the globe, with a January average of —59°F. In the tundra belt, as in that of Canada, precipitation is low, averaging about 8 inches annually, and in the temperate coniferous forest belt, also similar to that of northern Canada, there is an average rainfall of 15 inches east of the Urals and somewhat more to the west, where the forest is mixed with deciduous trees. Three great rivers—the Ob', the Yenisey, and the Lena —cross these zones from south to north, but, like the Mackenzie in Canada, remain frozen in their northern reaches many months annually, thus greatly limiting their usefulness for navigation. Russian energy and persistence in scientific studies of this forbidding region are notable, and have built up a considerable aggregate of information on the geology and oil resources. However, the only actual oil developments in the world to date under conditions of comparable severity are those of the United States-sponsored Canol Project and the Fort Norman field on the lower part of the Mackenzie River in Canada.

The third zone is a belt of steppes or prairies, and semi-desertic steppes to the south, with subtropical climates locally. The Asiatic portion of this southern belt includes desert areas with rainfall as low as 4 inches annually.

Sedimentary Areas and Oil Fields

The area of sedimentary formations essentially unaltered and capable of acting as host rocks for petroleum in the territories of the U.S.S.R.,

Editors' Note: A request for a full-length article by a Soviet authority was conveyed in 1945 to the President of the Academy of Sciences of the U.S.S.R. by the American Embassy in Moscow, but no reply was received. The editors therefore asked Mr. Stebinger, who has long watched Russian developments, to undertake this very difficult task, and they wish to thank him especially. In addition to the incomplete figures officially issued since 1938, Mr. Stebinger has also made use of a great range of data from other sources.

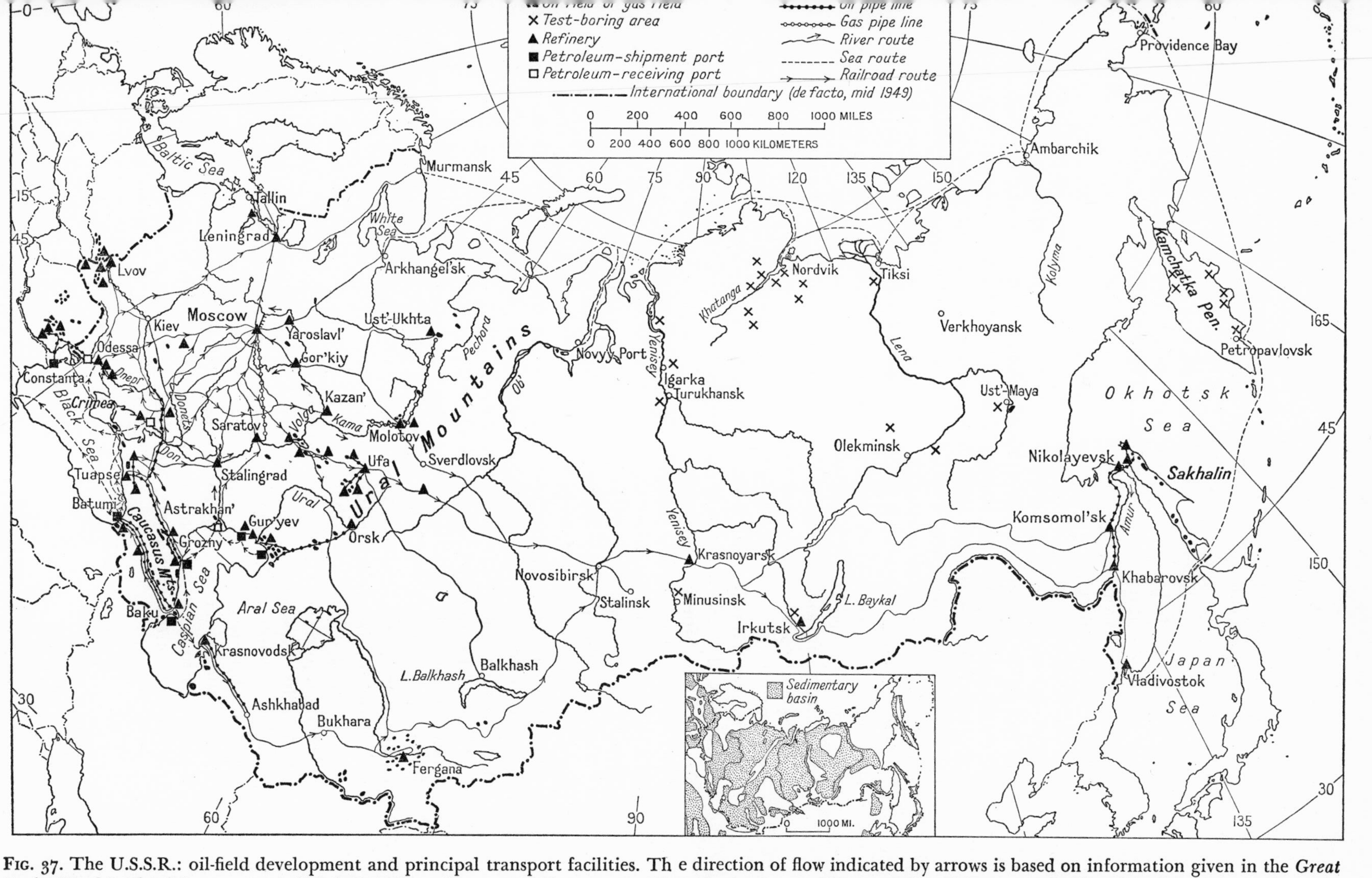

FIG. 37. The U.S.S.R.: oil-field development and principal transport facilities. Th e direction of flow indicated by arrows is based on information given in the *Great Soviet World Atlas* (Moscow, 1939), Vol. I, Part Two, Pl. 132 and Pls. 133–134. The other sources used in compiling this map are indicated in the acknowledgements in the footnote at the end of this chapter. Oil fields and facilities in the European part of the Soviet Union are shown in more detail in Fig. 38.

selected on the same basis as the other areas on our world map, comprise about 4.3 million square miles. This area is so great, and comprises so large a proportion of the world's total, that it invites comparison with other areas and political groupings, in order to obtain a true perspective on the various holdings of the world's undeveloped oil resources. The areas of sedimentary formations under control of the various important political entities on the basis of the prewar boundaries were as follows: British Commonwealth, 4.4 million square miles; U.S.S.R., 4.3; U.S.A., 2.5; Brazil, 2.1; French possessions, 1.5.

Certain of the political entities are especially favored in having more than their due proportion of the sedimentary areas. The United States is outstanding. With only 5% of the world's land, the continental United States has 12% of the effective sedimentary areas. The U.S.S.R. is also favored, having 20% of the sedimentary area, as against only 14% of the land.

The great sedimentary areas of the Soviet territories can best be described in relation to the main crystalline-rock shields of the continents that have limited them and controlled their sedimentation through the ages since Cambrian time. These shields include two great positive areas of Pre-Cambrian rocks: (1) on the northwest, the Fenno-Scandian Shield, constituting most of Norway, Sweden, and Finland; and (2) a much larger eastern area, constituting much of Central Asia. The latter is essentially a complex of shield areas centered on the 105th meridian, including areas south of the Siberian border in north China and Tibet, as well as the Angara and Anabar shields between the Yenisey and Lena rivers, and two shields in the Arctic Sea defined by outcrops in the Arctic islands as well as on the Siberian coast east and west of the 105th meridian.

Between the Fenno-Scandian Shield and the complex of shields in Central Asia there lies the great sedimentary area of European Russia, and the equally great area of western and southwestern Siberia. This inter-shield sedimentary area is very wide, about 3,150 miles on the south, and narrows northward to a width of about 1,075 miles on the Arctic coast.

This wedge-shaped area of dominantly sedimentary formations includes nearly all the U.S.S.R.'s proved oil reserves and development. It includes at least 21 well-defined petroleum-bearing basins (varying in extent from a few hundred square miles to over 100,000), whose number will probably be considerably increased as exploration and development proceed, and at least 41 individual oil fields, with daily yields ranging from a few thousand barrels to over 100,000 b/d, whose numbers also can confidently be expected to increase greatly in future years. Tectonic and structural trends in this huge area are in marked contrast. In its southern third, folding is very strong on a surprisingly regular northwest-southeast trend (Caucasus), extending for several thousand miles and apparent across a width of more than 1,000 miles over the flat plains of south European Russia and southwest Siberia. North of this Caucasus trend, and almost at right angles to it, lies the north-south uplift of the Urals' late Paleozoic folding, whose influence is evident in oil-field trends hundreds of miles from the outcrops of the older Paleozoic formations. This right-angle contrast of trends, on a very large scale, is one of the dominant features of the world's geology, almost as striking as the great Cordilleran trend extending nearly continuously the full length of the Americas.

In eastern Siberia there is another huge sedimentary area, divided into several separate lobes (not shown in detail on the map), fringing the central shield complex and extending from the meridian of the Lena River eastward to the Pacific. In this poorly explored region many more individual oil basins may be uncovered in the future. On the Lena River drainage, within a few hundred miles of the edge of the central shield, this area contains free oil showings in Lower Cambrian strata (several wells), which confirm the age range of Russian oil as extending from the Cambrian to the Pliocene, a range as great as that of North America.

The great inter-shield area comprising European Russia and west and southwest Siberia, with so large a proportion of the U.S.S.R.'s oil resources, deserves further comment. Its southern portion, dominated by the strong Caucasus trend in a belt wide enough to extend across the full north-south extent of the Caspian Sea for long distances both northwest and southeast, includes the first and most active oil developments in Russia. To the north the inter-shield area contains the other producing areas of European Russia and potential areas in Siberia (Fig. 37).

In the Caucasus region, the geanticline of the Caucasus Mountains, with a core of granitic and altered Paleozoic rocks, is overlapped and surrounded by a thick section of oil-bearing Tertiary formations unconformably overlying Mesozoic formations. This great uplift is surrounded by oil fields which have developed either (1) on strong folding in Tertiary rocks (for example, the Apsheron Peninsula on the east, on which Baku is located, and the Grozny Anticline on the north flank, second only to Baku and Maykop in daily output in 1944) or (2) on angular overlaps of Tertiary on the older rocks (for example, Maykop, with stratigraphic oil also on the north flank, whose first deep flowing well was obtained in 1909).[1]

Across the Caspian Sea, in line with the uplift of the Caucasus, there is strong folding in the Tertiaries, with important oil development at the Cheleken Islands and the Neftedag fields, near Krasnovodsk.

Many of the Caucasus fields, with their "high" structure and abundance of oil in unusually thick sections of Pliocene rocks, have close resemblances to those of California, notably in their very high yields of oil per field-acre. The Balakhany field on the Apsheron Peninsula, eight miles

[1] According to G. D. Hobson in *The Science of Petroleum,* p. 160, Russia's first gusher was completed in 1866 at a depth of 21 meters at Kudaho, in the Maykop region, 140 kilometers west of Maykop itself.

northeast of Baku, was Russia's first billion-barrel pool, the output having exceeded that amount as early as 1920.

Because of easy finding and prolific yields, Russian oil development was confined from an early date almost exclusively to the Caucasus region, thus excluding from attention other vast territories of promise. As late as 1939, nearly nine-tenths of all Russian production of oil came from this region—obviously a one-sided and non-strategic development of a vital resource. By 1945 the accumulated production at Baku had exceeded 4 billion barrels.

West of the Urals, in the heart of the Volga drainage basin, deep tests, beginning in the late 'thirties, appear to have discovered an area of superlative richness in a zone extending some 300 miles out from the Urals over a width of about 600 miles, equal to more than two-thirds the area of Texas. A Middle Devonian porous limestone appears to have given the most sensational results, and excited the Russians to call the region a "Second Baku," although the name from a technical standpoint is not very appropriate. (Such details of the oil geology as customarily would be widely published on a similar find in the United States were not published during the war period.)

Under a cover of undisturbed Cretaceous Tertiary and Permian formations, all of later age than the folding in the Urals to the east, there appears a thick section of Paleozoic rocks with oil-productive units in Carboniferous, Lower Carboniferous, and Devonian limestones that seem widespread and of great importance. Currently productive fields include: Molotov, Tuymazy (with 90,000 proven acres, and a porous zone in Middle Devonian that is also recognized at the Molotov field, 350 miles distant), Kuybyshev, and Sterlitamak. At Kuybyshev the total thickness of section into the Devonian is nearly 10,000 feet, and other drilling depths are comparably great. The best comparison with geologic settings elsewhere is with that of the West Texas Paleozoic limestone basin. Comparison with conditions at Baku is inapt, both as to the age of the producing section and the nature of the structure, and as to drilling conditions and other pertinent details. But, for the Russians under the stress of the war, the presence of flowing oil in great volume was sufficient to set the type. The official forecast for production in 1950 in this district, which may or may not be attained, is 250,000 b/d. This would be over one-third of all Soviet production, as forecast for that date, and a steady shift of operations to this center seems assured.

South of the above area, in the same position relative to the Urals but possibly somewhat deeper in the basin, is another region of great importance, called the Emba-Ural'sk district from the two river systems crossing it. Here Jurassic rocks are added to the producing section, and deep drilling is required. Structurally the area is characterized by broad folds on the north, with closer folding on the east parallel to the Urals; in the southern portion there are numerous salt domes over wide territory.

The western portion is the site of important gas developments. The north-south extent of the entire district is about 400 miles, extending down to the north shore of the Caspian Sea; the area is about four-fifths of that of California. The first flowing well was obtained in 1911, but by 1928 total production for the region was only about 5,500 b/d, indicating a comparatively slow rate of development for so much territory of promise.

East of the Urals, in the drainage area of the Ob' River, a geologic setting comparable to that of the rich area west of the Urals in the Volga drainage basin can be anticipated, with a thick section of Paleozoic rocks under an undisturbed cover of younger formations. As in the case west of the Urals, the deepest part of the basin probably lies within the first three hundred miles out from the Urals, and remains basinal in character along a trend over 1,200 miles in length. This immense region remains, or at least did so until very recently, practically unexplored and untouched by the drill, though it is of relatively easy access, compared to most of Siberia.

In the Far East, Sakhalin Island has long been an oil producer. In prewar years the Japanese obtained somewhat more than one million barrels annually from concessions in the northern (Russian) part which they operated under an arrangement with the Soviet government, which itself produced a like amount. Estimates of proved reserves for North Sakhalin suggest amounts of as much as 350 million barrels.

Oil-Field Development and Transportation

The main outlines of the regional development of the oil industry in the Soviet Union in comparatively recent years are indicated in Table 18 and Figs. 37 and 38. Fig. 37 also shows the natural-gas pipe lines, which form a vital link in the power network in the European part of the U.S.S.R.

The Caucasus area produced about seven-eighths of the total crude oil output in 1939 and about three-fourths of it in 1944. Most of the oil is refined there and sent north to the populous industrial and agricultural areas of the Donbas, Moscow, and Leningrad, although some is sent as crude. There are three principal routes.

The first of these is primarily a water route, across the Caspian Sea to Astrakhan' and up the Volga. Much of the oil cargo is entrained at Stalingrad or Saratov for the west and northwest, but some is transshipped at Astrakhan' into smaller craft which go up the Volga to Kazan, Gor'kiy, and Yaroslavl', and a few the whole way to Moscow, by the Moscow-Volga canal, a distance of over 2,000 miles from Baku. This route is icebound intermittently between Baku and Astrakhan', from 3 to 4 months between Astrakhan' and Saratov, and from 4 to 5 months above Saratov. Before the war about one-fourth of the oil produced in the U.S.S.R. was carried on the Volga. During the war, traffic was interrupted by the siege of Stalingrad, and damage was done at Yaroslavl'. Rebuilding has been

Table 18. The U.S.S.R.: production of crude petroleum by areas, 1939 and 1944.

	Production	
	1939 Actual	1944 Estimated
	(In million barrels)	
U.S.S.R. IN EUROPE [a]		
Caucasus and Crimea	190.9	175.4
Georgia (southeast slope of Caucasus)	0.4	1.5
Kura River basin (south of Baku)	1.1	4.9
Baku	154.2	143.8
Daghestan	1.4	6.8
Grozny	15.8	12.0
Maykop	18.0	6.4
Crimea	0.008	0.008
Emba-Ural'sk	5.1	10.2
Volga Drainage Basin	13.5	27.6
Kama River	0.6	2.4
Urals to Volga and Sterlitamak area	12.9	25.2
Pechora	0.7	3.5
Ukraine (Eastern)	0.008	0.2
Total, U.S.S.R. in Europe	210.2	216.9
U.S.S.R. IN ASIA [b]		
Turkmen	3.5	6.8
Bukhara-Fergana	1.5	9.1
Yenisey-Khatanga	—	1.5
Lena	—	0.1
Kamchatka	—	0.4
Sakhalin	3.4	6.0
Total, U.S.S.R. in Asia	8.4	23.9
Total, All U.S.S.R.	218.6	240.8

[a] New territory producing petroleum in the western Ukraine has been added since 1944.
[b] New territory producing petroleum in southern Sakhalin has been added since 1944.
Source: Computed from data on fields in *World Oil Atlas 1948,* pp. 251, 255.

pressed, and plans provide for control of the Volga navigation to enable large craft to make the full journey, and for full canal connection with Leningrad.

The second route is both a land and a water route: by pipe line and rail to Batumi and Tuapse, thence by water to the northern Black Sea ports, thence by rail to the industrial areas. This route carried more than one-fourth of the total output before the war. Destruction during the war was severe in the northern part of the area, which is still in process of reconstruction. There are also plans for expanding the inland waterways by two canals from the Caspian Sea to the Don, and for a Greater Dnepr scheme, which will give connections right through to the Baltic and White seas.

The third route is wholly a land route: by pipe line and rail from the northern Caucasus fields to the Donets basin, and thence by rail to the north. This route took perhaps one-sixth of the total prewar output.

Destruction was heavy during the war, and the present state is unknown. But the rail system is to be expanded.

The oil fields west of the Urals in the Volga drainage basin provided about one-twentieth of the U.S.S.R.'s production in 1939 and about one-tenth of it in 1944; and, as has already been mentioned, they are scheduled to supply one-third of the crude produced in 1950. The output has been transported by rail and local pipe lines to industrial cities in the Urals and western Siberia.

The oil fields of the Emba basin provided about one-fortieth of the U.S.S.R.'s production in 1939 and about one-twenty-fourth of it in 1944. Part of the output is shipped west by water through the Caspian Sea and up the Volga; part is sent northeast overland by pipe line to the Orsk refinery, thence by rail to the Urals and Siberia.

In Soviet Asia the principal producing areas are west and east of the great shield complex. In the arid southwest the Turkmen and Central Asia basins provided about one-fortieth of the U.S.S.R.'s crude in 1939 and about one-fifteenth in 1944. The output is shipped north by pipe line and by the Turksib railroad to the new industrial centers of west-central Soviet Asia.

The principal oil-field area in eastern Soviet Asia is the island of Sakhalin, which provided about one-seventieth of the U.S.S.R.'s production in 1939 and about one-fortieth in 1944. The crude is shipped by submarine pipe line and tanker to mainland, thence by the Amur River or rail to towns of eastern Siberia.

Between the oil fields of the Central Asia basin and those of Sakhalin there is an area of very small production some 2,500 miles wide from east to west and 1,800 miles from north to south. The Yenisey, Khatanga, Lena, and Kamchatka basins have produced a little oil in this vast region, but none of these basins provided any crude in 1939, and their combined output in 1944 was less than one-one-hundredth of the U.S.S.R.'s total. Their output is probably still too little to count. Rather, petroleum products have to be shipped in to northern Siberia in summer by the north sea route and transshipped up the rivers to the limits of navigation. The area is icebound five to six months of the year. Southern Siberia apparently depends for its oil supplies entirely on the Trans-Siberian railway.

The production of oil from shale appears to offer little hope of alleviation of the oil scarcity in Soviet Asia. The chief known oil-shale deposits in the U.S.S.R. are along the Volga, in the Leningrad district, and in Estonia, and other deposits are known in the Kuybyshev, Saratov, and Gor'kiy areas, and in the Chuvash and Tatar autonomous republics.

Present total production of oil shale and its products in the U.S.S.R. can be guessed at only, but may possibly be as high as 12 million tons of shale and over 2 million tons of shale oil. The Russians claim to have in operation a highly efficient thermal solvent extraction process that converts up to 95% of the shale organic matter into liquid fuel. Total known reserves of oil shales in the U.S.S.R. have been estimated at not less than 60 billion metric tons, but further exploration may of course greatly increase this figure.

A Comparison of Probable Ultimate Yields in the U.S.S.R. and the U.S.A.

Because of their great size and high rank among the nations, comparisons of the probable ultimate yields of petroleum that may be obtained in the U.S.S.R. and the United States appear in order. We have noted many similarities in the geologic settings of the petroleum basins of the two countries: all of the great types of petroleum occurrence are well represented in each, and the complete age range of the productive formations is also the same. The average thickness of the sedimentary series is roughly equal, with the result that in each case total volume of sediments is proportional to area. Counterparts of the high-yield areas of California and the Texas-Louisiana-Gulf Coast are readily found in the U.S.S.R. Counterparts of the great interior basins of the United States, with deep Paleozoic production, are also represented in Russia.

With no fundamental differences as to the types of occurrence, there is, however, a considerable difference in the size of the effective areas available to each, as indicated by the square miles of effective sedimentary area. As already mentioned, the Russian total, selected on the same basis, is 4.3 million square miles, compared to 2.6 million for the United States; giving a ratio of 1.7 to 1.0 in favor of Russia. Assuming, as we have, uniform and continuous operation of the petroleum-forming processes through the ages, the results as applied to two given areas of great size and varied types should be proportional to the square miles of area involved. On this basis, if we assume a 100-billion-barrel ultimate yield for the United States, the Russian total would be 168 billion barrels—with the qualification that the development data for the 100-billion-barrel yield from the United States are very much more complete than those for the larger total in the U.S.S.R., thus requiring the use of the United States as the unit of measure for the comparison.

The modern period of petroleum development in the United States began in 1859, with the Drake well in Pennsylvania, and the Russian period began twelve years later, in 1871, near Baku. From these beginnings up to January 1, 1949, the cumulated annual output for the United States has been about 37 billion barrels, and for the U.S.S.R. 6 billion. Another significant comparison is that of daily rates of output as of January, 1949: 5,608,000 b/d for the United States, and an estimated 610,000 b/d for the U.S.S.R. Thus the estimated current ratio favors the United States by about 9 to 1. However, for a few years, in the period 1898–1902, Russian daily output exceeded that of the United States. This flush period existed under the Czarist regime, with Swedish, Dutch, and British capital operating freely. The present large preponderance of United States' production has been built up only since that date.

A Note on the Sources for Fig. 37: The U.S.S.R. oil-field development and principal transport facilities. Information on sedimentary basins and on oil and gas fields from L. G. Weeks' "Developments in Foreign Fields in 1948," in *Bull. Amer. Assn. Petrol. Geol.,* Vol. 33, No. 6, June 1949, pp. 1074, 1075, 1101. Oil and gas pipe lines and refineries from *World Oil* atlas issues, May 1946, June 1947, July 1948, and July 1949. Data on water routes and ports from (1) *Great Soviet World Atlas,* Moscow, 1939, Vol. I, Part 2, Plates 132 and 133–134; and (2) S. T. Possony, "European Russia's Inland Waterways," in *U.S. Naval Inst. Proc.,* Vol. 73, No. 534, Aug. 1947, pp. 936–947. Railroads from *Great Soviet World Atlas* as cited above and from the National Geographic Society's maps, *Europe and the Near East,* June 1949, and *Union of Soviet Socialist Republics,* Dec. 1944.

11. EUROPE WEST OF THE U.S.S.R.

BY LYMAN C. REED *

EUROPE west of the Soviet Union is important from the petroleum point of view because the 386 million people living there constitute the world's second greatest market, surpassed only by the United States. In Europe a little more than one-tenth of the petroleum used is produced as crude within the area; the other nine-tenths has to be imported. Supplying the European market is one of the principal activities of the world's international petroleum trade. Here, however, we are concerned primarily with the production of petroleum and petroleum products within Europe. The theme of the present chapter is production, by areas, with brief notes on refining, transport, and supplementary fuels.

The Sedimentary Basins Favorable for Petroleum

In a very broad sense, Europe may be looked upon as having three distinctive geological regions. First, the Scandinavian countries of rather rugged relief, devoid of young rocks and sedimentary basins. Second, the north European countries, characterized by broad sedimentary basins and extensive areas of older rocks, occasionally rising to low mountains. Third, the Mediterranean region and countries of southeast Europe, characterized by high mountain chains of youthful topography bordered by narrow sedimentary troughs and interspersed by several small intermontane basins; here the rocks of both mountains and basins are of relatively young age.

The sedimentary basin areas of Europe are shown on Fig. 38. They may be divided into eight prominent and from eight to thirteen minor basins, depending upon the criteria used to define a separate basin. By the standards of other continents, these European basins are small, as is also the ultimate oil production that may be expected from them. Sedimentary basins are directly related to and tied in with tectonically active areas, and in Europe these are more complicated and more closely crowded than in most parts of the world.

Although there may appear to be little or no order in the arrangement of the European basins, they are, nonetheless, in harmony with the regional tectonics, as may be observed from detailed geological maps. The relation of a basin to the tectonics is often quite obvious, as in the troughs adjoining the Pyrenees, Apennines, Alps, and Carpathians. For some other basins, it is true, such as the North German and the Paris basin, the association of mountain uplift and basin sink is not clear, and their formation cannot be so obviously related to some structurally prominent compensat-

* Geologist, Standard Oil Company (New Jersey).

ing feature. Nevertheless, there are no basin areas whose origin appears to contradict present-day geological theories.

For some years it has been generally accepted that there is usually a definite relationship between the amount of marine sedimentation and the amount of oil present within a basin: normally, the larger the basin, the more oil to be expected. But there are many factors involved, each basin having certain determining factors, which may go far toward determining the quantity and distribution of accumulation.

The basins of Europe are classified in Table 19 in order of productivity, where commercial production exists, and in order of anticipated ultimate yield where commercial or non-commercial production has not yet been found or, having been established in the past (as in Thuringia), has at present come to an end. Only the first seven basins listed, together with the Transylvanian basin, which produces gas only, are, at this writing, commercially productive to more than the smallest extent.

Table 19. The principal sedimentary basins of Europe west of the U.S.S.R. (listed in order of importance).

Basins	Countries in which they lie
1. Carpathian	Rumania, Poland, Czechoslovakia
2. North German	England, Scotland, the Netherlands, Germany, Denmark, Sweden, Poland, Lithuania
3. Hungarian	Hungary, Yugoslavia, Rumania, Austria
4. Vienna	Austria, Czechoslovakia
5. Po Valley	Italy, Albania
6. Aquitaine-Rhône Valley	France
7. Upper Rhine Graben	Germany, France
8. Molasse	France, Switzerland, Germany, Austria, Czechoslovakia
9. Sicilian Central	Sicily
10. Duero-Ebro	Spain
11. Transylvanian (gas only)	Rumania
12. Thuringian	Germany
13. Paris	France, England
14. Portuguese	Portugal
15. Seville	Spain
16. Tajo	Spain

THE MAIN SEDIMENTARY BASINS

A few generalizations concerning the most important oil-producing basins of Europe west of the U.S.S.R. may be useful before considering the production of each European country separately.

1. The Carpathian Basin. This basin follows the outside arc of the Carpathians through Rumania and Poland, a distance of about 745 miles. The northwest continuation of this basin, along the edge of the Alps through Czechoslovakia, Austria, Germany, and Switzerland, is called the Molasse Basin.

Sedimentation, mainly marine, began about the Middle Oligocene, and continued actively through the Pliocene. Pronounced tectonic unconform-

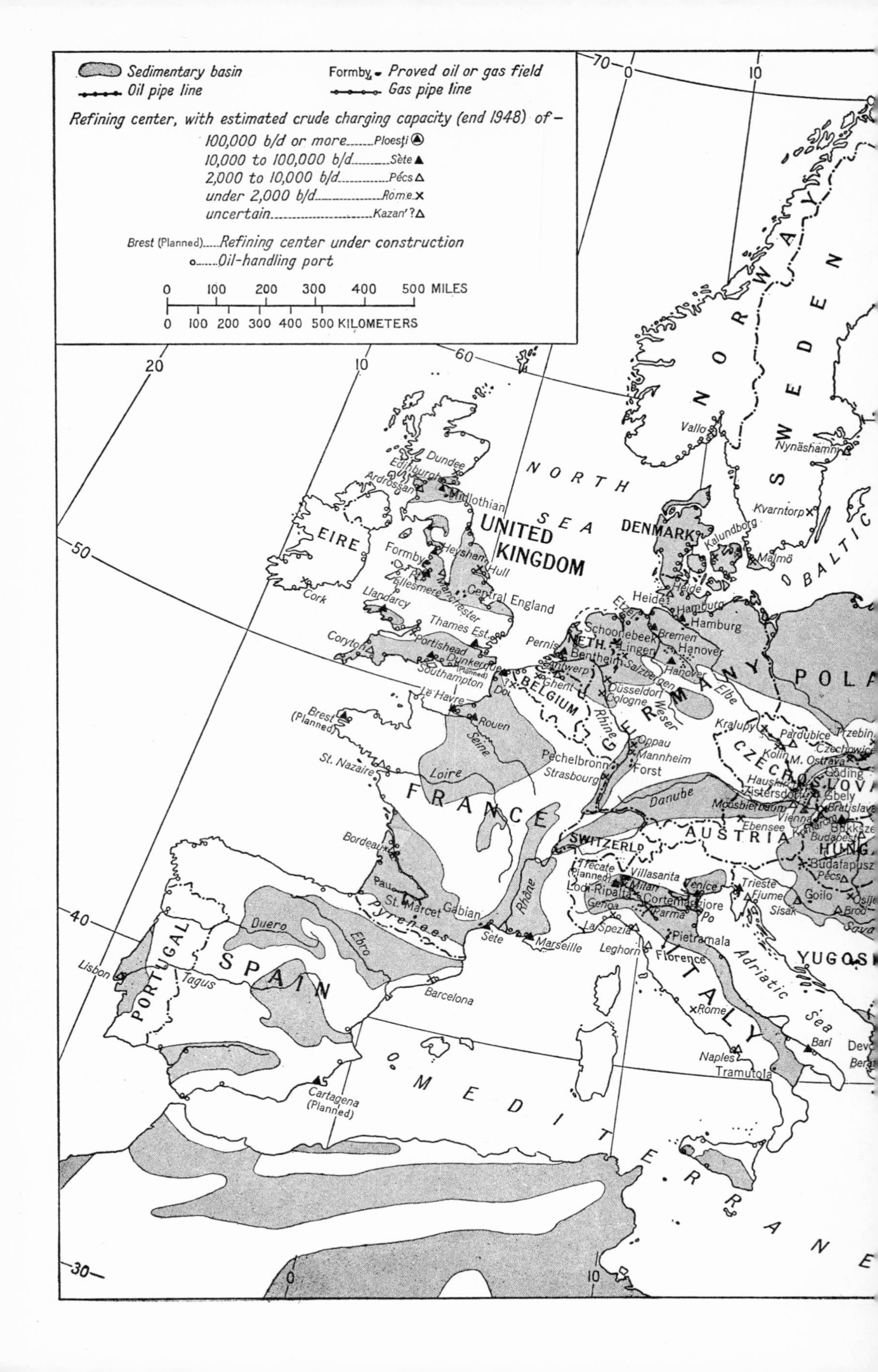

Sedimentary basin
Oil pipe line
Formby Proved oil or gas field
Gas pipe line
Refining center, with estimated crude charging capacity (end 1948) of—
100,000 b/d or more Ploeşti
10,000 to 100,000 b/d Sète
2,000 to 10,000 b/d Pécs
under 2,000 b/d Rome
uncertain Kazan'?
Brest (Planned) Refining center under construction
Oil-handling port
0 100 200 300 400 500 MILES
0 100 200 300 400 500 KILOMETERS
NORWAY
SWEDEN
NORTH SEA
UNITED KINGDOM
EIRE
DENMARK
BALTIC
NETH.
BELGIUM
GERMANY
POLA
CZECHOSLOVA
AUSTRIA
HUNG
SWITZERLD
FRANCE
SPAIN
PORTUGAL
ITALY
YUGOS
MEDITERRANE
Adriatic Sea
Vallo
Nynäshamn
Kvarntorp
Kalundborg
Malmö
Dundee
Edinburgh
Ardrossan
Midlothian
Formby
Heysham
Hull
Ellesmere
Manchester
Central England
Llandarcy
Thames Est.
Coryton
Portishead
Dunkerque (Planned)
Southampton
Cork
Heide
Hamburg
Etzel
Schoonebeek
Bremen
Lingen
Hanover
Pernis
Bentheim
Antwerp
Salzbergen
Ghent
Düsseldorf
Cologne
Elbe
Weser
Rhine
Le Havre
Rouen
Seine
Brest (Planned)
St. Nazaire
Loire
Oppau
Mannheim
Forst
Pechelbronn
Strasbourg
Kralupy
Pardubice
Kolin
M. Ostrava
Zistersdorf
Danube
Moosbierbaum
Vienna
Ebensee
Bratislava
Budapest
Pécs
Bordeaux
Pau
St. Marcet
Gabian
Pyrenees
Sète
Marseille
Rhône
Trecate (Planned)
Villasanta
Milan
Lodi-Ripalta
Cortemaggiore
Venice
Genoa
Parma
Po
Trieste
Fiume
Sisak
Goilo
La Spezia
Leghorn
Pietramala
Florence
Rome
Naples
Tramutola
Bari
Duero
Ebro
Tagus
Lisbon
Barcelona
Cartagena (Planned)
70
60
50
40
30
20
10
0

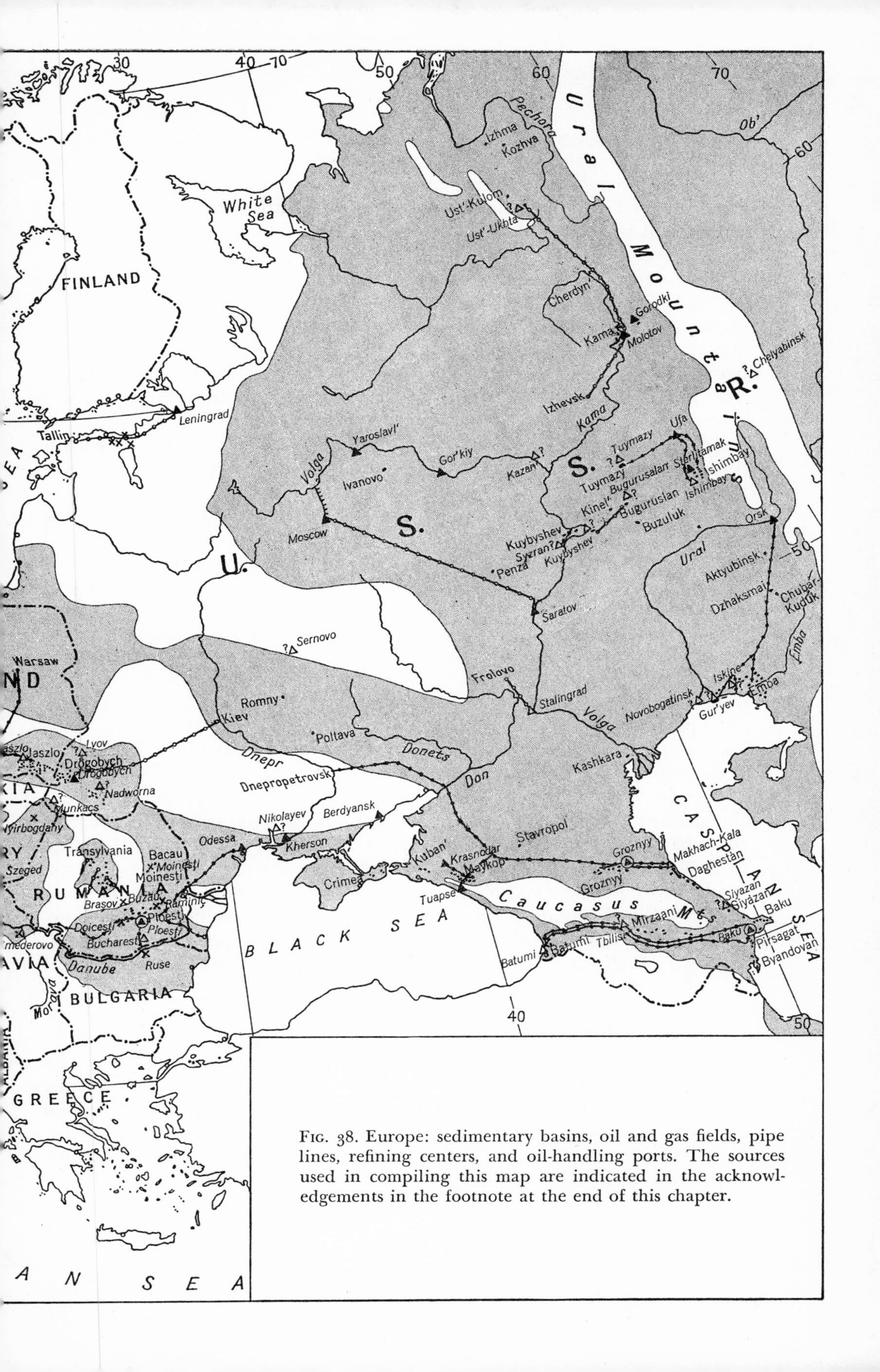

Fig. 38. Europe: sedimentary basins, oil and gas fields, pipe lines, refining centers, and oil-handling ports. The sources used in compiling this map are indicated in the acknowledgements in the footnote at the end of this chapter.

ities are present between Oligocene-Miocene and Miocene-Pliocene and Pliocene-Upper Pliocene. The last phase of folding and faulting, in Upper Pliocene, formed the structures as they are now, and accounted for much thrusting within the mountains and in the foothills. Maximum thickness of basin fill is estimated at from 26,000 to 40,000 feet. The Miocene itself may be over 16,000 feet thick.

The foredeep or trough portion of the basin forms a narrow rim, bordering and even underlying the Carpathians. In Rumania the deep part of the trough extends for only about 9 miles out from the mountain front, and the same situation probably prevails along most of the arc. Apart from the trough, the remainder of the basin is relatively shallow, shelf-like, and sloping gradually upward, away from the mountains. This is accompanied by wedging out of the older sediments.

The original trough of the basin must have occupied an area well within the present position of the Carpathians. Gravity measurements in Rumania show a decided minimum extending under the front range, and from this and other evidence it is concluded that the mountain mass thrust itself a considerable distance over the trough.

Evidence of severe thrusting is present in the belt extending outward from the front thrust range of the Carpathians. The thrusts within this belt are lubricated by salt (from the Aquitaine Oligocene formation) which often flows and distends into salt masses or stocks. The zone of thrusting (and folding) is confined to the trough portion of the basin, and the shelf area is undisturbed.

From the standpoint of petroleum, the zone of interest has been the deep portion of the basin, where it is folded and faulted; the Ploesti fields and those of Poland lie within this belt. No additional major fields have been found in these countries for more than ten years.

2. *The North German Basin.* This general basin area extends from Poland westward through north Germany, north Netherlands, and across the Channel into Scotland and England. It is characterized by the presence of many salt domes. Sedimentation began in the Permian and extended through the Triassic, the Jurassic, the Cretaceous, and the Tertiary in minor development. Almost the entire section is marine. Extensive deposits of salt were laid down within the Zechstein (Upper Permian) and, locally, in Rotliegend (Lower Permian). Maximum thickness of basin fill is probably in excess of 17,000 feet.

Several unconformities, mainly in the form of widespread transgressions, are found within the geological section throughout the basin. These transgressions are especially prominent in the younger beds, notably Upper Cretaceous and Tertiary, which have truncated older movements. The salt domes rising up from the deep Permian display well-defined periods of movement, the first period of growth being in Upper Jurassic and in Lower Cretaceous. These formations thin or wedge out against the flanks of the dome. In Upper Cretaceous time the salt pierced all overlying sediments, and renewed sporadic movements occurred in the Tertiary. The German geologists consider that the present accumulations took place during the Tertiary movements.

There are several subdivisions of the basin, the most important being the Hanover embayment, which contains the main group of oil fields and extends westward into the Netherlands. This subdivision is characterized by the presence of Jurassic and Lower Cretaceous, which generally pinch out along the north side of the trough. It may, therefore, be the presence of the Jurassic and Lower Cretaceous which make this the most prolific portion of the basin as a whole. Another feature of the Hanover embayment is a belt of anticlinal folding along its southern border where salt intrusions are generally absent. Near the Netherlands border this belt is purely anticlinal and has given rise to some relatively large fields.

The Heide group of fields in Schleswig-Holstein, in northernmost Germany, are in still another local embayment characterized by elongated salt ridges of Rotliegend salt (Lower Permian) in their cores, instead of Zechstein (Upper Permian), as in the other domes.

Minor subdivisions or embayments are the Munster Basin, the lower Rhine Graben between Belgium and the Netherlands, and several others. In Germany the southern limit of the basin is marked by the Harz Mountains and their western continuation, called the Osnabrück nose, to within the Netherlands.

Production of both salt domes and folded area is obtained over a large vertical range from Permian to Upper Cretaceous, the more important zones at present being Jurassic and Lower Cretaceous. The basin has not afforded as much oil as would be expected from its size, thickness of deposits, and apparently favorable grade of sediments.

3. The Hungarian Basin. This basin is situated within the arc of the Alps-Carpathians (see Fig. 38). It is fairly extensive, embracing both the Little and Great Hungarian Plains, which are partly separated by the ridge of old rocks along Lake Balaton.

The basin came into being during the Miocene, but the greater portion of the fill is comprised of Pliocene. The sediments are mostly brackish to marine. Maximum thickness is probably about 16,500 feet. Graben tectonics appear to have facilitated subsidence of the basin, and there is little or no folding that can be attributed to compressional movements.

Buried ridges, possibly horsts with relatively little sedimentary covering, form prominent features of the subsurface. These are flanked by deep areas (grabens), thus resulting in a generally irregular area of thick and thin basin fill. Besides irregularities resulting from the graben pattern, there are probably embayments having different facies and thicknesses, especially in the marginal areas.

Oil and gas from the six commercial fields, three in Hungary and three in Yugoslavia, are found in Lower Pliocene, and a minor amount is obtained from Miocene.

4. The Vienna Basin. The Vienna Basin is only about 110 miles long and 30 miles wide (Figs. 38, 39). It is an intermontane graben situated in the saddle at the junction of the Alps and the Carpathians.

Subsidence took place during Upper Miocene and Pliocene, and the sediments thereof (brackish to marine) attained a thickness of 13,000 to 17,000 feet. Sedimentation was probably closely related to that in the adjoining Hungarian Basin. The graben faulting shows up prominently in the subsurface along either side of the basin. The oil fields of Austria and Czechoslovakia are associated with this faulting, as well as with buried ridges reflected upward from the basement. Oil is from Pliocene, from Miocene, and from seepage, presumably, into porous pre-Miocene basement rocks.

The basin has afforded some relatively good fields. A new field, Matzen, was discovered in early 1949.

5. The Molasse Basin. The Molasse Basin, the westward continuation of the Carpathian Basin, follows the outer rim of the Alps through a portion of Czechoslovakia, Austria, Germany, and Switzerland. Sedimentation extended from Middle Oligocene to Upper Miocene, representing the last of the northward-migrating troughs of the Alpine geosynclinal system. Maximum basin fill was 15,000 to 30,000 feet, and the greater portion consisted of fresh-water deposits.

The trough of the basin originally extended considerably to the south, but the last great northward thrust of the Alps in Upper Miocene or Pliocene subsequently removed the sediments up to the northern edge of the deep trough. The present form of the Molasse is pictured as a gently-inclined shelf, dropping off steeply toward the south into a deep trough, of which only the northern portion remains intact. The shelf part of the basin was not disturbed by the Alpine thrust.

There is a small non-commercial gas field in the Miocene at Wels, in Upper Austria. A showing of heavy oil is present in Czechoslovakia at an old non-commercial field, another shallow showing in the northwest corner of Austria, and several seepages in Switzerland along the north side of the basin.

Altogether about 90 exploratory wells have been drilled.

6. The Po Basin. The Po Basin follows the outer edge of the Apennines, coinciding with the Po Valley and the Adriatic, and extends a short distance into Albania.

The basin of interest begins with the Oligocene or Miocene, and includes Pliocene and Quaternary. At least part of the original deep trough of the Miocene was situated back from the present front of the Apennine foothills. Subsequent extensive Apennine thrusting at or near the end of the Miocene resulted in much of the sediments being removed and tucked under. Gravity readings show an appreciable minimum anomaly under the foothills.

The Pliocene trough of deposition is situated in front of, and laps

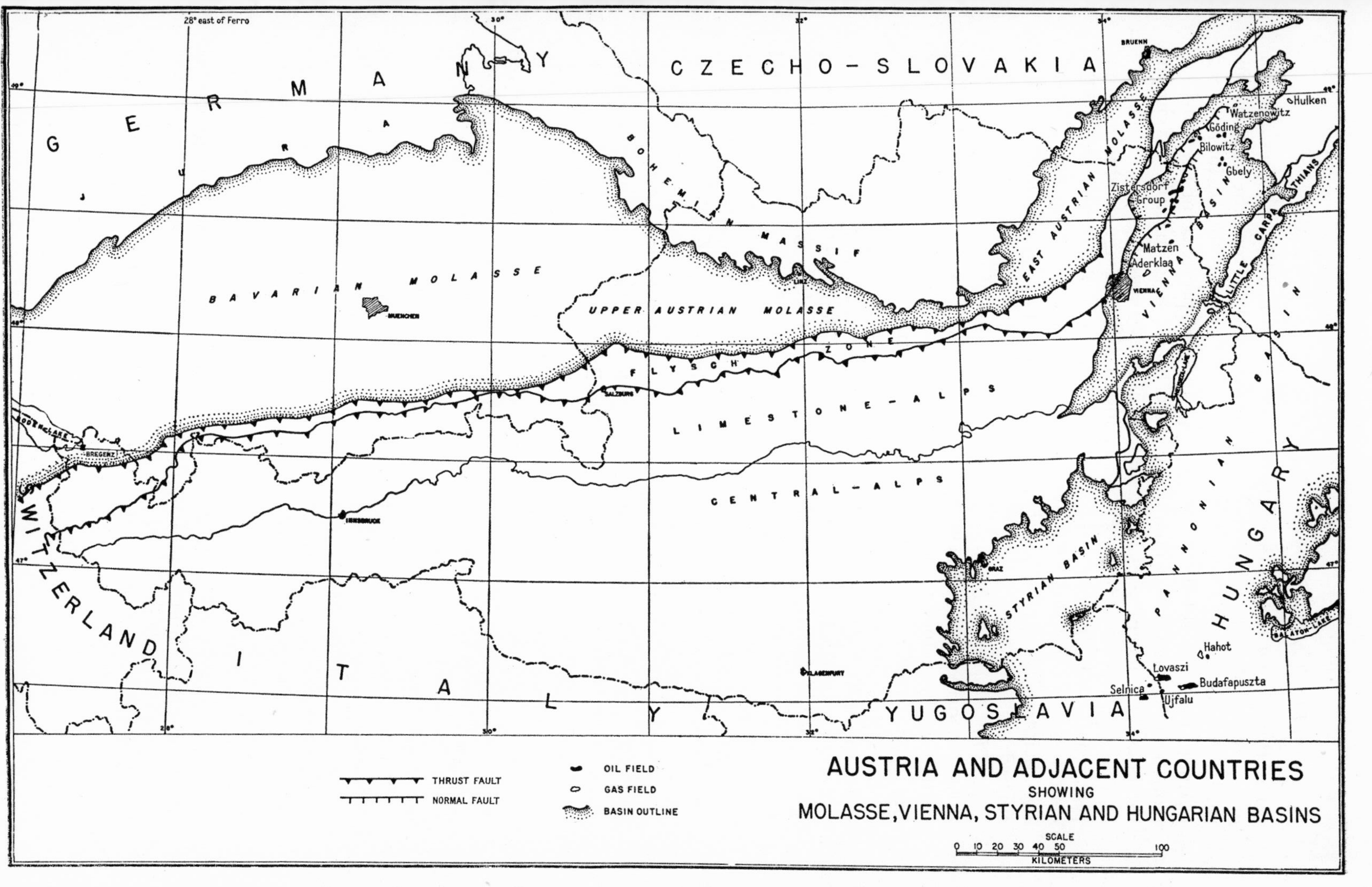

FIG. 39. Austria and adjacent countries: basins, oil fields, and gas fields in mid-1949.

upon, the present front line of the Apennine foothills. It occupies a width of 15 to 30 miles. This trough is clearly recognized from the gravity picture. Subsidence and deposition is still taking place in the basin, as noted by more than 3,000 feet of Quaternary and the presence of the Adriatic trough. The basin sediments are mainly brackish to marine; maximum thickness of Miocene plus Pliocene may exceed 30,000 feet.

The Apennine foothills of Cretaceous to Miocene age are highly complicated, mainly by much imbrication thrusting. Out in the basin Miocene is folded and thrusted, overlain by gently folded Pliocene, decreasing in intensity upward in the younger beds. The Quaternary is undisturbed.

7. The Aquitaine Basin. The Aquitaine Basin of southern France lies directly north of the Pyrenees. It represents the last outward-migrating trough in the sequence of the geosyncline. The deep portion of the basin lies to the south, partly tucked under the Pyrenees, and the shallower foreland rises northward.

In the initial stage of the Aquitaine Basin, Middle Cretaceous, followed by Upper Cretaceous and Eocene, were deposited upon folded and somewhat eroded Jurassic. During the Eocene orogeny the basin sediments were strongly folded and faulted, the intensity decreasing outward from the Pyrenees. The formations of the foreland area rise progressively toward the north. The subsequent Miocene transgression truncated folds and overlapped most of the basin with its flat-lying beds.

THE MINOR SEDIMENTARY BASINS

The minor sedimentary basins—Upper Rhine Graben, Sicilian Central, Duero-Ebro, Transylvanian, Thuringian, Paris, Portuguese, Seville, Tajo, and other smaller ones appearing on the basin map—are rated according to this author's personal opinion. Factors entering into the judgment are: form of basin, structural characteristics, geological history, character and thickness of sediments, size of basin, and results of exploration and development to date.

Petroleum Developments in the Separate European Countries

The individual countries of Europe, faced with the economic problem of supplying themselves with petroleum, have not only canvassed the possibilities of increasing their own crude production, but also the supplementary resources of natural gas, shale oil, and synthetic petroleum from coal; and they are giving especial priority to the development of petroleum refining within their borders. In the following survey all these features will be briefly scanned.

The sedimentary basins in Europe west of the Soviet Union overlap international boundaries in many places, but for reasons of practicality brief summaries of the oil developments are presented by countries rather than by the natural basinal units, the countries being reviewed in order of their total oil output. We shall give, for each country separately, a brief account of the geological, exploratory, production, and reserves data; data on refineries and shale-oil development; and statistics of net imports or exports. The latest dates for which information has been available are indicated after each appropriate item.

The countries of Europe west of the U.S.S.R. may be divided, as regards petroleum production, into three groups: first, those producing a greater amount than they consume; second, those that have commercial production, but insufficient in quantity for their own internal needs; thirdly, those countries that have no commercial production. The three groups are shown in Table 20. Based on crude-oil shipments to Europe from all sources of supply, and including oil in transit and for military use, the total imported European consumption amounted in 1948 to an average of 1,063,500 barrels daily.

GROUP I. COUNTRIES PRODUCING A GREATER AMOUNT THAN CONSUMED

These are, in order of their total production to January 1, 1949, Rumania, Austria, Hungary, and probably Albania (Table 20). They include about 33 million people, and their level of petroleum consumption is not high in comparison with the industrial countries in the west. Before the war Rumania was the only country of this group that was producing an exportable surplus; most of it went up the Danube to the countries bordering that river, and the larger part was delivered in Germany. During the war European production was pushed, and in the postwar years there have been exportable surpluses in Hungary and probably in Austria and Albania. These countries, being now in the east European trading sphere, dispose of their surpluses among the other member countries in the east.

Rumania, Austria, and Hungary have refinery capacity sufficient to treat domestic crude, but only Rumania has a sizeable industry. Announced plans for expansion are on a moderate scale.

Rumania. Rumania is the most petroleum-prolific of all the European countries west of the U.S.S.R.—a total output of about 1.2 billion barrels of oil has been produced since 1860—and, though in a marginal position, Rumania is properly considered as falling within the realm of the oil-rich "mediterranean region" centered on the Middle East. Rumanian territory includes a substantial part of the Carpathian basin, all of the Transylvanian gas basin, and a small part of the Hungarian basin.

The oil-producing sediments of Rumania in the Carpathian basin flank the outer Carpathians in their broad arcuate sweep of over 1,000 miles from the Danube in Austria northeastward, southward, and then westward to the Iron Gates gorge of that same river in southwestern Rumania. Oil

Table 20. European countries, west of the U.S.S.R.: production of crude petroleum and crude-charge refining capacity in relation to requirements in 1948.

Group I: Countries producing a greater amount than consumed.
Group II: Countries having commercial production, but insufficient for internal requirements.
Group III: Countries having no commercial production.

	Cumulative Production to 1–1–'49	Beginning Year	Peak Year	Daily Averages in 1948			
				Production	Internal Requirements	Surplus or *Deficiency*	Refinery Capacity
GROUP I	(In thousand barrels)			(b/d)	(b/d)	(b/d)	(b/d)
Albania	11,832	1933	1947	980 [a]	...	...	1,000
Austria [b]	51,383	1935	1944	15,604 [a]	12,000	3,604	20,010
Hungary	42,134	1937	1943	9,990 [a]	7,400	2,590	24,840
Rumania	1,187,386	1857	1936	80,600 [a]	39,100	41,500	126,250
Total, Group I	1,292,735			107,174	...	...	172,050
GROUP II							
Czechoslovakia	4,124	1919	1942	500	...	...	6,000
France	13,986	1918	1933	989	161,100	*160,111*	236,100
Germany [c]	92,761	1880	1940	13,000	45,300	*32,300*	56,355
Italy [d]	3,443	1865	1932	177	58,100	*57,923*	81,300 [d]
Netherlands	5,412	1944	1948	9,435	56,100	*46,665*	60,000
Poland	277,438	1874	1909	2,702 [a]	...	...	7,300
United Kingdom	4,132	1919	1943	923	309,900	*308,977*	115,150
Yugoslavia	2,147	1935	1944	1,100 [a]	...	...	8,000
Total, Group II	403,443			37,826	...	...	570,205
GROUP III							
Belgium & Luxembourg	—	—	—	—	36,500	*36,500*	9,165
Bulgaria	—	—	—	—	...	...	...
Denmark	—	—	—	—	24,500	*24,500*	530
Eire	—	—	—	—	9,800	*9,800*	600
Finland	—	—	—	—	8,600	*8,600*	—
Greece	—	—	—	—	14,000	*14,000*	—
Iceland	—	—	—	—	3,000	*3,000*	—
Norway	—	—	—	—	27,200	*27,200*	—
Portugal	—	—	—	—	13,500	*13,500*	7,000
Spain	—	—	—	—	24,300	*24,300*	700
Sweden	—	—	—	2,750 [e]	61,400	*58,650*	13,900
Switzerland	—	—	—	—	17,100	*17,100*	—
Total, Group III	—	—	—	—	239,900 [f]	*237,150* [f]	31,895 [f]
All Countries	1,696,178	—	—	145,000	...	...	774,150

[a] Estimated.
[b] Production is confined to the Soviet zone.
[c] Includes the U.S., British, and French zones.
[d] Includes Trieste, which has a refinery capacity of 15,000 b/d.
[e] Shale oil, not included in totals.
[f] Not including Bulgaria.

Note: The basic data for estimating the internal requirements for each country are: indigenous crude oil production; consumption; imports and exports of the industry.

Sources: Data on cumulative production from *World Oil,* Vol. 129, No. 4, July 15, 1949, p. 47; daily averages compiled by the Coördination and Economics Department, Standard Oil Company (N.J.).

production in the Rumanian portion of this basin is almost exclusively from Pliocene rocks; the earlier stages of the Tertiary down to the Oligocene yield only minor quantities.

The important fields of Rumania are concentrated in the small Ploesti * district, only 9 by 30 miles in extent, where some 25 fields are present. The oil-field structures in the Ploesti district are confined to the deep trough-like portion of the basin, and may be regarded as grouped in three belts, corresponding to the changing tectonics outward from the Carpathians. The innermost belt is characterized by long overthrust structures, often exposing Oligocene. The fields (Campina, Runcu, etc.) in this belt are old, mainly in the stripper stage, and produce from Miocene and Oligocene. The next belt is that of compressed, thrusted anticlines within the Pliocene. Salt of Aquitanian (Oligocene) age has flowed up through the thrust planes, acting as a lubricating medium, and often expanding to sizeable salt stocks. This belt is probably greatly reduced in width through being tucked under thrusts from the first belt. The famous Moreni-Baicoi trend is within this zone. The features distinguishing the third belt are more gentle folding and thrusting, and absence of salt penetrating the Pliocene section. Fields within this area are Aricesti, Bucsani, Margineni, Boldesti, and Ceptura. No structures are recognized beyond this belt, and it is thought that this may represent the outer rim of the deep portion of the trough, where the shallow shelf portion begins.

To the northeast of Ploesti on the east fringe of the Carpathians, the old fields of Bacau furnish a minor quantity of oil, and are commercial only because they can be worked in a very primitive manner. These small producing areas were opened about 1860 or earlier.

The Transylvanian basin is within the arc of the Carpathians. The marginal area of the circular basin is formed by non-folded sediments dipping gently basinward. Next is a zone of diapir folds with salt domes, then a zone of prolific gas-bearing domes in the basin's central portion. There are seven dry gas fields, and production is from Upper Miocene. First gas production was in 1913, and in 1946, the latest year for which figures are available, production amounted to 22.5 billion cubic feet.

In 1924 the government took over the subsurface rights of all unleased lands, which meant that outlying areas as well as some lands within the producing area had to be dealt for with the government. The government and the industry never arrived at an equitable mining law which would have encouraged exploration, and consequently very few new fields were found. Bucsani in 1933 was the last major discovery. On June 11, 1948, Rumania nationalized the oil industry.

In 1948 crude production is estimated to have been more than twice the domestic requirements, and refining capacity about one-and-one-half

* Typographical difficulties have made it impossible to follow the usage of the map in Fig. 38, in accenting Rumanian proper names and placing the diacritical mark beneath the *t*, as in Ploesti, Constanta.

times as great as production. The Ploesti area contains all the refining facilities, except a minor center at Bucharest, and scattered small plants on the lesser fields. Expansion is contemplated in plant at the Targul Mures gas field.

The output of the Ploesti refineries is transported by pipe lines, west to the Hungarian frontier at Turnu Severin, south to the Danube port of Giurgiu on the Bulgarian frontier, and southeast to the Black Sea port of Constanta. It has been reported that a pipe line has also been laid from Ploesti to Odessa. Gas pipe lines from the Targul Mures fields serve the principal towns of Transylvania.

Austria. Parts of three principal basins lie within Austria: the Vienna, the Molasse, and the Hungarian, with their subdivisions (Fig. 39). All of Austria's commercial oil production comes from the Vienna basin, along the faults and buried ridges on the west side of the graben, where fields have been producing since 1930. Oil is mainly from Upper Miocene, although some is present in the Eocene basement just below the contact with Miocene. A new field, Matzen, was discovered in early 1949, bringing the total to 11. The fields are about drilled up, and showed a decline from 1947: the daily average for 1948 was 15,604 b/d, but they are being pulled heavily, and their future decline is likely to be sharp.

The Matzen discovery is the southward continuation of the producing area. An anomaly was known from prewar gravity work, and drilling was undertaken in 1947–1948 by Soviet interests. Accumulation, also in Miocene, is thought to be controlled by a buried hill somewhat like Aderklaa, and is situated east of the west side faults which are so closely associated with the other fields.

The Soviet Oil Administration has carried out considerable exploration, core drilling, and seismic work within the Vienna basin and southeast into the Styrian embayment, but it is now confining its efforts to the Vienna basin.

The other possible oil areas of Austria are the Molasse basin and the Styrian embayment. The Styrian area, to the southeast of Vienna and near the frontier of Hungary and Yugoslavia, is an embayment of the Hungarian basin, and closely related also, structurally and in age, to the Vienna basin. Only one well has been drilled. The basin has been mapped by gravity and seismic survey.

In the Austrian portion of the Molasse about 30 exploratory wells have been drilled, but without finding commercial production.

As war reparations, the U.S.S.R. took over those companies having German interests, and this resulted in their holding the major portion of the oil area. Until the Austrian Peace Treaty is concluded, the Russians will continue at their discretion exploration and development within their occupation zone. Meanwhile, Austria is not giving out exploration concessions in the remainder of the country until the political situation has cleared.

Before the war Austria was dependent on imports of crude and refined petroleum from Rumania. The greatly expanded crude production since then now leaves a surplus for export.[1] In 1948 crude production exceeded domestic requirements by about one-third, and refining capacity was about one and one-third times the crude production. Except for two small plants, one at the Zistersdorf field and one at Ebensee on the Upper Danube, the refineries are all in the Vienna area. Proposed construction would increase the capacity by one-half or more, but would not much affect the concentration. The industry is primarily dependent on rail transport.

Hungary. The boundaries of Hungary include more than half of the main portion of the Hungarian basin, and its western and northern embayments. Hungary has three commercial fields, all in the Budafapuszta area in the western part of the main basin. Two fields, Budafapuszta and Lovászi, are of major size. They are gentle anticlines, with some 300 to 600 feet oil column. Oil is found in lenticular Lower Pliocene sands, and accumulation depends upon the presence of porous facies. The third producing area is on the Hahot anticline, which is also fairly large and gentle, but of which only a small portion is productive. A group of eight wells gives a small amount of oil from Miocene just above the basement, the Pliocene being barren. A mile or so northwest along the structure is a gas area of nine wells producing from the Lower Pliocene sands, the same productive zone as in Budafapuszta and Lovászi. A small amount of non-commercial oil was found at Ujfalu, in the Budafapuszta area, but the field has now been abandoned.

Extensive carbon-dioxide gas deposits are present within the basin, on buried ridges often associated with igneous dykes. One area is at Mihály, 62 miles north of the oil fields, where the CO_2 gas deposit is about 9 miles long. The other CO_2 area is at Inke, about 19 miles southeast of the fields. Some CH_4 gas (methane) has been found at Inke and Görgeteg. Forty-five exploratory wells have been drilled on seismic and gravity anomalies, but found only CO_2 and CH_4 gas.

Beginning in 1934, exploration and development was initiated in Hungary by Standard Oil Company (N.J.) interests on an 8,000,000-acre M.A.O.R.T. (Magyar Amerikai Olajipari, R.T.) concession south and west of the Danube. On September 24, 1948, these holdings were expropriated and taken over by the government.

Drilling and development has been carried out east of the Danube by the Hungarian government at the Bükkszék field, discovered in 1937 but never a commercial producer; and exploratory drilling during the war by a German-controlled company, the M.A.N.A.T., in their concession in southeastern Hungary apparently had little success. Natural gas is found in northeastern Hungary near Debrecen.

[1] Although Austria's production should be sufficient for the requirements of the country, the situation is complicated by the fact that little or none of the crude or products gets out of Soviet hands after supplying the needs of their occupation zone. Consequently, imports must be made to the American, British, and French zones.

Before the war Hungary imported petroleum from Rumania, mainly in the form of "artificial" (partially refined) crude; in 1938 the quantities imported were nearly five times as great as the Hungarian crude production. In 1948 crude production exceeded domestic requirements by about one-third, and the refining capacity was equal to about two and one-half times the crude output. The refinery with the largest capacity is near Komárom on the Danube above Budapest, but proposed construction would raise the capacity of the Budapest area to about the same level. The third center, at Pécs in southern Hungary, near the coal, may also be expanded. The crude oil is piped from the Budafapuszta fields to Budapest by a six-inch line.

Albania. A small part of the southeast extension of the Po basin lies in Albania. The Miocene oil of the Devoli and other small fields at Pahtoro are within this area.

Exploration was initiated by the Italians during the first World War, and the first oil well was discovered in 1918. Between 1923 and 1938, the Albanian government had granted seven or eight concessions to American, British, French, and Italian interests. However, all except the Italian companies became discouraged with the results of exploration, and eventually forfeited their rights.

The Italian companies transported most of the crude by pipe line to a tanker anchorage at Valona and thence to Bari and Leghorn, for treatment at hydrogenation plants. The asphaltic quality of Albanian crude oil is unsuitable for ordinary refining methods, and there is only a skimming plant at Berat, near the fields.

Production reached a peak in 1942, with production variously estimated at one and one-half to two million barrels from about 550 wells, and again in 1947, but production for 1948 is estimated at only 980 b/d.

After the second World War, following the termination of Italian occupation, the Albanian government assumed rights over all concessions and all production. For a time at least, Yugoslavia replaced Italy as the principal market for Albanian petroleum products.

GROUP II. COUNTRIES HAVING COMMERCIAL PRODUCTION, BUT INSUFFICIENT FOR INTERNAL REQUIREMENTS

These states, the home of 265 million people, include the most populous in Europe west of the U.S.S.R.: Germany, Great Britain, Italy, France, Poland, Yugoslavia, Czechoslovakia, and the Netherlands, in order of population size; or in the order of their total crude production to January 1, 1949, Poland, Germany, France, Netherlands, Great Britain, Czechoslovakia, Italy, and Yugoslavia. The crude produced was everywhere inadequate for the needs of modern industrial states. Only in western Germany and the Netherlands was it as much as one-fifth of requirements in 1948; in France, Italy, and Great Britain, it was less than one per cent of requirements.

On the other hand, these countries have natural advantages for the import of petroleum. All except Czechoslovakia have ocean ports, and the central European states have inland water transport as well as a dense rail net. All, being industrialized or planning to become so, have potential exports of manufactures. These advantages were temporarily canceled by the fact that all except Sweden were belligerents, and all except Czechoslovakia suffered war damage.

In the countries where war damage to industry and transport was especially severe—France, Germany, Poland, and Yugoslavia—and in Czechoslovakia for different reasons, the supply of petroleum in 1948 was less than in the prewar period. In other countries where coal supplies were short, petroleum products, and especially fuel oils, were being used in greater volume than before the war.

The refining industry was fast recovering in 1948 in western Europe, and in the Netherlands, Sweden, and the United Kingdom, it was operating on a larger scale than in 1938. Further reconstruction and expansion are planned by all the western countries in this group: the planned capacity for 1952–1953 ranges from an increase of about one-half over the prewar capacity in the Bizone of Germany, to a thirteen-fold increase in Italy, and a thirty-fold increase in Sweden. The combined program would make these countries net exporters of petroleum products instead of net importers, as they all are at present.

Information about current and proposed operations in Czechoslovakia, Poland, and Yugoslavia is somewhat fragmentary. The first two are apparently aiming at a moderate self-sufficiency, and Yugoslavia at a greatly enlarged production.

Poland. The territory of postwar Poland contains the northern part of the Carpathian basin and a broad expanse of the North German basin. Production comes from the Jaszlo area in the Carpathian basin, and before the war the adjoining Carpathian fields, then also Polish, furnished a larger amount.

In the Polish part of the Carpathian basin, Pliocene, which furnishes so much of the oil in Rumania, is absent. Another difference is that there are three extensive overthrust sheets composed of Flysch (Cretaceous to Oligocene) which seem to have extended outward for a considerable distance, and may cover much of the Miocene trough portion of the basin such as that noted in Rumania. Structural conditions are really complicated in the thrust sheets, one piled on top of another, and the search for oil is relatively blind. Oil occurs mainly in Oligocene series, and is found in each of the three sheets. The outer thrust contains the most important fields, but the middle thrust affords the greatest number.

The Polish fields date back to 1856 although production records go back only to 1874. Peak production of 41,000 b/d was attained in 1909; in 1938 the rate was only 10,000 b/d. The incorporation of eastern Poland, including the famous Boryslaw field, into the Soviet Union has meant a reduction of about 75 per cent in Poland's production. Estimated production for 1948 amounted to about 2,700 b/d, and cumulative production, including all production in the pre-World War II area up to 1939, to about 277,-438,000 barrels.

Present plans are to increase appreciably the search for new fields, and already discoveries have been claimed. But, judging by past records, discoveries may come slowly; between 1913 and 1939 only one minor field was found. However, the northern part of present-day Poland includes a large undeveloped sedimentary area, distinct from its present productive belt, extending to the Baltic Sea; and this region, practically unexplored geologically, is underlain by strata which include thick salt members. As a direct eastward extension of the wide North German Plain with its many oil fields, this part of Poland may prove productive, because of the similar age-range of the formations, namely, Permian to Cretaceous inclusive.

The refining industry in postwar Poland, though small in comparison with that of 1938, has a capacity well in excess of the domestic supply of crude. In 1947, it is estimated, the crude production was equal to a little more than one-third the internal requirements, and refining capacity was equal to nearly three times the crude production. Probably similar conditions continued in 1948. Five or six small refineries along the pipe line from the Jaszlo fields to the city of Kraków and then to Cieszyn (Tečin) on the Czechoslovak border, produce gasoline and other petroleum products. There is a products pipe line from Kraków to Warsaw.

The National Plan for Economic Reconstruction, covering the four years 1946 through 1949, provides for moderate increases in the production of crude oil, liquid gas and gasoline, and natural gas, and for the production for the first time of compressed gas, synthetic gasoline, methanol, and carbide.

Germany. The northern half of postwar Germany is included in the North German basin; in the southern half a much smaller area is occupied by the Thuringian basin and the eastern half of the Upper Rhine graben. Almost all of the current production comes from the North German basin, and about half the total from the Hanover district.

In the discussion concerning the North German basin, it was noted that the two producing provinces of northwest Germany are, on the one hand, the folded belt along the south and southwest side of the basin, and on the other, the area of salt domes covering the remainder of the basin. The whole region extends some 180 miles east to west, and about the same distance southward from the Danish frontier.

The folded belt extends from the Harz Mountains northwest toward the Netherlands, crossing the frontier in a zone which extends from Bentheim gas field to Schoonebeek field. From the standpoint of the sediments present, this belt may be thought of as being in the southern part of the Hanover embayment. There are very few salt extrusions within the belt, and westward the anticlines increase in number and the belt becomes wider. The area is of great importance in view of the fields (Emsland, west of the Ems River) developed within the past few years. In the first part of 1949 a new discovery, Ruhlertwist No. 2, promises to open up a new field just northwest of Dalen and east of Schoonebeek. Two modern seismic outfits are assigned to the Emsland actively searching for more fields.

The Hanover basin with its many small oil fields is restricted to the region south of a line trending northwest along the Aller and Weser rivers almost to Bremen.

The other oil areas of northern Germany are near Hamburg and, farther northwest, near Heide in Schleswig-Holstein, close to the North Sea and some 45 miles from the Danish border. Apart from the four oil fields and one gas field of the folded Emsland area, all the other fields to the north are associated with salt tectonics, with piercement domes or salt ridges. Of the 200 or so known domes, only 23 so far have been commercially productive. Out of this number, only about half have ultimate yields of more than 1,000,000 barrels, and only four have produced over 7,000,000 each to date.

Although the salt domes by and large have not given prolific production, the discovery in August 1949 of Suderbruch, an interdomal structure, situated about six miles east of Steimbke field, opens up a new type of prospect which may afford larger fields than heretofore found within the salt basin. Several interdomal structures are already known from geophysical work in various parts of the salt-dome area. Two of them have been drilled, but only the Suderbruch well was productive. From present information the interdomal structures (*Swischenhoch*) appear to be deeply buried gentle swells or "highs" which have not been pierced by salt.

The other oil-producing area of Germany is to the south, in the Upper Rhine graben, on the German-French frontier. Two fields, Weingarten and Forst-Weiher, are on German territory. The Upper Rhine graben is a very pronounced feature, although only about 30 miles wide and 150 miles in length. The graben began sinking during the Eocene, and continued through Oligocene into Miocene. Production is found in Triassic, Jurassic, and Oligocene.

A limited amount of exploration has been in progress for several years in the German portion of the Molasse Basin. No fields are present in this area, and only one exploratory well was drilled during the past few years.

Showings of gas are present in one or two wells drilled in the Munster embayment; a small amount of geophysical work is in progress, but prospects are poor for development. This embayment of the main basin is represented by Upper Cretaceous overlapping upon Carboniferous.

The discovery rate in Germany was slow, until pressed by military needs. From the time of Germany's first oil well in 1874 (there was production from pits as early as 1856) up to 1930, only six fields were developed, and the daily production was less than 2,000 barrels. But in 1934 the Nazi government nationalized unleased lands, gave out concessions, and subsidized exploration drilling, and in 1935 new fields began coming in. Between 1935 and 1945 there were 23 discoveries, and production rose correspondingly, a peak being reached in 1940, with 20,000 b/d. However, in succeeding years, production dropped rapidly because the fields had been pulled too heavily. No new discoveries were made from 1945 up to the end of 1948, although several fields have been extended appreciably by short stepouts.

Nearly all German production is in the British occupation zone, with a small amount in the French zone and some exploratory work being undertaken in the United States zone. There is no production in the Soviet zone, but several hydrogenation plants are located in it. It is noteworthy that none of the proposed revisions of German frontiers assigns any of her oil-productive territory to her neighbors.

The German oil-producing industry comprises sixteen operating companies, only two of which are controlled by foreign capital. Six of the companies control 75% of the production. The government controls the subsurface oil rights, and issues concessions of moderate size to interested companies. Royalty and working requirements are not onerous.

Germany had a total production of 635,240 barrels for 1948, amounting to a daily average of about 13,000 barrels. Totals have increased since 1945, and the present trend indicates that appreciable advances may take place within the next few years.

The German crude-oil refining industry, although heavily damaged during the war, far exceeds in capacity the production of the German fields and the imports of crude petroleum. In 1948 the crude production of western Germany (the Bizone) was equal to less than one-third of internal requirements, but the refining capacity was more than four times the crude production. The largest refining center is at the entrance to the Elbe River system, in the Hamburg district; it operates almost entirely on imported crude. The next largest center is the Hanover area, where capacity present and planned could refine all the locally produced crude and a larger volume of imported crude. The third largest refining center, in the Bremen port area, has no locally-produced crude. The refining centers in the Rhineland and in eastern Germany are mainly devoted to hydrogenation of heavy oils and coal, and to the production of synthetic products from coal or water gas. Their future status is still somewhat uncertain, but it appears that the Soviet Zone supplies its petroleum requirements by hydrogenation of local lignite.

There are no trunk pipe lines; the petroleum products are transported by river, canal, and rail.

Large deposits of low-grade oil shale [2] have been located in Württemberg,

[2] The notes on oil shales are abstracted by kind permission of the author from "The Oil Shale Deposits of the World and Recent Developments in their Exploitation and Utilization," by W. H. Cadman, M.B.E., being a paper presented to the 7th International Congress of Pure and Applied Chemistry, London, 1947, and reprinted in *Institute of Petroleum Review* (London), Vol. 1, No. 11, Nov. 1947.

Hesse, Bavaria, Baden, and the Rhine provinces. Commercial production began during World War II, and a retorting plant was installed at Frommern, Württemberg, to treat 1,100 tons of shale and produce some 40 tons of oil daily. The Germans have experimented with various processes for treating oil shales, notably with methods for underground extraction from low-grade beds, and probably could, in course of time, exploit successfully all their extensive deposits.

France. Six sedimentary basins are represented in France, three of them—the Aquitaine, the Rhône, and the Allier—in their entirety. A large part of the Paris basin, half of the upper Rhine graben, and the western tip of the Molasse, are all in France.

France's small oil production is obtained in the basins of the Upper Rhine graben (Péchelbronn field) and of Aquitaine (St. Marcet and Gabian), and total daily production for 1948 was estimated at only about 1,100 barrels.

The Péchelbronn field is an 8½ mile strip of scattered producing area along the French side of the Rhine graben. Production is from Oligocene, Jurassic, and Triassic, and shaft or mining operations initiated many years ago account for almost half the field's production. Although discovered as long ago as 1813, this field still furnishes the greater part of France's production. The crude is refined locally, at Merkwiller. It would seem that by now the French side of the Rhine graben should have been exhaustively explored, without hope of finding any new fields; but in 1948 alone 22 exploratory wells were drilled, and three of them found oil, although in extensions of the known Péchelbronn area. And in mid-1949 a deep well found flush production in Muschelkalk (M. Trias) yielding about 500 b/d. This, however, remains to be evaluated by additional deep drilling.

In the Aquitaine basin of southern France two fields have been proved: St. Marcet, in the foothills of the central Pyrenees; and Gabian, some 120 miles to the east, near Béziers, and some 25 miles from Frontignan, near Sète on the Mediterranean, where there is a refinery.

The St. Marcet field is a diapir-like structure characteristic of the Pyrenees foothills portion of the Aquitaine basin. Approximately 20 wells have been drilled, of which about half are Cretaceous gas wells, and two wells produce some oil from below, in Middle Jurassic. Daily production at the end of 1948 averaged 40 barrels of crude and 21 million cubic feet of gas. The gas is transported by pipe line to towns in the Pyrenees area and as far as Bordeaux. The old Gabian field produces from Lower Triassic dolomite; production is now only about 2 to 3 b/d.

Three large concessions cover the southern part of the Aquitaine basin. That of the Société Nationale des Pétroles d'Aquitaine (S.N.P.A.) covers about 9,266 square miles. Numerous structures are present in this territory, many of the diapiric type with salt or anhydrite cores, and many of these have been drilled without success. A discovery was reported in early 1949 at Lalongue, near Lembeye, in the Pau region, but this also turned out to be non-productive. S.N.P.A. is exploring with both gravity and seismic parties, and completed five exploratory wells in 1948.

The Régie Autonome des Pétroles (R.A.P.) concession of some 770 square miles, in which are located the St. Marcet fields, is enclosed, except on the south, by that of S.N.P.A. Besides gravity and seismic parties, about ten rigs are being used for exploration and exploitation, and 11 exploratory wells were completed in 1948. About six structures are known on this concession and have been tested, without finding production outside of the St. Marcet field.

The Société Nationale des Pétroles du Languedoc Méditerranéen (S.N.P.L.M.) concession of about 3,200 square miles lies to the east of the other two, in the Rhône delta area, and the old Gabian field is within the western part. Eight rigs are occupied with exploratory drilling, and a number of wildcats have already been completed. Some showings were obtained, but so far no new discoveries have been made.

S.N.P.A., R.A.P., and S.N.P.L.M. are state-controlled companies. Standard Française des Pétroles (Standard Oil Company, N.J., Atlantic Petroleum Company, Gulf Oil Corporation, and French interests) has applied for a concession in the Bordeaux region.

In 1948 crude production was equal to less than one per cent of the internal requirements, but the refining capacity was equal to one and one-half times the domestic requirements. The planned rebuilding and new construction of refineries would increase this capacity by about three times in 1952–1953. The refining industry is distributed among the ports convenient both for receiving the imported crude and for shipping products by inland waterways as well as by rail—Le Havre and Rouen on the Seine, Marseille and Sète on the Mediterranean, St. Nazaire and Bordeaux on the Atlantic. The refineries destroyed at Dunkerque, Brest, and Bordeaux are being rebuilt. Exports of products go to French possessions overseas.

The French bituminous shale industry is more than 100 years old. The richest deposits lie in central France, at Autun (Saône-et-Loire) and at St. Hilaire (Allier). There are also extensive low-grade deposits in the south and the east, and some in Vendée in the west. At Autun some 850 tons of shale can be handled daily, producing 5 to 6% by weight of oil, and at St. Hilaire about 1,000 tons daily, by specially devised plants that extract nearly 100% of the potential maximum.

The reserves in the Autun area are estimated at some 70 million tons, and in the St. Hilaire area at some 30 million. The reserves of lower-grade shales in the south are unknown, but the outcrops indicate from half a billion to one billion tons. The low-grade (less than 11 gallons of oil per ton) shale in the east, at Creveney, some 90 miles southwest of Strasbourg, is estimated at between 2 and 3 billion tons, and methods of underground distillation *in situ* are being studied.

The Netherlands. All but the southernmost strip of Netherlands territory lies in the North German basin. The sedimentary and structural conditions of the North German basin in Germany extend into the northeastern part of the Netherlands, where the folded belt and salt dome area are clearly represented, at least for a short distance inland from the frontier. The structural feature—the Harz Mountains and its northwest continuation in the Osnabrück nose—forming the south boundary of the basin in Germany, plunges northwest, thus permitting the Netherlands' portion of the basin to extend farther south and lap onto its original shoreline of the Brabant massif of northern Belgium.

Exploration in the Netherlands was initiated in 1932, and greatly intensified during the war years under German occupation. A regional gravity survey, refraction, seismic, and core drilling were carried out. Deep drilling led to the discovery of the Schoonebeek field in 1943. Active development of the field has been continued since the war, and extensive exploration is being carried out in various parts of the country.

This field is on the east plunge of the Coevorden anticline, one of the structures within the folded belt. The producing sand (Valendis of Lower Cretaceous) is eroded from the higher part of the anticline by the blanket-like transgression of Upper Cretaceous, thus leaving it "bald." The structure is gentle, somewhat faulted, and has an oil column of about 325 feet. The oil is 25° API, paraffinic, and very viscous. Limits of the field are established except to the west, where it may be further extended. Production for 1948 averaged 9,435 b/d.

A well completed in 1948 found gas at 9,055 feet in Zechstein (Upper Permian) on the Coevorden structure, about 9 miles west of Schoonebeek field, and in August 1949 shallow gas at 3,790 feet was found near De Wijk' about 25 miles west of Schoonebeek. These are the only discoveries since Schoonebeek.

Operations in the Netherlands are carried out by the Nederlands Aardolie Maatschappij (N.A.M.), a company jointly owned by B.P.M. (Royal Dutch) and by Standard Oil Company (N.J.), the former acting as operator.

Before the war the crude production was nil, and three-fourths of the imports were in the form of refined products. Since the war production has been a small source of supply, and imports of crude have increased. In 1948 the crude production of the Netherlands was equal to about one-sixth of internal requirements, but the refining capacity exceeded these requirements by about one-tenth. The whole refining industry is concentrated in the Pernis plant near Rotterdam, which has a crude-charging capacity of 60,000 b/d, to be increased to 90,000 b/d with an expanded range of chemical products.

Great Britain. About half the area of the island of Great Britain forms part of the area of sedimentary basins: the southern part of England and Wales fall within the Paris basin; and the Midlands and North of England, with southern Scotland, fall within the North German basin.

Production comes mainly from fields in the English Midlands, and to a slight extent from Dalkeith in southern Scotland.

Britain had produced a small amount of oil, from one to three thousand barrels yearly, from 1919, the year of discovery of the Hardstoft, Derbyshire, field; but the more concentrated efforts toward development of indigenous British petroleum resources dates from the early nineteen-thirties, when a systematic study of oil-field possibilities was undertaken by the geological staff of the Anglo-Iranian Oil Company.[3] Largely as a result, the Petroleum Production Act of 1934 nationalized oil rights, and allowed exploration licenses to be taken out on encouraging conditions over substantial areas, free from the former necessity of dealing with a multitude of private landowners.

Prospects ranged over a variety of separate geological provinces. In the southern counties, which form part of the Paris sedimentary basin, the objective was the Mesozoic; in the Midlands and the North of England, and in the Scottish Lowlands, which together form the northwestward extension from Germany and the Netherlands of the North German sedimentary basin, there were a number of possibilities.

Exploratory drilling under the new Act began in 1935, and the first objectives were certain prominent anticlines in the southern counties, but results were negative. Next the Permian Magnesian Limestone of northern Yorkshire and the Oil Shale group of the Calciferous Sandstone Series of Scotland were tested, and gas fields were proved at Aislaby in Yorkshire and near Dalkeith in Scotland. Meanwhile geophysical work had been proceeding in Nottinghamshire and Lincolnshire, where an unconformable cover of Permian and Mesozoic rocks shielded from view the nature and attitude of the Carboniferous strata beneath, and in March 1939 a well was drilled at Eakring, in the former county. This boring struck oil in the following June, in a sandstone in the Millstone Grit at 1,912 feet, and on test produced at a rate of some 12 tons daily. In the same year in Lancashire a small strike of oil was made at a depth of 125 feet, in Keuper Waterstones below a cover of glacial drift, at Formby, near Liverpool, where seepages of oil and gas had been noted in a peat bog.

The outbreak of war in September 1939 brought a decision to concentrate on the Eakring area, and additional fields were discovered nearby at Kelham Hills (1941), Duke's Wood (1941), and Caunton (1943), all in Nottinghamshire.

Very extensive seismic, gravity, and other surveys were later undertaken, and have been carried out, in most of the other promising regions of Great Britain, but so far their results have been very disappoint-

[3] The following paragraphs are in large part based on "The Exploration for Oil in Great Britain and its Economic Consequences," being the Abbott Memorial Lecture, 1946, of University College, Nottingham, by G. M. Lees, F.R.S., Chief Geologist, the Anglo-Iranian Oil Company, and author of the chapter "The Middle East" in the present volume.

ing. Production, which reached a maximum of 839,000 barrels in 1943, has slowly but steadily declined since, to some 700,000 barrels in 1944, some 530,000 in 1945, some 410,000 in 1946, some 351,000 in 1947, and some 323,000 in 1948: and the figure for 1948 would have been considerably lower but for the successful application of secondary-recovery methods. Up to the end of 1948, cumulative British production amounted to some 4,130,000 barrels, of which Duke's Wood had produced some 1,884,000 barrels, Eakring 1,157,000, Kelham Hills 858,000, Caunton 154,000, Formby 59,000, Dalkeith (Scotland) 18,000, and Hardstoft 9,500.

All the producing British fields are operated by the D'Arcy Exploration Company, Ltd., a subsidiary of the Anglo-Iranian Oil Company, with the one exception of the Dalkeith field, which is operated by the Anglo-American Oil Company, Ltd. The Anglo-American Oil Company has carried out gravimetric and refraction seismic surveys, and holds extensive prospecting licenses, and up to 1941 surveys were also carried out by the Gulf Exploration Company, Ltd., and by Steel Brothers and Company, Ltd. At the present time Imperial Chemical Industries hold prospecting licenses covering 169 square miles in Yorkshire, and hope to develop gas production for local operations.

From all the surveys that have been undertaken, very valuable information has been derived as to Britain's non-petroleum mineral resources: indeed, in the long run this knowledge is likely to provide the most profitable return, to the country as a whole, for the effort expended. One example is the discovery of a completely new coal field north, east, and southeast of Lincoln, and there are excellent prospects for new coal fields elsewhere. There has also been an important discovery of potash salts of Permian age in north Yorkshire, though the deposits at present proved lie at a considerable depth. In addition, the contribution to geological and scientific knowledge has been notable.

In 1948 the domestic production of crude petroleum was equal to about 0.1% of requirements, while the refining industry had a capacity of slightly more than one-third of requirements. The large refining centers are situated at ports, with the greater capacity on the western side of the island, but the plans for refinery construction will make the Lower Thames and the Firth of Forth the leading centers in size. The output of refinery products planned for 1952–1953 is slightly more than six and one-half times the prewar average.

Only the Scottish shale fields have been commercially exploited. These fields extend about 15 miles, from below the Firth of Forth in Midlothian into West Lothian. The seams of rich oil content extend over an area of some 75 square miles. The shale deposits have been mined for the production of oil since 1851, extraction thus antedating by seven years the sinking of the first United States petroleum well.

The Scottish oil shales occur in the Calciferous Sandstone Series of the Lower Carboniferous, lying between the Old Red Sandstone and the Carboniferous Limestone. The series are some 5,000 feet thick, with recurring outcrops, which results in large areas of oil shale lying within easy mining reach. The shale beds vary from a few inches to 15 feet in thickness, the workable seams being from 4 to 12 feet thick. Yields of crude oil vary from about 16 to 40 gallons per ton, and contain an average of about 2% by weight of sulphur. The crude oil yields motor spirit, kerosene, gas oil, diesel oil, light lubricating oil, paraffin wax, fuel oil, and coke. It is particularly suited for diesel oil, the yield being as high as 50% by volume.

In 1946 some 1⅓ million tons of oil shale were produced, from 12 mines and from open-cast workings. Good use was made of by-products and of spent shale; the latter, mixed with lime, being converted into bricks. Reserves of Scottish oil-shale are estimated at between 480 and 880 million tons.

The extensive deposits of bituminous shales in Dorset, southwestern England, and Norfolk, East Anglia, contain such a high percentage of sulphur as to make the cost of removing it prohibitive, though parts of the seams have long been used locally as a solid fuel. Elsewhere in the British Isles oil shales have been found in small quantities only.

Czechoslovakia. Small parts of the Carpathian and Molasse basins, and the northeastern half of the Vienna basin, are included in Czechoslovakia. The Czechoslovak fields are situated in the Vienna basin, under the same structural control as the Austrian fields. However, they are on the eastern updip end of the graben, and consequently shallower in depth, and this may account for their lower productivity. Several small fields in the Carpathian basin near the Polish border apparently ceased producing before the war.

The Czechoslovak oil industry is now operated under the Ministry of Industry; it has not been possible to obtain information concerning operations. The 1948 production is estimated to have averaged about 500 b/d; it is also estimated that this production of crude petroleum was equal to less than one-tenth of requirements, and that the refining capacity was equal to about three-fourths of requirements. This capacity was far in excess of the supplies of crude available. Before the war petroleum imports consisted mainly of crude, but since the war products predominate. The principal refineries in order of size are at Pardubice on the Elbe system, Bratislava on the Danube system, and Kolín on the Elbe system. Before the war it was proposed to build a refinery in Slovakia.

The bituminous shale of Permian age located in the Kladno coal basin is mined, together with the underlying coal, and used as fuel in glass works and for gas making. It has not been used commercially for oil production. The extensive Middle Miocene beds of Cypris oil shale in western Bohemia, between Falknov and Cheb (Eger), had not, up to 1946, been utilized either for fuel or for oil production.

Italy. The Po Valley basin occupies a large part of northern Italy, and extends southeast through the Adriatic. Up to now Italy's meagre crude-oil production has been almost entirely from the foothills of the Apennines, derived from Cretaceous to Pliocene seepage oil in structurally and stratigraphically complicated areas. There are about ten of these small fields, which together produced only 177 b/d in 1948, although production was first recorded nearly ninety years ago, in 1860.

Gas in the Po Valley has assumed major importance since the discovery of Lodi in 1943, and subsequently Ripalta in 1948 and Cortemaggiore in 1949, all controlled by the Azienda Generale Italiana Petroli (A.G.I.P.). These are large anticlinal structures of 12 miles or more in length, and it seems likely that other structures of this magnitude will be found as exploration progresses. These anticlinal gas fields are situated in the western part of the Po Basin; production is from Lower Pliocene or Upper Miocene, at depths of 4,000 to 5,250 feet. The fields of Lodi and Ripalta produce dry gas, whereas the Cortemaggiore gas is accompanied by condensate or light crude amounting to 100 to 200 b/d in some of the first wells to be completed in 1949. This may mark the turning point of Italy's oil production.

The other source of gas is from the numerous small Quaternary fields in the eastern part of the Po Valley; these are operated generally by independent producers. The gas is shallow, 300 to 1,300 feet in depth, with correspondingly low volume and pressure. Lenticular sand conditions rather than structure (the Quaternary is apparently almost flat) seem to determine accumulation. Production of gas in 1948 is estimated at 4.6 billion cubic feet, or 12,600,000 cubic feet per day. This is an increase over 1947, due chiefly to better outlets rather than to new discoveries.

The western deep gas production up to now is all from the Lodi field (Ripalta and Cortemaggiore are not yet in the exploitation stage) which has about 15 wells completed, although only about half are producing, alternately. The daily average in December 1948 was 3,710,000 cubic feet; production may be increased as additional pipe-line facilities are completed.

The shallow gas fields in the eastern Po Valley produced at the rate of about 8,890,000 cubic feet per day in 1948, although some of this total may be attributed to gas from various areas in the Apennine foothills. The relatively greater production of the eastern area is a temporary question of outlet, and the situation may be reversed within the next few years. Much of the gas is bottled for use in automotive transportation.

In 1948 crude production was equal to less than one per cent of requirements, but refining capacity (excluding Trieste) was equal to about five-sixths of requirements. The largest capacities are now at Bari and Venice, on the Adriatic, but proposed construction, totaling about 150,000 b/d, is planned mainly in port areas on the western side—La Spezia,

Leghorn, and Naples—and near Turin, connected by pipe line to Vado near Genoa. These when completed could supply an enlarged domestic and export market.

Numerous beds of bituminous shales occur in the Alpine Trias in Northern Italy near Besano, south of Lake Lugano, and small quantities of oil are distilled; and bituminous shales in the Cretaceous, from which good asphalt can be made, have been located near Morraro in the Trentino. In Sicily, bituminous shales containing from 9 to 19% by weight of hydrocarbons have been reported from many districts. Distillation tests are said to have been successful.

Yugoslavia. The southern part of the Hungarian basin and very small portions of the Carpathian and Po Valley basins are within Yugoslavia. Commercial production is considered to be limited to three fields—Peklenica, Selnica, and Lendva—situated in the Hungarian basin along the western trend of the Hungarian fields. Production is from the Pliocene, as in Hungary.

Peak production of 2,000 b/d was reached in 1944, during the German occupation, under great pressure and by the use of reckless methods. It is reported to have been only about 1,100 b/d in 1948. However, in 1939 it was only 20 b/d, and to date a cumulative total of about 3,105,000 barrels has been produced.

The petroleum industry was nationalized in December 1946. The state's current five-year program aims at making the country self-sufficient, but the past history of oil-finding in Yugoslavia indicates that this may be difficult to accomplish within a short period.

The lack of recent information prevents an estimate of requirements in 1948. Before the war, however, these appeared to be about 3,000 b/d, of which the greater part was imported as crude from Rumania. There were small refineries along the Sava, Morava, and Danube rivers, but perhaps all were demolished or inoperative in 1948, except one of 2,200 b/d at Sisak. Since the war the acquisition of Fiume has provided a refining center at an Adriatic port where crude from overseas may be used. If the five-year plan, calling for a crude output of about 9,500 b/d, is fulfilled, a parallel expansion of refining capacity is to accompany it.

The Trieste Area. Trieste has long been one of Europe's major commercial as well as strategic ports. Political considerations apart, it is a logical site for petroleum storage, as well as for transit trade. In what ways the large expansion of petroleum production that is planned by Yugoslavia might affect neighboring Trieste is at present scarcely predictable, but the Trieste territory's present refining facilities may be called upon to the full, or expanded. At present (spring, 1949) however, while the largest refinery, the 9,000 b/d plant of the Aquila S.A. Tècnico-Industriale, is in operation, the 2,500 b/d plant of the Sòcietà Americana del Petròlio is shut down.

GROUP III. COUNTRIES HAVING NO COMMERCIAL PRODUCTION

Of these thirteen countries, only five—Finland, Ireland, Iceland, Luxembourg, and Norway—are lacking in sedimentary basin areas favorable for furnishing oil; but, with a combined population of some 86 million people, all except Spain are relatively small, and for the most part lack the raw materials of heavy industry.

Some have small refining industries, but most of the petroleum they use is imported after refining, and between 1938 and 1947 the volume of imported products doubled. None of these countries has sufficient refining capacity to supply the domestic requirements, and only in Belgium, Spain, and Sweden have building plans for substantially expanded refining capacity been announced.

Sweden. The southwest tip of Sweden is within the North German sedimentary basin. A deep well was drilled in this area in 1946–1947, but without favorable result. However, Sweden produces about 2,750 b/d of shale oil, by mining operations in several provinces in the southeastern part of the country.

It is estimated that Sweden possesses about one billion tons of oil shales minable by open pits, and a total reserve of some five billion tons, in localities conveniently situated for rail or water transport. Swedish scientists have been responsible for great inventiveness in oil extraction—in particular, the Ljunström method of electrically heating the shale underground by means of bore-holes placed a few feet apart. Three months are required to heat the shale before vapor is produced; vaporization then proceeds for at least two months; and then for many months, while the area is cooling down, the ingenious Swedes use the tropically heated soil for the production of luxuriant vegetables. Swedish scientists have also played a leading part in investigation of the uranium content of shales, and have estimated that the Swedish deposits contain more than one million tons of uranium. (Both the U.S.A. and the U.S.S.R. have deposits of similar or larger size and of similar origin, and deposits of similar origin are also known to exist in Belgium, Czechoslovakia, Denmark, France, Germany, and Norway. They may also exist, according to the Swedish experts, in virtually every country in the world.)

In 1948 the shale oil produced was equal to less than one-twentieth of Sweden's petroleum requirements; the refining capacity, using both shale oil and imported crude, was about 14,000 b/d, or less than one-fourth of those requirements. Some three-fourths of these were therefore met by importing products. Proposed refinery construction would raise the Swedish crude-refining capacity to about 175,000 b/d by 1952–1953. The present center is at Nynäshamn, a port south of Stockholm on the Baltic Sea.

Denmark. All Denmark lies within the North German basin. Prospecting rights for the whole country are held by one company, the Danish

American Prospecting Company, which is a subsidiary of the Gulf Oil Corporation. Active geological and geophysical work was begun in 1938, suspended during the German occupation, but resumed after the war.

The first deep test, Vinding No. 1 in west-central Jutland, was spudded on July 21, 1947, drilled to 7,985 feet, and abandoned in December of the same year. The second deep test was Gassum No. 1, in east-central Jutland, some 80 miles northeast of the abandoned Vinding site, and at the end of 1948 a depth of 8,816 feet had been reached. The company has undertaken considerable shallow drilling with portable rigs, confirming the existence of a number of shallow salt domes that had been indicated by geophysical exploration; rock salt has been reached on the Vejrum, Suldrup, and Hornum domes in northern Jutland, in the same region as the two deep tests. The company has maintained a considerable geological organization in the country, including a gravimeter party, a refraction-seismograph party, and a stratigraphic laboratory. But to date there has been no commercial discovery.

Denmark has one small refinery, at Kalundborg, and two small blending plants at Copenhagen. The crude-refining capacity is only about 530 b/d, and the country must import almost all its requirements as finished products. The national economic plan for 1952–1953 includes a refinery, to operate on imported crude and produce a considerable proportion of the country's needs.

Belgium. Northeastern Belgium lies within the western end of the North German basin. Belgium has no commercial production, but considerable refining facilities: in 1948 the refining capacity was equal to one-fourth the internal requirements. Prewar capacity was over 11,000 b/d, but most of the refineries suffered great damage, and the Belgo Petroleum plant at Terdonck was completely destroyed. Capacity was back to some 7,500 b/d by the spring of 1948, and by the spring of 1949 had climbed to some 9,000 b/d.

The primary center is the port of Antwerp, and a second is at Ghent. Prospective additions would increase capacity at Antwerp about five times by 1952–1953.

Portugal. The western part of Portugal includes a sedimentary basin along the coast, north and south of the Tagus. The sediments are Tertiary to Jurassic in age. Oil shales in the lower Lias exist, but the seams are thin, and consequently development would be difficult.

The Anglo-Portuguese Oil Company carried out prospecting activities for a number of years, and shallow drilling operations near Torres Vedras, some 25 miles northwest of Lisbon. Non-commercial oil showings were obtained, and the company suspended operations at the end of 1947.

Beginning in 1947 the Companhia Portugueza do Petróleo, a joint enterprise of the Portuguese government and of the Swedish Axel Johnson shipping interests, undertook geological and geophysical investigations but up to the latter part of 1949 had done no drilling.

Portugal's one refinery is at Lisbon, operated by the Sociedade Anónima Concessionária da Refinação do Petróleos em Portugal (S.A.C.O.R.), in which the state holds a one-third interest. In 1948 its reported capacity of 7,000 b/d was equal to about half the internal requirements; according to current plans it will be increased to 10,000 b/d.

Spain. Of the three basin areas in Spain, the Duero-Ebro, the Tajo, and the Seville, attention is presently directed chiefly on the Duero-Ebro basins fronting the Pyrenees.

C.A.M.P.S.A. (government-owned company) and C.I.E.P.S.A. (50% Socony-Vacuum, 50% private Spanish interest) are carrying out a modest geophysical and exploratory drilling campaign. In 1948 C.I.E.P.S.A. completed an 8,200-foot dry wildcat in the eastern or Ebro basin. The rig then went to Burgo de Osma, in the Duero basin, in Soria province, where a well was started in December 1948.

Farther north, in Burgos province, C.A.M.P.S.A. continued drilling. A third active corporation, A.D.A.R.O., a government organization that operates Spain's mineral industries, continued drilling near Chiclana de la Frontera, south of Cadíz, in the Cadíz-Seville basin, but operations later were temporarily suspended, pending modernization of the equipment.

In addition to these organizations, the Instituto Nacional de Geofísica has undertaken some electrical prospecting.

In 1948 the refining capacity on the Spanish mainland was equal to only about 700 b/d, or less than 3% of internal requirements. In the Canary Islands, however, a refinery with a crude capacity of 10,000 b/d can now supply about two-fifths of Spain's internal requirements, and a proposed refinery at Cartagena, on the Mediterranean, with a capacity of 25,000 b/d could supply the domestic market at a somewhat higher level of consumption.

Considerable deposits of shale have been found in several provinces, and large-scale distillation began in 1948.

Switzerland. Most of northern Switzerland lies within the Molasse Basin, the western continuation from Germany. Sedimentary conditions are similar to those in Germany, except that in Switzerland, along the north side of the west half of the basin, oil seepages and folding are present. Both folding and seepages are related to the presence of the Jura Mountains. About seven exploratory wells having geological importance were drilled before and during the war, and, as a wartime measure, some thought was given to possible mining of the oil-impregnated sands found west of Geneva.

Switzerland makes much use of hydroelectric power and of coal, but imports of petroleum products amounted to 8,700 b/d in 1938, to nearly 15,000 b/d in 1947, and to more than 17,000 b/d in 1948. The Swiss have no crude-refining capacity, although there are several small specialized processing plants.

Bulgaria. The northern half of Bulgaria is within the southern limits of the Carpathian basin. While Bulgaria has no present commercial production, prospecting is being undertaken in several areas. The Bulgarian oil industry was nationalized in December 1947, and the country's three small refineries are government-operated. Their capacity at the beginning of 1949 was estimated at 1,200 b/d.

Greece. Greece has no indigenous oil production. Sedimentary rocks are present within the territory, although not under sufficiently favorable conditions to warrant the term of sedimentary basin as applied in oil exploration. Sediments that accumulated during the late Eocene, Oligocene, and early Miocene were highly disturbed by Alpine orogeny. Subsequent deposits of late Miocene and Pliocene within local faulted areas also suffered from much folding and volcanic activity. The overall section of the Tertiary sediments is thought to be thin.

Numerous seepages of petroleum, asphalt, and gas are present within Greece, and at least five localities of oil shale are known. An oil seepage on the island of Zanta was mentioned by Herodotus in 484 B.C.

The first concession for oil development dates back to 1865. Hand pits were dug and several exploratory wells drilled during the subsequent years, but without finding commercial production. No exploratory activity has been reported since just before World War II.

No refining capacity existed in 1948, but the national economic plan includes the erection of refineries to process about 8,000 b/d by 1952–1953, which, it is estimated, will then be equivalent to one-third of the internal requirements.

Finland. With no oil fields nor sedimentary basins, Finland must import all its petroleum and petroleum products. In 1938 crude oil imports averaged somewhat over 600 b/d, or about one-ninth of the total requirements of roughly 5,500 b/d. Information about recent years is incomplete. Apparently the imports now consist entirely of products, which are obtained from both the west and the east. In 1948 imports were estimated at about 9,250 b/d, of which nine-tenths came from the west and one-tenth from the east.

Iceland. This island in the North Atlantic has been claimed as part of Europe, and also as belonging to North America. It has extensive coal deposits, but no oil fields and no refineries, and imports its petroleum requirements in finished products form. In 1945 the imports of the four main types of liquid products, supplied almost entirely by the United States, totaled 277,000 barrels. After the war ended, imports, especially of fuel oils, rose sharply, and Venezuela became the principal supplier. In 1948, imports of the four main types of products totaled 958,000 barrels, of which 902,000 came from Venezuela. Gas and fuel oils accounted for 673,000 barrels in the total import.

Ireland. Ireland has no oil fields nor any sedimentary basin. One refinery, with a capacity of 600 b/d, was in operation in 1948 near Cork, but no additional refineries are indicated in the national plan for 1952–1953.

.S.S.R. Kazakstan. New derricks in the South ield of the Emba basin. This is one of the largest regions in the Soviet Union, covering an area ndred thousand square kilometers (38,600 square and is estimated to contain a billion tons of oil. .)

S.R. Kuybyshev. The pipe line at the Kuyby- refinery, east of the Volga. (Sovfoto.)

76. U.S.S.R. The Saratov-Moscow gas pipe line. These pipes, insulated with bitumen and covered with a casing of wood, are ready to be dropped into a trench on the bed of the Oka. (Sovfoto photo by D. Minsker.)

77. South Sumatra. Sungei Gerong refinery of Standard Vacuum P.M., prewar view. (N. V. de Bataafsche Petroleum Maatschappij.)

78. South Sumatra. Sungei Gerong. February 1948. Many of the refinery workers come to work in their own boats. (Standard Oil Company, N.J., photo by Corsini.)

Luxembourg. With no oil fields nor sedimentary areas, Luxembourg imports its requirements. It has no refineries.

Norway. Norway has no oil fields nor, unlike Sweden, has it commercial shale deposits, but it is a considerable and increasing consumer of petroleum products, especially fuel oils. Its refinery at Vallo, on Oslofjord, has a capacity of 700 b/d, or only about one-fiftieth of the estimated requirements in 1948. The long-term program for 1952–1953 contemplates increasing imports of products, but not the development of refining on any extensive scale.

Europe's Requirements and Capacities

From the information given in Table 20, a general picture of the situation in 1948 emerges, in spite of the gaps in the information about parts of eastern Europe. If the countries for which information is lacking—Albania, Bulgaria, Czechoslovakia, Poland, and Yugoslavia—are left out of account, the deficiency of production below requirements averaged about 795,000 b/d in 1948 for the countries of all three groups taken together. The gap to be filled by imports from outside the region was probably larger than this figure, however, since a substantial portion of the surplus available in the countries of Group I may well have been exported to the Soviet Union rather than to petroleum-deficient countries in the rest of Europe.

While an average daily import of 800,000 to 850,000 barrels represents a large absolute demand for petroleum, it is relatively small in comparison with the rate of use in the United States, which, on a per-capita basis, was from thirteen to fourteen times that of Europe in 1948. The potential expansion in European demand, whenever economic conditions permit, is obviously very great.

If one compares the output of crude petroleum in 1948 in the twelve producing countries of Europe west of the U.S.S.R. with that in the last prewar year, 1938, and again with that in the war years of peak production, 1942 and 1943, it appears that the efforts to meet the deficiency by expanding crude production have not resulted in a significant increase in output for the area as a whole; the gains made in Hungary, Austria, the Netherlands, and Germany have been more than counterbalanced by the decline in Rumania.

The petroleum deficiency in Europe west of the U.S.S.R. must be met in the main by imports from outside that area. The reorientation between 1938 and 1947 in sources of supply and in proportions of crude and refined products has been set forth in Figs. 2–5 in Part Two.[4] In 1948 imports were for the first time preponderantly from the Middle East, rather than from Venezuela; and the Middle East is designated as the

[4] The utilization pattern of 1938 and the direction of future developments as regards both Europe and the world as a whole are discussed in Part Four, Chapters 3 and 4, of this volume.

principal source of the expanded imports contemplated in the coming years.

In anticipation of a substantial increase in the proportion of crude imports, Europe's actively expanding refining industry made rapid progress in 1948. At the end of the year the combined crude capacity of the European countries west of the U.S.S.R., as shown in Table 20, totaled 777,530 b/d, or almost three times as much as the crude imported from extra-European sources in 1938. By 1953 their combined capacity may be more than twice as great as that for 1948.

A Note on Sources for Fig. 38: Europe: sedimentary basins, oil and gas fields, pipe lines, refining centers, and oil-handling ports. Information on sedimentary basins, oil fields, and refining centers supplied by the Standard Oil Company (N.J.). Pipe lines from *World Oil* atlas issues, May 1946, June 1947, July 1948, and July 1949. Oil-handling ports from Hurd's *Ports of the World,* London, 2nd edition, 1948, in which it is noted that "it has been impossible to obtain any reliable up-to-date information for some German or for any Japanese or Russian ports." Information on Estonia also from Pullerits, A., ed., *Estonia: Population, Cultural and Economic Life,* Tallinn, 1937, p. 98 and appendix pp. 3–7.

12. INDONESIA, BRITISH BORNEO, AND BURMA

BY I. SWEMLE *

The Oil Province of Southeast Asia as a Whole

IN 1939 the total production of crude petroleum in Southeast Asia amounted to approximately 10 million metric tons (76 million barrels), or about 3½% of world production for that year. In the Southeast Asia total about 80% came from Indonesia [1] while British Borneo furnished nearly 10%, and Burma a slightly larger amount. In Table 21, production during 1948 and the cumulative production, from the beginning through 1948, are summarized.

The areas in Southeast Asia where oil is being recovered, and those where additional sources may be developed, are scattered over a very large expanse (Fig. 40). The producing fields on the mainland, in the valleys of the Irrawaddy, and those in the western part of the island of New Guinea are more than three thousand miles apart. The rest are distributed over the vast Indonesian archipelago, compromising some thousand islands and a land area of about one million square miles—stepping stones between Asia and Australia. The largest islands—Borneo, Sumatra, New Guinea, and Java—possess the chief oil-producing areas.

Since the present work is primarily geographical, the geology will be dealt with only briefly. The reader who wishes more may be referred to the bibliography at the end of the book.

The oil-bearing formations of the important producing areas of Sumatra, Java, and Borneo belong exclusively to the Tertiary, as does the only basin so far evaluated in the western part of New Guinea. During Lower Tertiary time a number of geosynclinal basins came into existence along the marginal part of the Asiatic continental block, which had been stabilized by pre-Tertiary orogenesis. Subsidence of an oscillating character was maintained in these basins during the Tertiary until Plio-Pleistocene time, when the main folding occurred in Java, Sumatra, and southern Borneo. In western New Guinea a period of folding of middle Miocene age is indicated by a widespread unconformity.

The thickness of the Tertiary deposits in the central parts of the well-

* Geologist, Bataafsche Petroleum Maatschappij.

[1] In this paper the general term "Indonesia" is used to denote all territories formerly known as the Netherlands Indies. Since this paper was written, the United States of Indonesia have been recognized by the Netherlands; the new state's territories comprise all the lands formerly known as the Netherlands Indies, except Netherlands New Guinea.

Table 21. Indonesia, British Borneo, and Burma: production of crude petroleum in 1948 and cumulative totals as of January 1, 1949.

	Year of Discovery	Production In 1948 b/d at year-end	Production In 1948 year's total	Cumulative to 1/1/49
		(in thousands of barrels)		
Sumatra				
Northern basin	1893	—	—	156,114
Central "	1941	—	—	243
Southern "	1901	81.0	25,439	435,607
Java	1896	1.5	475	120,026
Netherlands Borneo				
Eastern basin	1897	9.2	3,185	257,192
Tarakan	1906	7.8	2,670	158,454
Netherlands New Guinea	1936	3.9	137	146
Ceram	1897	—	—	8
Total, Indonesia		103.4	31,900	1,127,790
Sarawak (Miri)	1911	1.1	342	70,811
Brunei (Seria)	1929	61.0	18,246	82,295
Total, British Borneo		62.1	18,588	153,106
Burma	1889	—	300	288,000
Total, Southeast Asia		165.5	50,794	1,568,896

Source: *World Oil,* Vol. 129, No. 4, July 15, 1949, p. 263; and other sources.

known basins ranges between about 20,000 and 50,000 feet. The latter figure applies to the geosynclines of Borneo, where the Paleogene reaches thicknesses of a mile or more, in contrast to the basins of Sumatra and Java, where strong subsidence started later; in Java and Sumatra Lower Miocene does not reach the great thickness found in Borneo.

Production is, with few exceptions, confined to sands and sandstones of Upper Miocene and Pliocene age, with the Upper Miocene by far the more important. Only in Tarakan is most oil derived from Pliocene beds. In South Sumatra the Lower Miocene holds some highly productive sands. The structures in western New Guinea are so far the only ones in the archipelago where oil is produced from limestone of Upper Miocene age, rather than from sands.

The greater part of the oil produced comes from relatively shallow depths on gentle to moderately steep folded anticlines, most of it from depths between a few hundred and less than 3,000 feet. Although a few important fields produce from depths of nearly a mile, deep tests in the producing structures have had rather disappointing results. Practically no attempt has been made to discover accumulations in stratigraphic traps.

The forms of the structures in North Sumatra and Burma are much alike. The Chindwin-Irrawaddy basin is the northern end of a large Tertiary basin in which the North Sumatra oil fields occupy a southern position. The Chindwin-Irrawaddy oil fields occur in the first structures

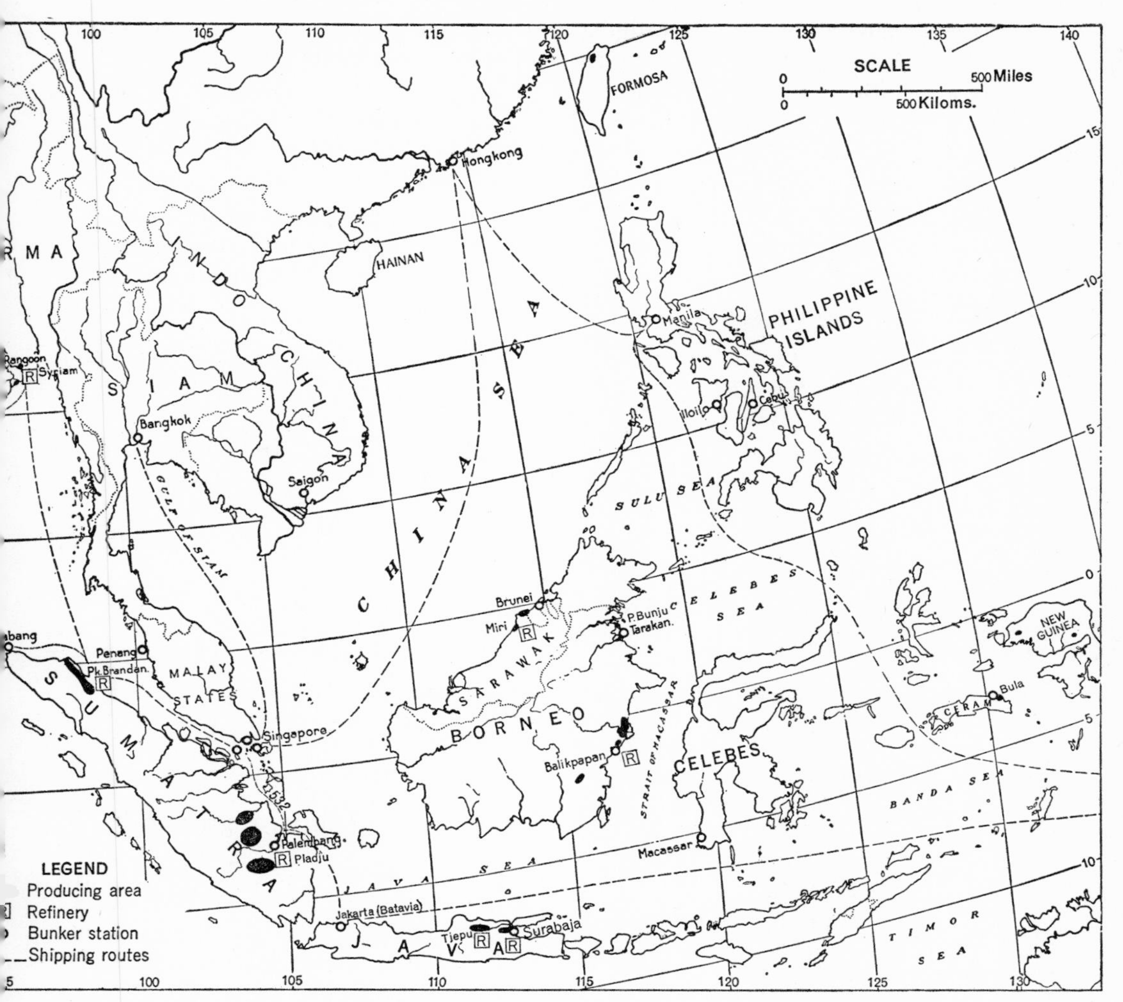

o. Southeastern Asia: oil-field areas, refineries, bunker stations, and shipping routes. This map was pre- by the Bataafsche Petroleum Maatschappij and is based on the prewar state of the petroleum and ing facilities. The sizes and shapes of the producing fields are indicated in Figs. 41 to 44 and Fig. 46.

east of a great western syncline which borders the Arakan Yoma on the east. The producing anticlines, save for the almost symmetrical Yenangyaung anticline, are asymmetrical, usually with the steeper limb to the east. In Burma more unconformities occur in the Tertiary series than in North Sumatra; there is a great thickness of Oligocene and Eocene beds, with a thinner Miocene development. While in North Sumatra oil production is limited to Upper Miocene beds, in Burma oil occurs in both the Oligocene and Miocene. In the Yenangyaung field the greatest production has been from Lower Miocene and Upper Oligocene sands; in the Singu-Lanywa fields Oligocene sands are productive. The oil sands of the Burmese fields are very inconsistent; the bulk of production comes from shallow depths ranging to about 3,300 feet.

The types of oil obtained in the different basins vary considerably, and even within the basins themselves. The main North Sumatra field, Rantau, for example, yields a light paraffinic, highly aromatic oil with a gravity of 44.4° to 51.2° API. The yield of gasoline is 65% and fuel oil 30%. Burma and Assam oils are high in kerosene and wax, and comparatively low in gasoline, the reverse of the North Sumatra oils. The types of oil from the Indonesian fields are indicated below.

NORTH SUMATRA. Light paraffinic, highly aromatic oil, gravity 44.4° to 51.2° API.

SOUTH SUMATRA.

Djambi: Light paraffinic, 47.0° to 48.0° API. In one field, shallow heavy asphaltic oil of 22.5° API.

Kluang: Light paraffinic.

Talang Akar, Pendopo, Djirak: Paraffinic to very paraffinic, 37° API.

Talang Djimar: Heavy paraffinic, 27.6° API.

Suban Djerigi: Light paraffinic.

JAVA.

Tjepu: Paraffinic to heavy paraffinic, 35° to 45° API.

Surabaja: Heavy asphaltic, 25° to 33° API.

BORNEO.

Barito (Tandjung): Paraffinic to heavy paraffinic.

Balik Papan: At shallow depths heavy asphaltic, at greater depths lighter and more paraffinic.

Tarakan: Heavy asphaltic oil of 18° API, fuel; at greater depths some lighter paraffinic oil.

CERAM. Heavy asphaltic, 22.5° API.

NEW GUINEA.

Klamono: Asphaltic, 18.8° API, used as fuel after light flashing-off process.

Wasian-Mogoi: Paraffinic, 47.6° API.

Throughout their development the important oil-producing territories of Southeast Asia—Indonesia, British Borneo, and Burma—have each had their own mining legislation. Thus, in spite of similarity in various natural circumstances, each of these territories has shown distinctive development.

Indonesia

In general development Java is especially favored. Java and Madura supported a population of about 48 million, out of a total population of 70 million for the whole of Indonesia in 1939. The fertile Javanese countryside has been described as "an unending village surrounded by its laboriously tilled little fields and groves, save where the plantations intervene." Sixty-five per cent or more of the adult population are engaged in agriculture, and some fifteen per cent in industry.

Of the other great islands, none of them so richly endowed as Java for agriculture, Sumatra appears to offer the best climatic and soil possibilities. Its advanced development in comparison with the islands other than Java is partly due to its position on the Straits of Malacca, one of the great avenues of world trade, and partly to its mineral resources, particularly oil and coal. Yet Sumatra, more than three times as large as Java, had only about 8 million people in 1939, and the greater part is covered by dense virgin forests.

In Borneo and New Guinea the mountainous interiors are even more difficult to penetrate, and remain to this day inaccessible jungle; much of the coastal land is mangrove swamp. Borneo supports only a little over two million people, and New Guinea about a million.

The Indies have an equatorial climate, with little variation in either temperature or length of day. Although the temperature remains consistently high, it is not so high as might be expected for the latitude, on account of the widespread interpenetration of the land by the sea; but pleasantly cool temperatures are found only at a sufficient altitude. All the oil fields are situated in the lowlands.

The greater part of the archipelago has no arid season, and the terms "dry" and "wet" are to be understood in a relative sense: the easterly and westerly monsoon currents with which they are associated are less important as rain-makers than the local convectional currents. The intense heat of the mornings from sunrise on produces thunderstorms, and in this form a sufficiency of rain is received nearly everywhere, even during the "dry" monsoon. In the larger islands there are always wet mountainous districts sufficiently near the lowlands to allow irrigation in the plains.

The Development of the Oil Industry

In 1871 an enterprising man, Jan Reerink, tried for the first time to obtain commercial production by drilling in the neighborhood of oil seepages at the foot of the volcano Tjerimai, south of Cheribon in West Java; but after five years he had to give up without having been successful.

In 1880 a tobacco-planter, A. J. Zylker, found oil seepages near his plantation northwest of Medan in North Sumatra, and, with the consent of the government, in 1883 he secured the Telaga Said concession from the Sultan of Langkat. In 1885 the first production was obtained from a depth of only about 400 feet in the Toenggal No. 1 well.

This first productive well in Indonesia led to the foundation of the "Royal Dutch Company for the Working of Petroleum Wells in the Netherlands Indies" (which later altered its name to Royal Dutch Petroleum

Company), the nucleus of the world-wide Royal Dutch/Shell Group. Formed in 1890, the Royal Dutch took over the Zylker concession, and in the first year produced 1,200 metric tons of oil from Toenggal No. 1.

Other entrepreneurs had become active, and within fifteen years after the first oil had been struck in North Sumatra, the basins in South Sumatra, East Java, eastern Borneo, and northeastern Borneo were all producing. All the important basins had been opened up, with the exception of those in central Sumatra and southern Borneo and that of western New Guinea, which gave the first promising results just before the outbreak of World War II.

In this early period eighteen different companies were active in the various basins. Most of them confined their activities to a single area each, but the Royal Dutch began to get a firm footing everywhere, both by its own exploration and by participation in other companies. In 1907, after some years of difficult negotiations, it amalgamated with the Shell Transport and Trading Company, which operated in eastern Borneo, and the Royal Dutch/Shell Group came into being. This Group had control of all production in Indonesia, except in part of East Java, where the Dordtsche Petroleum Maatschappij remained independent until 1911, when it also joined the Royal Dutch/Shell Group.

After the amalgamation in 1907, the Bataafsche Petroleum Maatschappij (the B.P.M.) was founded as the operating company for the Royal Dutch Shell Group in the Indies. In 1911 the B.P.M., then the only operating company in Indonesia, held 19 concessions in Sumatra, 18 in Java, and 7 in Borneo, with a total area of about 1,250 square miles (3,200 square kilometers), but the bulk of the production came from a relatively small number of fields. In North Sumatra, where the production of the original Telaga Said field had fallen considerably, the Perlak field in Atjeh, situated about 100 miles northwest of Pangkalan Brandan, had become the main producer, and in South Sumatra the bulk of the production came from the area round Muara Enim, in the southwestern part of the basin. In Java the production was concentrated near Tjepu and Surabaja, while in eastern Borneo the original fields near the Mahakam River delta produced more than any other area. In northeastern Borneo one important field on the island of Tarakan remained the only producer. In 1911 the total production of Indonesia amounted to 1,700,000 metric tons, of which North Sumatra produced 22%, South Sumatra 20%, East Java 10%, eastern Borneo 34%, and Tarakan 14%.

A vigorous industry had come into being, able to compete in the world market. It owed this position especially to a far-seeing commercial management, and to a Mines Act favorable to the industry. All concessions were granted for 75 years; and as entire structures were covered by the concessions, no competitive drilling could arise within an oil field and the operating company could exploit the fields as economically as possible. In later years especially, when oil-field engineering became more scientific, it was a great advantage that the Indonesian fields were opened up in their entirety by one company.

The next landmark was the advent of important competition for the B.P.M., when Americans became interested in oil operations in the archipelago. In 1912 the Nederlandsche Koloniale Petroleum Maatschappij, now called the Standard-Vacuum Petroleum Mij, and a subsidiary of the Standard Vacuum Oil Company, was founded and started prospecting. A small production in the neighborhood of Tjepu, East Java, never attained importance, but great success was achieved in South Sumatra, where in 1918 the rich Talang Akar–Pendopo area came into production, wholly in the hands of the Standard Vacuum's subsidiary. Just before the second World War it was yielding 50% of the whole South Sumatran production; it had then already produced about 21 million metric tons, and the annual production amounted to more than 2 million tons. In North Sumatra and Borneo much exploration was carried out by the Standard Vacuum Petroleum Mij, but without commercial production.

The coming of competition naturally promoted exploration. It has already been mentioned that in 1911 some 44 oil concessions had been granted in the Netherlands Indies, with a total area of about 1,250 square miles (3,200 square kilometers). In 1924, the last year that concessions were granted on the old conditions for a period of 75 years, the oil concessions in the Netherlands Indies amounted to 119, with a total area of about 2,500 square miles (6,400 square kilometers). Of the total production of 2,926,000 metric tons in 1924, North Sumatra produced 6%; South Sumatra, 17%; Java, 9%; eastern Borneo, 36%; and Tarakan, 32%. As compared with 1911, the decline of North Sumatra and the great increase in production, absolute as well as relative, from the rich Tarakan field are striking. The American company accounted for 5% and the Royal Dutch/Shell Group, through the B.P.M., for 95% of the entire production.

After World War I new political and economic viewpoints, which were gaining ground with the government and in the legislature, began to make their influence felt in the oil industry. The economic position of the industry was considered too one-sidedly favorable, and a larger share in the profits for the state was advocated. This tendency made itself felt in particular when the exploitation of the mineral riches of the province of Djambi, in South Sumatra, came up for discussion. In 1904 this area had been closed to private mining by the government, and held in reserve. It was known that the oil basin of South Sumatra stretched into Djambi. When, during and immediately after World War I, the government considered the exploitation of the Djambi area, acrimonious debates were held in the legislature, and finally it was decided to form a new company, the Nederlandsch-Indische Aardolie-Maatschappij, customarily called the N.I.A.M., in which the Netherlands Indian government became a partner of the B.P.M. on a fifty-fifty basis. The N.I.A.M. obtained the exploitation rights for the greater part of Djambi, for a smaller area in North Sumatra, and for the small island of Bunju, near Tarakan in northeastern Borneo. Djambi production began in 1924 and increased steadily; in 1940 nearly 8 million metric tons had been produced, and the annual production amounted to 1,370,000 metric tons from four fields.

Although the government received considerable revenues from its partnership in the N.I.A.M., no similar contracts were entered into for new exploration except an extension of the Djambi area to include the coastal plains of Djambi and of part of Palembang, which were added to the N.I.A.M. exploitation rights in 1941. In 1928 a change in the granting of concessions was made for the first time, and large areas were granted on contracts differing considerably from the old conditions. The most fundamental differences are that the contracts run for 40 years; the contracting company is under an obligation to drill; it has the right to return its area by sections in case of negative results; and the state, in addition to the royalty due for concessions, receives a highly progressive profit share, which usually is as high as 20% of the net profit.

Before World War II, the area held by the various oil companies under exploitation contracts totaled about 9,000 square miles (23,000 square kilometers), held by B.P.M., the Standard Vacuum, and the N.P.P.M. (Nederlandsche Pacific Petroleum Maatschappij). The N.P.P.M., a subsidiary of the Standard Oil of California and Texas groups, entered the field in 1931, started extensive exploration work in the coastal plains of central Sumatra and in western Java, and shortly before the war realized commercial production in central Sumatra.

The latest development before World War II was the creation of the Nederlandsche Nieuw Guinea Petroleum Maatschappij (N.N.G.P.M.) in 1935. In this company the Royal Dutch/Shell Group (through B.P.M.) and Standard Vacuum Petroleum Maatschappij each have 40% interest, and the Far Eastern Investment, Inc. (Caltex) has 20%. Exploratory work has been carried out in an area covering 100,000 square kilometers (38,610 square miles) in western New Guinea, and the first production from the Klamono field was being transported at the end of 1948 at a rate of 4,000 b/d. A major effort will still be necessary to put New Guinea on the map as another big producer.

Under contracts concluded since 1928, important fields have been discovered in the basins already yielding oil in the other islands. Of these we would mention the Rantau field of the B.P.M. in North Sumatra, and the great additional production in the Talang Akar–Pendopo area of Standard Vacuum and in the Talang Djimar field of the B.P.M., both in South Sumatra, while in the Barito area of southeastern Borneo the B.P.M. obtained commercial production.

In 1939 the B.P.M. share in the total production of Indonesia amounted to 72%, and that of the Standard Vacuum to 28%. It is interesting to note that after 1930 eastern Borneo yielded first place to South Sumatra. The old Borneo fields declined greatly, without important production being realized from new areas, while in South Sumatra the production of rich new fields, both in Djambi and in Palembang, far exceeded the decline of the old fields.

Besides the important basins discussed above, a small production has been obtained from the island of Ceram, in the eastern part of the archipelago. The accumulated production amounts to only about one million metric tons, less than 1% of the total production of Indonesia, but the Ceram production is interesting in that, while the oil is found in shallow Plio-Pleistocene sands, it probably formed originally in beds of Triassic age, migrating subsequently to its present position.

The petroleum industry of Indonesia fell into the hands of the Japanese early in the war (1942), and Japan's petroleum needs to fight the war in the Pacific were largely drawn from that source. It was, however, an impossible task for the Japanese to rehabilitate the industry, which had been subjected to the scorched-earth policy of the Netherlands Indian government at the time of the Japanese attack. The Japanese had therefore to employ makeshift arrangements in the fields as well as at the refineries. It is estimated that approximately 105 million barrels of oil were produced by the Japanese during the period 1942–1945. The output in the entire Indies, including British Borneo, is estimated to have been around 65,000 b/d in 1942; 130,000 b/d in 1943; 75,000 b/d in 1944; and 17,000 b/d in 1945. The industry suffered great damage as a result of the war; and since the end of the war the political instability, especially in Java and Sumatra, has been responsible for the fact that Indonesia's prewar position has not yet been recovered. Restoration measures after the war were concentrated chiefly in Borneo and in southern Sumatra.

In the fields of the Borneo mainland and Tarakan (an island northeast of Borneo) drilling was resumed and repair strings were set the task of improving existing wells. Many factors adversely affected the rehabilitation work—lack of materials, of skilled labor, and of transportation. In June 1949 this area had reached about half its prewar production. Balik Papan's refining capacity, from two trumbles built from scrap material available on the spot, is still small, but a big refinery building program is under way.

In South Sumatra, however, the fields were not reentered until 1947 and during the course of 1948, although it had been possible to work on the reconditioning of the refinery sites at Pladju (B.P.M.) and Sungei Gerong (Standard Vacuum) since October 1946. They also were damaged, but are now running at full prewar capacity. The rehabilitation of the fields is a big affair, but progress is satisfactory, and there is a regular increase in production.

In North Sumatra the Pangkalan Brandan area is not yet accessible because of political unrest. Until recently the same was the case with the

B.P.M.'s refinery and fields on Java (Tjepu). They are now again accessible (December 1948), but the refinery has been completely destroyed. Rehabilitation of the fields is in hand.

In New Guinea operations were resumed in 1946, and in December 1948, as stated above, the Klamono field was brought into production. Several new tests are being drilled in the Vogelkop area, and geological and geophysical field surveys are again in progress.

To summarize: in spite of unsettled political conditions and the restricted scope of operations in Indonesia, considerable progress has been made in the rehabilitation of the petroleum industry in that country since the war. During 1948 approximately 32,000,000 barrels of oil were produced, and at the close of the year the average production was 105,000 b/d, as compared with a prewar rate of 165,000 b/d. Geological and geophysical field exploration surveys were resumed in 1947, and exploration drilling has been resumed also, but no new discoveries of importance have so far been reported.

PROBLEMS OF GEOGRAPHICAL ADAPTATION

The internal transport possibilities of the different islands match their general state of development. Java, with many through roads and railways, and independent of river transport, leads the way; then comes Sumatra, with a few separate railway systems and a through highway; next Borneo, without railways and with separate road systems; and finally New Guinea, with no modern transport facilities at all, until very recently.

The Javanese rivers are not very useful as waterways because of the abrupt gradient of the river beds, and as early as the seventies of the last century it was necessary to supplement the traditional buffalo by railways to transport agricultural products from the interior to ports on the northern coast, notably Batavia, Cheribon, Semarang, and Surabaja. A rail network links all the important places, and crosses Java's mountain backbone. The distance of 530 miles between Batavia and Surabaja can be covered in 11½ hours. Practically all Javanese towns have good roads, and in the present century Java has been provided with a network of motor roads which make most centers accessible by car.

In Sumatra a number of navigable rivers in the southern half are important for transport of products to the river ports of the east coast—Palembang, Djambi, and Rengat—where transshipment into ocean-going vessels takes place. Three different Sumatran railway systems have developed, which are not yet linked: one system serves the southern part of the island and was important for opening up this area; a second small system serves the coal fields in the interior near the west coast; and a third system serves the most important producing areas along the northeast coast—the northern part of this system, in the province of Atjeh, was originally built as a strategic railway. There are good roads in the northeast, all along the west coast, and across the southern part of the island. Before World War II one could drive by car from Kutaradja in the northern tip of Sumatra to Telukbetung in the extreme south, a distance of 1,300 miles.

The vast island of Borneo has four navigable rivers of importance: two relatively short ones, which serve the most highly developed areas of the east coast; the long and winding Barito, of major importance for the agricultural district of Kapuas Murung in southeastern Borneo; and a fourth, the Kapuas, which serves the best developed plantation area of Borneo, on the west coast.

There are no railways. Borneo has three widely separated road systems, one leading to the harbor of Pontianak on the west coast, another leading to the harbor of Bandjermasin at the mouth of the river Barito, and a third, less developed, system along the middle east coast, with the oil harbor of Balik Papan as center. The linking up of the last two systems is well advanced.

New Guinea has no navigable rivers of importance, except the Mamberamo, and New Guinea cannot boast modern roads; the jungle trail still prevails. However, the N.N.G.P.M. is building roads now. The first of these roads is finished, and links the Klamono field with the harbor of Sorong. Its length is thirty miles, with an eight-inch pipe line along the road. Tremendous difficulties with continuous rainfall had to be overcome in building this road through virgin jungle.

Most Sumatra and Borneo fields are situated in areas of low hills or hillocks, with elevations ranging from about 175 to 250 feet; so there are no special problems of adaptation in the location of producing fields. The normal procedure for reaching a new field in the jungle includes, first, surveying a good trail, with as few river crossings as possible, and going around bigger hills; then cutting a broad trail some 50 yards or more wide so that the future road will receive sufficient sun to remain dry; and then building the road proper with the help of bulldozers, grading machines, and so on. The necessary paving is mostly supplied from boulder beds in nearby rivers. Road building is done by preference during the dry season.

For drinking water in the field-camps, filtered and boiled river water is used. Malaria is fought by cutting all shrub, by good irrigation, and by drenching stagnant water with oil in order to destroy the mosquitoes' breeding places. In New Guinea regular spraying of DDT by airplane around the settlements has brought down the number of malaria cases from 25 per 100 workers in 1936 to 7 per 100 in 1949.

The situation of the refining and transport centers of the petroleum industry was decided by both historical and geographical factors. For Java the case was simple. When in 1888 the first production had been obtained in a field south of Surabaja, a refinery was built in the immediate vicinity, at Wonokromo, and in 1890 this plant began to supply illuminating oil to the home market. After production had been obtained in the province of Rembang, the refining center of Tjepu was established in 1894, in the neighborhood of the producing fields, which were able to market their products in Java during the period of development of the industry. Thus in both cases distribution was effected direct from the refinery. When, on account of the increase of production and the differentiation of products supplied, the home market became too small, Surabaja, the largest shipping port of the Indies, became the logical port of loading for export. The Wonokromo refinery lies in the immediate neighborhood of this port, and Tjepu is situated at a distance of only 80 miles, and connected with it by a railway. Up to 1930 the Tjepu products were transported to Surabaja by rail tank cars; after that, by pipe line. The Wonokromo refinery has been repaired and is operating, but the Tjepu refinery is still inoperative because of the serious damage.

In Sumatra and Borneo the oil industry had been dependent upon export from the beginning, as these areas had no hinterland capable of absorbing the production.

In North Sumatra (Fig. 41) a primitive refinery was completed on the river Babalan, near the fishing hamlet of Pangkalan Brandan, in 1892. This refinery depended on the production of Telaga Said, the only field producing at the time. The choice of the site was decided by three factors: the necessity of building on ground not liable to inundation at high tide, the desirability of remaining as close as possible to the productive field, and the desirability of combining the refinery and the port of shipment for overseas transport. Pangkalan Brandan seemed to satisfy these conditions, for here a somewhat higher sand ridge extended along the swampy banks of the Babalan; the channel in the river was sufficiently deep; and although there was a bar across the river mouth, ships drawing 8 feet could pass at high tide. Large ships were not used for the export of products to Penang and Singapore, on the opposite coast. After a few years, however, the shipping facilities proved to be inadequate. As the need arose for shipment to India, Indochina, and China, larger tankers were required, for which the Babalan channel was too shallow. Aru Bay was chosen for a new port, and in 1897 the loading site of Pangkalan Susu was completed, 7 miles from the Pangkalan Brandan refinery and connected with it by pipe lines. The refinery remained at Pangkalan Brandan, and all the fields pumped their crude oil to it through pipe lines. The distance to Perlak is about 100 miles. As mentioned before, the North Sumatra fields and the Pangkalan Brandan refinery are still, in mid-1949, not yet accessible to the operators.

In South Sumatra three companies, at first independent, each built a small refinery. For the Sumpal field in northeast Palembang a refinery was built in 1898 at Bajunglentir in the jungle on the tortuous Lalang River, navigable up to that point for ships with a length of 300 feet. The company operating in southwest Palembang, in the neighborhood of Muara Enim, decided at once to establish its refining and loading center on a larger waterway, at Pladju on the right bank of the Musi River, 6 miles downstream from the town of Palembang. Just below Pladju the coastal swamps of the

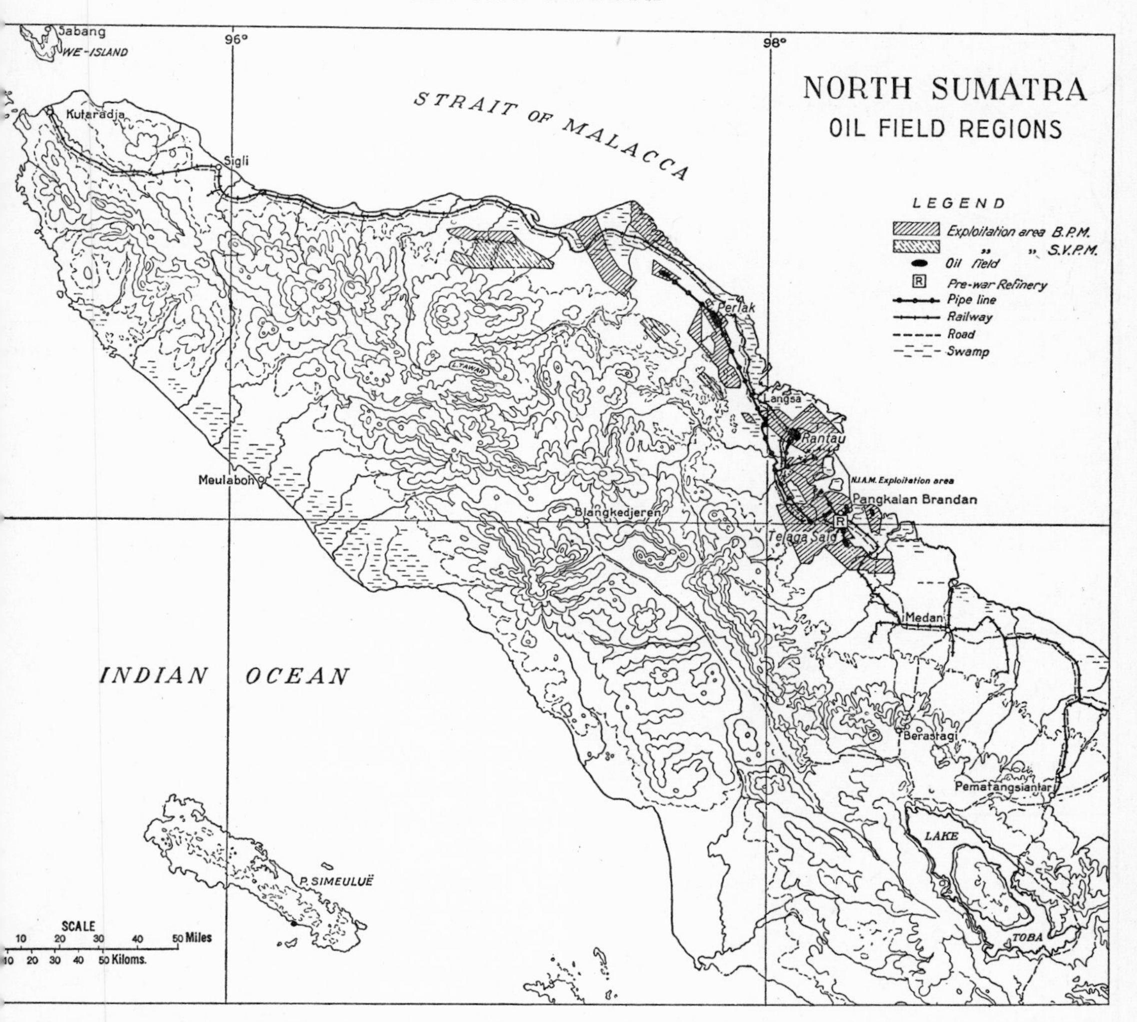

41. North Sumatra: oil-field regions and facilities. Map prepared by the Bataafsche Petroleum Maat-ppij, showing oil-field development in 1941. The altitude contours shown denote 100 meters (broken), 500 meters (first solid line), and then rise by 500-meter intervals to 3,500 meters. These contours are n in color in the plates of the *Atlas van Tropisch Nederland* (Batavia and Amsterdam, 1938).

delta begin, and the river, which at Pladju is over half a mile wide, is navigable up to that point for larger sea-going vessels. A 100-mile pipe line connected the oil field with the refinery. The company exploiting the Babat field in central Palembang also established its refinery on the Musi, between Palembang and Pladju, and laid a pipe line extending beyond Palembang town.

After the Royal Dutch obtained control of all production and processing plants in South Sumatra in 1906, Pladju was expanded to become the only

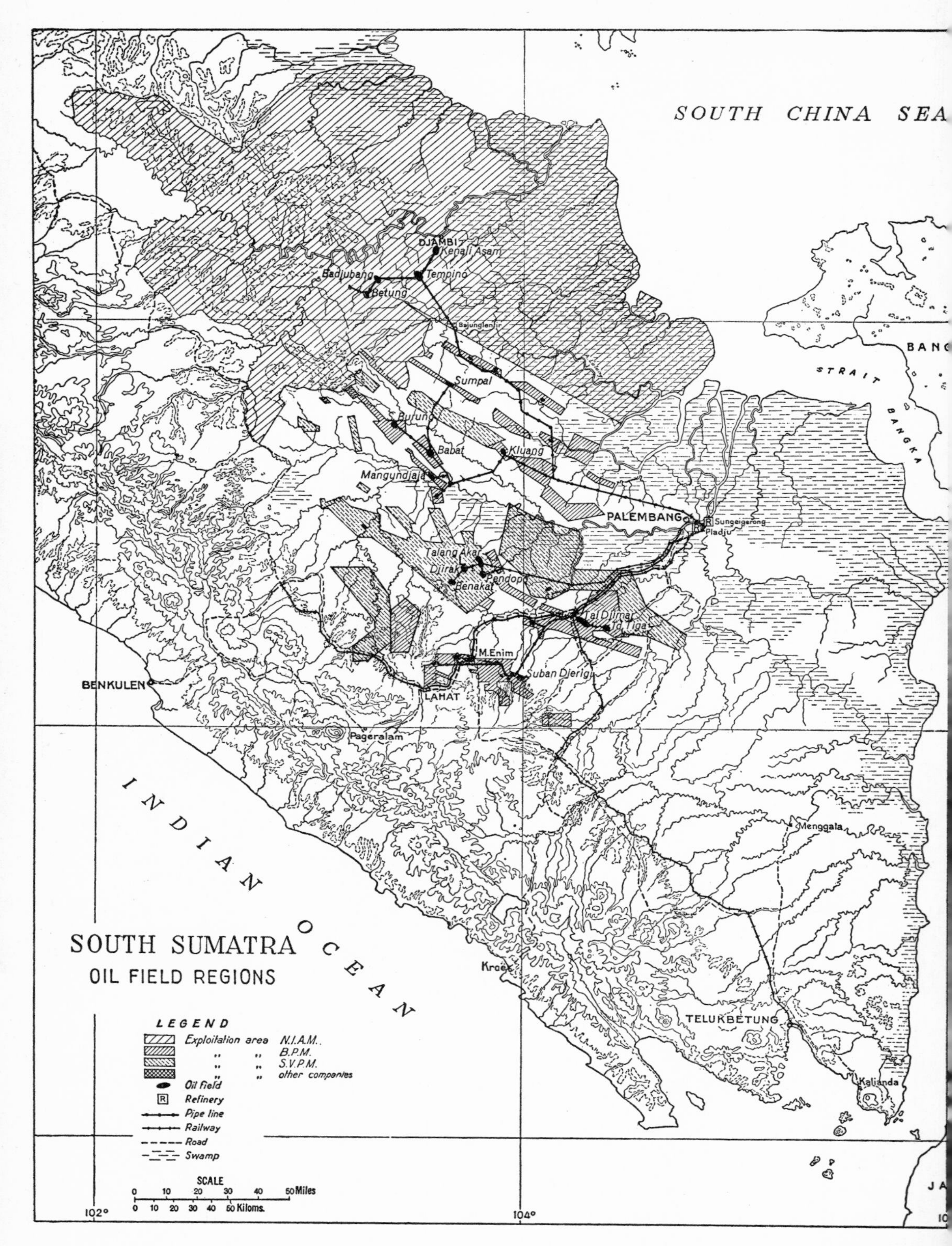

FIG. 42. South Sumatra: oil-field regions and facilities. Map prepared by the Bataafsche Petroleum M schappij, showing oil-field development to mid-1949. The altitude contours follow the same progress as in Fig. 41. The entire area of South Sumatra east and southeast of the oil fields is less than 100 meter altitude.

. Indonesia. Modern transport by tractor and trailer.

ght). Northeast Borneo. Tarakan oil field, prewar view.

81. East Borneo. Pipe lines near Balik Papan.

82 (lower right). Yenangyaung, Burma. Well-digger in diving dress.

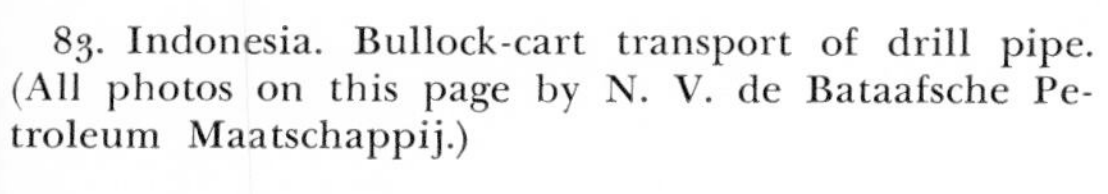

83. Indonesia. Bullock-cart transport of drill pipe. (All photos on this page by N. V. de Bataafsche Petroleum Maatschappij.)

84. New Guinea. Air photo of base camp, showing: (1) airfield; (2) houses for European personnel of th NNGPM; (3) office of the NNGPM; (4) camp of flying personnel; (5) Indonesian quarters; (6) fresh-water pump ing station; (7) church; (8) missigit-Mohammedan church; (9) warehouse. (N. V. de Bataafsche Petroleun Maatschappij.)

85. North Sumatra. Air photo covers part of map in Fig. 45. (N. V. de Bataafsche Petroleum Maatschappij.)

refining and loading center. When the N.K.P.M. had come into the picture, it established the Sungei Gerong refinery, next to that of Pladju, and separated from it only by a tributary of the Musi. Reconstruction and rehabilitation of the Pladju and Sungei Gerong refineries, situated in South Sumatra, have been largely completed, with an aggregate capacity of 110,000 b/d.

All fields in Palembang are connected with the refineries at Pladju or Sungei Gerong by pipe lines, and though these loading ports lie 45 miles from the mouth of the Musi, they can be used for all except the very largest modern tankers.

An important engineering job was the direct connection between the Djambi fields and the Pladju refinery by an all-welded 8-inch pipe line, which was completed in 1935. The length of 170 miles is not great according to American standards, but as it runs entirely through jungle, and largely through swampy jungle, and as six river crossings had to be undertaken with swings, this was a major job.

Along the whole pipe line runs a road, of which a length of 100 miles had to be newly constructed. The trail cut is from 50 to 60 yards wide. Over 100 million cubic feet of timber had to be felled and removed, while the earthwork for dikes through the swamps, digging through hills, and filling up ravines amounted to over 20 million cubic feet. Some 19 bridges had to be constructed; 300 culverts with a total length of 10,000 feet were employed; and the length of the six swings totaled 11,000 feet.

The entire pipe line was buried to an average depth of 2 feet 4 inches, after coating it with pipe asphalt and wrapping it round with coco fiber cloth soaked in bituminous oil. One reason for digging in the pipe line was to eliminate the difference between day and night temperatures, and another, to protect the line against damage from falling trees, elephants, and marauders.

On the east coast of Borneo, where the first, and still the most important, production was realized just above the delta of the Mahakam river, the pioneers saw that the difficult channel would be unsuitable for loading. So for the construction of a refinery they looked for a roadstead where the largest tankers could moor and load day or night, in all winds and all weathers. Such a "noble sheet of water" was found in the jungle-fringed bay of Balik Papan some 60 miles southwest of the productive fields. Here, in 1898, the first refinery in Borneo was erected, and in the course of years this location has developed into one of the most important centers of the oil industry in Southeast Asia, rivaling Pladju in Sumatra. The 35,000-barrel Balik Papan refinery, which was completely destroyed during the war, has been only partly reconstructed, with a capacity of about 8,500 b/d; construction of a new modern refinery at Balik Papan is at present under way.

The fields on the island of Tarakan, a few miles off the east coast of Borneo (3°20′N, 117°45′E) produce a heavy crude oil which, without

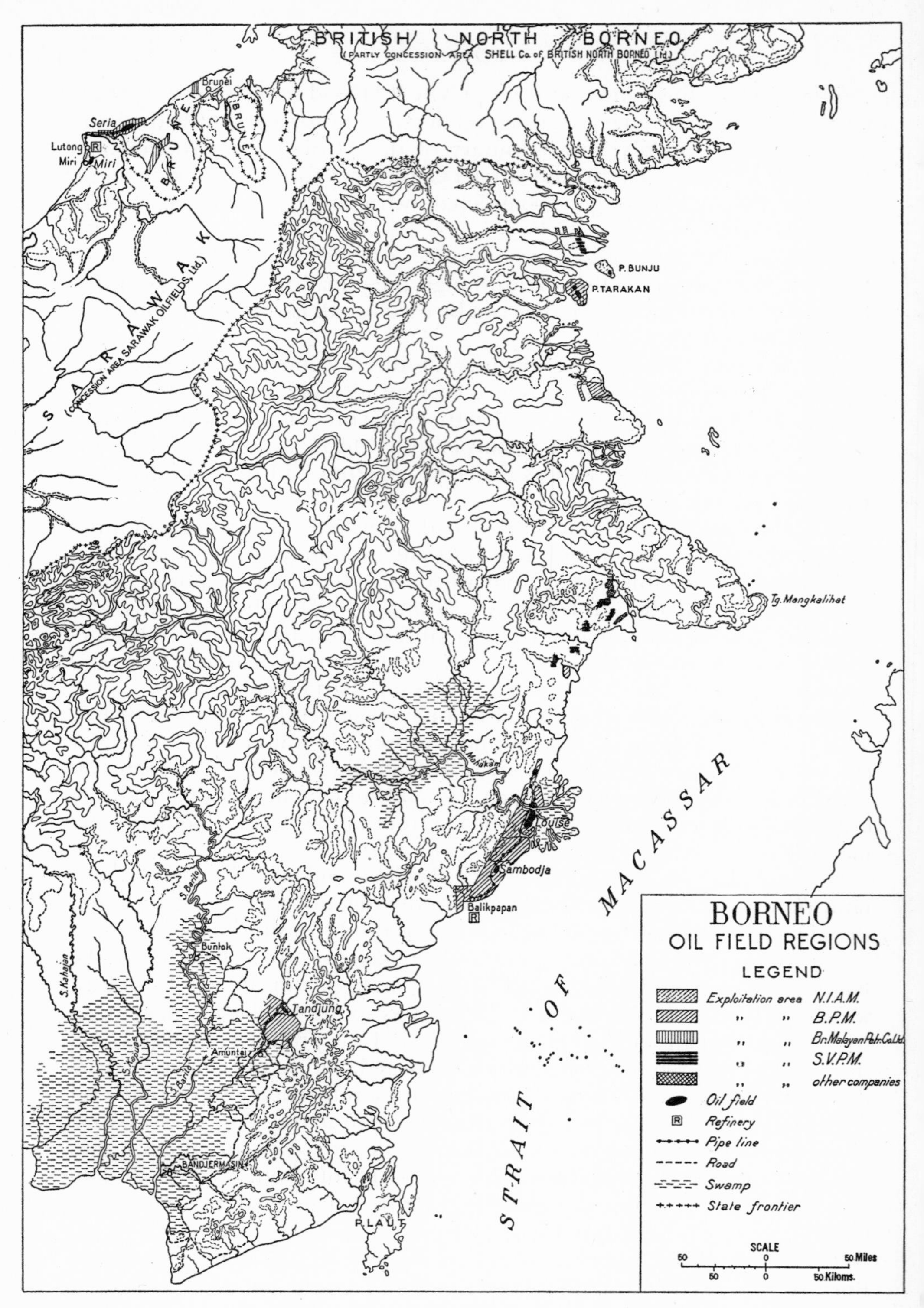

FIG. 43. Borneo: oil-field regions and facilities. Map prepared by the Bataafsche Petroleum Maatschappij, showing old-field development to mid-1949. The contours rise by the same progression as in Fig. 41, except that none is shown here above 2,500 meters, and all contours except the 100-meter have been omitted in British Borneo. An oil field in the B.P.M. concession, in the small group lying 160 miles north of Balik Papan, is indiscernible on this map.

refining, is suitable for fuel oil. Tarakan, therefore, has no refinery, and the loading is done from an open roadstead on the southwest tip of the island.

INTERNATIONAL COMMUNICATIONS

Before World War II Indonesia had excellent shipping and air communications for inter-island traffic. The Royal Packet Navigation Company (Koninklijke Paketvaart Maatschappij or K.P.M.) maintained regular and frequent services by passenger steamer and freighter throughout the archipelago. The Royal Netherlands–Indian Airlines (the Koninklijke Nederlandsch–Indische Luchtvaart Maatschappij or K.N.I.L.M.), the civil aviation company of Indonesia, operated an extensive net of inter-island air lines, unique in the tropics, with services several times a day between the communication centers of Java, and several times a week from Java to Sumatra, Borneo, and Celebes.

The Indonesian archipelago lies at the junction of many lines of international shipping. In the ports of Java ships meet bound for Europe, South Africa, the mainland of Southeast Asia, China, Japan, the Philippine Islands, the Americas, and Australia. Regular communication between Indonesia and the Netherlands was operated by the Royal Dutch Airlines (the Koninklijke Luchtvaart Maatschappij or K.L.M.) up to three times a week in both directions, before World War II. Air services both in Indonesia and between the Netherlands and Indonesia by the K.L.M. were resumed after the war, and it is hoped to make the service from Europe to Indonesia a daily one.

MAPPING BY THE OIL COMPANIES

It stands to reason that as the oil companies explored large jungle-covered areas, especially in the islands outside Java, they also carried out important mapping work. Although for the vast, and to a large extent still undeveloped, areas the mapping carried out by the government was on a comparatively high level, yet a more detailed mapping of the virgin areas into which the exploration for petroleum penetrated was out of the question. The topography of many inland regions came to be known more exactly only from the surveys of the oil geologists who penetrated into these areas. If then concessions were granted and roads were constructed which connected the developing oil fields with the refining centers, all this resulted from the activities of the staffs of the oil companies. The survey data are handed over to the government, which uses them for its official maps.

Although before 1934 a great deal had already been done in this field, mapping by the oil companies was greatly expanded and accelerated from that time on, when photographic surveys from the air were undertaken. The immediate cause of this air-mapping was the following. Originally the boundaries of concessions had to be staked out in the field and marked by boundary posts at intervals of 500 meters, within seven years after the concessions had been granted. In the island of Sumatra the total area of all

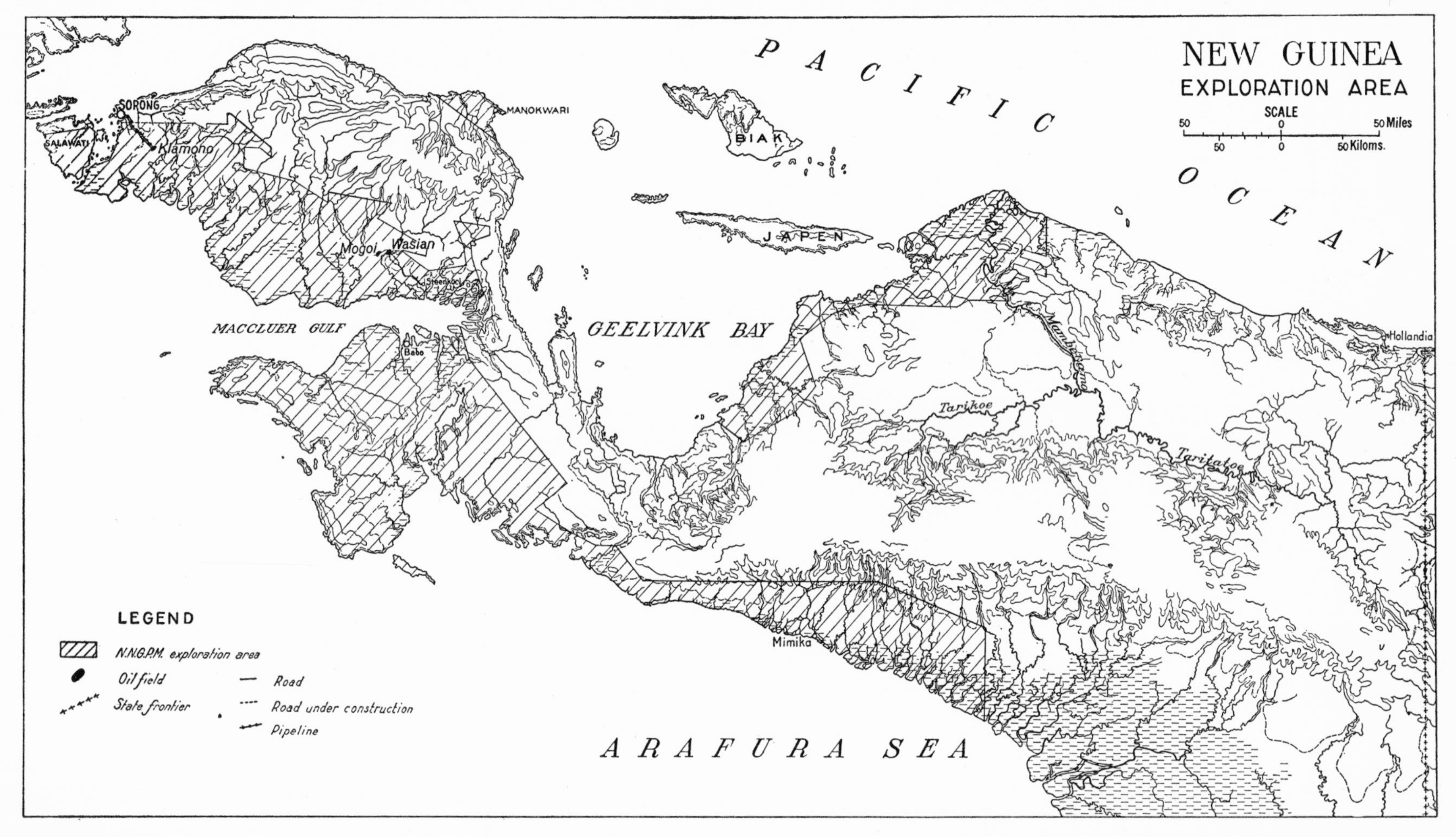

FIG. 44. Netherlands New Guinea: exploration area. Map prepared by the Bataafsche Petroleum Maatschappij, showing oil-field development in mid-1949. The altitude contours denote 100 meters (broken lines), 1,000 meters (first solid line), and then 1,000-meter intervals to 4,000 meters. These contours are shown in color in the plates in the *Atlas van Tropisch Nederland*. The high areas where contours are broken or missing were uncharted when the base map was drawn.

52 oil concessions granted before 1924 amounted to about 2,400 square kilometers (about 925 square miles) and the boundaries of these concessions had all been staked out in the field in the manner prescribed.

As a result of intensive exploration, however, in 1928 and 1931, an exploitation contract was granted to B.P.M. and Standard Vacuum, jointly, for twenty new areas in Sumatra up to a total of not less than 9,000 square kilometers (about 3,475 square miles). The total length of the boundaries of these scattered areas, mainly lying in virgin, often swampy, jungle amounted to several thousands of kilometers, and the staking out in the field of the boundaries in the old-fashioned manner as prescribed by the Mining Act would have been a costly work, involving much time and a large staff.

Therefore in 1932 the two oil companies interested proposed to the government to have the staking out on the ground replaced by air mapping. Air mapping would be appreciably cheaper than staking out on the ground, and would answer the purpose—the indisputable fixing of the boundaries of the fields—at least as well. Moreover, a much more complete and, in every respect, a more exact map of the whole of the fields and the intervening and adjacent areas could be made from the aerial photographs.

The government saw the great advantages of the new process, and in 1934 the photographic section of the Air Force Department was instructed to carry out the photographic flights. From the aerial photographs, taken on a scale of 1 to 20,000, the government topographical service made the boundary maps of the concessions by aero-triangulation, and for the oil companies the cost amounted to only one-fourth of that required for ground mapping. In all 17,000 square kilometers (about 6,500 square miles) were photographed, although the area of the concessions granted amounted to only 9,000 square kilometers (3,475 square miles).

Finally, the government topographical service used the photographic material as a foundation for maps of the areas in question on a scale of 1 to 50,000, to be newly issued in the place of the old maps dating from before 1920. As an illustration of the great difference in exactitude and detail attained in this way we reproduce here part of a map sheet of North Sumatra issued in 1917 underlying the same part issued in 1939 and made with the help of aerial photographs (Fig. 45 and Plate 85).

In another respect, too, air mapping proved important in facilitating exploration. For it has proved possible to trace valuable geological evidence, even in jungle-covered areas, by the stereoscopic study of adjoining and partly overlapping aerial photographs.

The success of the first air mapping in Sumatra led to further large commitments by the oil companies for the making of aerial photographs. Between 1935 and 1937 the exploration region of the N.N.G.P.M. in New Guinea, an area of 100,000 square kilometers (about 38,600 square miles), was photographed. The company constructed the necessary airfields, and the K.L.M., the civil aviation company in Indonesia, carried out the photo-

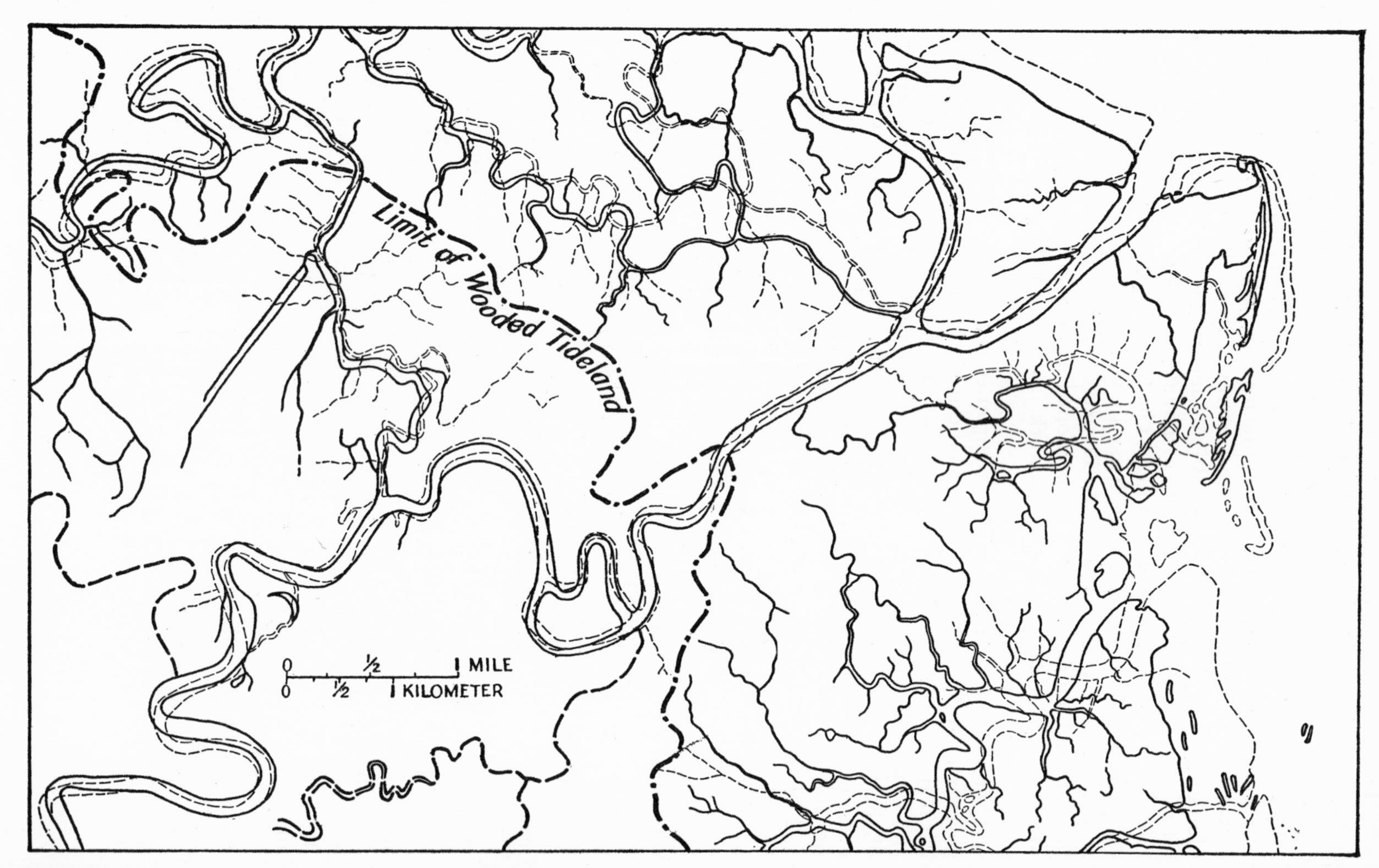

FIG. 45. North Sumatra: simplified drawing of part of the Government Topographical Map, showing section of tideland mapped in 1917 and remapped, after air surveys, in 1929. Broken lines show 1917 information, solid lines 1929 information.

graphic flights. In more than 2,000 flying hours, of which nearly 1,000 were useful mapping hours, more than 16,000 photographs were made, and an area of about 110,000 square kilometers (42,500 square miles) was mapped on a scale of 1 to 40,000, the greatest air mapping scheme ever carried out in such a primitive region. Plate 84 shows the base camp at Babo from which the work was directed.

Between 1934 and 1940 an area of about 175,000 square kilometers (67,500 square miles) was mapped from the air by the oil companies in Sumatra, Borneo, and New Guinea: an area one and a half times as large as the State of New York.

THE OIL INDUSTRY AND THE POPULATION

The people of Indonesia are mainly of Malay stock. They include some 140 groups differentiated in type, religion, and language: their civilization is generally described as Indonesian. Strong Hindu influences prevailed in the archipelago before Islamization took place. The Malay stock is predominant in Java and Sumatra; in Borneo we find Malays and Dayaks; and New Guinea is inhabited by Papuans (Irians) of Melanesian origin. The Papuans who live in the swampy jungle-covered lowlands are still living in their natural state, but contact with them during and after World War II proved that they could quickly adapt themselves to modern conditions. Those in the mountains, where the climate curbs the jungle from overgrowing everything else, have developed into peaceful tillers of the soil.

The most highly developed of the different Indonesian languages is Javanese; but in the whole archipelago the Malay language, spoken throughout the coastal districts by Malays, later immigrants than the original stock, is spoken.

Although much has to be done in this respect, literacy was increasing rapidly before World War II, when much was done for the development of these territories in the way of education, hygiene, living conditions, transport, and economic improvement in all spheres. The basis of education is the primary or village school, where instruction is given in the vernacular or in Malay. After six years the intelligent child may proceed to a secondary school or to a trade school where he can specialize in certain trades. And practical agricultural schools and technical schools are available. In 1938 there were 15,000 village schools and 3,000 popular libraries, for which books are provided from a central institute. The secondary schools lead up to institutions of university level. All this is in an uncertain state in the postwar period, but will certainly come back when once peace and order have returned in Indonesia.

In 1939 about 20 million persons were gainfully employed in Indonesia, including 14 million on agricultural estates and European plantations, and 2 million in industry, of whom some 50,000 found employment with the oil companies. In the industry itself indigenous labor plays a very important part; for instance, in 1939 the largest operating company, the B.P.M., employed about 35,000 Indonesians and 1,260 Europeans. In 1939 the three major companies, B.P.M., Standard Vacuum, and N.N.G.P.M., employed just over 2,000 European and American employees altogether.

Indigenous labor is used in every branch of the industry; practically all manual labor, skilled and unskilled, and many office functions are carried out by Indonesians. And Indonesians have been trained as drillers on the derricks. Transport is entirely looked after by Indonesians on river boats, as are all cars and trucks. In the refineries, the workshops, and the warehouses, the staff is also for the greater part indigenous. So far the Indonesians have displayed little interest in technical and chemical posts, as most of them join the ranks of state and municipal officials. It may be expected in the future, however, that many more Indonesians will find their way into other functions, not only administrative but also technical, for which the oil companies are opening up possibilities by founding the requisite training schools.

Part of the Indonesian staff is locally engaged in each island, but the greater number are from Java. This applies largely to the higher staff, but workmen also are recruited to a great extent from Java. Most Javanese return home after completion of their contract; those who remain assimilate in local communities. In Balik Papan much labor was recruited from south Celebes and Timor, as well as from Java. In both Sumatra and Borneo hundreds of Lascars were employed, especially as stevedores. The Chinese artisan is often encountered in the shops.

The oil industry is a good employer. Wages are as high as any paid, and the oil companies have seen to good housing for the staff. The industry also provides all kinds of schools, and much is done for the training of personnel. There are special trade schools in the main oil centers, where the Indonesian staff is trained for various positions, both in the lower and higher ranks. As regards intensity of work, high demands are made of the workers, but, on the other hand, the companies pay high wages compared with most other industries, and make excellent social and hygienic provision for their staffs.

The oil industry is, furthermore, of economic importance to thousands of Indonesians who find their living in and around the oil fields and refining centers as suppliers of vegetables and fruit, small food-vendors, artisans, taxi-drivers, servants, and in many other occupations. Many shops are run by Chinese and Indians.

In another respect, the oil fields are of importance to the Indonesians: where exploration leads to the opening up of the jungle by road construction, the population often follows.

HOME AND EXPORT MARKETS

The comparative development of home and export markets for petroleum products from 1913 to 1938 is indicated in Table 22. In 1938 the export value of oil products amounted to some 25% of the value of all Netherlands Indies' exports, whereas in 1929 the proportion was only about 13%. In 1938 oil products constituted by far the greatest bulk export, and Fig. 5 summarizes their principal destinations. In this respect Indonesia is of primary importance to the Far East—Australia, China, Japan—as the natural source of supply of liquid fuels.

Table 22. The Netherlands Indies: trend of export trade and domestic consumption of certain classes of petroleum products, by weight, in selected years from 1913 to 1938.

	Quantities exported and consumed in:					
	1913	1923	1927	1930	1935	1938
	(In thousand metric tons)					
Kerosene						
Export	382.5	205.3	235.4	515.6	579.6	720.0
Consumption	211.9	241.2	290.4	303.9	311.6	258.0
Gasoline						
Export	249.8	479.4	723.1	1,310.7	1,649.4	2,156.0
Consumption	15.6	60.9	135.1	191.8	188.2	153.0
Lubricating Oil						
Export	15.6	25.1	77.7	31.7	23.8	30.0
Consumption	2.9	13.7	23.1	28.1	18.6	20.0
Gas Oil and Fuel Oil						
Export	141.8	1,314.0	1,471.4	2,186.1	2,540.7	3,053.0
Consumption	377.2	586.9	907.2	1,149.0	715.4	1,100.0

Source: Statistics supplied by Bataafsche Petroleum Maatschappij (B.P.M.).

British Borneo (Sarawak and Brunei)

For many years petroleum has been known to exist in Sarawak; it was long obtained for lighting purposes from hand-dug pits and collected from the numerous seepages as a tarry residue with which to caulk canoes. The petroleum industry, however, cannot be said to have begun until 1909, when the Anglo-Saxon Petroleum Company received a concession from the Rajah to exploit the oil resources of his country. This concession they held until 1921, when they transferred it to the Sarawak Oilfields, Ltd., a subsidiary of the Royal Dutch/Shell Group, incorporated in Sarawak.

Oil was struck in 1910 near Miri in northern Sarawak, and the first well put in production yielded 12 tons daily. Once the existence of oil in commercial quantities was proved, development proceeded rapidly. In the Miri field, situated on a highly faulted asymmetric anticline on the coast, production is obtained from depths to about 3,000 feet in Upper Miocene sands. Annual production passed 100,000 metric tons in 1920; by 1926 it was over 700,000; and peak annual production of 760,000 metric tons was reached in 1929. But after 1930 a rapid decline in productivity set in, and in 1939 the field was nearing depletion, the production down to 200,000 tons a year.

In the meantime the British Malayan Petroleum Company, another subsidiary of the Royal Dutch/Shell Group, had discovered the Seria field, on a gentle faulted anticline along the coast of southern Brunei, the adjoining state. In 1924 the company secured a first lease for 50 years and in 1932 another, also for 50 years. Production in the Seria field started in 1932, with 177,000 metric tons for that year, rapidly increased, and in 1938 passed 700,000 metric tons. Heavy asphaltic oil is produced at shallow depths (500 to 2,000 feet) and light paraffinic oil from depths to nearly a mile. Part of the production is found offshore, the outer boundary of the field not having been established. Though extensive exploration has been carried out in other parts of Sarawak and Brunei, no commercial production has been attained.

In 1916 a small refinery was erected at Miri. It was moved in 1919 to Lutong, about 7 miles north of Miri, and was considerably enlarged later, to cope with the combined output of the Miri and Seria fields—almost 1 million metric tons in 1939. From a village, the town of Miri grew to be a flourishing community of a few hundred Europeans and several thousand Malays and Chinese, with bus services, motor launches, schools, a modern hospital, clubs for the Asiatic as well as the European staff, and golf links, tennis courts, and other sports grounds.

When the fields were reoccupied after the war, it was found that the Japanese had thoroughly wrecked the refinery and wells. Their rehabilitation was begun before V-J day, and the effectiveness of the measures is indicated by the fact that, by the end of 1948, production was at a rate three times greater than that of 1939, and a large expansion program was being undertaken.

During 1948 British Borneo (Sarawak and Brunei) produced a total of 20,000,000 barrels, equivalent to a daily rate of 55,000 b/d. The Lutong refinery, with a capacity of 25,000 b/d, has been rehabilitated, but it is reported that a new and larger refinery is to be built. Meanwhile, crude oil is being shipped to the Royal Dutch/Shell Group's refineries abroad.

The coast of Sarawak has no sheltered harbors available for large ships; consequently all oil has to be loaded through submarine pipe lines which run nearly three miles out to sea. Flexible hoses are provided at the seaward end, and these are hoisted on board and connected to the tankers' receiving lines.

Burma

In the oil industry of Southeast Asia, Burma, with about 17% of the total cumulative production, occupies a secondary position. Only two major fields, Yenangyaung and Singu-Lanywa, have been developed. But the oil industry in Burma is distinguished by a number of interesting features from that of the East Indian archipelago. The history of the remarkable

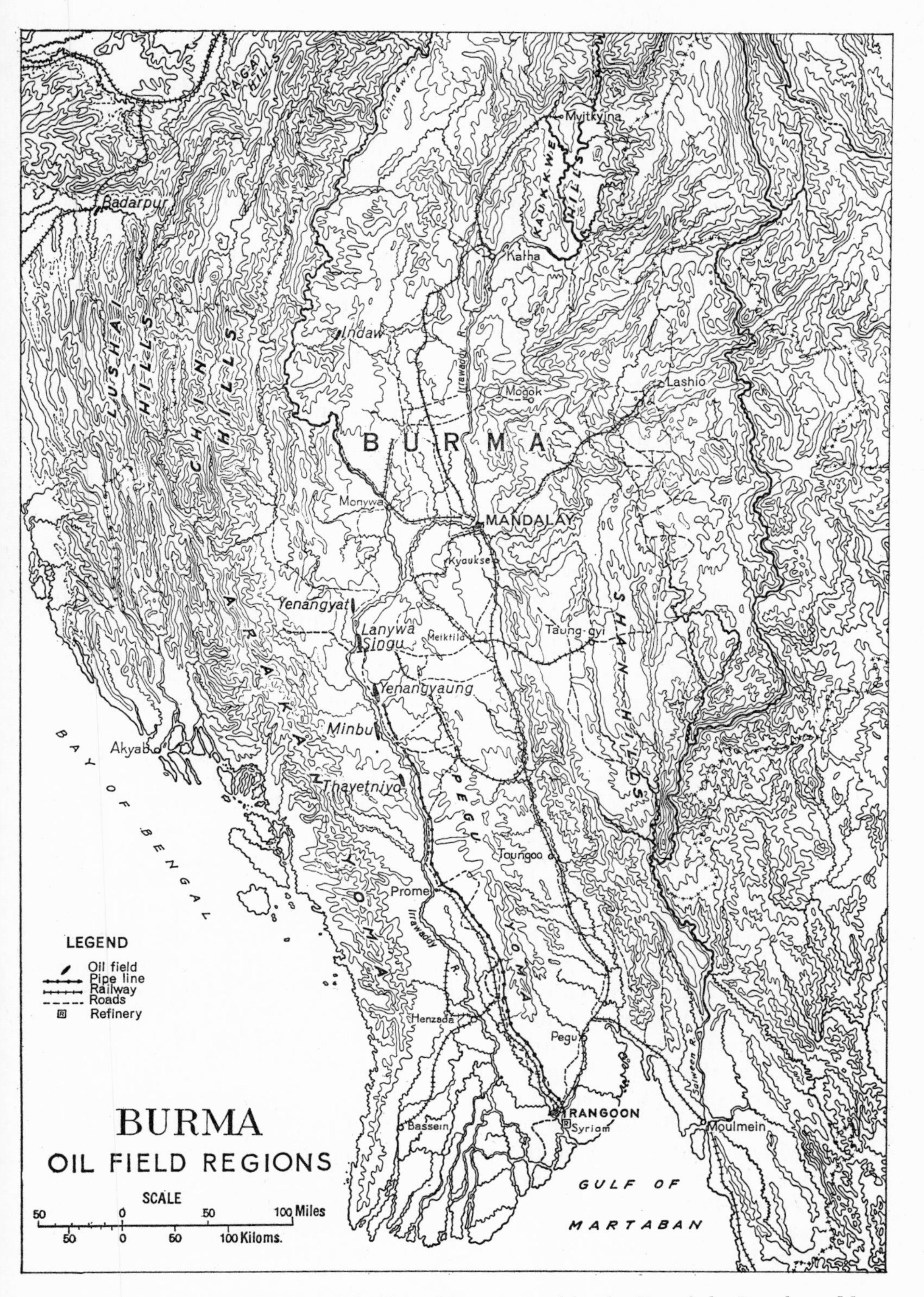

FIG. 46. Burma: oil-field regions and facilities. Map prepared by the Bataafsche Petroleum Maatschappij, showing oil-field development as it was before the war, with the addition of railroads and highways built to mid-1949. The altitude contours (all in solid lines) denote: 200 meters, 500 meters, 1,000 meters, and then 1,000-meter intervals to 4,000 meters. Two producing fields in Assam, in the Brahmaputra Valley north of Badarpur, do not appear on this map, and fields yielding a few barrels daily on Western Baronga Island, near Akyab, and near Kyaukpyu on Ramree Island, the largest of the island group farther south, are likewise omitted.

exploitation of the Yenangyaung oil field covers a longer period of time than that of any other field in the world. Most of the oil fields in Burma, though situated in the tropics, lie in a subdesert. They are the only fields in Southeast Asia where competitive drilling is known. Finally, the oil industry in Burma was, during its development, restricted to British enterprise, foreign influence having been unable to obtain any foothold.

The oil fields are, for the most part, in the great river basins of the Chindwin and the Irrawaddy (Fig. 46). Northward lie the great ranges of the Himalayas, to the north of the Assam basin of the Brahmaputra in northeastern India, where the Digboi field of Assam is situated. The Himalayan trend is represented in eastern Burma by the hills and plateau of the Shan states. The Shan hills continue farther south, forming the ranges of Lower Burma, which are a part of the Himalayan-Malayan arc, but much reduced in height. The central Irrawaddy basin is flat and covered with alluvium, but from these plains project the knife-edge ridges of the Pegu Tertiary rock. On the east are the Pegu Yoma, a broad range of hills very much dissected and densely wooded. On the west are the Arakan Yoma, also densely wooded but higher, and separating the Irrawaddy basin from the plains of Bengal. On the south the Irrawaddy basin opens out into its great delta, which is monotonously flat and almost everywhere covered by rice fields.

Burma has a monsoon climate, in which almost all the rain falls between May and September. Lower Burma is hot and damp, and has a high annual rainfall, exceeding 300 inches in places, but it is not particularly unhealthy. In the Arakan Yoma, however, and most of the Upper Chindwin and the foothills of the Shan ranges, blackwater fever is not uncommon in some valleys. The Shan hills are on the whole healthy, and the Shan plateau rivals Europe in climate.

The main oil fields of Burma lie in what is known as the "dry zone," a subdesert which Sir Edwin Pascoe of the Geological Survey of India likens to the "bad lands" of western Nebraska. The rainfall is from 25 to 30 inches, as compared with averages in Assam of 100 to 150 inches and a maximum of 450 inches; and the temperature reaches 117°. The ground is parched and sandy, with salt efflorescence in places, and the vegetation is comprised mostly of scrub, cactus, and other thorny plants.

The story runs that, centuries ago, captives of an ancient Burmese king started working the oil on the barren plateau near the east bank of the river Irrawaddy. It is certain that in the eighteenth century oil was known and used by the Burmese for their lamps, and two petroliferous tracts (Twingon and Beme) of the present Yenangyaung field were worked toward the end of the century. Hereditary rights to dig for oil were granted to twenty-four heads of families, known as "Twinzayos" (which literally means well-eaters, those who eat or obtain a living by possessing wells). When Burma was annexed by Britain, in 1885, the production from native wells amounted to some 10,000 metric tons a year: by 1900 it had increased to 34,000.

The first wells were plank-lined shafts, 5 feet square. In the nineteenth century depths of 300 feet were reached and later, slightly more than 400 feet; but it took nearly two years to attain such a depth. A shaft once completed, the oil diggers went down practically naked, lowered on a rope over a pulley. In a well about 250 feet deep it took 15 seconds to reach the bottom. After half a minute of frantic digging and loading of pots in an atmosphere saturated with gas, the digger was hauled up in two minutes; then he needed half an hour's rest before he was sufficiently recovered to go down again. The light by which the digger worked was provided by a mirror at the mouth of the shaft. Towards the end of the century digging was somewhat modernized by the introduction of a sort of diving dress for the digger, with air pumped down to him by a hand-driven pump.

The formation of the Burmah Oil Company in 1886 laid the foundation of the modern oil industry in Burma. The rights of the Twinzayos were recognized by the British government, and the Twingon and Beme reserves, containing 295 and 155 acres respectively, were demarcated. The minimum distance between well sites was fixed at only 60 feet.

The Burmah Oil Company started the first machine-drilled well in 1887, between the two reserves, but also acquired well sites within the reserves by purchase. Between 1888 and 1906 the whole of the production from machine-drilled wells was that of the B.O.C. Then new companies began to lease well sites from the Twinzayos, and competitive drilling in the reserves became very fierce; though outside the reserves practically no competitive drilling was possible, since most of the remainder of the oil-field area was leased by the B.O.C. The output of the Yenangyaung field passed its peak in 1918, with a yield of about 800,000 metric tons for that year.

In 1902 the Burmah Oil Company commenced drilling at Singu, farther north on the same belt of folds. Other companies, notably the Indo-Burma Petroleum Company, are at work there, too, but as the ground has been leased in square-mile blocks there is less feverish competition than in the Yenangyaung reserves. Singu, with its extension Lanywa on the opposite bank of the Irrawaddy, became the second most important field in Burma. The Lanywa extension is worked by the Indo-Burma Petroleum Company. The field is on a sandbank two miles long, reclaimed from the river between 1925 and 1928. The Singu-Lanywa fields have been steadily and carefully developed, and constitute an important reserve for Burma's oil industry; shortly before the war production surpassed that of Yenangyaung.

The other Burmese fields are of minor importance. They are found in the Irrawaddy basin below Yenangyaung, at Indaw in the Chindwin basin, and on the Arakan coast.

The Burmese output was refined at two refineries on the estuary of the Rangoon River belonging to the B.O.C. and I.B.P. The B.O.C.'s oil was transported by pipe line to the large Syriam refinery, while the I.B.P.'s was carried down the Irrawaddy by barges. In 1939 the crude oil produced in Burma amounted to 1,087,000 metric tons.

Production for war requirements was pushed actively in Burma during the early part of the war. As the Japanese invading forces pressed westward into the Irrawaddy Valley early in 1942, however, preparations were made ready for the demolition of wells, pipe lines, power houses, and refineries, with a view to rendering the facilities useless to the enemy. The scorched-earth process that was carried out in April 1942 resulted in the destruction of 3,500 derricks and some 700,000 barrels of crude oil. Despite this destruction, the Japanese succeeded in gradually building up production to several thousand barrels per day; but in April 1944, when they were finally driven out, the output was still small.

Very little rehabilitation work has been possible because of the state of civil war, which has spread over nearly the whole of Burma. Reconstruction of the 325-mile pipe line has been at a standstill, and work on the new refinery at Syriam suspended. The output of crude oil during 1948 averaged only 820 b/d.

We must not close this brief review of the Burmese oil industry without a further remark: though the American geologists, E. B. Andrews and T. Sterry Hunt, were the first to work out, in 1861, the anticlinal theory which has had so profound an influence on the better understanding of commercial oil accumulations, it was Dr. Thomas Oldham, the father of the Geological Survey of India, who, at Yenangyaung in 1855, first observed the association of oil with anticlinal structure.

13. OTHER AREAS IN AFRICA, ASIA, AND OCEANIA

BY D. DALE CONDIT *

THE survey of the Eastern Hemisphere can be concluded with a brief survey of areas of less importance that are either producing in a small way, or present some possibilities for oil. In this classification are included sedimentary basins in Africa other than Egypt, and various basins in Asia other than in the Persian Gulf region, Burma, and the Soviet Union, as well as basins in Oceania other than Indonesia. All of the excepted areas are described in preceding chapters.

Apart from the continental shelves, these areas of less importance include numerous scattered basins in Africa, two in Afghanistan, several in India and Pakistan, numerous basins in China, several in the Philippines, one in Formosa, several in Japan, and, finally, several in Australia. Oil fields have already been developed in India, Pakistan, China, Formosa, and Japan. Their aggregate output, however, amounts to less than .2% of current world production.

Africa Other than Egypt

Because of the magnitude of its central shield, Africa has a smaller proportion of sedimentary area of interest for petroleum than any other continent. The only known area of any promise in the interior of Africa is in western Uganda, near the border of the Belgian Congo, where oil seepages are present in the extensive Lake Albert graben, a down-faulted trough in the much older crystalline rocks—composed of Tertiary sediments. The remaining regions of any promise in Africa are mainly on coastal plains around its outer margins.

In northern Africa the broad sedimentary belt extends for over 2,000 miles westward from Egypt. In Libya, exploratory studies only have been carried out. In Tunis, Algeria, and Morocco exploration in proximity to seepage localities has reached the drilling stage, most of it under the auspices of the French government. In Algeria three small oil fields, with a nominal output, have been developed in Tertiary formations; and in Morocco four fields, yielding in all about 100 barrels per day during 1946, have been discovered. Pay sands range from Jurassic to Tertiary in age.

Along the west coast of Africa south of Morocco six areas merit exploration, mainly in Cretaceous and Tertiary formations; they are distributed at intervals over a lengthy coast line extending from French

* Geologist, Standard Vacuum Oil Company; formerly Chief Geologist, Whitehall Petroleum Corporation, India, and Oil Search, Ltd., Sydney, Australia.

West Africa to Angola. In Nigeria, where the sedimentary area of interest extends a long distance inland, in French Equatorial Africa, and in Angola, drilling for oil has been carried out, but without success.

The extensive Karroo basin in the Union of South Africa, at the south end of the continent, is filled with a thick section of Gondwana fresh-water deposits (Permo-Carboniferous to Triassic in age). Drilling for oil, some of it by the Union government, has not yielded encouraging results, and the area is of doubtful promise.

Gondwana beds, while unpromising for oil, have important coal measures and associated bituminous shales which, on distillation, yield oil. The oil shales are now being exploited in a small way at Ermelo, about 80 miles southeast of Johannesburg. Similar deposits also occur in the Gondwana of the Belgian Congo.

Along the east coast of Africa exploratory studies have outlined two areas of interest for oil, principally in rocks of Mesozoic and Tertiary ages. One area covers the greater part of southern Mozambique, and another, parts of Tanganyika, Kenya, Ethiopia, the former Italian Somaliland, and British Somaliland. While there have been no successful developments in any of these extensive basins, explorations are being pushed actively in Mozambique and Ethiopia by American companies. Eritrea, still farther north, and bordering on the Red Sea, has some possibilities.

In Madagascar, sedimentary rocks ranging in age from Permian to Tertiary extend along the west side of the island. Tar sand outcrops and oil seepages lend interest to the area as a possible oil producer. Shallow drilling, carried on at intervals for many years by the government, has not, however, yielded more than slight amounts of oil.

Asia Other than the Middle East, Burma, and the Soviet Union

AFGHANISTAN

No oil has been produced in Afghanistan, and no drilling for oil has been done. It is known, however, that the northernmost part of the country on the north side of the Hindukush Range is promising territory. Oil is now being produced in several fields on the U.S.S.R. side of the international boundary, in territory which constitutes a part of the same sedimentary basin. Remoteness from large markets, such as those of populous India, as well as difficult access, is a handicap which may long delay large-scale exploration and development.

INDIA AND PAKISTAN

In Pakistan and India, which until August 1947 were parts of British India, oil has been produced for many years in two widely separated Tertiary basins, one in the extreme northeast, in the part of Assam Province now in India, and the other in the extreme northwest in Punjab Province,

86. Assam, India. A general view of a small part of the Digboi field, giving some idea of the difficult terrain. (Government of India Information Services, Washington, D.C.)

87. Assam, India. An elephant and a tractor moving drill pipe, Digboi field. (Government of India Information Services, Washington, D.C.)

88. Canada. North West Territories. Norman Wells seen across the frozen Mackenzie. (Standard Oil Company, N.J., photo by Collier.)

Pakistan, a distance of some 1,400 miles west of the Assam field. The aggregate output in both regions has rarely exceeded 7,000 b/d, as compared with a postwar daily demand of 55,000 to 60,000 b/d for the two countries. In prewar years the deficiency was largely supplied by Burma, which until 1937 was a part of British India. Unfortunately, very little progress has been made toward restoring production in that country since the war.

All of the present production of India is derived from the Digboi field of Assam, situated in the mountain-locked upper Brahmaputra Valley, remote from the populous plains. Of the 43 million barrels (U.S. 42's) of oil produced in Assam to the end of 1948, more than 95% came from this one field. Though production began here 60 years ago, the field is still being actively produced.

During the war the output of Digboi, though at best averaging little more than 7,000 b/d, was of unusual importance because of its situation close to the scene of fighting. Ledo, the starting point of the memorable Ledo Road, which was constructed as a war measure across the mountains into Burma, is near Digboi.

Oil explorations including drilling are being pushed by British companies in the more accessible parts of India southwest of Digboi, and it seems probable that new discoveries will eventually result. Until recently, the Mining Rules have excluded non-British companies, but revisions in process of being enacted by both India and Pakistan may remove those restrictions.

The northern Punjab oil fields, of which there are four, namely, Khaur, Dhulian, Balkassar, and Joya Mair, derive their production from limestone reservoirs of Eocene age. Depths in Balkassar and Joya Mair, the two more recently discovered fields, vary from around 7,000 to more than 8,000 feet. While the oil from Joya Mair is heavy and asphaltic and difficult to refine, that from Balkassar is of better quality. Depths in Khaur and Dhulian, the two older fields, also are considerable, but the quality of the oil is good. Production in these fields has declined rapidly in late years, and they appear to be in an advanced stage of depletion. The cumulative production for the Punjab fields to the end of 1948 has been about 11 million barrels.

Since the war, British companies have carried on geological and geophysical surveys in the lower Indus valley. A deep exploration well recently drilled about 100 miles inland from the coast has been abandoned as a dry hole at a depth of 12,666 feet.

While both India and Pakistan require far more oil than either has been able to produce, the fuel problem of Pakistan is particularly aggravated by an almost complete lack of workable coal, with poor prospects for important coal discoveries. India, in contrast, has many important, widely distributed coal fields. Prospects for future oil discoveries in Pakistan, however, are promising because Pakistan includes within its boundaries practically all of the major Tertiary sedimentary basin in the west, namely, the Indus Valley–Punjab basin, and also much of the Bengal-Assam basin in the east.

SIAM AND INDOCHINA

The fold map of World Sedimentary Basins and Petroliferous Areas shows a basin occupying the adjacent parts of Siam and Indochina. Very little information is in hand regarding the geology of this basin. So far as is known, no drilling for oil has been done anywhere.

CHINA

To the Chinese goes the credit for being the world pioneers in the development of well-drilling technique. The history of their achievements in drilling to substantial depths in the exploitation of salt-water sands goes back at least one thousand years. At the scene of these operations in the Tzuliuching field, near Chungking, Szechuan Province, homemade outfits constructed mainly from bamboo and other homemade materials had attained depths of several thousands of feet long before John D. Rockefeller went into the oil business, and tools of this type are still standard equipment. Small amounts of natural gas, encountered in the drilling, were utilized in evaporating the briny waters in the extraction of salt.

A little oil also obtained in some of the Tzuliuching wells, together with oil seepages in widely scattered districts, has stimulated a search for oil. In late years the government has carried on exploration drilling, so far without success, though unimportant amounts of gas have been found near Chungking. The age of the formations ranges from Cretaceous to late Paleozoic.

The one important oil development in all China is the Yumen field in Kansu Province, some 1,500 miles northwest of Chungking, where the discovery well was drilled in 1939. During the war, despite the handicap of remote situation in this mountainous, sparsely populated desert region, a production of 1,000 b/d or more was gradually built up. The age of the oil sands is Eocene. Information in hand suggests that additional discoveries may be made in various parts of this same general region.

Oil seepages have long been known in two sedimentary basins in the extremely remote Sinkiang Province, 1,000 to 1,500 miles northwest from Yumen. During the war, oil was discovered at Wu Su with the aid of the Russians, who supplied equipment. So far as is known, no effort has been made to exploit this deposit.

Still other sedimentary basins in various parts of China are believed to have oil possibilities, among them parts of the Gobi Desert and the basin of Shensi. In Shensi small amounts of oil have been obtained in shallow wells, the first of which was drilled near a seepage during World War I.

During their occupation of Manchuria, the Japanese exploited coal and oil shale deposits in the Fushun district near Mukden. In the same general region, drilling for oil in areas of Jurassic rocks resulted in discovery of small amounts of oil.

PHILIPPINES

Next to be described are the several island chains off the southeast coast of Asia, including the Philippines, Formosa, and Japan, all of which are of interest for oil.

Though the Philippines have not as yet yielded oil in paying amounts, Tertiary sedimentary basins with seepages have been mapped on several of the islands. Drilling carried on intermittently for many years has yielded shows of oil, and the outlook for discoveries appears moderately promising.

FORMOSA

In 1895, when the Japanese took over Formosa following the Sino-Japanese war, the Chinese were already producing oil in the field now known as "Shukkoku" in the northern part of the island. Despite prolonged efforts in the search for further fields, with the drilling of dozens of wells on what appeared to be promising anticlinal structures, the Japanese never succeeded in discovering another oil pool. They did, however, develop numerous gas fields at widely scattered points, the gas in several instances being so rich in petroleum vapors that substantial amounts of gasoline are extracted.

The productive formations, of late Tertiary age, extend along the west side of the island. The total amount of oil produced since early days has probably not exceeded 1,000,000 barrels. To this figure should be added some 800,000 barrels of gasoline extracted from natural gas.

JAPAN

Japan produces oil from numerous small fields in three Miocene sedimentary basins on the northwest side of Honshu Island. Minor quantities are also produced in Hokkaido. The best years were from 1915 to 1917, when nearly three million barrels were produced annually. The output early in 1949 was at a rate of around 3,500 b/d. Total production since 1872, the year when the keeping of statistics was begun, has been about 93,000,000 barrels.

Oceania Other than Indonesia

AUSTRALIA

In Australia there are three large sediment-filled basins that may offer some promise for petroleum, two being in Western Australia, and the third, in the east, known as the Great Artesian basin and covering more than 200,000 square miles in Queensland, New South Wales, and South Australia.

Desert basin, the northern of the two basins in West Australia, with an area of 150,000 square miles, includes Devonian and Permian formations. Oil showings encountered in a well drilled for water resulted in drilling activity that has been in progress intermittently ever since 1921, without encouraging results. The southern basin, known as the Northwest basin, comprises formations from Permian to Tertiary which include important fresh-water-bearing sandstones. No showings of oil or gas have been encountered, however, in the numerous deep wells drilled for artesian water.

The Great Artesian basin is filled with a moderately thick section of late Paleozoic, Triassic, Jurassic, and Cretaceous formations in which are important water-bearing sands that have been tapped by thousands of artesian wells. Chance showings of oil and gas, encountered locally in wells in the eastern part of the basin, have led to serious oil prospecting which, to date, has not been successful.

The discovery of natural gas in a water well at Roma, Queensland, more than twenty years ago led to considerable drilling in that district, the results of which were mostly discouraging. In later years the search shifted to the region 80 to 100 miles north of Roma where a small Australian Company drilled two tests into formations of Permo-Carboniferous age, one of which encountered non-flammable gas consisting of 70% carbon dioxide. Despite this discouragement exploration work has been resumed since the war, and a major British company is preparing to drill a deep test at a point about 15 miles farther north.

A discontinuous narrow strip of Tertiary overlap extends along the south coast in the states of Victoria and South Australia. Shows of oil in an artesian water well at Lakes Entrance, east of Melbourne, led to drilling of numerous shallow wells that have yielded a little oil. Prospecting elsewhere has met with discouraging results.

The search continues actively in various parts of Australia with the financial aid of Commonwealth and State governments. Meanwhile, oil is being derived in a small way from rich oil-shale deposits in New South Wales, and plans are being considered for the manufacture of oil from coal, of which there are vast supplies.

The Australian territory of Papua, in the eastern part of the island of New Guinea, has several large Tertiary sedimentary basins which are being prospected actively. In 1948 a well was abandoned at a depth of 12,621 feet without obtaining any oil shows. Drilling is in progress in three other wells.

NEW ZEALAND AND NEW CALEDONIA

Both the North and South islands of New Zealand appear to present some promise for oil in Cretaceous and Tertiary strata. Drilling in early years around oil seepages in Taranaka District yielded small but encouraging amounts of oil. Extensive geological surveys, followed by deep drilling, in various districts, carried out in recent years have not, however, resulted in any oil discoveries.

The island of New Caledonia, situated northwest of and in alignment with New Zealand, has oil seepages in strata of Eocene age. The scant information available does not indicate that the outlook for discoveries is promising.

14. PETROLEUM IN THE POLAR AREAS

BY WALLACE E. PRATT

THAT petroleum, a product of the decomposition of the organic residue of marine life, might be expected to occur in an environment so inhospitable as the polar areas are today, may come as a surprise to the reader. Yet when we recall that petroleum has been a normal result of common earth processes through much of geologic time, and that the present polar ice caps are geologically recent phenomena which only very lately brought to an end a long sequence of equable climates at the poles—palm trees flourished as far north as Spitzbergen—when we reflect on these facts, we realize that the polar climates of today in nowise preclude possibilities that in a geological yesterday important petroleum resources may have come to exist near the poles.

But any discussion of petroleum in the polar areas must be prefaced by the admission that these resources are but little known. Polar exploration in general is far from complete, and the petroleum resources of the greater part of these areas are hardly explored at all. Except in Soviet Asia, where the Russians claim to have proved large petroleum resources, the search for oil in the polar areas has hardly begun.

Some Contrasts between the Arctic and the Antarctic

The polar regions of the earth, alike in the frigid climate common to both, appear in striking contrast in other geographical aspects. At the North Pole a depression lying between the continents is occupied by a body of water which, incorrectly known as the Arctic Ocean, is in reality a partly land-locked sea; it has appropriately been referred to as "the Arctic Mediterranean." At the South Pole is a lofty continent, Antarctica, greater in area than Australia. Antarctica is remote and separated by oceanic deeps from every other continent, inhospitable, devoid of land-animals other than a few species of insects, and all but devoid of land plants; and no important route of world commerce traverses the Antarctic. The Arctic, on the other hand, is the permanent home of a variety of animal species, including a large human population, and, with the advent of aerial transport, the position of the Arctic on the shortest direct route between the principal cities of Asia, Europe, and America promises to make it a crossroads of the world. Whereas economic activities extended into the Arctic centuries ago, and have expanded continuously, world trade has hardly been made aware of the Antarctic.

If, then, petroleum were to be developed in the Arctic areas it would find an expanding local market, and it might well become a strategic source of fuel for world air commerce. Petroleum in the Antarctic, even if it were found in abundance, would be exploited commercially only if the demands of distant centers of population should become sufficiently insistent to justify costs and hardships greater than such demand has ever justified in the past.

If commerce in general has failed so far to exploit the resources of the Antarctic, the petroleum industry has been even less concerned, if that is possible, with this vast unknown area; no exploration whatever for petroleum has been undertaken in the Antarctic. Yet Antarctica is a great continent of more than 5,000,000 square miles. Of the other continents, all but one, Australia, are commercial sources of petroleum: the chances are good, therefore, merely on the basis of probabilities, that commercial petroleum resources exist in Antarctica too. Moreover, we know that Antarctica's rocks consist in large part of marine sediments, the usual environment of petroleum; and we know that throughout geologic time glacial conditions have been the exception and not the rule in Antarctica. The present ice cap developed only in late Pliocene time. A thick series of Paleozoic rocks, including Cambrian, Silurian, Devonian, and coal-bearing Permo-Carboniferous strata, outcrops in the lofty horst which forms the western margin of the Ross Sea. On the peninsula of Graham Land, west of Weddell Sea, is a younger series of rocks in which Jurassic, Cretaceous, Oligocene, and Miocene beds of marine origin have been identified, and this area appears to be related geologically to the southern part of South America, where oil has been discovered in rocks of corresponding age.

Professor F. Alton Wade, geologist on the second Byrd Expedition, 1933–1935, and again on the U.S. Antarctic Service Expedition, 1939–1941, says: [1] "There is no reason for concluding that oil does not exist in some of the Antarctic formations. To the contrary, there are good reasons to assume that it does." In the same article Dr. Wade quotes Sir Hubert Wilkins: "From an examination of the ice-free surfaces of the United States sector of the Antarctic, bounded on the east by Hearst Land and on the west by Marie Byrd Land, it seems that the Pacific sector of the Antarctic might hold the greatest pools of oil yet to be tapped on earth."

In the Arctic, the wide expanse and great volume of the marine sedimentary rocks which compose the coastal areas and the continental shelf lead us at once to expect to find petroleum, and the prospect is enhanced by the favorable geologic structure into which these bedded rocks have been warped by earth movements. In Permian times geologic folding produced a great geosyncline extending along the entire Arctic coast of Asia, and subsequent (Cretaceous) folding can be detected in the rocks of Spitzbergen, northern Greenland, and Grinnell Land.

In this favorable setting tangible evidences of petroleum are, in fact,

[1] Wade, F. A., "Oil in Antarctica," *Oil Weekly,* Vol. 121, No. 5, Apr. 1, 1946, International Section, p. 5.

so conspicuous that they impressed themselves upon early explorers, intent though these men were upon quite other objectives. An entry in the diary of Alexander Mackenzie[2] under date of August 2, 1789, when he was engaged in his pioneer exploration of the great river which subsequently came to bear his name, appears to record his observation of seepages of petroleum, despite the fact that he identifies the material in question as coal (a natural mistake at a time when petroleum was little known): "We set out at three this morning with the towing line. . . . When we came to the river of Bear Lake . . . in our progress we experienced a very sulphurous smell, and at length discovered that the whole bank was on fire for a very considerable distance. It proved to be a coal mine to which the fire had communicated from an old Indian encampment. The beach was covered with coals and the English Chief [the title Mackenzie had bestowed upon his Indian guide] gathered some of the softest he could find, as a black dye; it being the mineral, as he informed me, with which the natives render their quills black."

This passage is believed to refer to one of the copious seepages of petroleum in the general vicinity of Fort Norman, Northwest Territory; seepages which led to the discovery, more than 100 years later, of the Fort Norman oil field.

Altogether, an area of 1.5 million square miles north of 60°N appears to be generally favorable for the occurrence of petroleum in commercial quantities. About one-third of this area lies in the Western Hemisphere, and the portion in Alaska alone exceeds 200,000 square miles. As a comparison with the Arctic, it may be noted that the area similarly favorable for the occurrence of petroleum within the continental United States, exclusive of Alaska, is also of the order of 1.5 million square miles.

Petroleum in the Arctic of the Eastern Hemisphere

In the Eastern Hemisphere the million square miles of Arctic territory favorable for petroleum is made up of several distinct areas, distributed along the belt of Arctic coastal plain which stretches eastward from Arkhangel'sk to Bering Strait over a distance of 3,000 miles. Guided by seepages of petroleum and natural gas, and by exhaustive geological and geophysical surveys, intensive exploration has been under way in this region since 1934, according to the published reports of the Soviet government.

Fringing the eastern margin of the old continental shield of crystalline rocks which makes up most of the land mass of the Scandinavian countries, stretching northward from Moscow to the Barents Sea, and thickening eastward as far as the axis of the Ural Mountains, there is encountered a great wedge of Paleozoic sediments of a character generally favorable for the occurrence of petroleum.

Beyond the Ural axis, filling the foredeep along its eastern flank, occupying the broad valley of the Ob' River, and persisting northeastward beyond

[2] Quoted by Aubrey Fullerton in the *Edmonton Journal,* Apr. 16, 1921.

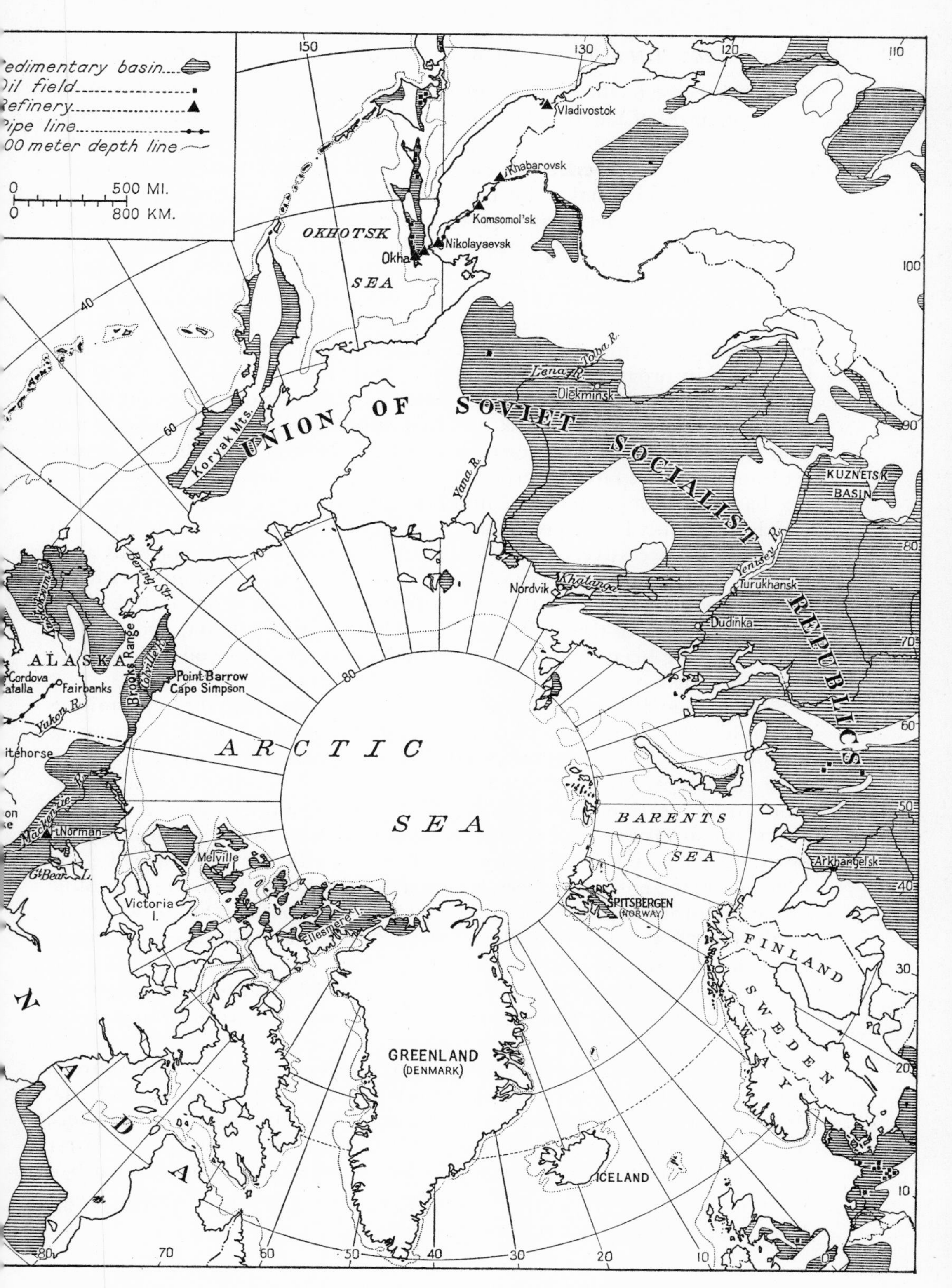

47. Petroleum in the Arctic and adjoining areas, as developed in mid-1949. Information on sedimentary ıs and oil-field areas from the Standard Oil Company (New Jersey), and on pipe lines and refineries from *ld Oil,* July 15, 1949, and from other sources.

the Arctic coast to the broad continental shelf beneath the Arctic waters, this same series of Paleozoic rocks is again in evidence. This series extends as far eastward as the Yana River, and, together with Jurassic and Cretaceous rocks which overlie the Paleozoics east of the Urals, forms a thick succession, filling a major downwarp in the earth's crust.

In extreme northeastern Siberia, again, extending out under the Bering Sea to form a wide continental shelf, is a large promising area of Cretaceous rocks which are tilted off the eastern flank of the Koryak Mountains, and thrown into favorable attitudes to form traps for petroleum. This province is the northward extension of the petroleum-bearing rocks of Kamchatka Peninsula, where exploratory and development activities have been in progress many years.

Along the Yenisey River at its estuary, and for 300 miles southward from Dudinka to Turukhansk, petroleum seepages have been noted, and near them test wells have been drilled. Near Nordvik on Khatanga Gulf, six hundred miles farther east, other seepages are found, and further drilling has been done. Here are numerous salt domes—geologic structures which commonly house petroleum accumulations elsewhere over the earth —intruding successively beds of Devonian, Jurassic, and Cretaceous ages, each of which contains units of petroliferous character and appears to be of likely source rocks. Southwest of Nordvik, oil has been found along the entire course of the Tolba River, a tributary of the Lena; and producing wells have been drilled near Olekminsk and within 200 miles of the Sea of Okhotsk.

Siberia, then, like other parts of the U.S.S.R., appears to possess promising potential sources of petroleum. The evidence of these possible resources is replete in rocks of very wide areal distribution, as well as throughout the geologic column, from Cambrian upward into the Eocene. Even the remote Arctic portion of Siberia, the territory here under inquiry, exhibits abundant surface manifestations of these resources, as the enthusiastic reports of Soviet scientists demonstrate. Farther south, for a distance of 2,500 miles along the entire southern boundary of the U.S.S.R., from the prolific oil fields of the Caspian Sea region west to the Carpathian arc in Rumania on the one hand, and east to the frontiers of India and China on the other, there are imposing evidences of vast petroleum resources. And in the Far East, on Kamchatka Peninsula and on Sakhalin Island, there are important producing oil fields, together with large, promising sedimentary basins still unexplored.

In the face of this wide distribution of potential petroleum sources, Soviet production is woefully small in relation to her needs, and Siberia, in particular, appears to suffer from inadequate supplies of liquid fuels. The total consumption of petroleum products in Siberia for the year 1941 is reported at slightly less than 30 million barrels,[3] or about 80,000 b/d

[3] Shanazarov, D. A., "Petroleum Problem of Siberia," *Bull. Amer. Assn. Pet. Geol.*, Vol. 32, No. 2, Feb. 1948, p. 154.

—and this entire supply had to move into and across Siberia from outside sources over the Trans-Siberian Railway. This movement of petroleum products must constitute an almost intolerable burden, and so act as a sharp spur to the development of internal sources of liquid fuels for Siberian consumption. Yet, as far as is known, Arctic Siberia still produces no petroleum.

Among the factors which have seriously impeded the development of oil fields in Siberia, according to the reports of Soviet scientists, is the permanently frozen character of the earth's crust to great depths. Wells drilled at Nordvik on the Arctic coast reached a depth of 2,000 feet without passing out of the frozen zone,[4] and the temperature in the frozen zone in these wells was recorded at —14°F. But it seems unlikely that this problem, or any related problem, will long prevent the forward-looking Soviet leaders from developing the petroleum resources of northern Siberia, so sorely needed by the expanding industrial economy of this great empire.

Petroleum in the Arctic of the Western Hemisphere

Typical of the surface manifestations of petroleum in the Arctic are seepages near Cape Simpson, east of Point Barrow on the coast of northernmost Alaska, examined by geologists of the United States Geological Survey more than 20 years ago, and described by them as follows: [5] "Seepage No. 1 occurs near the inland face of this ridge. . . . Here, in an irregular area several hundred feet in diameter, the moss is soaked with petroleum, which also slowly seeps from the gentle slope. Seepage No. 2 is . . . 3 miles almost due south of Seepage No. 1. . . . Here the residue covers several acres. The main petroleum flow moves southward down the slope for 600 or 700 feet, to a lake."

The discovery of these remarkable seepages led the U.S. government in 1923 to segregate some 35,000 square miles of the surrounding territory as Naval Petroleum Reserve No. 4. No attempt was made to develop this reserve, however, until the latter years of World War II, when further geological exploration was undertaken, both by the United States Navy and the Department of the Interior. In the course of this further exploration, numerous other impressive seepages of petroleum were found along the northern coast of Alaska as far east as the Canadian boundary, a distance of 300 miles.

Most of these oil seepages come to the surface at points underlain by rocks of Tertiary age, occupying the central portion of a great depositional basin which constitutes much of the land area of Alaska north of the Brooks Range, and may extend farther north, under the Arctic Sea. The Tertiary rocks are believed to attain a thickness of several thousand feet, and beneath them are even thicker sections of both Upper and

[4] *Ibid.*, p. 192.

[5] Paige, S. S., Foran, W. T., and Gilluly, J., "A Reconnaissance of the Point Barrow Region, Alaska," *U.S. Geol. Surv. Bull.* 772 (1925), p. 23.

Lower Cretaceous rocks, as revealed by outcrops in the northern foothill belt of the Brooks Range to the south. Jurassic and Triassic rocks are also known to be present, locally at least, beneath the Cretaceous series, and are themselves more widely underlain by Devonian limestones. Each of these series of beds is of such a character as to make it a possible petroleum source.

The oil seepages which overlie Tertiary rocks appear to be petroleum residues, such as might result from the evaporation of a more liquid oil. They cover considerable areas and represent large tonnages of asphaltic material. Seepages of a different character, active flows of light oil accompanied by natural gas, have been noted at a number of places where Cretaceous rocks are exposed at the surface. These live seepages escape directly from the broken edges of the Cretaceous beds and are dissipated, whereas the heavy residues form more or less permanent bodies.

Near a group of seepages from Cretaceous rocks on the Colville River, within the limits of the Naval Petroleum Reserve in the Umiat District, the drilling of a test well was undertaken by the United States Navy in the summer of 1945. Operations have now been carried on fairly continuously at Umiat for several years, and several wells have been completed, including a small shallow producer. These wells are located near the crest of a promising anticline 150 miles southeast of Point Barrow and 350 miles northwest of the city of Fairbanks.

Northern Alaska, the slope from the Brooks Range to the Arctic coast, is a bleak, tundra-covered waste land, dotted with thousands of small lakes. There are no harbors on the Arctic coast. Only for two or three months of the year at most is the sea free of ice and the beach at Point Barrow accessible to ordinary shipping. Yet men with significant experience in this region express confidence that the development of petroleum resources there can be carried on efficiently throughout the year in healthful, reasonably comfortable surroundings. Lieutenant-Commander William T. Foran, who, because of his earlier experience in geologic studies in Alaska, was chosen to lead a geological reconnaissance party sent to northern Alaska by the U.S. Navy in 1944, and to whom the writer is indebted for the accompanying chart (Fig. 48), showing temperatures and hours of daylight on the Navy's petroleum reserve in this region, makes the following statement: [6] "The plains area with the highest latitude on the continent (the Arctic coastal plain) is characterized by a climate not nearly so severe as the oil-producing regions of northern Montana and west central Alberta. The Arctic area is less subject to extreme low temperature, despite the fact that the mean annual temperature is 9° Fahrenheit and permanent frost persists to a depth of 625 feet below the surface."

This problem of permafrost strongly impresses most observers of oil-field operation in the Arctic. The preoccupation of the Russians with this aspect of the problem of developing the petroleum resources of Siberia has

[6] Personal communication.

already been noted. As the surface mantle of the earth's crust on the northern coast of Alaska is commonly ice-bound several hundred feet down, water lines, sewage lines, often even oil lines, must not only be buried but must also be heated to prevent congealing. Oil wells themselves may require warming through steam injection far beneath the surface, although shallow wells in Alaska are reported by Commodore Greenman, Director of Naval Petroleum Reserves, to be producing oil in a normal manner from sands within the permanently frozen zone. In an oil field, where fuel is abundant and cheap, it is to be anticipated that permafrost will not prove an insuperable barrier to development. A readily available source of heat, such as the liquid fuels oil fields afford, would likewise dispel much of the hardship and exposure that normally attend life in the Arctic.

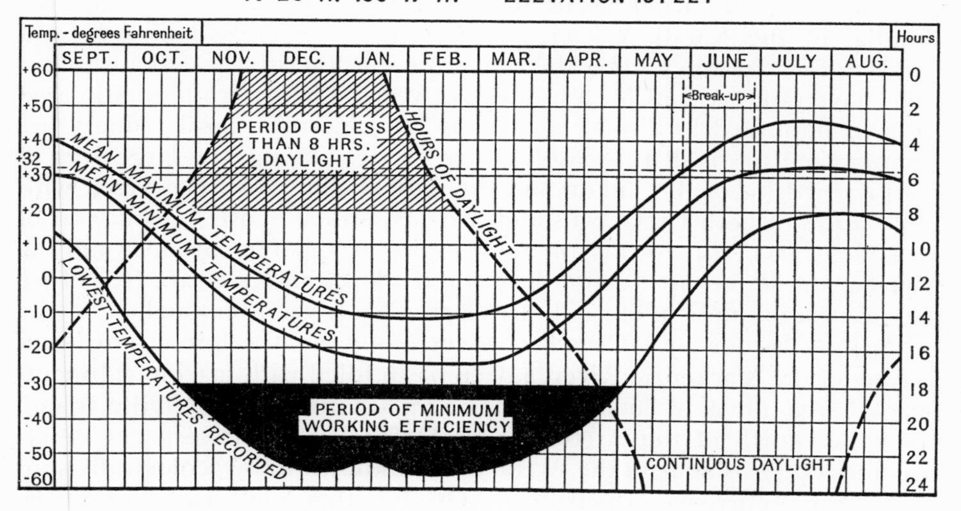

FIG. 48. A work feasibility chart for Naval Petroleum Reserve No. 4. Temperature data from U.S. Weather Bureau, 25 to 30 years' record. Hours of daylight data from U.S. Naval Observatory.

If commercial oil fields were to be developed on the Arctic coast of Alaska it is probable that only a minimum of staff and plant would be maintained at the site of the wells. If the crude oil were sufficiently fluid at prevailing temperatures, it would move through pipe lines to Fairbanks or another center of population for refining. If the crude oil were too viscous to permit this movement, it would be subjected to a simple cracking process at the source and the fluid fractions be moved out through pipe lines, the residual fractions being utilized on the ground as fuel.

The construction and maintenance of a pipe line across the Brooks Range between Point Barrow and Fairbanks is a formidable undertaking, although a reasonably direct route about 500 miles in length and at less than 3,000 feet maximum altitude appears to be available. The local market at Fairbanks (population 5,000) is inadequate to absorb the output of a major oil field, and pipe-line facilities beyond Fairbanks for another 300 miles, to a deep-water port on the south coast of Alaska, would be essential. The present total consumption of Alaska would not equal the volume of oil which would have to move through such a pipe line in order to amortize its cost: Alaska would have to export oil products before the Point Barrow region could become a commercial source of petroleum.

Elsewhere in Alaska, extensive areas favorable for the occurrence of petroleum have commanded the interest of oil producers for many years. On the Alaska Peninsula there are seepages of petroleum from rocks of Jurassic age, distributed over a distance of more than 100 miles, and the promising territory stretches along this coast over a total length of 350 miles. Several unsuccessful deep wells have been drilled on Alaska Peninsula, but the exploration has been entirely inadequate. Farther east, near Katalla, in the Katalla-Yakataga region which extends eastward from Cordova for 250 miles along the accessible south coast of Alaska, shallow wells have produced oil for a number of years. Their output is estimated to have amounted to a total of 300,000 barrels, and a local refinery has been erected to distill this gasoline-rich crude. Finally, the large area of the lower basins of the Yukon and the Kuskokwim rivers is covered with thick series of both Cretaceous and Tertiary strata, gently folded, and of a character favorable for petroleum. Despite its apparent promise and the fact that the waters at the mouth of the Kuskokwim River are free of ice the year round, no exploration of this area has been attempted.

Turning to northern Canada, we encounter, in the lower (northern) reaches of the Mackenzie River, a deep basin filled with marine sediments over an area (north of 60°N) of 160,000 square miles. On Melville Island, 500 miles north of the Arctic Circle, Stefansson noted seepages of petroleum. Other islands in the Arctic Sea are known to be geologically suitable to the occurrence of petroleum. On the other hand, eastern Canada, made up largely of the crystalline rocks of the ancient continental shield, Iceland, of volcanic origin, and Greenland, with its thick ice-cap, hold little promise of petroleum resources. Only northernmost Greenland, beyond the limits of the ice-cap, seems worthy of consideration.

The only actual oil field in the Arctic of the Western Hemisphere is situated near the settlement of Fort Norman on the Mackenzie River at 65°N. Here are oil wells which have long supplied fuel for trappers, river boats, mining operations, and airplanes. For years a pipe line from the refinery in this field to Great Bear Lake carried the fuel oil that made possible the activities of Eldorado Mines. This enterprise developed extensive uranium-ore deposits, which were finally taken over by the Canadian government. Thus petroleum in the Arctic has contributed to the final success of man's long struggle to command the energy of the atom, which ought eventually to supplant petroleum as a source of power.

In the emergency of total war Arctic oil again served importantly. In 1942 Japanese naval operations in Alaskan waters rendered the usual method of transporting fuel oil and aviation gasoline to Alaska by ocean-going tankers so hazardous that it seemed imperative to develop an alternative source of supply. Accordingly, U.S. Army engineers, in cooperation with our Canadian ally, undertook to increase the output of the Fort Norman field sufficiently to meet the anticipated requirements for military operations in Alaska. This effort was successful in that under U.S. Army control the Canadian company which had discovered and developed the oil field expanded the volume of its output from 800 to 5,000 b/d. Other contractors laid a 500-mile long pipe line from Fort Norman to a refinery which the Army hastily erected at White Horse, near the Alaskan boundary, whence products could move along the newly completed Alcan Highway to Fairbanks.

The petroleum at Fort Norman comes from a limestone of Devonian age, lying between beds of bituminous shale. The size of the area proved for production and the thickness of the producing formation are such as to define an oil reserve which, were it situated in the United States, would constitute a major oil field. The discovery well at Fort Norman was drilled in 1920 under the direction of Mr. T. A. Link, a geologist of long experience in western Canada. In recent years both the drilling of oil wells and the refining of the oil have proceeded without serious hindrance throughout the winter months. Permafrost complicates operations, but is by no means an insoluble problem. The oil and its products retain their fluidity at the lowest winter temperatures. Mr. Link and his fellow-workers assert that the Arctic climate, although it makes for higher costs, is no real barrier to the development of oil fields.

The Fort Norman oil field is situated north of the middle point of a great sedimentary basin of unusual promise for petroleum production—a basin which extends several hundred miles farther north, until it is lost beneath the waters of the Arctic Sea, and an even greater distance south along the Rocky Mountain Front, beyond the southern boundary of Canada. North of Fort Norman this basin is marked by oil seepages at the surface and by rocks of a character favorable for the occurrence of oil beneath the surface. Far to the south are the important oil fields of the prairie provinces of southern Canada. Between Fort Norman and the oil fields of southern Canada are other oil seepages, including the tremendous deposits of petroleum residue that constitute the Athabaska Tar Sands, the largest known accumulation of petroleum on earth. This whole region, immense as it is in area, is a unit as a potential petroleum province. It is to be expected that other oil fields will be discovered and developed between southern Canada and Fort Norman, and farther north. These anticipated developments might justify a pipe-line transportation system as far north as Fort Norman, and thus afford access to the world's markets.

The Fort Norman oil field is owned by the Canadian company that originally discovered and developed it. Continuation of the emergency development work performed by the owners as contractors for the United States Army is not justified by the present demand for petroleum products in the surrounding area. Like Alaska, northwestern Canada must export its oil before it can command markets large enough to justify the capital expenditures required for the full development of its petroleum resources. Except for the local consumption of a few hundred barrels daily by mining operations, airplanes, and river transport in the Mackenzie valley, there is now no commercial outlet for Fort Norman oil. However, the demand for petroleum products in the Arctic promises eventually to attain larger proportions. International commerce would be so greatly facilitated by rapid, direct transport between the centers of population in North America, Europe, and Asia that the eventual establishment of a network of airways across the north polar region seems inevitable. This prospect of mankind encroaching ever farther into the Northland lends to indigenous resources of petroleum in the Arctic—potential supplies of heat and power—the character of vitally significant assets.

). Alaska. U.S. Naval Petroleum Reserve No. 4. Point Barrow, 71°15′ N, 156°15′ W, in the left center of the
hotograph, and the Arctic Sea coast on the right. The ice pack is visible in the distance. (U.S. Navy photo.)

90. Gulf of Mexico. The oil well at sea. Air photo of large platform, so designed that, if oil is found, up t 35 wells can be drilled directly from it. Photo shows 136-foot derrick, drilling rig, truss-type substructure, an other equipment. Platforms of this nature have been set up as much as 60 miles off the coast, on the Conti nental Shelf in the Gulf of Mexico. (McCarty Company, and Emsco Derrick and Equipment Company.)

91. Texas. Redfish Reef, Galveston Bay. Pumper tying his boat to one of the producing wells to make periodical inspection. Windmill on well generates electricity for light which shows well location at night. (Standard Oil Company, N.J., photo by Bubley.)

15. PETROLEUM ON THE CONTINENTAL SHELVES

BY WALLACE E. PRATT

AS OUR knowledge of the origin of petroleum has increased, competent observers have come to suspect that the continental shelves of the earth may house great stores of this indispensable resource. President Truman, aware of this growing conviction in the minds of his technical advisors, dramatized its possibilities when, on September 28, 1945, he proclaimed that "the Government of the United States regards the natural resources of the subsoil and sea bed of the continental shelf, beneath the high seas but contiguous to the coasts of the United States, as appertaining to the United States, subject to its jurisdiction and control."

What is the character of the continental shelves of the earth? What makes them a favorable environment for the generation and accumulation of petroleum? What relative importance may be anticipated for these possible resources in comparison with those of the land areas of the earth?

The Character of the Continental Shelves

The continental shelf is the sea floor beneath the belt of marginal shallow waters that encircles the continents. It is the submerged periphery of the great platforms on which the continents stand in relief. The deep oceanic basins are at present overfilled, so that their waters rise above their rims, to flow out over the lower surface of the continental platforms. If the oceans were confined to their deep basins they would cover only about 64% of the earth's surface, and the land areas would amount to 36% of the total. However, the lands at present comprise but 28% of the earth's surface, and the continental heights constitute only 21%, leaving nearly 15% of the earth's surface as a great shelving plain which intervenes between these heights and the oceanic basins proper. It is the outer portion of this plain, covered by the waters of the oceans, which is termed the continental shelf.

It is commonly stated that the continental shelf is "arbitrarily" limited to the sea floor beneath coastal waters not deeper than about 100 fathoms, or 600 feet. It is worth while to realize that this limitation is not, in fact, arbitrary. The great plain, whose outer submerged portion forms the continental shelf, is a distinct unit in the earth's crust. The inland edge of this plain, at an elevation of about 600 feet above sea level, marks the mean level of the land-and-water surface of the globe. The seaward edge, at a depth of about 600 feet below sea level, marks the brim of the deep oceanic basins. Although it is submerged at present, it is still the true margin of the continents. In addition, it is logical to differentiate the continental shelf

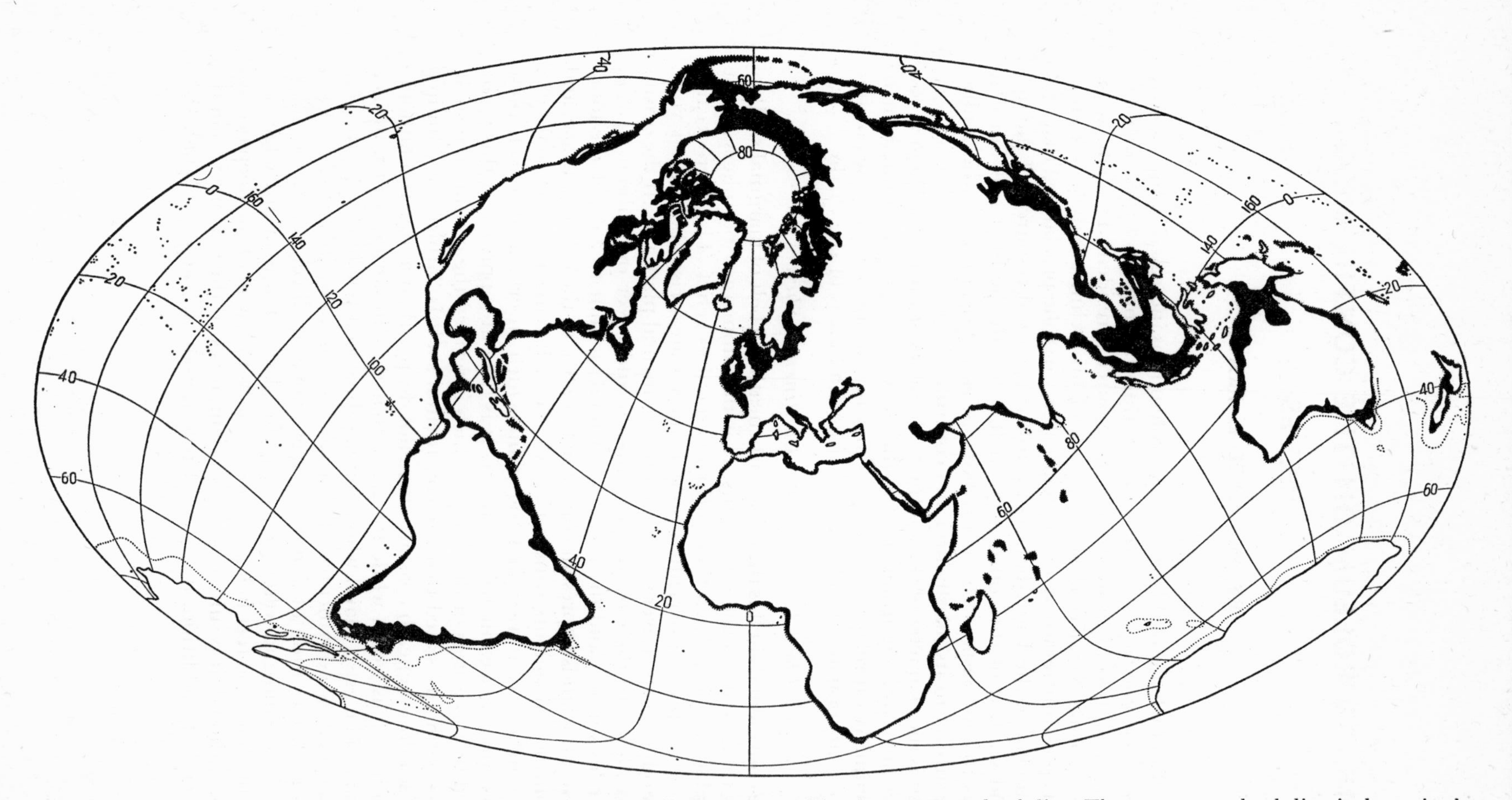

Fig. 49. Sketch map, showing the continents and the continental shelves, as measured by the 200-meter depth line. The 2,000-meter depth line is shown in the neighborhood of Antarctica, to emphasize the contrast with the shallow margins in the Arctic. The continental shelves are shown also in the fold map (Pl. I).

because its lower edge marks the depth limit of effective wave and current action on the sea floor, and the approximate limit to which sunlight penetrates marine waters.

Beneath the landward portion of this great plain are situated the natural reservoirs from which has come the bulk of all the petroleum so far discovered on earth. The question we ask ourselves as we contemplate the continental shelf is, therefore, "What are the prospects for petroleum under the adjacent submerged portion of this same plain?"

If we were to ignore the Antarctic continent we might speak of the continental shelf in the singular. All the other continents lie within the confines of an essentially continuous belt of shallow water—the continental shelf. But Antarctica has its own individual shelf; hundreds of miles of oceanic depths must be traversed to reach it, whatever the line of approach.

The aggregate area of the continental shelves is roughly 11 million square miles.[1] Of this total area, about one million square miles is contiguous to the coasts of the United States, including Alaska. Among the continents, Africa exhibits the least expanse of continental shelf. The eastern margins of North and South America have wide continental shelves, whereas their western coasts plunge with relative abruptness to oceanic depths. Similarly, the eastern edges of the continents of Asia and Australia, together with the East Indian archipelago, are bathed by wide stretches of shallow continental-shelf water.

The northern coasts of each of the three continents—North America, Europe, and Asia—which encircle the North Pole are marked by wide shelves. Indeed, the most conspicuous development of the continental shelves of the earth is to be found in the four great mediterranean regions: the Arctic mediterranean (often spoken of as the Arctic Ocean, but more appropriately designated as the Arctic Sea), the American mediterranean (the waters of the Gulf of Mexico and the Caribbean Sea in the irregularly down-warped basin between the continents of North and South America), the Asiatic mediterranean (the island-studded and essentially land-locked seas lying between the continents of Asia and Australia), and the classic mediterranean seas of Europe and the Near and Middle East. Within these four regions lies more than 50% of the total continental-shelf area of the earth.

Three of the great mediterranean regions of the earth are also major sources of petroleum; and the fourth, the Arctic mediterranean, still largely unexplored, manifests impressive surface evidences. Elsewhere the continental shelves also appear to bear a significant relationship to petroleum resources on land. Even the relatively narrow continental shelf of the western coasts of the Americas expands locally and takes on important proportions in the vicinity of existing oil fields, as, for example, in the region of the Los Angeles Basin in Southern California. Viewed from the adjacent

[1] Kossinna, E. v., "Die Erdoberflache," *Handbuch der Geophysik,* Vol. 2, Berlin, 1933.

continental shelf the Los Angeles Basin becomes only one lobe of a larger basin, much of which is submerged beneath continental-shelf waters.

It is obvious that these fringing terraces, encircling the continents, are in part built up of the debris of soil and rock fragments weathered and eroded from the surface of the adjacent land, and carried into the sea by flowing water and by wind. In part, again, they consist of the residue of marine organisms and the chemical precipitates which sink to the bottom and are buried upon the floor of the sea. This blanket of sediments is in general very thick, although in areas of a stable earth crust, where there has been little or no foundering of the continental margins, the veneer of sediments remains thin, or may even be entirely lacking.

The Continental Shelves as a Petroleum Province

In order to comprehend the promising character of the continental shelf as a petroleum province it will be helpful to recall some of our basic conceptions on the occurrence of petroleum, namely: (1) that great oil fields are usually found in rocks formed from sediments deposited in deep marine basins, characteristically downwarped in the mobile sectors of the earth's crust; (2) that petroleum appears to have originated from the organic matter of former marine life, through the transformation which takes place when this organic matter accumulates in the stagnant concentrated brines which occupy the deeper parts of these basins, below the general level of the surrounding sea floor; here the organic matter is preserved in an environment so foul that it excludes at once all scavenging life which would otherwise devour it, and all oxygen by which organic matter accumulating on normal sea floors is oxidized; (3) that the natural reservoirs which contain the accumulations of oil in the earth's crust usually take the form of porous rocks—sandstones or limestones—closely associated with impervious fine-grained shales of high organic content.

With these facts in mind it becomes clear that several attributes stamp the peripheries of the continents as a fundamentally superior environment for the generation and accumulation of petroleum. Among these factors is the coincidence of the shelf with the contact zone between the low-specific-gravity rocks of the continental platforms and the heavy rocks of the ocean floor on which the continents float. This contact zone is the locus of most of the earth movements which adjust and compensate, by thrust, subsidence, and uplift, the stresses built up in the earth's crust. As a result of these movements we have in the vicinity of the continental shelf the sharp downwarps of the sea floor and the accompanying rapid sedimentation which combine so effectively to entomb and preserve from oxidation organic residues from which petroleum may be generated. The shelf is also characterized by the most extensive development of Mesozoic and Cenozoic sediments, beds which house much the larger part of the petroleum so far discovered. Another favorable aspect of the environment of the continental shelf is the concentration of marine life and the resulting wealth of organic residue in the waters overlying it. This concentration of life is accentuated by the upward deflection of ocean currents impinging on the steep continental slopes at the outer margins of the shelf, and the consequent overturning and upwelling of bottom waters laden with the inorganic plant nutrients upon which, in the final analysis, all marine life is sustained. Still another favorable factor is the inherent tendency of continental-shelf processes efficiently to classify coarse and fine sediments, to segregate the coarse sediments on the highs, and to sweep the fine sediments, which normally consist largely of organic residue, into the adjacent downwarps of the shelf surface. In this way porous reservoir beds are formed, ideally placed to facilitate the migration into them of any petroleum which may be generated in the adjacent fine-grained organic source beds.

Relative Importance of the Possible Resources

Our search for petroleum has been confined almost entirely to the land —to the sediments deposited in former seas which spread over the present land areas during periods of temporary submergence. Twenhofel [2] has estimated the total area of these sediments at 45 million square miles, and their volume at 45 million cubic miles. Only about 80% of these rocks, according to him, are of marine origin; the rest were laid down in fresh or brackish water, and are therefore not promising source rocks for petroleum. Moreover, a large proportion of the sediment deposited on the present land area has been so squeezed and indurated during the recurrent periods of intense folding, thrusting, and buckling which have characterized the development of the earth's crust, as no longer to retain whatever petroleum may formerly have been present. Taking into account this factor, Weeks [3] has estimated the total area of the possibly oil-bearing sediments on the land surface of the earth at about 15 million square miles, and their volume at 20 million cubic miles.

The area of the sediments on the continental shelves has been estimated by Twenhofel at 10 million square miles, and their volume at 30 million cubic miles. The area of the adjacent continental slope, where the fine organic sediments that sweep out to sea across the general surface of the continental shelf finally come to rest, Twenhofel places at 15 million square miles, and their volume at 40 to 50 million cubic miles, or more.

Altogether then, in or adjacent to the continental shelf we have a total volume of 70 to 80 million cubic miles of sediments. Some part of this volume, however, like the sediments of the present land surface, has been indurated by intense pressure and so rendered unpromising for petroleum, although, because of the younger average age of the shelf sediments, this change must be less complete than that of the sediments of

[2] Twenhofel, W. H., *Treatise on Sedimentation,* Baltimore, 1932, p. 290.

[3] Weeks, L. G., "Basins and Oil Occurrence," an unpublished report to Standard Oil Company (N.J.).

the present land surface. Perhaps we shall make a sufficient allowance for these indurated beds if we reduce the foregoing total to 50 to 60 million cubic miles.

According to these rough estimates, then, the continental shelves and slopes of the earth should include an accumulation of sediments, similar in general character to the sediments we have long been exploring over the land areas of the earth, of an aggregate volume of more than fifty million cubic miles. This volume is more than double the estimated volume of favorable sediments within the land areas.

We are already engaged in a preliminary way in exploiting the petroleum resources of the continental shelves. Off the coasts of Texas, Louisiana, Florida, and California extensive geophysical surveys have been carried out, and numerous wells have been drilled into continental shelf sediments; petroleum is already being produced from the continental shelf adjacent to California, Texas, and Louisiana. In our search for oil on the land, our operations have brought us, in one region after another, all over the earth, to the seacoast, and it is clear that petroleum resources do not stop at the water's edge.

The extension of our established land operations out over the water is the logical development of the present technique of winning petroleum. These operations may reasonably be expected gradually to quicken their pace and expand their scope until oil from the continental shelf supplies a significant part of our total national consumption. These operations do not, however, promise a comprehensive solution of the problem of recovering the petroleum resources of the continental shelves. The solution will not be forthcoming until the time is at hand when our natural resourcefulness fails longer to contrive a preferable source of liquid fuels. In view of the large volume of our known petroleum resources beneath the land areas of the earth, the prospect of almost unlimited energy from subatomic forces within the next few decades, and the even more imminent promise of liquid fuels from coal at costs but little higher than the prevailing cost of gasoline distilled from petroleum, this contingency appears to be remote. We may never find ourselves called upon to attempt the recovery of more than the most available fraction of whatever petroleum the continental shelves may hold. If, however, the occasion to develop fully the petroleum resources of the continental shelves ever arises, the technical problems incident to their development will surely be solved under the spur of an adequate incentive.

16. THE MAJOR AREAS OF DISCOVERED AND PROSPECTIVE OIL

BY EUGENE STEBINGER

IN PART I of this volume it was noted that the fundamental pattern of oil occurrence, as it exists today over the earth, was established by the distribution of the marine sediments through the geologic ages since Cambrian time, and reference was also made to the world-wide extent of geologic field studies since the beginnings of the science. The knowledge acquired, in so far as it has appeared in print, with respect to the distribution of marine formations is summarized on the world map in Part I showing the areas of promise for oil. It should be noted that, in defining these areas, two important eliminations have been necessary. The first of these comprises all marine sediments altered and disturbed to such an extent that they are incapable of retaining the petroleums in important quantities, and this requirement eliminates extensive areas generally marginal to those shown on the map. The second elimination comprises all thin marine sediments existing as a veneer of surface-cover over Pre-Cambrian or other altered rocks. This requirement also eliminates considerable areas on all the continents, marginal to those indicated, as well as extensive completely isolated areas.

The producing oil-field areas are shown in generalized red outlines on the map, and their production record by countries and by broad regions is stated in Table 1, in Part I of this book.

The Mediterranean Basins

In the Western Hemisphere the outstanding feature of the geographic distribution of petroleum is its relative concentration in the essentially mediterranean region that surrounds the waters of the Gulf of Mexico and the Caribbean and lies between the main land masses of North and South America. This basin includes the great oil fields of the Gulf Coast of the United States and those of Mexico, as well as those of Colombia and Venezuela.

The detailed accounts of this area in preceding chapters have brought out the features that characterize it, and the reasons for its dominating position as the great accumulation center of the two continents. The bulk of its oil is youthful, that is, it was generated in and is trapped in rocks of mainly Cretaceous and Tertiary age. The reservoirs lie in great downwarps of these periods, which permitted enormously thick prisms of the sedimentary rocks to pile up at rapid rates. Salt domes abound in the Gulf Coast of the United States and also in parts of Mexico, piercing the oil-bearing rocks and creating favorable traps for oil accumulation. Stratigraphic traps also characterize this petroleum province, and two of the greatest of our known oil fields occur in stratigraphic traps in this area. The position of this petroleum province midway between the Canadian shield of ancient rocks in North America and the Brazilian shield in South America is very probably significant, if the broader cycles of sedimentation and resedimentation from prior sediments since Cambrian time are considered.

Proved reserves in the Caribbean–Gulf of Mexico region, as of January 1, 1949, amounted to over 20 billion barrels, and total proved reserves for the remainder of North and South America, to about 18 billion barrels. If we consider the relatively limited land area in the Caribbean-Gulf region, as compared with the remainder of the two continents, the great relative richness of these lands is evident.

When we turn to the Eastern Hemisphere, the outstanding feature of the distribution of its petroleum sources is their preponderance in the area surrounding the group of inland seas that comprise the eastern Mediterranean, the Black and the Caspian seas, and the Persian Gulf. This area on its northern margin includes the rich fields of Rumania, with their salt-cored anticlines; the great fields flanking the Caucasus in the southern U.S.S.R.; and, lastly, the great group of fields in the Persian Gulf area, comprising those of Iran, Iraq, Kuwait, Saudi Arabia, and Bahrein. Proved reserves in this broadly conceived "Middle East" province are nearly equal to those of the entire remaining area of the continents. These Middle East reserves were estimated to total about 36 billion barrels as of January 1, 1949, compared to about 42 billion barrels for the remaining area of the world.

Comparisons with the Caribbean–Gulf Coast province in the Western Hemisphere are striking. The age range of the productive formations is similar, namely, Mesozoic through Tertiary. In both provinces sediments are very thick, and include clastic rocks as well as limestones, with pooling in both sands and limestones; and in both provinces salt domes are numerous. Furthermore, each province is in a central position with respect to shield areas of Pre-Cambrian rocks. For the Middle East there are three such shield areas: the Scandinavian shield on the northwest, the African shield on the southwest, and the shield of Central Asia on the east. These two great centers of oil accumulation, each in a central position between adjacent continents, are well placed for the transport of their petroleums to their corresponding territories.

Sedimentary Areas on the Various Continents

The pattern of the sedimentary areas of North and South America, considered as a whole, is characterized by a belt of sediments extending north-south along each continent. In North America the belt is continuous from the Arctic Sea to the Gulf of Mexico. It comprises a series of geosynclinal troughs, lying for the most part east of the North American cordillera. At one place or another, oil-bearing sediments of practically every age from Cambrian to Tertiary are included; at many points the total volume of the sediments is very great, equaling if not exceeding the maxima from Cambrian to Tertiary inclusive found elsewhere on the globe. In South America an analogous belt of sediments continues, with only minor interruptions, the length of the continent, from the Caribbean to Tierra del Fuego. The overall thickness of the sediments, particularly the Paleozoic portion of the section, is somewhat less than in North America, but the continuity of oil-bearing facies is equally persistent, and the developed oil fields, present at intervals along its entire length, are comparable in reach, although not in intensity of development, to the oil fields of North America.

The fundamental factor producing such similar results on the continents of the Western Hemisphere was the presence in North America of the Canadian shield, in South America of the Brazilian. These two positive areas supplied sediments to the profound troughs on the west throughout much of earth history, from Cambrian to Tertiary, while the North and South American cordilleras were developing as a continuous belt through orogenies beginning in Paleozoic time and extending through the Tertiary.

Turning to the map of the Eastern Hemisphere, it is evident that the relatively simple pattern of distribution of the sedimentary areas in the Western Hemisphere is not repeated. Instead of two primary shield areas there are four, excluding the Antarctic continent.[1] Nevertheless, a rude alignment, more or less east-west, is evident in the land area of the Old World north of the Equator. This alignment, again, is due to two factors. The first is the areal influence of three of the Pre-Cambrian shields in controlling the sweep of the immense volumes of sediments that came from them through the ages. The second factor is the influence of the extended orogenies of the late Mesozoic and Tertiary in developing the structural basins flanking the Alpine systems of Europe and Asia; these systems occupy an east-west zone bounded roughly by the parallels 30°N and 50°N.

The influence of the Pre-Cambrian shields is most evident in the very large area of potentially petroliferous formations, ranging from Cambrian to Tertiary in age, which covers much of Russia and Siberia north of 50°N; in the aggregate this province comprises the largest area of sediments to be found on the globe. For the most part the strata are little disturbed from their original depositional attitude. They were formed by long-continued deposition on the northern flank of the shield of Central Asia and on the south and southeast flanks of the Scandinavian shield. In the same manner the 4,500-mile band of Mesozoic and Tertiary forma-

[1] Sedimentation peripheral to the Antarctic shield appears, through the ages, to have been more or less completely isolated from that of the other continents. It has as yet been only partly explored.

tions flanking the African shield on the north and northeast and extending southwest in the Persian Gulf area was formed.

Earlier chapters in this book have described the geology and oil developments of North America, the Caribbean area, the cordilleran sector of South America and the lowlands to the east, the parts of Oceania lying north of Australia as well as the adjoining portions of southeast Asia, the U.S.S.R. territory, and the Middle East province centered on the Persian Gulf. These areas, which include about 90% of the world's proved oil reserves, are the sites of most of the current exploratory and development activities of the world's oil industry.

The Marine Areas of the Continental Shelves

An attempt to indicate what proportion of the world's potential petroleum area has been effectively explored to date must also consider the prospects on the continental shelves that surround the land areas of the globe.

These shallow-water shelves extending out to the 100-fathom (600-foot) depth limit, usually with very gentle bottom slopes, which lie above the steep continental slopes that surround the continents and plunge down to the abyssal depths of the oceans, are essentially portions of the continents. Their aggregate area amounts to millions of square miles, considerable portions of which border sedimentary areas on the lands and are favorable for petroleum exploration. The more favorable of these shelf areas are underlain by simple down-dip extensions of adjacent coastal-plain sediments of Tertiary and Cretaceous ages, comprising the most productive units of the geologic column.

To date, very limited development operations extending seaward on the continental shelves have been carried out more or less successfully in several localities in the world. Beginning in 1890, California operators have exploited a number of fields to distances extending several thousand feet offshore, into open waters of the ocean; since 1929 the Seria field off the coast of Brunei has produced over 18 million barrels; since 1932 the Argentine Government operations in the Comodoro Rivadavia field have been extended seaward for distances comparable to those off California; more recently, beginning in 1938, several fields have been developed off Texas and Louisiana; and in 1949 exploratory drilling had begun off Mexico. Extensive marine drilling in enclosed waters near the coasts has, in addition, been carried out in water as deep as 120 feet, notably along the east side of Lake Maracaibo in Venezuela and at numerous localities on the Texas and Louisiana coasts. The practicability of offshore operations both in the open seas and in more protected waters has thus been established.

In recent years the government of Iran has conceded extensive water areas in the Persian Gulf to the operators of the adjacent land areas; and agreements are being sought on the national boundaries of lands beneath the marginal waters in other parts of the Persian Gulf. The Venezuelan and Trinidad governments have agreed to boundary definitions on their intervening marine areas in anticipation of developments there. In the United States, the legal problems surrounding State, as against Federal, ownership of marine areas, and the technical problems of actual operations, have been under discussion for many years.

If the technical problems of operating in marine waters down to 600-foot depths, or even to 300 feet, can be solved, the areas of immediate world-wide interest are considerable. These would include the shallow-water areas adjacent to territory already productive. Notable among these are: (1) the entire area of the Persian Gulf, adjacent to prolific fields in Iran, Iraq, Kuwait, Saudi Arabia, and Bahrein; (2) the area off the Gulf Coast of the United States and Mexico, extending from 30 to as much as 200 miles offshore; (3) the marine areas between eastern Venezuela and Trinidad adjacent to prolific fields in each country; (4) practically all of the area of the Caspian Sea; (5) a comparatively narrow strip off the coast of southern California; (6) extensive areas surrounding the East Indies; (7) the marine area off the Argentine coast extending south from 38°S and reaching as much as 300 miles seaward. In every instance these areas of immediate interest would comprise terranes underlain by Tertiary or Cretaceous formations, or by both.

PART IV

ASPECTS OF UTILIZATION

1. THE AVAILABILITY OF PETROLEUM—TODAY AND TOMORROW

BY KIRTLEY F. MATHER *

AS WE turn to the distressingly difficult task of arranging a peaceful world, in which men may use the rich resources of our bountiful earth for the welfare of all mankind, petroleum continues to hold its leadership among the natural resources of the earth. Both as fuel and as raw material for chemical industries petroleum will hold the center of the stage for many years to come. Hardly any other substance illustrates so fully the manner in which science and technology may be combined to achieve the utmost success in contributing to human efficiency and comfort.

A Non-Renewable Resource

Petroleum is a non-renewable resource; it is in the category of nature's stored capital, not of man's annual income. It is, of course, true that the geologic processes responsible for oil pools are continuing to operate today as in the past. On the sea floor off the coast of southern California, for example, there are broad hollows where the tissues of marine animals and plants are now accumulating in mud and ooze at depths of 200 or 300 fathoms. The conditions are closely similar to those that recurred repeatedly during the Paleozoic era in Oklahoma and Texas, when the oil of certain rich oil fields was being generated. But millions of years would have to elapse before that organic material could be transformed into petroleum, stored in the interstices of overlying sandstones, and made available by crustal movements for recovery from wells that might be drilled by some future inhabitants of earth's surface.

In relation to the feverish haste of mankind's insatiable demands, the

* Professor of Geology, Harvard University.

Acknowledgements: This is a revision of an article published in *Science,* December 19, 1947, by arrangement with the Officers of the British Association for the Advancement of Science. The paper was originally presented as an illustrated lecture on August 29, 1947, during the Association's meetings in Dundee, Scotland. I am deeply indebted to R. G. Watts of the Magnolia Petroleum Company, W. H. Wilson of the Humble Oil Refining Company, George Krieger of the Ethyl Corporation, and Roy Stryker of the Standard Oil Company (New Jersey) for information used in the preparation of this paper. The revised data pertaining to proved reserves of petroleum in countries other than the United States are based upon statistics published in the 1949 edition of *Twentieth Century Petroleum Statistics,* prepared by DeGolyer and MacNaughton. Data on production in 1948 have been obtained from *World Oil: Review-Forecast Issue,* February 15, 1949. The statistics pertaining to exploratory wells drilled in the United States in 1943 were compiled by F. H. Lahee and the Committee on Exploratory Drilling of the American Association of Petroleum Geologists.

creative processes of nature's laboratory operate very slowly. For all practical purposes our planet must be reckoned as a storehouse of such minerals as petroleum, not as a factory in which that substance is generated year by year, or even millennium by millennium. Mother Earth has made available a cupboard richly stocked with a vast amount and a great variety of goods indispensable to us in an age of science and technology, and among these stores we find petroleum. Each year we go to the shelves of that cupboard and take away a few packages of the goods stored thereon; if we keep going long enough, some day some one will find that the cupboard is bare. Indeed, petroleum is now being used at such a rate, in relation to its total amount in the earth's crust, that its complete exhaustion is, from a geological viewpoint, alarmingly imminent.

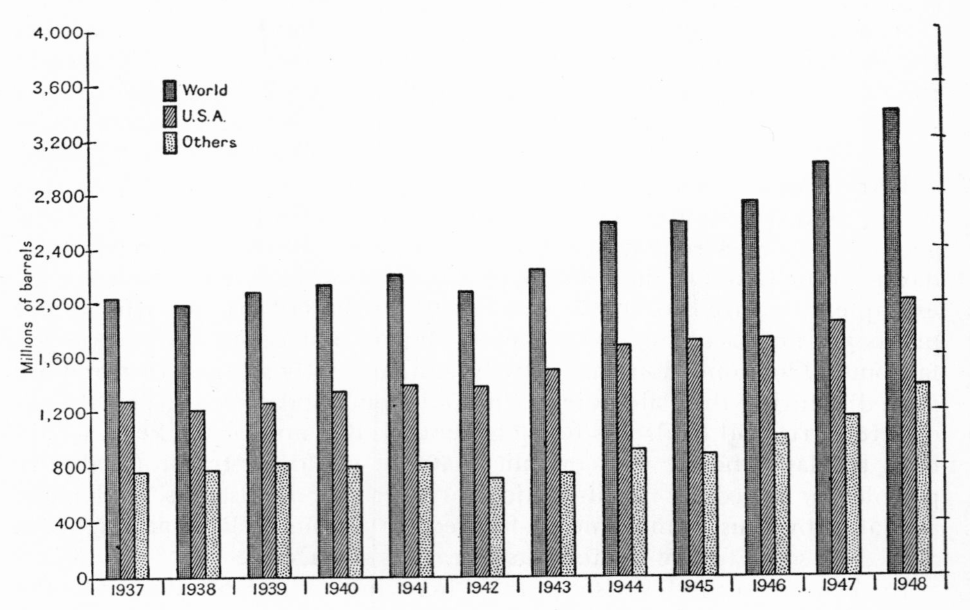

FIG. 50. Annual output of crude petroleum from 1937 to 1948: in the world as a whole, in the United States alone, and in all other producing areas. Source of data: *World Oil, Review-Forecast Issue,* February 15, 1949, p. 221.

Between 1859, when the first oil well was drilled in the United States, and January 1, 1949, the production of petroleum from all parts of the world totaled more than 58,000,000,000 barrels. Of that total, slightly more than 37,000,000,000 barrels were produced in the United States. As shown in Fig. 50, the annual production both of the world as a whole and of the United States has increased each year since 1938, with the sole exception of 1942, the first year of American participation in the war; and there is every indication that these increases will continue for the next few years. Indeed, it is a conservative estimate that the world's oil production will average more than 3,500,000,000 barrels per year during the decade 1947–1956.

Such figures begin to take on real significance when we note that the recent estimates made by competent petroleum geologists give a figure of slightly more than 78,000,000,000 barrels (78,322,000,000) for the proved reserves of petroleum throughout the world on January 1, 1949. It would, however, be far too simple and quite erroneous to make the easy arithmetical calculation and announce that the world's oil will be exhausted in 26 years. The proved reserves are only a fraction of the actual reserves. Whether the fraction is one-half or one-tenth, however, is a matter of great moment, requiring careful consideration.

The Search for Petroleum

The particular combination of specifically defined geological conditions requisite for an oil accumulation of economic value is now well known to petroleum geologists. Every oil field fulfills the specifications in one way or another, and the failure to find oil in an exploratory borehole can almost always be explained by an observable lack of one or more of the requisite geological conditions.

The first requirement is an adequate source of supply of the organic materials from which petroleum may be formed by natural processes of a biochemical and geochemical character. Source beds are sedimentary rocks, most commonly shale, but in some instances limestone or sandstone, in which the tissues of plants or animals were buried, before they were completely oxidized, while the sediments were accumulating on a sea floor, lake bottom, or stream bed. Certainly almost all, and very probably all, adequate source beds are of marine origin.

In the second place, there must be a suitable reservoir rock in fairly close proximity to the source beds. Reservoir rocks must be both porous and permeable, in order to provide space for the oil in the interstitial voids and to permit its movement, first from the source beds into the interstices, and later from them into a borehole. The best reservoir rocks are sandstones and limestones, with a porosity of 5% to 25% of their volume and with permeability ranging from 100 or 200 to 1,000 or 2,000 millidarcies. It is not essential that a reservoir rock be marine in origin; lacustrine, fluviatile, or eolian sandstones will do just as well. In fact, a volcanic ash deposit, a tuff, or an agglomerate, might be quite satisfactory.

Thirdly, there must be some sort of trap to prevent the upward migration of the oil, as it seeks to ascend to the surface because of its lesser density than that of water. The trap may be (1) "structural," produced by folding or faulting of the rocks or a combination of the two, or by the intrusion of salt or igneous rock into the sedimentary series; or (2) "stratigraphic," resulting from the pinching out of the reservoir rock updip, or from excessive lateral reduction in porosity or permeability within the reservoir bed, or formed by the unconformable deposition of relatively impermeable beds across the eroded edges of tilted strata that include a reservoir rock.

Finally, the sedimentary system containing source beds and reservoir rocks must not have been unduly metamorphosed, either as a result of deep burial, horizontal compression during mountain-making movements, or igneous intrusions on a large scale. Moderate pressures stimulate geochemical changes conducive to the evolution of petroleum from its source materials, but excessive pressure destroys petroleum. Gentle folds are desirable to provide structural traps, and to tilt reservoir rocks so that the oil can move upward into stratigraphic traps, but closely compressed folds do not yield oil. The greatest depth from which oil has thus far been produced is slightly more than 14,000 feet, although several holes have been drilled deeper than 15,000, and at least one exceeds 20,000.

Although geological and geophysical surveys at the earth's surface permit the discovery of places beneath which one or more of these four requisites for an oil field are known to be met, no one can guarantee in advance of drilling that oil will be encountered at depth. Satisfactory criteria for the identification of source beds as such are not yet known. Sedimentary rocks are notorious for the facility with which they change in textural, lithologic, and mineralogic characteristics, when traced laterally from place to place. Not until the drill has yielded information concerning the texture and thickness of potential reservoir rocks within the favorable structure itself, and has demonstrated the actual presence of petroleum, can one begin to calculate the quantity of proved reserves for any oil field.

Proved Reserves and Total Resources

The relation between proved reserves and the total supply of petroleum in the ground is best indicated by the history of oil fields in the United States. Almost every minor variant of the geological conditions essential for an oil pool is found somewhere within the extensive petroliferous area of the United States. Accurate records of nearly all drilling operations, and of the resulting production of oil and gas, are available over a long period of years; and American geologists have specialized in the computation of known and probable reserves for each of the many oil fields at various stages in their development, from the drilling of the "discovery well" to the exhaustion and abandonment of the older fields.

The broad picture is indicated in Fig. 51. During the thirteen years from 1936 to 1948, inclusive, there was intensive search for new oil fields, the results of which made possible the increase in annual production from approximately 1,000,000,000 barrels in 1936 to more than 2,000,-000,000 in 1948. In spite of the removal of this vast quantity of oil, the proved reserves were greater at the end of each year than at the beginning, with the sole exception of 1943, when there was a relatively small amount of exploratory drilling because of the limitations upon steel for well-casings imposed by the diversion of large quantities of steel to shipbuilding and military equipment. It will be noted, however, that the curve entitled "Proved Reserves at End of Year" rises steeply for four years, from 1936 through 1939, and from 1940 to 1947 shows a definite tendency

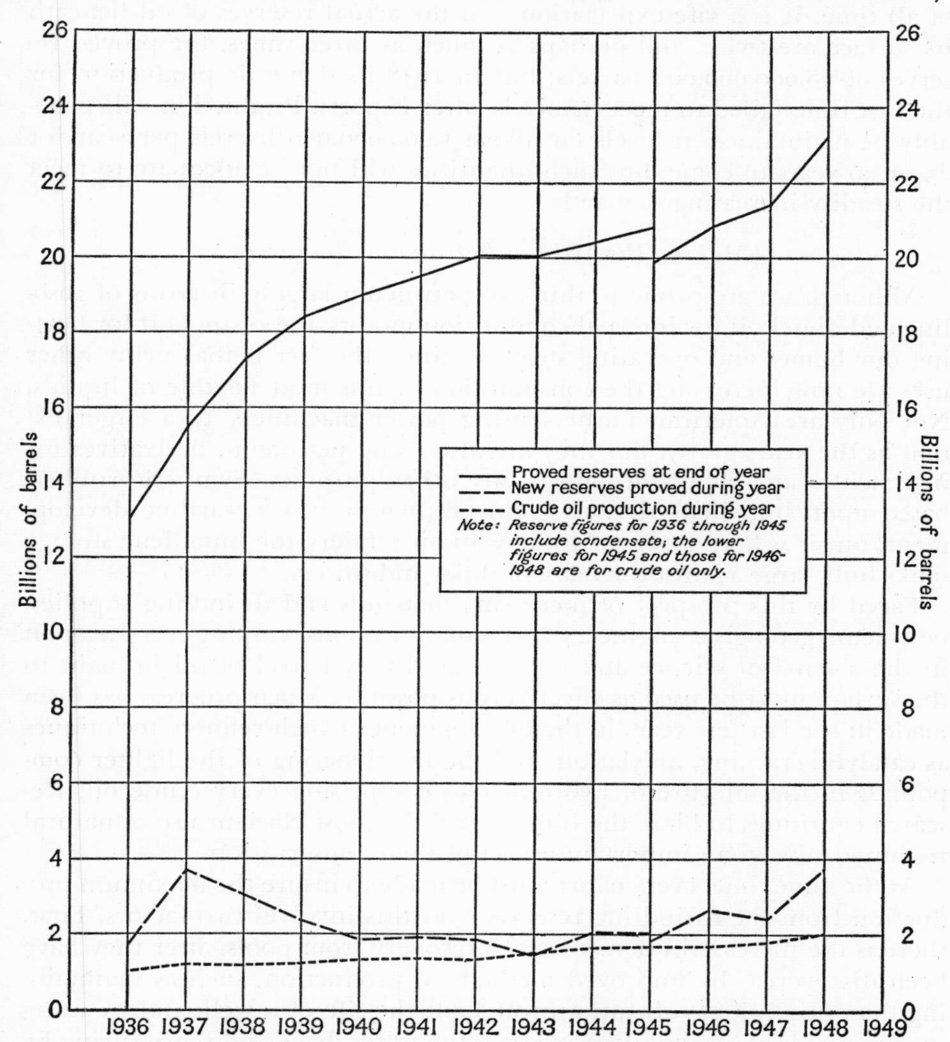

FIG. 51. United States: annual crude-oil production compared with changes in proved reserves, 1936–1948. Source of data: DeGolyer and MacNaughton, *Twentieth Century Petroleum Statistics 1949,* p. 1.

to level off. The peak of that curve will probably be recorded in the early 1950's, and thereafter it will descend toward zero.

This, of course, is the inevitable story for every oil-producing region in the world. Although for a time a locality or a nation may profit greatly from the rich stores of oil within its bounds, sooner or later those stores will be exhausted. We are dealing with a non-renewable resource. We are burning up the capital provided by nature.

The United States is now enjoying the greatest petroleum production of all time. It is a safe expectation that the actual reserves of oil beneath its surface are twice, and perhaps as much as three times, the proved reserves of 28,000,000,000 barrels; but in 1948 its domestic production for the first time failed to meet its needs, after exports. Production will probably be maintained at levels far above 1,000,000,000 barrels per year for 25 or 30 years to come, but such quantities will not be adequate to meet the steadily increasing demands.

Making the Most of Petroleum Resources

Although we are prone to think of petroleum largely in terms of gasoline and diesel oil for internal combustion motors and of fuel oil for heating our homes and operating steam engines, the fact is that many other uses are now increasing the consumption of this most flexible of liquids. Not only are American farmers using power machinery to a larger extent as the years go by, but they are also using petroleum derivatives for insecticides and fungicides and many other purposes. Synthetic rubber, based upon butadiene derived from petroleum, was a wartime development, but it is here to stay. At many an oil refinery the butadiene storage tanks bulk large against the maze of pipes and towers.

Faced by this prospect of increasing demands and dwindling supplies, petroleum geologists, engineers, and economists must employ every weapon in the arsenal of science and technology. Every barrel of oil brought to the surface must be used as effectively as possible. Much progress has been made in the last few years in the development of such refinery techniques as catalytic cracking, alkylation, and the fractionating of the lighter compounds in the mixture of hydrocarbons comprising every crude oil. Research continues to blaze the trail toward the most efficient use of natural resources, always an important aspect of their conservation.

At the same time, every effort must be made to insure the maximum production from the dwindling reserves, and this involves two factors. First, there is the increase in percentage of recovery from pools, after they have been discovered, by improved methods of production, such as maintaining the optimum gas-oil ratio throughout the life of a well, and by using such methods of secondary recovery as the water drive and repressuring by gas injection. Second, there is the discovery of new pools, either in new localities or by deeper drilling in old fields. This is the function primarily of the petroleum geologist, and involves continuing research as well as the application of procedures already well-established by long practice.

The importance of geology and geophysics is now well recognized throughout the entire industry in the United States, if not also in every other country. The statistics pertaining to the exploratory wells drilled in the United States in 1943 may be cited as illustrative of the reasons why geologists and geophysicists have established a secure place for themselves, and may expect to continue their activities in the search for petroleum for many years to come. In that year 3,843 exploratory wells were drilled. The location of 3,242 of these is known to have been selected on the basis of geological or geophysical surveys, or a combination of the two. Of these scientifically located wells, 626 proved to be productive—that is, approximately 20% were successful. Of the remainder, 523 are known to have been located by non-technical methods. Only 23 of these, or less than 5%, were successful. The method of determining the location of the other 78 wells is unknown; 6 of them were productive.

It is not to be expected, however, that this excellent record of geological and geophysical achievement can be maintained in future years. Less favorable locations must be tested with the drill, if every oil pool, no matter how small, is to be discovered. A smaller percentage of successful wells to total exploratory tests is inevitable as the search becomes concentrated upon stratigraphic traps and the doubtfully effective structural traps.

Accepting the fact that almost all the large oil pools within the land area of the United States have by this time been discovered, attention at the moment is being given to the possibilities of oil production from the submerged portion of the continent, known to geologists as the continental shelves. The coastal plain of Louisiana and Texas continues far out beneath the waters of the Gulf of Mexico, and the present position of the shoreline has no relationship whatsoever to the occurrence of oil beneath its surface. From the standpoint of geology there is no reason to doubt the presence of salt domes beneath the marginal waters of the Gulf, quite similar to those beneath the adjacent shore.

Petroleum engineers have already had considerable experience with drilling operations in shallow water. Hundreds of wells have been drilled in Lake Maracaibo, Venezuela, near its eastern shore. But off-shore drilling in the Gulf of Mexico is a very different proposition from that in the quiet waters of Lake Maracaibo, close to shore. Drilling platforms must be high enough above the water to escape damage from storm waves, and, out in the open waters of the Gulf where hurricanes occur almost every year, that means a height of at least 30 feet above high-tide level. The engineering problems involved in constructing stable bases for drilling operations are obviously very difficult.

Surveying techniques to determine favorable drilling sites depend largely upon the portable seismograph and the gravity meter. Adaptation of their use for underwater surveys has been greatly facilitated by certain of the ingenious devices developed for military purposes during the war. Similarly, radar has proved most satisfactory for precise location of points occupied by the surveyors—an extremely difficult operation when one is out of sight of land.

This extension of the search for oil to the continental shelf may confidently be expected to add a few billion barrels to North American reserves, but at best it will postpone by only a few years the time when in the United States production will lag far behind the consumption of petroleum products. It is therefore appropriate to look next at the prospects for meeting the future needs of the United States by importing oil from other countries.

The United States and the World's Problem of Supply

As of January 1, 1949, United States interests controlled about 75% of the more than 10,000,000,000 barrels of proved reserves in South America, more than 40% of the 32,696,000,000 barrels of proved reserves in the Middle East, and nearly 30% of the 1,300,000,000 barrels of proved reserves in the Far East (not including Asiatic portions of the U.S.S.R.). Thus, if political factors are favorable, the United States may draw upon nearly as many barrels of foreign oil reserves for its domestic needs as are presently available within its own boundaries.

But the United States does not exist in a geographical vacuum. It is but one among many nations having needs and rights with respect to petroleum. It is therefore imperative to consider the reserves from a world point of view. The broad features of the world picture are shown in Fig. 52, based upon the best estimates of proved reserves made by competent geologists as of January 1, 1949. In all probability, a similar graph made ten years hence would show a decrease in the proved reserves for the United States and an increase for every other country (with the possible exception of Venezuela) inasmuch as discovery and development have already advanced so much further in the United States than elsewhere. One could almost venture the guess that, whereas the actual reserves beneath the United States are probably not much more than twice the amount shown as proved reserves, in the other countries they average at least three times the amounts shown for them.

Most amazing is the position of Kuwait, the tiny country at the head of the Persian Gulf, exceeded only by the United States in the volume of its proved reserves of petroleum. I am, however, assured by Mr. E. L. DeGolyer, who has studied the petroleum potentialities of the Middle East at close range, that the estimate of 10,500,000,000 barrels for that country is very conservative. Significant also is the position of Saudi Arabia, coequal with Venezuela in third place, and that of Iran and Iraq as the fifth and sixth countries of the world, when the nations are listed in the order of their petroleum potentialities.

The most important inference, however, that should be drawn from this picture is based upon a comparison of the two percentage columns shown for each country. Whereas the United States possesses only a little more than 30% of the world reserves, its production had been running well over 60% of the world production each year until 1948 and 1949, when it was about 59%. Venezuela and the Soviet Union are the only other countries, of those possessing more than 1,000,000,000 barrels of proved reserves, in which their percentage of world production in 1948 exceeded their percentage of world reserves. The deduction is obvious: if present trends continue, in ten to twenty years from now the United States will be a "have-not" nation, so far as petroleum is concerned; the "have" nations will be Kuwait, Saudi Arabia, Iran, Iraq, the Soviet Union, and the

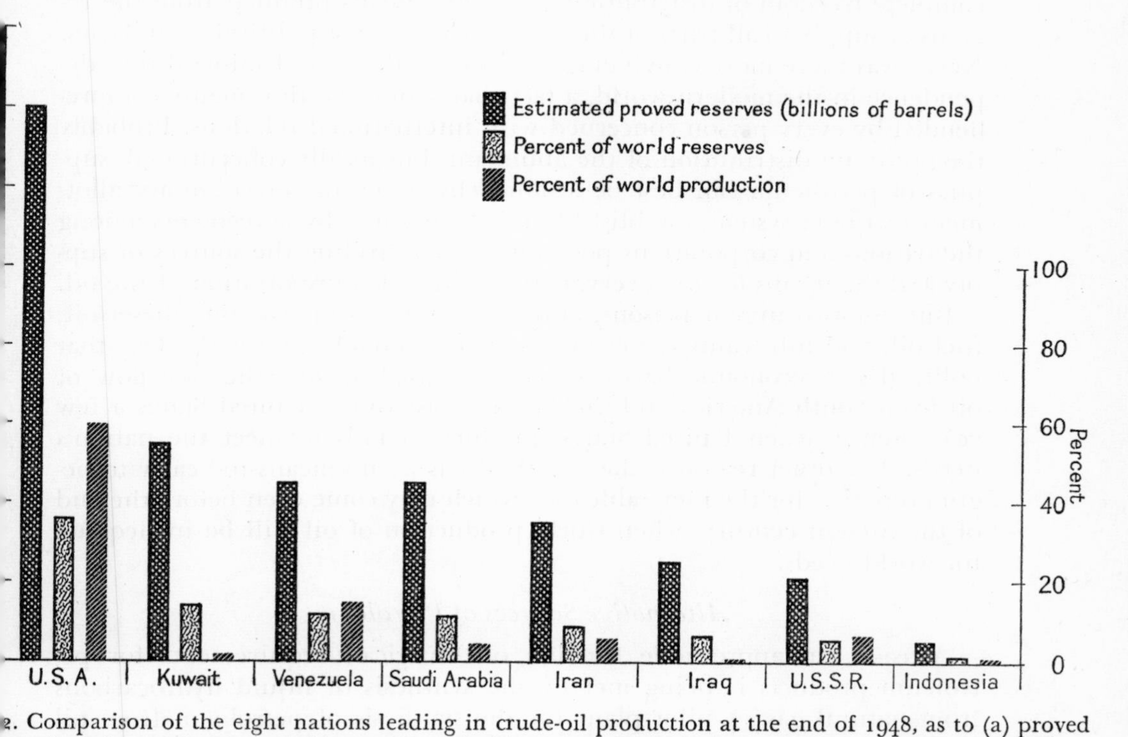

2. Comparison of the eight nations leading in crude-oil production at the end of 1948, as to (a) proved es of crude petroleum in billions of barrels; (b) percentages of total proved world reserve; (c) per-ges of total world output. Source: DeGolyer and MacNaughton: *Twentieth Century Petroleum Sta- 1948,* p. 4.

East Indies. It is no accident that practically all of the potential oil fields in these countries, apart from the Soviet Union, are under the control of American, British, and Dutch corporations.

The World Problem

This raises the question of the adequacy of the world's supply of oil for meeting the future world demands for petroleum products. In 1948, world consumption of petroleum was running at an annual rate of approximately 3,400,000,000 barrels. That was 4% of the world's proved reserves, and something like 1% of the world's actual reserves. Demands are certain to increase greatly in the next few years, not so much as a result of the expected increase in world population, as because of increased consumption by people who have not hitherto been large users of petroleum products.

Even so, it would appear that world petroleum reserves are quite adequate to meet world needs for half or three-quarters of a century to come. But to use those reserves in the most efficient way, there must be almost complete freedom of distribution of the oil and its products from the regions of supply to all parts of the world, regardless of political boundaries. Never was there more convincing evidence of the fact of mineral interdependence in the modern world, a fact that should be thoroughly comprehended by every person concerned with international relations. Probably the optimum distribution of the abundant, but locally concentrated, supplies of petroleum can best be attained by some sort of voluntary allotment or quota system, established and maintained by agreements among the nations and corporations possessing or controlling the sources of supply and the means for discovery, extraction, and transportation of the oil.

But, for two urgent reasons, alternative sources of gasoline, diesel oil, fuel oil, and lubricants are even now being sought. One is the fear that political and economic barriers may be raised against the free flow of oil from South America and the Middle East to the United States a few years hence, when United States' production fails to meet the nation's needs. The other reason is the fact that it is by no means too early to begin preparing for the inevitable day, which may come even before the end of the present century, when world production of oil will be inadequate for world needs.

Alternative Sources of Petroleum

Already, an appreciable fraction of American requirements for petroleum products is being met by the synthesis of liquid hydrocarbons from natural gas. A pilot plant for the synthesis of petroleum from oil shale has been put into operation at Rifle, Colorado, by the U.S. Bureau of Mines, and methods of synthesizing petroleum products from coal, developed by German scientists before and during World War II, are being checked and improved in the research laboratories and demonstration plants of governmental bureaus and private corporations in the United States. The success of the oil-shale industry in Scotland, and the great quantity of synthetic gasoline that helped so materially in providing power for the German war machines, are sufficient guarantee that science and technology have provided an alternative source of supply, long before the petroleum resources of the world have even begun to approach exhaustion.

These alternative sources are almost unbelievably abundant. In the United States alone there is enough oil shale and coal available to provide the equivalent of 2,000,000,000 barrels of crude oil each year for at least 1,000 years. Similarly, if known techniques of producing petroleum products from coal by chemical synthesis were applied to the coal reserves of Great Britain, all British requirements for such products could be met for at least several hundred years.

But such application of technical knowledge is not by any means an easy and simple matter. Processes now available are too expensive to permit competition with natural petroleum products except under extraordinary conditions, such as those that make the Scottish and Swedish oil-shale industries economically feasible. The necessary equipment is extremely complicated, very costly, and requires extensive plants, and its installation is not yet practical on a sufficiently large scale to permit such a substitution of mineral-fuel sources in the immediate future.

Nevertheless, there is every reason to expect the gradual replacement of natural petroleum products by synthetic products in the more distant future. As the supplies of crude oil are exhausted, this alternative source will take their place. Thanks to science and technology, the mineral fuels stored within the earth will prove adequate for all human needs, for as long a time as they are needed. Long before the oil-shale and coal are exhausted, still other sources of energy, such as the atomic energy released by nuclear fission, will be available to do the work that men want done.

Far more difficult than the technical problems involved in meeting the physical needs of humanity are the psychological and spiritual problems that retard the process of learning how to live in a world community. Perhaps knowledge of the fact that our small world is a world of potential abundance, but also of inescapable interdependence, as illustrated by such an inquiry as this concerning one typical non-renewable resource, may help to accelerate the learning process.

2. GEOGRAPHICAL ASPECTS OF PETROLEUM USE IN WORLD WAR II

BY THE OFFICE OF THE ARMY-NAVY PETROLEUM BOARD OF THE JOINT CHIEFS OF STAFF

DURING World War II the armed forces of the United States were spread through the land and waterways of a large part of the world. As approximately half of the total supply tonnage was made up of petroleum products, the geography of petroleum was of primary importance to our military tactics and strategy.

We fought geography in the form of great distance, high mountain ranges, and vast oceans. Enormous expenditures were laid out to insure adequate and timely distribution of oil from our Gulf Coast fields to the east-coast refineries and the New York port of embarkation. A source of great worry to many people was the problem of transporting Mid-Continent and Gulf Coast oil to augment the inadequate supplies available on the west coast for dispatch to the Pacific, and our petroleum help to China was controlled by the elevation of the Himalayas and the tortuous defiles of the Burma Road.

Let us review the activities in setting up strategic storage and distribution of petroleum products in the several theaters of war.

The Mediterranean and European Theaters of Operations

Logistics in the Mediterranean and Northwest European theaters of operation involved a variety of petroleum products: high-octane gasoline and special lubricants for the air forces; 80-octane gasoline, engine oil, gear oils, and greases for the ground forces; heavy fuels for combatant and transport vessels; and kerosene for civilian heating and cooking.

From November 1942 to April 1945 the importation of petroleum products through the Mediterranean totaled 16.2 million tons. Of these, approximately 15% had to be handled in packages. The Mediterranean operations required a total of 5.7 million 55-gallon drums, 23.9 million standard 5-gallon fuel cans, and 10.1 million 5-gallon fuel cans called "flimsies," a light-weight type of British can. These containers alone cost a hundred million dollars. The Normandy-German operations required an additional 35.0 million 5-gallon "Ameri-can" and "Jerrican" fuel cans, and 1.5 million 55-gallon drums. The remaining 85% of the petroleum tonnage was received in bulk, and about seven hundred tankers were used for this job.

Practically all of the petroleum products used by the United States armies were produced in and transported from the United States or Venezuela, but in the later stages of the conflict a small percentage was obtained from the Middle East, after the Mediterranean was safe in Allied hands.

At the end of 1943, approximately 767 miles of military pipe line were in operation in Africa: by the end of 1944, 689 miles of this line had been removed and relaid in Italy. During the same period the Army dismantled in Africa a bolted steel tankage of 481,000-barrel capacity, and moved it to Italy to provide bulk storage. During the campaign that followed the invasion of southern France, one 6-inch pipe line and one 4-inch pipe line—a total length of 1,507 miles of line—were laid from Marseille, up the Rhône River valley, thence to Saarebourg, and across the Rhine to Mannheim. This system supplied petroleum fuels received in southern France from 127 tankers shuttling from the United States to the Mediterranean, and delivering there an average of 43,000 b/d. A total of approximately 13.3 million barrels were pumped through the Rhône Valley pipe lines in support of the American and French forces during the operation ending September 15, 1945.

In addition, the naval operations in the Mediterranean theater also required fuel. During a nine-months' period, 1,143,000 barrels of Navy Special fuel oil accounted for about 10% of the total petroleum supplied to the theater during that time.

It is well to remember that at the beginning of the African campaign the Mediterranean was denied to us, and shipment to our troops was a major problem. Little by little, sea, air, and ground gains were made, until eventually our ocean supply lines were established from loading terminals at New York and Philadelphia, Galveston and Houston, Aruba and Curaçao, to landing points at Bizerte, Tunis, Casablanca, Algiers, Oran, Naples, and Marseille, as well as Liverpool, London, and Bristol in England, and Cherbourg and Le Havre in France, and Antwerp in Belgium. From New York the supply ships traveled 3,150 miles to Bristol, 4,000 miles to Marseille. From the Gulf it is a tedious 5,000 miles to London, 5,500 miles to Marseille. From Venezuela it is 4,000 miles to Bristol, 4,200 miles to London. The story of those sea lines—the tanker losses, the savage battles against weather and submarine menace, the tenacious and indefatigable efforts our forces exerted to protect the sea lanes—all that is too well known to need elaboration here.

Preparatory to the invasion of France through Normandy, 23 million tons of petroleum products were transported across the Atlantic and stockpiled in England. This enormous stock pile supplied our forces during the pre-invasion period, and was drawn upon as a supplementary source throughout the operations.

All petroleum requirements had to be imported, and distributed in containers for the assaulting troops until bulk storage could be captured or erected. As French, Belgian, and Dutch ports were liberated, direct bulk shipments to them from this side of the Atlantic became the principal source of supply.

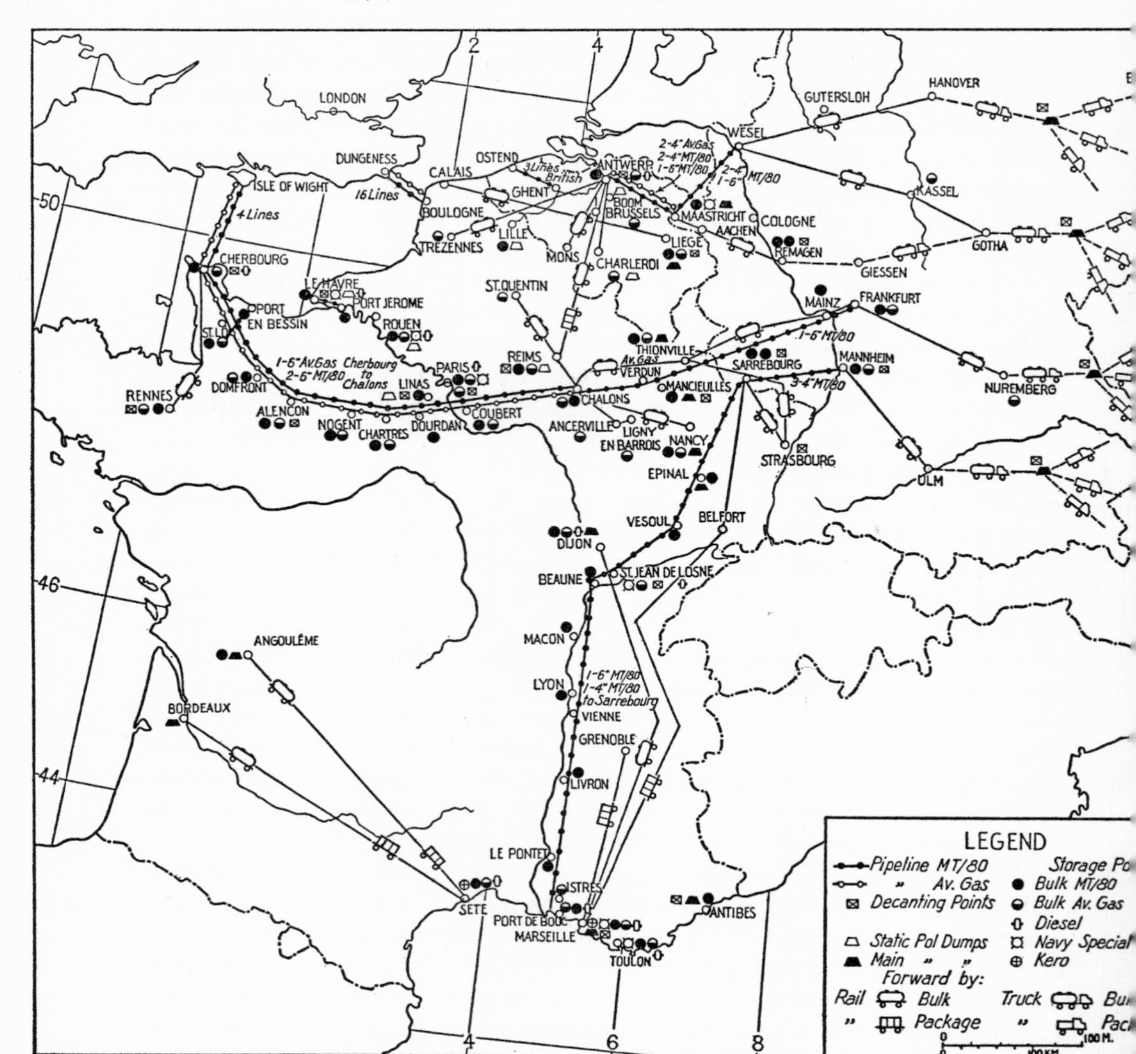

FIG. 53. European Theater of Operations: military petroleum distribution system, June 6, 1944, to M 1945. A static POL dump was a rear supply point at which incoming shipments were received and held f sue to main dumps. A main POL dump was a forward supply point from which Petrol, Oil, and Lubri (British terms) were provided to military users. The information shown on the map was supplied b Army-Navy Petroleum Board of the Office of the Joint Chiefs of Staff.

To assure a continuous supply of essential petroleum products, and to avoid the danger of interrupted delivery because of attack on our tankers by air, submarine, or surface craft, the distribution of bulk petroleum from England to the Continent was supplemented by PLUTO (Pipe Line Under The Ocean). This unique pipe-line system under the English Channel was a wholly British achievement, carried out by British engineers in cooperation with the British Navy. Under a pressure of 1,500 pounds per square inch, the undersea pipe lines of the system carried about 120 million gallons of gasoline to the Anglo-American armies between August 12, 1944, and May 8, 1945, and a million gallons daily were reaching France through it on VE Day. Sixteen of the lines ran from Dungeness to Boulogne, about 30 miles, and four ran from the Isle of Wight to Cherbourg, about 75 miles, but these four lines were abandoned before VE Day.

At D plus 90, the pipe line in France consisted of two 4-inch lines extending from Port-En-Bessin to St. Lo. This was named the Minor System. Three 6-inch lines from Cherbourg were routed through St. Lo, via Alençon, Chartres, Dourdan, and across the Seine to Coubert, the northern terminus. This was named the Major System. Two of the lines were utilized for 80-octane motor gasoline, and one for aviation gasoline.

It had been planned to utilize the port of Le Havre for a discharge port. However, it was found so demolished that it could not be used; and the major pipe-line system from Cherbourg had to be extended to Châlons-sur-Marne. From this point, aviation gasoline was transported via rail through Thionville to the Rhine River at Mainz, in Germany. Eventually, one 6-inch motor gasoline pipe line was continued from Châlons-sur-Marne to Mainz. The total length of pipe in the major system, then from Cherbourg to Mainz, was 1,908 miles.

A lesser, but important, pipe line was installed from Antwerp to Maastricht, Holland, then to Wesel, Germany as the armies pushed on; 190 miles of four 4-inch pipes and one 6-inch pipe were laid from Antwerp via Maastricht to Wesel. The initial lines from Antwerp to Maastricht were completed in 48 days. For this job, three liberty ships stood by in the United Kingdom, loaded with construction materials to be shipped to Antwerp as soon as the port was open.

Thus in Europe we had three pipe-line systems from sea ports to the Rhine: (1) from Antwerp to Wesel; (2) from Cherbourg to Mainz; (3) from Marseille to Mannheim—the three totaling 3,508 pipe-line miles. Connecting pipe lines made possible a continuous flow of gasoline from Liverpool to Frankfurt-am-Main, some 43 miles east of the Rhine. Through July 1945 over ten million tons of petroleum products in package and bulk had been discharged on beaches and at ports in northwestern Europe and southern France, and had been moved forward by pipe line, motor transport, rail, and barge to forward dumps in support of the German campaign.

The bulk distribution of motor gasoline and aviation gasoline to the ground and air forces on the Continent constituted the big problem of petroleum logistics. Over 5,000 tank cars and 10,000 tank trucks were utilized to supplement the pipe-line transport, for obviously pipe lines cannot be constructed at the speed with which General Patton raced across France. Rail lines were used as soon as they could become operational, although Herculean effort and ingenuity were necessary to restore them. Large quantities of petroleum were transported by motor truck along the 700-mile Red Ball Highway, the longest one-way traffic road in the world. At times, such as September 1944, when General Patton outran his gasoline supply near the Siegfried Line, airplanes were used to transport gasoline to the forward elements of his mechanized equipment. And to keep pace with the rapid advance of our army east of the Rhine, once it was breached, 3,660 aircraft of the Troop Carrier and Bomber Command were enlisted.

In France, Belgium, the Netherlands, and Luxembourg, before the war, indigenous crude production, never exceeding 500,000 barrels a year, was virtually negligible, and the refinery throughput was largely imported. During the war refineries had been closed, damaged, or dismantled and shipped east by the Germans. Tank cars, tank trucks, and all other petroleum equipment had been confiscated and despatched to Germany or Eastern Europe as part of the Nazi war machine. The Allied invasion forces moved into liberated areas bereft of petroleum supplies even for transportation and industry. A considerable quantity of enemy stock was captured, but, as it was practically all below United States military specifications, it was allocated to essential civilian economy; for in addition to supplies for our own forces, civil petroleum requirements, for essential purposes, had become a military responsibility.

The Caribbean Theater

Although we are disposed to look upon this theater as having been more or less inactive, neither the military nor the civil population felt that way about it. The tanker convoys, of necessity originating from ports within the Caribbean, were easy prey for German submarines.

A constant reminder of the war was the oil from tankers sunk by submarines that covered many of the fine beaches and low areas on all the Caribbean islands, and as far west as the east coast of Florida. German submarines surfaced in the harbor area in Aruba and Curaçao and fired on the refineries, and tankers at anchor in the harbors of some of the Caribbean islands were sunk by German torpedoes. However, the apparent geographical advantage to the Axis in the concentration of the oil fields around the Caribbean proved the enemy's ultimate undoing, for it permitted us to establish a series of bases from which anti-submarine forces could operate most effectively, eventually freeing this area of the Nazi menace.

Early in the war, movements to the Pacific area were made by tankers traversing the entire sea route, via the Panama Canal. But before the end of the war, pipe lines were built across the Isthmus of Panama, effecting great economy in tanker use. These trans-Isthmian pipe lines were built by the Navy, and consisted of two 20-inch lines, one 10-inch line, and one 12-inch line, extending from Cristobal to Balboa, about 36 miles.

As far as possible, the petroleum products required by the forces in Santa Lucia, Trinidad, the Guianas, Brazil, and Ascension were refined in Aruba, Curaçao, or Trinidad, while bases such as Puerto Rico, Cuba, and to some extent the Panama area were supplied with products refined in the Gulf Coast area of the United States. Our forces in the Caribbean required practically every known petroleum product; paradoxically, the list included some cold-weather grades of lubricants such as grade 1120 aviation lubricant for Navy blimp squadrons on anti-submarine patrol work. On the Atlantic side effective distribution was mandatory for

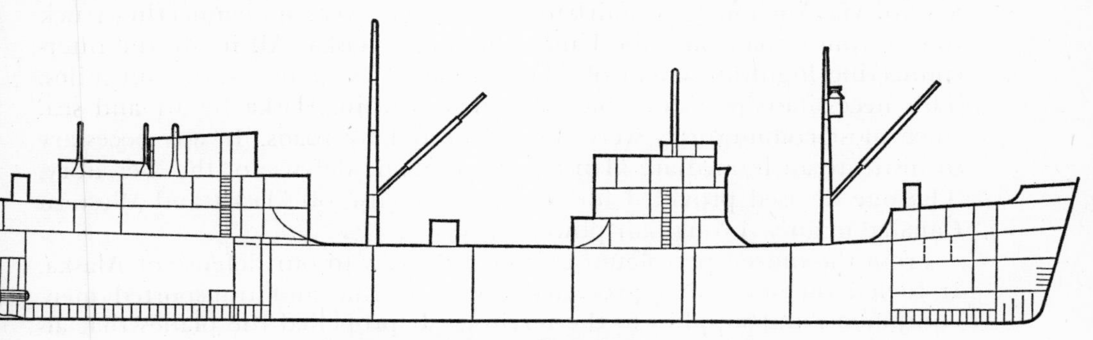

TYPICAL BT TANKER

GENERAL CHARACTERISTICS OF MOTORSHIP TARANTELLA (TI-M-BT2 DESIGN)

gth, overall	325'2"
n, moulded	48'2"
t, loaded	19'1"
s tonnage	3160
dweight tonnage	3933

Machinery	DIESEL
Horsepower, shaft	1400
Speed, knots	10
Cruising range, miles	8630
Crew, approximately	34

LIQUID CARGO	BARRELS
TANK NO. 1	4946
TANK NO. 2	5372
TANK NO. 3	5342
TANK NO. 4	5282
TANK NO. 5	5242
TANK NO. 6	5108
TOTAL	31,292
DRY CARGO	BALE CU. FT
FRAME NO. 9 TO 23	4301

;4. Cross section of a "BT" tanker, typical of the small T-1 tankers used during the war. The BT is l 31,000-barrel tanker with a raked stem and a cruiser stern. The propelling machinery consists of a engine geared to a single screw, and the machinery is located aft. The diagram and descriptive notes been supplied by the U.S. Maritime Commission.

Cuba, the Caribbean chain of islands, the Guianas, and Brazil; on the Pacific side, from Mexico through Central America to Santiago, Chile, including the Galapagos Islands. Some petroleum supplies were required at bases high in the Andes.

The geography of the numerous countries in this theater necessitated the solving of many special and intricate problems of supply. For example, supplying aviation gasoline for a photo-mapping squadron in the interior of Venezuela involved difficult methods, as did supplying motor gasoline to infantry units in the tropical jungle of the interior of Panama, and supplying petroleum products to Navy craft controlling the waterways between the Caribbean islands.

The Alaskan Theater

By the spring of 1942 the Japanese were firmly entrenched on Attu and Kiska in the Aleutians, and their submarines were active off our western coast. Our main military garrisons were in southern Alaska and the Aleutian Islands. Most of the petroleum, food, and manufactured items required for the support of the population, including the military garrisons, were imported by water from the United States, mainly through Seattle. There is only one primary railroad in the Territory, running from Seward via Anchorage to Fairbanks, and there were no connecting truck routes from Canada and the United States to Alaska. All in all, the interconnecting highway system of Alaska totaled less than a thousand miles. This necessitated considerable movement within Alaska by air and sea, since most communities were not connected by roads. It was necessary to institute an immediate plan for the strategic defense of the Territory. The one devised provided for the joint defense of Alaska and Western Canada, mainly through air power.

From the outset, petroleum provided the key to our defense of Alaska. It fueled the ships that protected the coast line and transported men, equipment, and supplies to the territory. It propelled the planes that attacked the Japanese installations in the Aleutians and later in their home islands. It was a source of motor power for military equipment. It heated hangars, barracks, and homes. Without this supply all operations would have been crippled, and Alaska would have been indefensible.

To make petroleum products available our supply problems encompassed ports, storage, roads, and pipe lines, as well as the availability of the products themselves. In the spring of 1942 there existed a shortage of crude on the west coast of the United States, a shortage of tankers to move products from the United States to Alaska, and a shortage of shipbuilding facilities for the construction of additional tankers. The situation was further aggravated by ship losses from submarine warfare, and a shortage of shuttle tankers and barges for the coastwise movement of petroleum products. In Alaska the ports of entry were incapable of handling large volumes of products, and there was a shortage of bulk tankage, both in the port areas and in the interior, for the petroleum products required by the Army and Navy. The available storage which existed at major ports was sufficient in quantity only to supply the prewar civilian requirements of approximately 113,000 barrels a month of all products, whereas in 1942 the demand for storage for the military alone amounted to 100,000 barrels monthly.

All of these shortages, in the face of the Japanese threat to the west coast of the United States, made it necessary that Alaska and northwestern Canada be as far as possible self-sufficient in petroleum. The Canadian government placed the crude resources of the Norman Wells field in the Northwest Territories at the disposal of the joint United States–Canadian Board

92. Trooz, Belgium, 1945. Class III decanting point. Thousands of gas cans standing beside tank cars at the 813th QM Gasoline Distributing Company's decanting point. The gasoline was pumped from the railroad cars into the empty cans. (U.S. Army Signal Corps photo.)

3. China, 1945. Chinese soldiers unloading pipe, the first to be flown to Mangshih air field from India via C-47. (U.S. Army Signal Corps photo.)

94. Greenland, December, 1941. Pipe line to move fuel oil from ships in harbor to Greenland Base Command
(U.S. Army Signal Corps photo.)

95. Alaskan defense oil depot. (U.S. Army Signal Corps photo.)

of Defense. The program called for putting the Norman Wells field into operation, transporting a refinery all the way from Texas to Whitehorse to provide 4,500 b/d capacity, connecting this refinery with the Norman Wells field by the Canol pipe line, and building three other pipe lines: one from the refinery to the port of Skagway, a second between Whitehorse and Fairbanks, and a third from Carcross to Watson Lake Airfield, Canada. Between February 1944 and July 1946 an average throughput of 1,000 b/d was provided by this supply system.

By developing several ports of entry the problem of winter-blocked ports was solved; and by constructing huge bulk petroleum storage at these ports, the problem of reserves was solved. The port of Anchorage was developed because of its proximity to Fort Richardson and Elmendorf Airfield. It is blocked by ice in the winter, has excessive tides, and was obstructed by silt; but it had to be utilized. The port of Seward was developed as an alternate port of entry to southern Alaska. It is ice-free, open to navigation the year round, and less rough. The petroleum demands of southern coastal Alaska were met by the use of small shuttle tankers of 16,000-barrel capacity, based at Anchorage and serving all coastal points readily reached by water.

The problem of transport was solved by building the pipe lines already mentioned, by utilizing the Alaska railroad, and by developing established roads or building new ones. The construction of the Alcan Highway, 1,400 miles long, from Port St. John in British Columbia to Big Delta, Alaska, traversed an almost unsettled region, formerly accessible only by dog team, plane, or, during the summer, by river. In December 1943, Army-operated vehicles hauled approximately 10,500 tons of petroleum products and miscellaneous supplies over this highway. But in the interior of Alaska there were many stations that could not be reached by road. The need of supplying these points by air resulted in new techniques, such as airlifts, which later in the war made history in the China-Burma-India, European, and Pacific theaters. Pilots fought snow, sleet, and fog to carry petroleum products to the remote areas.

The Pacific Theater

The strategy and tactics of warfare in the Pacific involved a different story with respect to petroleum. The single factor that shaped the history of petroleum in the Pacific more than anything else was the distance between the sources of supply and the points of actual consumption. For example, to supply the needs of a vast carrier force, one-way trips as long as 8,500 miles had to be made by our tankers. These tankers formed a virtual pipe line across the Pacific Ocean. In midsummer 1945, some 350 tankers were making the long runs from the West Indies, Balboa, and San Pedro. The run from California to Leyte, Philippine Islands, is 7,028 miles via Guam and Hawaii. From New York to Leyte via the Panama Canal the distance is 11,500 miles and from the Gulf Coast to

Leyte via the Panama Canal it is 11,025 miles. These distances involved a one-way sailing time of 27 to 42 days, and a total turn-around time of 69 to 100 days.

In contrast to the situation in the European theater, where storage facilities were usually available, whether in England or on the Continent, storage facilities were nonexistent on pin-point Pacific Islands. The fleet was constantly on the move. Forward bases varied with varied operations, and even the fleet base had to be movable. To meet these conditions, old tankers, affectionately known as "dirty Gerties," and barges or any other type of craft available were assembled into a floating tank farm. The only essential characteristics common to these craft were that they could hold oil and stay afloat. Except for escort vessels, any craft in this heterogeneous assemblage which could move under its own power had in tow one or more oil craft incapable of doing so. Fleet oilers were the tank trucks of the Navy, by which petroleum was moved from the floating tank farm to rendezvous with the combat fleet. Admiral Halsey's famous Third Fleet, of over 100 fighting ships, utilized a group of 28 fleet oilers for this purpose.

Before the Philippine Islands invasion the amount of Navy Special fuel consumed in the Pacific was 150,000 b/d. As the second battle of the Philippine Sea began and increased in intensity, more than 300,000 b/d were consumed, but the greatest sustained demand for Navy Special was during the Okinawa campaign, when for a period of one month the daily consumption of our Pacific fleet averaged about 290,000 b/d.

The China-Burma-India Theater

Supplying liquid fuel to the Allied Army Forces in the China-Burma-India theater was made extremely difficult by the very inadequate transportation facilities. Most of the petroleum products came from the Persian Gulf, except that lubricants in cans came from the United States. As enemy action had closed all of the normal supply lines to the forces in China, the establishment of new channels was of utmost importance.

Supplies of drummed gasoline were shipped by rail from Karachi on the west coast of India to upper Assam (as soon as the tactical situation permitted, supplies were collected in Calcutta and Chittagong, rather than Karachi) and thence by air to Kunming, with an escort of a few P-38's to attend to enemy activity en route. While the distance, 525 miles, was not great, the route was over terrific country, from the Mishmi Hills (12,000 to 14,000 feet) to the Hump (17,000 to 18,000 feet). Operational plans called for 20,000 tons of cargo per month to be delivered to Kunming by this method, and about half of that quantity consisted of petroleum products in drums and packages. To accomplish this task 320 aircraft—75% continuously operable—8 airfields in upper Assam, and 4 airfields in Kunming were required. The planes went by air over the most dangerous kind of terrain, a conglomeration of towering mountains and deep narrow valleys.

Although aircraft transportation was used so intensively, the shortage of cargo planes and the large quantities of gasoline consumed by the planes in transit made this method of supply both inadequate and costly. For example, to deliver 4.2 tons of supplies a plane consumed 3.5 tons of fuel. The Ledo-Burma Road was therefore scratched out of the side of hills by Chinese labor with antiquated tools, and by U.S. troops working indefatigably in all kinds of weather. At strategic points along the road bolted tanks, from 1,000 barrels to 10,000 barrels in capacity, were erected, and gasoline service stations were installed to supply fuel to the never-ending stream of trucks loaded with all kinds of badly needed supplies for the China theater. These stations obviated the necessity of each truck's carrying as parts of its load sufficient gasoline to make the entire trip. Later, parallel to part of the Ledo-Burma Road, pipe lines were laid by U.S. Army Engineers, attached troops, and native labor. The lines were buried in trenches at some places and were exposed at others; they were slung over deep canyons on steel cables, and sunk in the swiftest streams.

The first part of the pipe-line system, a 6-inch line, 750 miles long, from Calcutta to Tinsukia, in the upper valley of the Brahmaputra, was completed in August 1944. Another 6-inch line, 570 miles long, from Chittagong to Tinsukia was completed in March 1945. Railroads and barge lines were also utilized to transport petroleum to Tinsukia. The Assam area requirements were filled from this point, and large quantities were transported direct to Kunming by air. One 6-inch pipe line and two 4-inch pipe lines were laid the distance of 308 miles from Tinsukia to Myitkyina, in the upper valley of the Irrawaddy; from storage at that point the requirements of the North Burma area were drawn. Finally two 4-inch lines were laid from Myitkyina to Kunming and to the airports in that vicinity. Plans called also for the construction of a 6-inch line between these two places, but the China Theater Command decided it was not needed.

In all these ways supplies went through, and the Allied Forces were kept from being choked off.

3. WORLD PATTERNS OF CIVILIAN UTILIZATION

BY JOHN W. FREY *

MODERN transportation and mechanized industry demand oil, and no country with industrial ambitions can hope to realize them without it.

Obvious as these facts are, it took the cataclysm of war to bring the significance of petroleum utilization into focus. Between the world wars, automobiles, airplanes, oil-burning ships, diesel locomotives, and oil-burning tractors made great headway, with resultant enormous increases in the civilian use of petroleum products. The building up of oil demands in many parts of the world was to a great extent concealed, however, by the oil companies' competition for markets; it was taken for granted that oil would flow to areas where it was needed. But when in World War II military requirements and the loss of tankers curtailed the availability of oil to civilians, only primitive societies could take the reduction without serious inconvenience and loss. Postwar rehabilitation similarly demands oil.

The Types of Petroleum Products and Some of their Uses

Most petroleums as they come from the ground have little or no value for direct utilization, although some crudes are so light and volatile that they will burn in gasoline engines, and, at the other end of the scale, some crudes are so heavy and low in volatiles that they may be used directly as road oil or fuel oil. Most crude petroleums, however, lie between these extremes, except for the components that exist as gases at atmospheric pressure.

Just as the crude petroleums are in nature complex mixtures of hydro-

* Director, Division of Marketing, American Petroleum Institute.

Note on Sources: Many of the data for this chapter were taken from standard references, such as: *Petroleum Facts and Figures,* published by the American Petroleum Institute; the *Minerals Year Book* of the U.S. Bureau of Mines; *Automobile Facts and Figures,* published by the Automobile Manufacturing Association; and several studies on petroleum made by the U.S. Tariff Commission. The petroleum problems of Europe were checked against the U.S. Administrative Branch reports to the Congress completed in January 1948: (1) Chapter G, "Petroleum," in *Commodity Report, European Recovery Program;* (2) *Detailed Analysis of the CEEC Estimates for the Petroleum Requirements of the Participating Countries and their Dependent Overseas Territories, Fiscal 1949 through Fiscal 1952.* The present writer, as a joint editor of *A History of the Petroleum Administration for War,* reviewed many unpublished war records concerning petroleum. A number of unpublished reports have been valuable as references. Not a least source, but more difficult to classify, has been contributions by friends in the oil industry who criticized the first draft.

carbons, so, too, are the refined products. None of the common petroleum products is a single chemical compound; each is a mixture of hydrocarbons, and each mixture boils within a fairly definite temperature range.

In the old refineries using shell stills, the gases that boiled at low temperatures were dissipated into the atmosphere, and the fraction that boiled at about 95° to 450°F was condensed as gasoline. The temperature was again increased, to boil off a range of components collectively known as kerosene. A still higher range of temperatures yielded gas oil; heating further resulted in the heavy distillates for lubricating oil stock; and the very end of the batch left a residue of heavy fuel oils, asphalt, or coke. Although the shell still has been superseded, the principle remains, and it is convenient to classify the refined products according to range of boiling points in the following summary.

Motor Fuel and Associated Products. One of the light distillates that is fluid at ordinary temperatures at the top of the refining tower's fractionator is gasoline. This product as used in automobiles and tractors is usually a blend of straight-run, cracked, and natural gasolines, frequently with additives such as benzol or tetra-ethyl lead. Aviation gasoline of high-octane number requires a superior base stock of cracked gasoline, plus alkylate made from processed refinery gases, plus isopentane, also a gas product, plus the anti-knock additive (tetra-ethyl lead), plus inhibitors, such as anti-gum. During the war the tremendous requirements for aviation gasoline necessitated the taking of the cream of light distillates, which not only reduced the yield of automobile gasoline, but degraded the quality; and the military requirement of 80-octane "all-purpose" gasoline, probably unnecessarily high, was also a light-distillate robber.

The associated products have an important role to play. Certain trades need products similar to gasoline, but with narrower boiling ranges than ordinary motor gasolines have. For these, raw straight-run gasoline is redistilled variously to make solvents and naphthas. During the war the demand for toluene, as the base of TNT, was so great that a number of oil refineries were equipped to extract toluene from distillates, or to synthesize it. In 1939 the United States production was 25 million gallons, virtually 100 per cent from coal; by 1944 production had increased to 208 million gallons, of which four-fifths was from petroleum.

Kerosene. This group of products, which boil at a higher temperature range than motor fuels and are much less volatile, were in the early days of the industry used principally as illuminants, but are now used mainly for cooking and space heating. Raw kerosene from the fractionator is chemically treated to remove compounds that would, when burned, yield objectionable by-products.

Middle Distillates. The group boiling at temperature ranges averaging about 150°F higher than kerosene were known as gas oil in earlier days, and used to enrich artificial gas by the addition of heat units and hydrocarbons contributing incandescence to open gas flames. Gas oil is still used

Table 23. Chief types of petroleum products obtained from crude petroleum.*

- HYDROCARBON GASES
  - Liquefied Gases
    - Metal Cutting Gas
    - Illumination Gas
  - Petroleum Ether
    - Laboratory Ether
    - Motor Priming Ether
  - Alcohols
    - Isopropyl
      - Solvents
      - Acetone
    - Secondary Butyl, Secondary Amyl, Secondary Heyxl
      - Lacquer
      - Solvents
  - Other Synthetics
    - Benzene — Chemicals, Explosives, Pharmaceuticals
    - Toluene — Explosives, Toluidine, Saccharin
    - Xylene — Explosives, Dyes
    - Naphthalene — Dyes, Perfumes
    - Anthracene — Dyes
    - Resins — Lacquers, Varnishes, Paints
  - Gas Black
    - Rubber Tires
    - Inks
    - Paints
  - Fuel Gas
  - Light Naphthas
- LIGHT DISTILLATES
  - Naphthas
    - Light Naphthas
    - Intermediate Naphthas
    - Heavy Naphthas
      - Gas Machine Gasoline — Domestic Illumination Naphtha
      - Pentane — Candlepower Standardization Naphtha
      - Hexane — Laboratory Naphtha
      - Chemical Solvents — Drug Extraction Solvent
      - Aviation Gasoline
      - Motor Gasoline
      - Commercial Solvents
        - Rubber Solvent
        - Fatty Oil Solvent (Extraction)
        - Lacquer Diluents
      - Blending Naphtha
      - Varnishmakers & Painters Naphtha
      - Dyers & Cleaners Naphtha
      - Turpentine Substitutes
      - Soaps
  - Refined Oils
    - Kerosene
      - Lamp Fuel
      - Stove Fuel
      - Motor Fuel
    - Signal Oil
      - Railroad Signal Oil
      - Lighthouse Oil
    - Mineral Seal Oil
      - Coach & Ship Illuminants
      - Gas Absorption Oils
- INTERMEDIATE DISTILLATES
  - Gas Oil
    - Carburetion Oils
    - Metallurgical Fuels
    - Cracking Stock
    - Household Heating Fuels
    - Light Industrial Fuels
    - Diesel Fuel Oils
  - Absorber Oil
    - Gasoline Recovery Oil
    - Benzol Recovery Oil
- HEAVY DISTILLATES
  - Technical Heavy Oil
    - White Oils
      - Technical
        - Emulsified Spray Oils
        - Bakers Machinery Oil
        - Candymakers Oil
        - Fruit Packers Oil
        - Egg Packers Oil
        - Slab Oil—Candy and Baking
      - Medicinal
        - Internal Lubricant
        - Salves
        - Creams
        - Ointments
    - Saturating Oils
      - Wool Oils
      - Twine Oils
    - Emulsifying Oils — Cutting Oils
    - Electrical Oils
      - Transformer Oils
      - Switch Oils
    - Flotation Oils — Metal Recovery Oils
  - Wax
    - Candymakers Wax
    - Candle Wax
    - Laundry Wax
      - Detergent Wax
      - Iron Wax
    - Sealing Wax
    - Etchers Wax
    - Saturating Wax
      - Cardboard Wax
      - Match Wax
      - Paper Wax
    - Chewing Gum Wax
    - Medicinal Wax
    - Insulation Wax

- Lubricating Oil
- RESIDUES
  - Lubricating Oil
    - Light Spindle Oils
    - Textile Oils
    - Transformer Oils
    - Household Lubricating Oils
    - Compressor Oils
    - Ice Machine Oils
    - Meter Oils
    - Journal Oils
    - Motor Oils
    - Steam Cylinder Oils
    - Compounded Oils
      - Water-Soluble Oils
    - Valve Oils
    - Turbine Oils
    - Tempering Oils
      - Heat Treating Metals
    - Floor Oils
    - Transmission Oils
    - Railroad Oils
    - Printing Ink Oils
    - Black Oils
    - Grease Oils
      - Compounded Greases (For General Lubrication)
        - Gear Grease
        - Axle Grease
        - Switch Grease
        - Cable Grease
        - Cup Grease
  - Petrolatum Grease
    - Petrolatum
      - Technical
        - Metal Coating Compound
        - Lubricants
        - Cable Coating Compound
      - Medicinal
        - Petroleum Jelly
        - Compounded Products
        - Salves
        - Cold Cream
        - Skin Cream
        - Vanishing Cream
        - Wrinkle Remover
        - Massage Cream
        - Rouge
        - Lipstick
  - Residual Fuel Oil
    - Wood Preservative Oils
    - Gas Manufacture Oils
    - Boiler Fuel
      - Merchant Marine Fuel
      - Naval Fuel
      - Railroad Fuel
      - Industrial Fuel
    - Metallurgical Oils
  - Still Wax
    - Roofing Material
  - Asphalts
    - Liquid Asphalts, Binders, Fluxes
      - Road Oils
      - Roofing Saturants
      - Emulsion Bases
    - Steam Reduced Asphalts
      - Briquetting Asphalts
      - Paving Asphalts
      - Shingle Saturants
      - Paint Bases
      - Flooring Saturants
    - Oxidized Asphalts
      - Roof Coatings
      - Waterproofing Asphalts
      - Rubber Substitutes
      - Insulating Asphalts
  - Coke
    - Carbon Brush Coke
    - Carbon Electrode Coke
    - Fuel Coke
- REFINERY SLUDGES
  - Sulfonic Acid
    - Saponification Agents
    - Demulsifying Agents
  - Heavy Fuel Oils
  - Sulfuric Acid
    - Fertilizers

* Table reproduced by courtesy of *National Petroleum News* and the Reinhold Publishing Corporation.

to enrich artificial gas, although the trend is to crack residual oil. The principal use of middle distillates is for home-heating and diesel-engine fuel; within the refinery they are important as charging stock for crackers. The middle-distillate products range from those almost as light as kerosene to the dark and more viscous oils that resemble residual. All fuel oils, distillate and residual, are designated by numbers, from No. 1 at the light end to No. 6 at the heavy residual end.

Heavy Distillates. The boiling range of the heavy distillates averages about 150°F higher than that of the middle distillates just described. This group is the source of most lubricants, as well as technical and medicinal oils, paraffin, and petrolatum. Some heavy distillates serve as cracking stock or fuel oil, however, and not all lubricating-oil stocks are heavy distillates.

Where this stream is used for lubricants, the first fractionating process is only the beginning of another series of fractionations and treating processes, virtually an industry within an industry. The range of lubricants extends

---

FIG. 55. The components of aviation gasoline and how they are made. As indicated by the vertical ruling, the diagram is divided into three parts dealing, respectively, with the production of base stock, with the production of blending agents, and with the blending.

Base stock. The petroleum as it comes from the well is usually separated there, or at the gathering station in the field, into natural gas and crude oil. The top line of the diagram illustrates the treatment of the natural gas, first, in a gas absorption plant to free the lighter hydrocarbons for fuel gas, and then in a condensation plant, to condense the heavier hydrocarbons into base stock for aviation gasoline and into natural gasoline suitable for ordinary motors. The second line of the diagram illustrates the distillation of gasoline-range crude petroleum in a pressure still to produce the useful fractions, ranging from gases to solids. Of these products, straight-run gasoline is suitable as a base stock for aviation gasoline, and gas oil, the fraction next heavier than kerosene, provides the base for the process illustrated in the third line—thermal cracking—to produce a motor gasoline with somewhat higher octane rating than straight-run gasoline. Gas oil also serves as the basic material for catalytic cracking to produce base stock for aviation gasoline. Lines 4, 5, and 6 illustrate three alternative methods of catalytic cracking: the Houdry, a fixed-bed type; the thermofor, a moving-bed type with the catalyst in the form of pellets; and the fluid catalyst method, with the catalyst in the form of a very fine dust that behaves like a liquid under pressure.

Methods of producing blending agents. The 7th, 8th, and 9th lines of the diagram illustrate the production, from refinery gases, of three principal blending agents, viz. iso-octane or codimer, produced by polymerization followed by hydrogenation (line 7); alkylate iso-octane, produced by the single process, alkylation, using sulfuric acid as catalyst (line 8); and cumene, produced from benzol and petroleum gas, by a similar process, using hydrogen fluoride as catalyst.

Blending. Straight-run and catalytically cracked base stocks are mixed in various proportions with the above or other blending agents (notably toluene and triptane), and with tetraethyl lead, to make aviation gasoline of 100-octane, or higher, rating.

[Description based on the article, "High Octane Fuel," by C. A. Scarlott, in the *Westinghouse Engineer,* Vol. 4, No. 6, Nov. 1944, pp. 162–168, where the diagram originally appeared. The processes depicted are discussed by Mr. Heroy in Part Two of this book, in the section "Refining."]

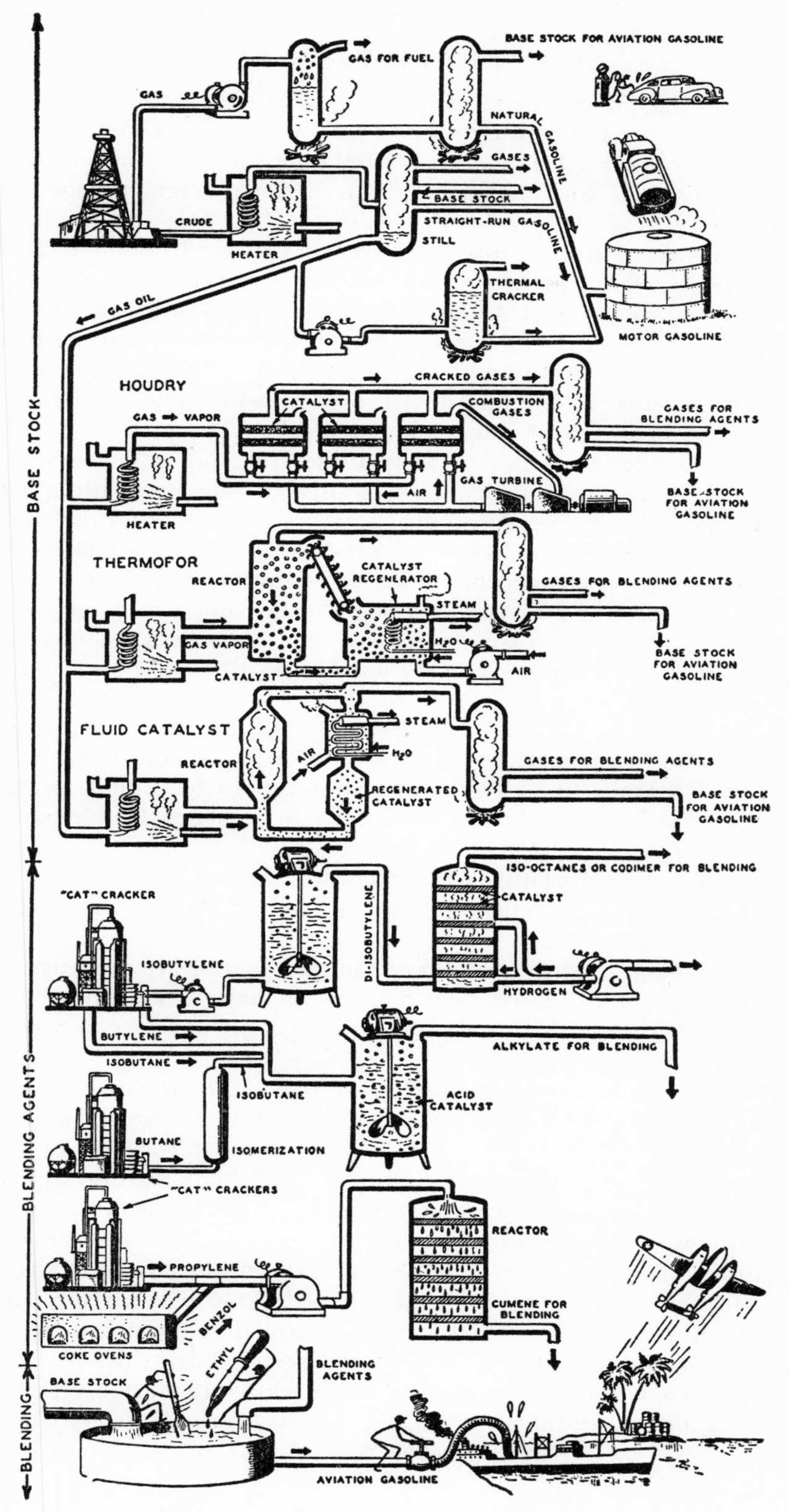

FIG. 55. See legend on opposite page.

from those so light that they are applied in small fractions of a drop, to products so heavy and viscous that they are shoveled with a scoop. Certain of the white oils are freed from all color and odor for use by bakers, candy makers, egg-packers, fruit-packers, pharmacists, and cosmetic manufacturers.

Lubricating oils and inorganic soaps are the principal raw materials of the grease maker. Some greases contain additives, such as mica, graphite, asbestos, animal fats, vegetable fats and fibers. There are combinations with special characteristics to meet hundreds of technical requirements. The development, from the simple "axle grease" of horse-and-wagon days to those greases used in the high-speed, high-pressure, high-temperature, silent gears of modern machines is a long story.

Paraffin, a by-product removed from lubricating oil by low temperatures or solvent action and filtration, is another industry within itself. Paraffin preparations range from those containing a substantial percentage of oil to highly clarified and deodorized products with a variety of melting points. The etcher preparing his plates for printing, the dentist taking the impression for an inlay, the pathologist preparing a thin section for examination, the manufacturers of candy, candles, matches, butter cartons and papers, even the cook in the kitchen sealing jams and jellies—all use paraffin.

Some clarified form of another by-product of lubricating-oil manufacture—petrolatum—is found in most salves, ointments, and skin creams. Dark colored petrolatum is used in many ways: in veterinary preparations, in rust inhibitors, and as an ingredient in rubber and cable-coating, to give examples.

Both paraffin and petrolatum were important during the war in protecting from the elements shipments of machinery, guns, ammunition, food, and other supplies.

Residual Products. At the bottom of the fractionator are the least volatile of petroleum products: residual fuel oils, asphalts, cokes, and sludges. In areas where little residual fuel oil is used, road oil and coke may be the end-products.

Residual fuel oil, sometimes called No. 6 fuel oil, is used primarily for steam locomotives and ships, stationary boilers, gas manufacture, and heat in industrial processing.

Petroleum asphalts, now more extensively used than natural asphalts, are obtainable as liquids, emulsions, semi-solids, and solids. Their waterproofing qualities are outstanding, whether in roofing, paint, or floor materials; and asphalt in highway construction is ubiquitous. Certain asphalts may be mistaken for rubber, in insulating wires and electric appliances. During the war the civilian consumption of asphalt was cut low by military demand for building materials and the paving of air fields.

Petroleum coke is used as fuel, and as the source of carbon electrodes for batteries, metallurgical furnaces, and carbon brushes in electrical machinery. Petroleum sludges are raw materials for a number of chemicals, especially the sulphonates.

Patterns of World Utilization

In industrialized countries the full catalogue of petroleum products is utilized; but from the world standpoint petroleum is essentially a source of power, heat, light, and lubrication. Therefore, motor fuel, kerosene, fuel oils, and lubricants are the products on which to concentrate in arriving at utilization patterns. Of the world's prewar annual production of petroleum—almost 2 billion barrels—the estimated world consumption of the four principal products accounted for nearly nine-tenths (Table 24). All other products amounted to less than one-tenth, for refinery losses, stock changes, and other factors, enter into a calculation of net balance.

Over a period long enough to average seasonal variations, petroleum products are consumed at approximately the rate they are produced. They are produced as needed (except the relatively small stores to meet seasonal fluctuations); and in order that they shall be available almost everywhere people live, there must be a constant flow of large volume in transit and in process.

In 1938 North America was the only continent where production and consumption were almost in balance. South America consumed only about one-fourth as much as it produced, and Asia about half its production. Europe as a whole consumed almost 1¾ times as much as it produced; Africa, some 18 times as much as it produced; and Oceania's consumption was virtually all imported.

Table 24. Production of crude petroleum, and estimated consumption of four leading products, by continents, in 1938.

	Crude Production [a]	Estimated Consumption of:				
		Motor Fuel	Kerosene	Fuel Oil	Lubricants	Total of These Oils
			(In millions of barrels)			
North America	1,277.4	552.1	58.5	469.1	25.7	1,105.4
South America	245.1	14.1	3.0	39.5	1.0	57.6
Europe [b]	263.2	156.0	61.8	173.3	23.3	414.4
Africa	1.6	9.0	3.5	11.7	0.9	25.0
Asia [c]	200.4	18.7	17.5	43.3	4.5	83.9
Oceania [d]	—	15.2	2.2	7.1	0.7	25.3
Undistributed	—	1.0	0.6	0.5	0.1	2.2
World Total	1,987.7	766.0	147.0	744.6	56.2	1,713.8

[a] Does not include related fuels (natural gasoline, benzol, power alcohol, and synthetic mineral oils from coal and shale).

[b] Includes all the U.S.S.R.

[c] Includes the East Indies.

[d] Australia, New Zealand, the Fiji Islands, the Hawaiian Islands.

Source: *Petroleum Facts and Figures, 1941,* pp. 16, 20–23. Because the figures in the source have been rounded, the totals given above do not check in the first decimal place. The production figures are those of the U.S. Bureau of Mines, and the consumption estimates those prepared by V. R. Garfias, R. V. Whetsel, and J. W. Ristori, of Cities Service Oil Company, as given in *Petroleum Facts and Figures,* Amer. Petrol. Inst., 7th ed., 1941, pp. 16, 18.

NORTH AMERICA

In 1938 the United States and Canada consumed about 63% of the world's oil. While the intensity of utilization in the United States was greater than that of Canada, each had a high per-capita rate of consumption, and used motor fuel in greater quantity than any other group of products. In Mexico, on the other hand, fuel oil was dominant. For almost ninety years North America was self-sufficient in petroleum, and a source of exports. But in 1948 the United States became a net petroleum importer.[1]

During the war military demands made it necessary not only to produce greater quantities of all products but, also, many new products. By the last year of the war production in the United States had increased about one million barrels per day. At the same time it had been necessary to decrease the civilian consumption of oil for heating and of gasoline for passenger automobiles.[2] When the war came to a close, gasoline consumption immediately started upward, and in 1947 averaged 2,177,500 b/d, as compared with 1,828,800 b/d in 1941. This rise is traceable in part to tremendous increases in agricultural demand. There were in 1948 about 3 million tractors on farms, as compared with about 1.6 million in 1941, and in addition about 1.9 million trucks were serving farms, an increase of 62% in the same period. A high percentage of farm tractors burn gasoline, although many use diesel fuel, tractor fuel, or kerosene.

During the years between the two world wars the consumption of kerosene in the United States was relatively steady until 1933, when increases, caused principally by range burners, raised the consumption from 105,500 b/d in 1933 to 190,300 b/d in 1941, and 280,800 in 1947. This latter gain has been largely due to the installation of space heaters. ("Space heaters," a trade expression, include vaporizing pot-type oil burners that use kerosene, range oil, or No. 1 heating oil. Although "oil burners" also are used for heating space, that trade expression refers to "gun" or "wall" types that burn No. 2 and higher fuel oils.)

In the United States there is a heavy concentration of oil burners that

[1] There would be a somewhat different statistical picture if bunkers in foreign trade, for both U.S. and foreign vessels, were reported as exports; this method is used by a number of European countries. All oil imported into the United States is so reported; on the other hand, bunkers are reported as part of the domestic demand. The imports of fuel oil and the fuel-oil bunkers are almost in balance—in round figures, 60 million barrels a year.

[2] In 1941, of a total gasoline production in the United States of 1,904,000 b/d, passenger cars are estimated to have consumed 1,217,000 b/d, and trucks and buses 390,000 b/d. In 1944 the total production had increased to 2,008,000 b/d, but passenger-car consumption was reduced to 663,000 b/d, a decline of 45.6%, while trucks and buses had increased to 403,000 b/d. The total military and export demand for gasoline, 127,000 b/d in 1941, is estimated to have increased 452% by 1944, while the total civilian utilization, 1,777,000 b/d in 1941, had declined 26.5% by 1944. The consumption of gasoline by scheduled airlines in the United States increased from 80.7 million gallons in 1941 to 231.6 million in 1946.

consume middle-distillate oil, for which the total demand has almost doubled since 1941. During the war it was necessary to control the consumption of this product because of the increased military demand for diesel and fuel oil (military consumption of diesel oil increased from 2.6 million barrels to 22.9 million annually), and its increased use by industry for the manufacture of essential military material. In 1941 the civilian heating-oil requirements of the United States were estimated at 331,000 b/d. There were 2,135,000 domestic oil burners in use at the beginning of 1941, but by the beginning of 1948 this number had jumped to 3,650,000, and there was barely enough distillate oil available to keep them going.

Another phenomenal increase in the consumption of distillates has come about during the past decade through the extensive adoption of diesel locomotives. At the beginning of 1948 diesel engines operating on the railways consumed about 21.5 million barrels of oil annually, compared with 2.7 million in 1941, and it is estimated that by 1953 the horse power of diesel locomotives, and hence their diesel-oil consumption, will be doubled. Stationary diesel engines as of January 1, 1948, totaled 6.8 million horse power, and marine engines, 3.3 million, and both types have been increasing rapidly.

Residual fuel oil, by reason of its viscosity, which requires preheating for atomizing, is used only in large burner installations. Most of such oil, as produced by the refinery, is of the type known as No. 6. Large commercial vessels generally use this type of fuel, but the Navy during the war required a lighter product known as Navy Special, a considerable quantity of which was made by blending residual fuel oil with distillates. The comparative utilization of distillates and residual fuel oil before and at the close of the war is summarized in Table 25.

Table 25. United States: estimated annual consumption of distillate and of residual fuel oils in 1941 and in 1945.

	Distillates		Residuals	
	1941	1945	1941	1945
	(In million barrels)			
Heating other than range oil	120.9	121.3	46.6	43.9
Range oil	4.4	7.5	...	...
Vessels	11.0	14.1	56.7 [a]	100.4 [a]
Manufacturing and mines	10.4	19.1	78.0	91.2
Railroads	4.9	14.5	80.6 [a]	112.3 [a]
Oil-company fuel	1.1	1.1	54.2	57.3
Gas and electric companies	5.2	6.8	33.6 [b]	34.5 [b]
U.S. Army, Navy, Coast Guard	3.0	30.4	25.7	97.5
Tractor fuel	—	4.0	...	...
Miscellaneous	12.4	12.8	5.6	5.2
Total of items listed	173.3	231.6	381.0	542.3

[a] Residual fuel oil consumed on vessels and railroads as steam-boiler fuel.

[b] Residual fuel oil consumed by gas and electric companies includes that used in the manufacture of gas, as well as that used for power.

Source: *Petroleum Facts and Figures, 1947,* pp. 34, 35.

Statistics on the end use of lubricants leave much to be desired, but it seems that more than half the total consumption in peacetime is industrial, and the remainder primarily automotive. The consumption of industrial lubricants reflects the intensity of industrial activity.

The growth in the utilization of no petroleum product is more dramatic than that of liquefied petroleum gas (commonly known as LPG), the sales of which in 1935 amounted to only 76 million gallons, but by 1941 had jumped to 462 million, and by 1945 were over 1,250 million. About 40% is burned in homes; about 35% is used in the manufacture of chemicals, including components for synthetic rubber; and the remainder is employed in the manufacturing of city gas, as industrial fuel, and as fuel for internal combustion engines. Utilization in homes, where it had been making great headway, was restricted by the war, because LPG was so important in advancing the military program, especially in the manufacture of synthetic rubber and in the production of components for aviation gasoline. The shortage of steel for the manufacture of containers for shipping and distribution (insulated pressure tank cars and steel bottles) also held back the development. With the postwar decline in consumption of aviation gasoline and the better availability of steel, LPG is forging ahead in cooking, water heating, and refrigeration.

SOUTH AMERICA

More than four-fifths of the land and population of South America are found south of the equator, but more than nine-tenths of the petroleum is produced north of it, where there are three important sources of the world's supply: Venezuela, Trinidad, and Colombia. These three Caribbean countries have a huge exportable surplus. To the south, Ecuador, Peru, and Bolivia have a smaller surplus for export. In Argentina, the largest consuming country in South America, the consumption and domestic production in 1938 were estimated at 29 and 17 million barrels, respectively. In 1948 Argentina used about 33 million barrels, of which about two-thirds came from indigenous production.

In the prewar period, more automobiles, better highways, and "packaged" power (gasoline or diesel engines built into self-contained units, e.g. electric-generator units, arc welders, air compressors, shovels, drills, wood-working machinery) were steadily increasing the consumption of motor fuel; but the war, with its shortages in tankers and tremendous increase in military demand, had a restricting effect, which has not been entirely overcome.

In volume, kerosene is only about one-fifth as important as gasoline; of the South American total of about 3 million barrels in 1938, Argentina consumed nearly half, and Brazil nearly one-third. Vast areas of South America have little artificial illumination, and wood- and charcoal-burning have barely been challenged, beyond the well-lighted cities.

Fuel oil is much more important in South America than either kerosene or gasoline; the 1938 consumption of about 40 million barrels was 13 times that of kerosene, and nearly 3 times that of gasoline. The fuel oils' primary job is to turn the wheels of ocean and river shipping, and to generate heat and electricity for mining and other extractive industries. In a continent with little coal production, the comparatively low cost of transporting fuel oil makes it loom large in industrial uses, as well as in transportation.

If there were no other measure of industrialization, the volume of lubricants used would tell that Argentina, Brazil, and Chile rank first as industrial states. Of the total consumption of lubricants in South America in 1938, about one-half was used in Argentina, one-quarter in Brazil, and about one-tenth in Chile.

EUROPE

Before discussing the use of petroleum products in Europe, we should remind ourselves that in the utilization of mechanical energy Europe led the way for most of the nineteenth century, and that the growth in the application of such energy was directly based on coal. In most countries in Europe, petroleum is still not a large factor in the total energy utilized.

Except for the U.S.S.R., Rumania, Poland, and Czechoslovakia, the countries of Europe in 1938 produced either no petroleum, or less than 5% of national requirements. However, there was a wide range in the per-capita consumption of liquid fuels. In the northwestern and Scandinavian countries the per-capita consumption of motor fuel was 5 to 8 times as large as in Spain and Italy. In eastern Europe the per-capita consumption was low, even in Rumania.

In industrial Europe oil was chiefly important because of its role in transportation, and, except in the U.S.S.R., fuel oil and motor fuel were almost equally important. About three-fourths of these liquid fuels were used by automobiles, ships and boats, railways, and the military. Automobiles, with a consumption of about 40% of the total of liquid fuels, were the largest users. Fuel oil was much used by European ships and boats, especially those engaged in the coastwise, river, and canal trades, but the full significance of fuel oil in the European economy is obscured by the omission in the statistics of bunker fuel taken by European ships in non-European ports. On the other side of the ledger, the fuel-oil figures of Gibraltar include oil for more than European use. Since the war new oil-burning ships have replaced many coal burners sunk during the war. Ships vital to Europe's economy now use bunker fuel oil produced, for the most part, in the Caribbean and Persian Gulf areas.

Outside of the U.S.S.R., where the kerosene consumption in 1938 was estimated at 42 million barrels, the European consumption of kerosene was low. For all the rest of Europe it was estimated at about 20 million. In many countries kerosene is the primary source of rural and small-town illumination, and is extensively employed in cooking. While information is not available concerning what percentage of the kerosene is used to produce power, many European tractors and marine and stationary engines are built to use it.

Europe ranked first among the continents in the use of lubricants in the prewar period, with a total of about 23 million barrels annually, or about 40% of the world consumption (these figures include the U.S.S.R., estimated at 9.5 million), but the consumption of lubricants in Europe has been reduced by the devastation of war.

During the war, non-Axis countries dependent on tanker imports presented serious problems, especially during the first two years; and those countries dominated by the Axis and blockaded by the Allies lost virtually all petroleum products. Germany, as the dominant power, did everything possible to maintain her own civilian economy; and, with the advantage of supplies obtained by capture and from satellite countries, especially Rumania, German civilian consumption was not at first drastically reduced. By early 1943, however, the basic civilian allowance was reduced to 50% of that of 1941, and the total effect by the end of 1943 was about a 40% cut in the consumption of petroleum products,[3] as compared with the prewar figures. The subsequent rapid deterioration of petroleum supply incident to the bombing of refineries and synthetic

[3] In most German cities, taxis could be obtained only for specific emergencies, and doctors, who previously had been permitted to use automobiles, were denied them. Little alcohol was available for admixture to gasoline, as more indigenous food production had to be used directly for human consumption. Benzol, which in the prewar period had also been added to gasoline, was more important for other war uses than for transportation. By late 1943 a program was instituted to convert motor vehicles from liquid fuel to gas and solid fuel. Earlier in the year, about 40% of the buses were converted from gasoline or diesel oil to city-gas or solid fuels, and the percentage of converted trucks was higher.

Long-distance haulage by trucks was particularly hard hit, because most were diesels, and the supply of diesel fuel for them was very short, although it was stretched by blending with gasoline. The conversion of diesels to solid fuel was more difficult than that of gasoline engines; and when converted, their efficiency was reduced, because of the lower heat value of gas generated from solid fuel, and the frequent stops to clean out generators and to refuel. Moreover, the supply and distribution of solid fuels for gas generators, especially charcoal and wood, became increasingly difficult.

By early 1943, about 100,000 vehicles were equipped to burn a liquefied mixture of butane and propane that was produced by the synthetic industry and delivered in steel bottles. However, the competing needs for synthetic rubber, aviation gasoline, and other war products, and a reduction in the amount of steel available for bottles, curtailed further expansion of the use of these gases for motor vehicles, long before bombing had seriously interfered with synthetic production.

Although the industrial use of fuel oil was cut in half, Germany's 130,000 farm tractors received a fuel cut of only 20% in 1943, but their use for road haulage or for stationary power was prohibited. One of the most critical problems was the shortage of lubricants, even though Germany, long a manufacturer of lubricating oils, had augmented that capacity by producing synthetic lubricants. Rationing was applied, and much of the used oil was re-refined.

96. High-sea filling station and carrier. A U.S. Navy tanker, laden with fuel oil and gasoline for the fleet, is photographed as she steams along in a rough sea, accompanied by an Essex-class carrier in the background. (U.S. Navy photo.)

97. Fueling at sea. Aware of the vastness of Pacific spaces, the Navy long ago sought to perfect its technique for fueling at sea. After Pearl Harbor, task forces steamed out for weeks and months, fueled at sea as a matter of course. Calling for the smartest seamanship, fueling at sea would be hazardous even in calm waters were it not for the trained crews to whom the operation is entrusted; and in heavy seas there is always danger. Water breaks high on the sides of a tanker as she pumps fuel to the light carrier *Independence*. Slack on the hose astern has been taken up, but the hose forward is still dragging in the water. (U.S. Navy photo.)

98. A six-million-gallon tanker of the wartime merchant marine. Each such tanker carried enough gas to las an A-ration holder 35,000 years. The War Shipping Administration's own tanker fleet of more than 900 vessel plus those it allocated to the armed services, moved to Europe the gasoline and other oil that played such a par in the defeat of Germany. The W.S.A.'s oversea tanker fleet was made up primarily of T-2 or Standard typ tankers, similar to the above, of from 15,900 to 23,000 dead weight tons, and of emergency tankers with con verted Liberty ship hulls, having a dead weight tonnage of 10,600 or 10,800. W.S.A. tankers going overseas ha a skeleton deck on top, on which were carried unboxed airplanes processed so that the spray would not hu them. By the end of the war about 90 per cent of the smaller aircraft used by the U.S. Army Air Corps oversea were so transported. The W.S.A. also built tankers similar to the T-2 for the Navy; these, known as the Cimarron type, are 553 feet long and have a dead weight tonnage of 18,300. (War Shipping Administration photo, b courtesy of U.S. Maritime Commission.)

plants became evident, even in vital air operations, many months before the end of the war.

The postwar patterns of petroleum utilization in Europe are like those of a kaleidescope in constant rotation. In 1947 the coal production of Western Europe was roughly 100 million tons less than in 1938, a decline of about 17%, but petroleum consumption was 10 million tons greater than in 1938, a rise of about 27%. Obviously, shortage of coal accounts for part of the increased oil consumption. Economic rehabilitation, however, requires total energy production and utilization at a level higher than that of 1938. There are many incidental problems, but their insistence should not obscure the fundamental relationship between the capacity to produce, the standards of living, and a high per-capita utilization of energy.

Looking ahead to 1952, the European Recovery Program calls for almost doubling the petroleum consumption of 1938, although the pattern of utilization is based on curtailment to essential uses only, with 45% of the total as fuel oil. In spite of the fact that Europe needs more automobiles, trucks, tractors, and diesel locomotives, the recovery program, as far as petroleum is concerned, indicates only moderate increases in these consuming devices. On the other hand, it indicates marked increases in the use of petroleum in steam locomotives, under stationary boilers, and as industrial fuel. Gasoline consumption by personal automobiles is to increase little, but truck transport, farm tractors, barges, and fishing boats will more than absorb the savings that result from continued curtailment of the use of automobiles.

AFRICA

The estimated per-capita consumption of 8 gallons of petroleum products in Africa in 1938 was about one-fifth of the world average, but within Africa the variations were enormous. Of a motor-fuel consumption of about 9 million barrels, almost one-half was used in South Africa, and about one-third in countries touching the Mediterranean, leaving about one-sixth for the vast regions between. And the concentration of automobiles, with resultant motor-fuel consumption, at a number of mining centers and areas of specialized agriculture in the interior, impairs the validity of average consumption figures for some political divisions. During the war, aviation in Africa became very important, and civilian aviation now requires considerable gasoline, though in a small number of centers.

The period of the great explorers of the African interior coincided approximately with the early years of the petroleum industry in the United States. During that period, while kerosene became the oil for the lamps of much of the world, Africa remained dark. Even now, when in the industrial countries electricity has displaced kerosene as the principal illuminant, the interior of Africa, except the areas noted for motor fuel, has little light. Of the African consumption of 3.5 million barrels of kerosene in 1938—less than 3% of the world total—two-thirds were used by Egypt and South Africa. Little wonder that an empty kerosene can is a valued possession in many parts! Incidentally, the need that empty cans used for cooking should be non-toxic caused a problem during the war, because of the shortage of tin. Terne plate, a possible substitute, is toxic.

The peripheral location of African industrial and commercial activities shows strongly in the consumption of fuel oil and lubricants. Of a total of more than 11½ million barrels of fuel oil in 1938—less than 2% of the world total—French West Africa, the Union of South Africa, and Egypt consumed four-fifths. Of the less than 900,000 barrels of lubricating oil—about 0.5% of world consumption—one-third was used in South Africa, and two-fifths in the Mediterranean countries.

In Egypt, the only country in Africa that has significant crude-oil production and refining, the 1938 production of 1.6 million barrels of crude (about 4,300 b/d) was equal to less than one-third of the Egyptian consumption. On the other hand, the refining capacity of about 9 million barrels per year (about 24,700 b/d), operated largely on foreign crude, made Egypt an exporter of petroleum products.

ASIA

Occasionally someone with a penchant for arithmetic shows that if the per-capita consumption of oil in Asia were increased to equal that of the United States, the world would face an astronomical demand figure. Asia could, of course, profit by a greater use of power; a modest doubling of the consumption would produce revolutionary economic changes.

The slight use of motor fuel in Asia as a whole is related to the fact that the vast interior has poor highways, or none. Because of this, it cost hundreds of dollars per barrel to fly aviation gasoline over the Himalayas into China during the war, and the small Chinese petroleum production required some 1,500 miles of high-cost haulage by trucks, river rafts, and railways; moreover, there was a large shrinkage of gasoline in transit, due to its use as fuel *en route*. In the total of 18.7 million barrels of gasoline used in Asia in 1938, the Japanese civilian consumption amounted to 7.8 million barrels, or more than two-fifths of the total. British India came next, with 3.0 million barrels.

The old story of lamp oil in Asia may be pointed up by the fact that in 1938 the consumption of kerosene in India, China, and Iran still was more than twice that of gasoline.

Compared with other petroleum products, fuel oil is relatively important. Asia's consumption was in 1938 about four-fold that of Africa, and a bit larger than that of South America. Ocean shipping accounted for at least two facets in this situation: the use of fuel oil for ships' bunkers; and low ocean freights, which made for fuel-oil use in many power plants on or near the sea coast.

As might be expected from their industrial development, Japan and

India used the greater part of the lubricants: 1.8 million barrels and 1 million, respectively, in the Asiatic total of 4.5 million barrels consumed in 1938.

In no country of Asia (except the Soviet Union, which is included with Europe in this discussion), or of Oceania, have supply and consumption been nearly balanced. The few countries with large production have little consumption, and the larger consumers have little or no production. In 1938 Iran and Iraq each produced about eight times as much as was consumed there; Brunei and Sarawak, about six times; the Netherlands Indies and Burma, each about five times. Among the large consumers, British India produced about one-seventh of its requirements; and Japan, the largest Asiatic consumer, produced only about one-tenth of its civilian requirements. How much it imported for military consumption and storage was not publicly reported.

It is, of course, well known that by her conquests Japan did not immediately obtain rich stores of petroleum products. Because her victims' abandonment of their facilities was accompanied by demolition, the South Pacific producing area captured by Japan yielded only about 12 million barrels in 1942, but by 1944 the output had been brought up to about 46 million. But during the later course of the war, the benefits of rehabilitation were first diminished and later eliminated by military action. The Japanese government began, as early as 1923, to consider a national liquid-fuel policy to make the country independent of foreign sources, but not until ten years later was a definite program instrumented to encourage the low-temperature carbonization of coal in the home islands, and to increase the shale-oil production at Fushun. However, the former was an important source of liquid fuel during the war, from 1940 to V-J Day.

Australia and New Zealand had no significant commercial production in 1938, but consumed about 22 million barrels. The annual per-capita consumption of Australia was about 2.1 barrels, and of New Zealand, about 2.7. The pattern of consumption is very much like that of the United States and Canada, except that central heating and railway dieselization have not been much developed. About two-thirds of the petroleum utilized is in the form of gasoline.

The Upward Demand for Petroleum

The job of supplying oil to maintain the world's increasing intensity of petroleum use—a higher rate of increase than before the war—is gigantic. Looking ahead we can see that consumers will want, before 1952, about 4 billion barrels of petroleum a year.[4]

When World War I came to a close in 1918, petroleum was consumed at the rate of about 1.5 million barrels per day. Today the rate is more

[4] These figures are from a paper, "The Place the Middle East will Occupy in World Markets," given by C. J. Bauer, February 15, 1948, and published in *Mining and Metallurgy*, Vol. 29, No. 500, August 1948, p. 436.

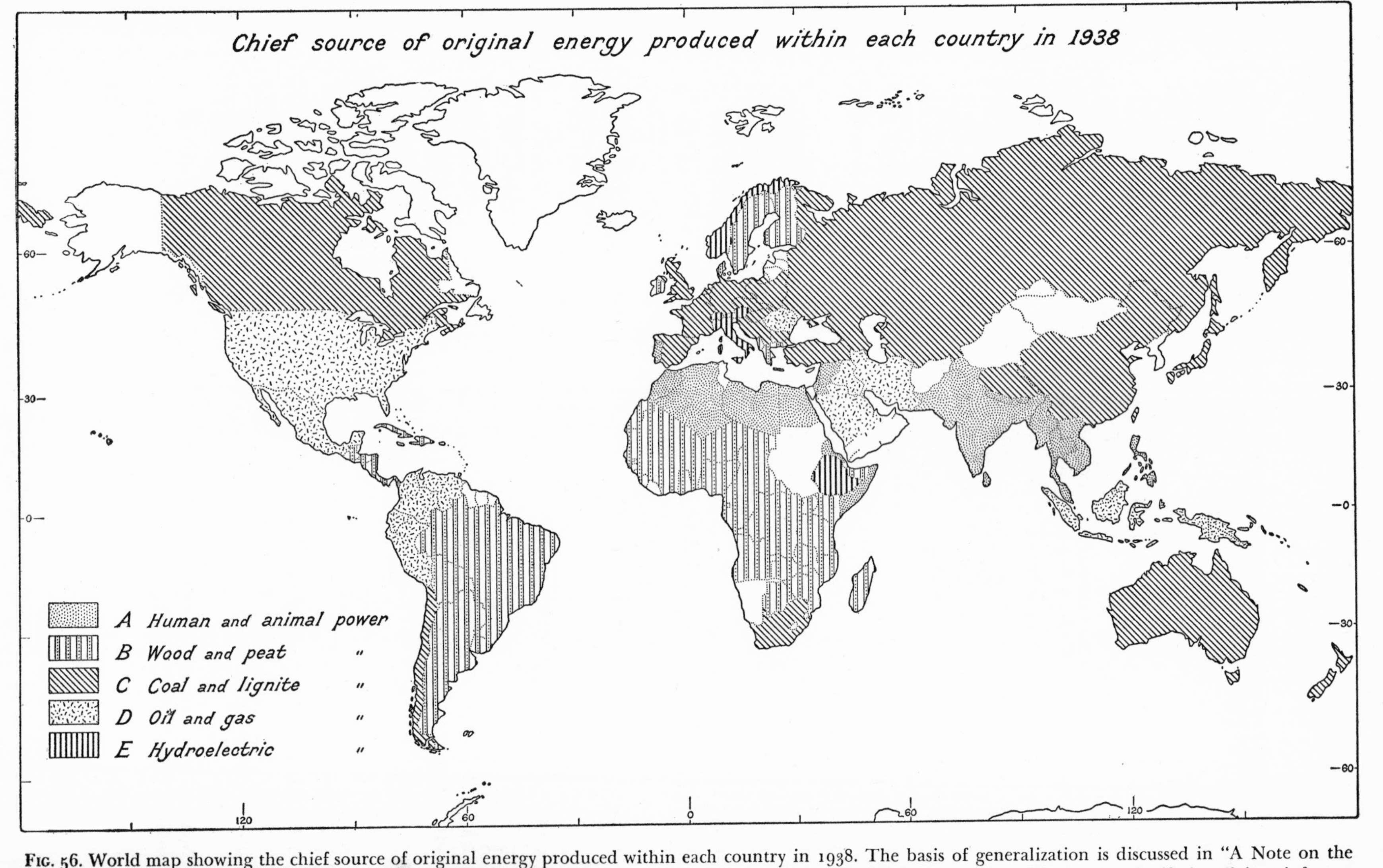

Fig. 56. World map showing the chief source of original energy produced within each country in 1938. The basis of generalization is discussed in "A Note on the Comparative Position of Oil and Gas among the World's Chief Sources of Energy," appended to Chapter 3 of Part Four. Blank areas signify insufficient informa-

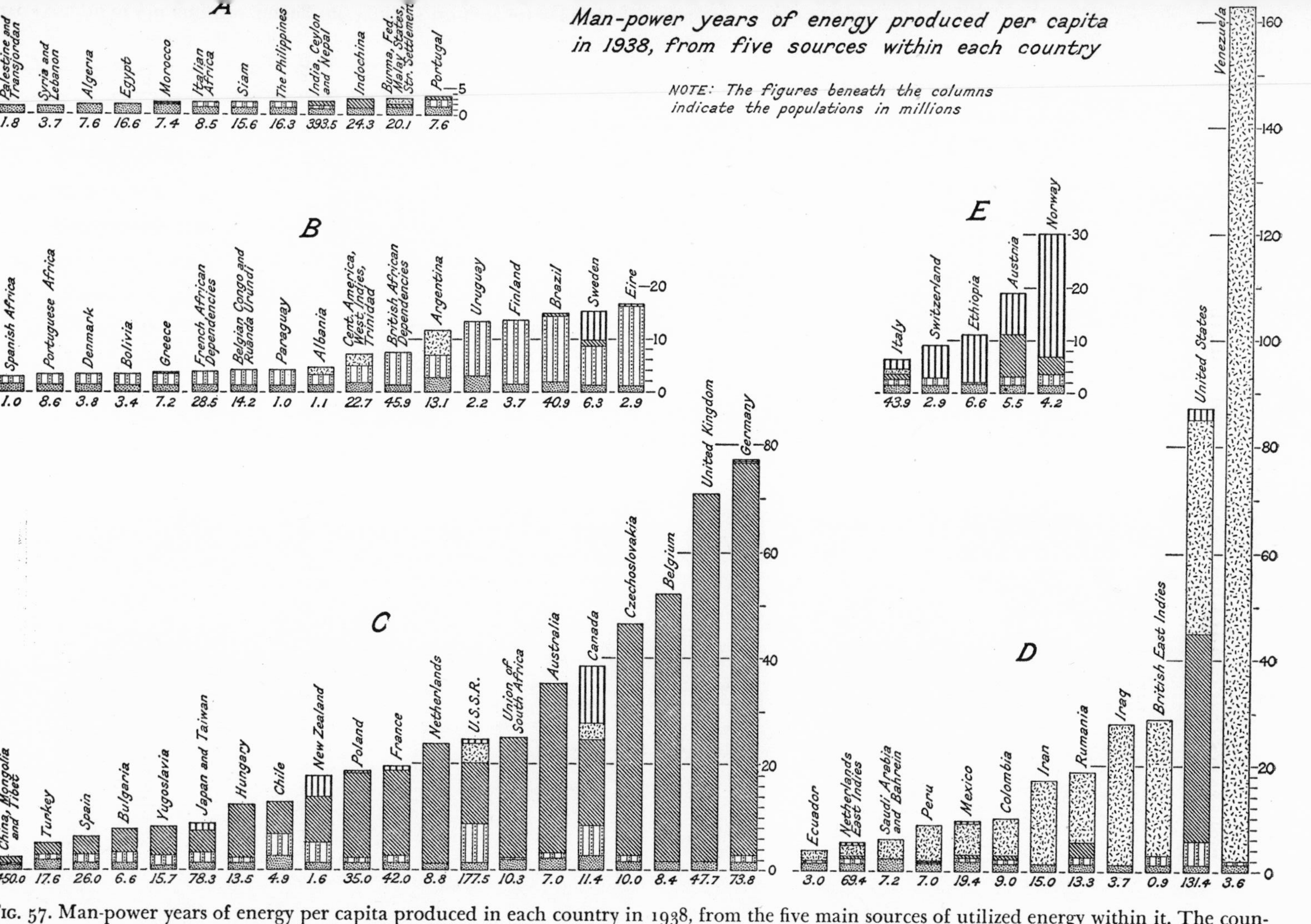

FIG. 57. Man-power years of energy per capita produced in each country in 1938, from the five main sources of utilized energy within it. The countries are grouped according to their chief source of original energy, as shown in the world map, Fig. 56.

than six times that figure, and for 1951 a daily consumption figure nearly eight times that of 1918 has been estimated. No other source of energy has grown, and promises still to grow, so rapidly in volumes used.

An extensive pattern of petroleum utilization is more than an accompaniment of high living standards; it is a basic source of the energy release that is the essence of the modern world.

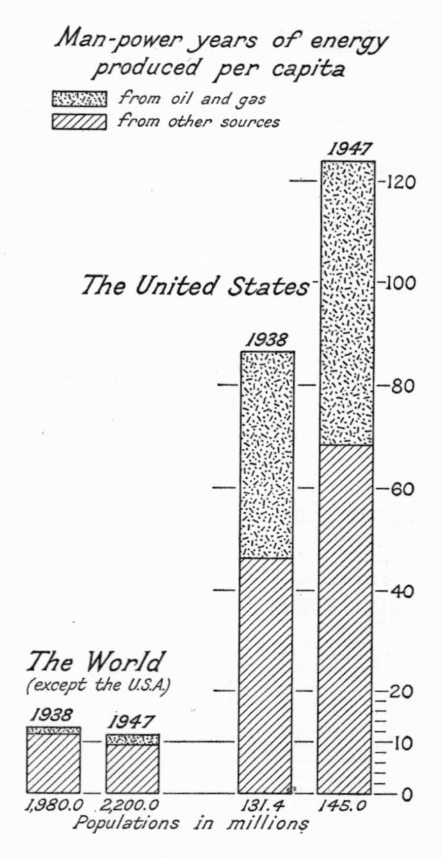

FIG. 58. Man-power years of energy produced per capita in 1938 and in 1947 from (1) oil and gas, and (2) from other sources (human and animal labor, coal, lignite, hydroelectricity) in the United States and in the rest of the world.

Editors' Note on the Comparative Position of Oil and Gas among the World's Chief Sources of Energy

The map in Fig. 56 and the bar diagrams in Figs. 57 and 58 were prepared by the American Geographical Society from data supplied by Professor W. T. Thom of the Department of Geology, Princeton University. They are outgrowths of a major project that he has long had in mind: a measurement of the rates and kinds of power production (apart from atomic power) potentially available in different areas of the world.

The computations that are diagrammatically represented in these charts measure the original energy produced within each country in relation to its population, although they do not indicate the way in which the energy is utilized. To measure utilization adequately, Professor Thom considers that quantitative means must be found by which allowance can be made for all the following factors in each country: (1) international shipments of petroleum, coal, and vegetable fuels, and international transfers of gas and electric current; (2) the degree of efficiency in the conversion of the energy from each type of fuel into mechanical work; (3) the availability of essential metals and other industrial materials, and the degree of command over the complex technical facilities, knowledge, and organization through which mechanical work may be effectively applied. Short of this, however, his present per-capita ratios give a rough measure of the comparative position in raw energy output held by each country in the stage of development it had reached in 1938, and of the rank of oil- and gas-produced energy among the kinds of energy produced in each.

As a basis of computation, Professor Thom assumed that a population unit of ten million people can be productive of five million man-years of work *per annum*. He further assumed that one horse, mule, or camel has the work-capacity of eight manual workers, and that horse-power years of fuel or hydroelectric energy may be converted to man-power years by a conversion factor of approximately 12 men per mechanical horse-power. The animal, fuel, and electrical energy figures have been calculated to horse-power years of work and then to man-power years, which have been divided by the national populations to give per-capita ratios.

From the results of these computations, the countries have been grouped, in the map in Fig. 56 and the bar charts in Fig. 57, according to their chief type of energy produced in 1938. The Group A countries, primarily dependent on human and animal power, have the lowest ratios of energy output per capita. The Group B countries, primarily dependent on wood and peat, come next, and their energy ratios are also fairly modest. The Group C countries, primarily dependent on coal, show a great range, and the Group D countries, producing predominantly oil and gas, show an even greater range. The "great powers" of 1938 are distinguished in the upper ranges of these two groups (i.e. in the ranges with per-capita energy ratios of 20 man-years or more) by the combined factors of large populations and large outputs of coal. The decisive advantage possessed by the United States and by the Soviet Union in also having large proportionate outputs of oil and gas is striking. The countries producing most of their energy in the form of hydroelectric power (Group E) are a relatively small group, with an especial advantage for light industry.

Fig. 58 summarizes, by energy ratios for 1938 and 1947, the United States' growing command over work energy, greatly augmented as it has been by the very large U.S. production of oil and gas, and the concurrent position in the rest of the world, where oil and gas form a much less considerable element in the sources of energy to be drawn upon. A map showing the utilization of petroleum products by the countries of the world in 1938 is presented in Fig. 59.

4. A STATISTICAL SURVEY

BY ANASTASIA VAN BURKALOW *

STATISTICAL data on the consumption of petroleum are surprisingly incomplete even as regards the amounts used in different parts of the world; and in terms of the uses to which it is put and its influence on man's economic, social, and political life, our information is still less adequate, and must often be obtained indirectly from statistics on related topics. A fully satisfactory treatment of all aspects of the geography of petroleum consumption is therefore impossible. The data we do have, however, are sufficient to reveal certain significant patterns and trends, very generally for the early decades of the petroleum era, but with increasing detail for more recent years.

Chief Statistical Sources

The chief sources of data dealing more or less directly with the use of petroleum are: (1) government reports on petroleum utilized; (2) government statistics of annual production, imports, and exports, from which the general order of magnitude of petroleum utilization can be estimated; and (3) compilations combining government statistics and estimates based on all other available data.

GOVERNMENT REPORTS ON PETROLEUM UTILIZED

For most parts of the world there are no official published statistics of petroleum consumption. Since World War I a few countries have made such information available, but usually in incomplete form—the total consumption (unclassified) of petroleum,[1] or the consumption of refined products,[2] or of motor fuel only,[3] or of industrial fuel and lubricants.[4]

* Department of Geology and Geography, Hunter College.

Acknowledgement: In the preparation of this paper the author received assistance from her colleagues on the staff of the American Geographical Society, of which she was at that time a member.

[1] Denmark, for example, has listed the consumption of petroleum (total and per capita figures) in its statistical yearbook (*Statistisk Aarbog*) since 1926, giving annual values from 1922 on, and average values for the periods 1906–1910, 1911–1915, and 1921–1925.

[2] Venezuela publishes the consumption of refined products within the country, by product and by company (*Anuario Estadístico de Venezuela*).

[3] For example, Canada, by individual provinces (*The Canada Year Book*), Hungary (*Annuaire statistique hongrois*), and Czechoslovakia (*Annuaire statistique de la République Tchécoslovaque*), the latter giving only the motor fuel consumed by autobuses run by the state railways.

[4] For example, Argentina (*Estadística Industrial*), which gives the industrial consumption of petroleum fuels and lubricants by provinces and territories and by groups of industries.

For the United States the figures published by the Bureau of Mines [5] assume that annual domestic consumption of petroleum and its products may be represented by production plus net imports (or minus net exports), plus the difference between refinery and primary bulk terminal stocks at the beginning of the year and these stocks at the year end. For this purpose data are available on production since 1859 [6] and on exports since 1864 [7]; imports, being very small, were not listed systematically until 1914; [8] refinery stocks and refinery production by product were first listed in July 1917, when the Bureau of Mines began, with the cooperation of the refiners, to collect these statistics.[9] Therefore, beginning with 1918 [10] there have been available the data for the calculation of petroleum utilization by the method outlined above. However, two items remain unknown—changes in stocks held by consumers, jobbers, and other secondary storage holders, and losses resulting from evaporation, leakage, and other causes after the products have been sold.[11]

In addition, the United States Bureau of Mines publishes estimates of the country's monthly "demand"—that is, the disappearance of primary supplies—by product; the annual consumption of gasoline, kerosene, and fuel oils by states; the consumption of these products and of lubricating oils by type of use; and the percentage of the total annual supply of energy furnished by petroleum in the United States.[12] Reports of state authorities on monthly and annual consumption of gasoline by type of use, compiled by the Public Roads Administration,[13] also include itemized data on non-highway uses of gasoline where such information is available.

Governmental data gathered primarily for other purposes sometimes record the use of petroleum products. In the Census reports for 1920, 1930, and 1940 there is information by states on the number of automobiles, trucks, and tractors on farms, and the percentage of farms having such equipment,[14] and information on the number and percentage of

[5] See the annual summary, *Minerals Yearbook,* published since 1932. It succeeded the annual volume, *Mineral Resources of the United States,* published by the U.S. Geological Survey from 1882 to 1925 and by the U.S. Bureau of Mines from 1925 to 1931.

[6] See the *10th Census of the United States, 1880,* Vol. 10, p. 278 (data for fiscal years ending June 30); and *Mineral Resources of the United States,* 1882 on (data for calendar years given from 1885 on).

[7] See the *10th Census of the United States, 1880,* Vol. 10, p. 278 (data for fiscal years ending June 30); and *Mineral Resources of the United States,* 1892 on (data for calendar years).

[8] *Mineral Resources of the United States, 1913,* p. 1080.

[9] See Hopkins, G. R., "Petroleum Statistics, 1935–38," *U.S. Bureau of Mines, Economic Paper 20,* 1940, p. 1.

[10] *Mineral Resources of the United States, 1918,* p. 1,128.

[11] *Ibid., 1922,* p. 397.

[12] Annual summaries published in the *Minerals Yearbook* and in the *Mineral Market Report.* Current information on some of these topics published in the *Monthly Petroleum Statement.*

[13] See its periodical, *Public Roads,* and its *Highway Statistics, 1945.*

[14] *14th Census of the United States, 1920. Vol. V. Agriculture: General Report and Analytical Tables,* Table 9, p. 514. *15th Census of the United States, 1930. Agriculture:*

dwellings using petroleum products for lighting, cooking, and heating.[15] The latter is given for the whole nation, for each census region, and for each state, in terms of the entire population and of urban, rural-non-farm, and rural-farm population groups, for large cities and metropolitan districts, and for counties as regards lighting equipment and cooking fuel. Special studies by government agencies give additional details on these matters,[16] and on the use of automobiles in different parts of the country by people of different income groups and in places of different sizes.[17]

In interpreting the statistics of individual nations, it must be remembered that they vary in nature and quality. Some desirable test questions are: How were the data collected? Are they for fiscal or calendar years? Are amounts used by the government, especially for military purposes, included in the totals (as they are in the United States), or omitted? Are natural gasoline and benzol included, or do the figures cover only crude petroleum and its products? Often the answers to such questions are not readily available, but the possibility of variation in procedure from country to country must be kept in mind.

GOVERNMENT STATISTICS OF ANNUAL PRODUCTION, IMPORTS, AND EXPORTS AS A BASIS FOR ESTIMATING RETAINED PETROLEUM

Annual production of petroleum, if any, plus imports, minus exports, represents the amount of petroleum retained within a country. This may be more or less than the amount consumed during the year, for it disregards not only the two unknown items listed above, but also changes in producers' and refiners' stocks. The results are, probably, fairly accurate approaches to true consumption for two types of countries: (1) those that derive their petroleum products entirely from imports, which will normally be adjusted fairly closely to current demand, and (2) petroleum producers who use most of what they produce, exporting relatively little; the United States and the Soviet Union are the outstanding examples of this second type. In 1938 the United States used about 90% of the petroleum it produced,[18] while, according to the compilations of Garfias, Whetsel, and

Vol. IV, General Report, Table 17, p. 535. *16th Census of the United States, 1940. Agriculture, Vol. I,* State Table 11; *Agriculture, U.S. Summary,* 2nd Series, p. 11.

[15] *16th Census of the United States, 1940. Housing, Vol. II,* Parts 1–5.

[16] See Kirk, H., Monroe, D., Pennell, M. Y., and Rainboth, E. D., Family Housing and Facilities, Five Regions (*U.S. Dept. of Agric., Misc. Pub. 399*), 1940; also Kirk, H., Monroe, D., Brady, D. S., Rosenstiel, C., and Rainboth, E. D., Family Expenditures for Housing and Household Operation, Five Regions (*U.S. Dept. of Agric., Misc. Pub. 457*), 1941.

[17] Monroe, D., Brady, D. S., Constantine, J. F., and Benson, K. L., Family Expenditures for Automobile and other Transportation (*U.S. Dept. of Agric., Misc. Pub. 415*), 1941; and reports in *Public Roads* (published by the Public Roads Administration, Federal Works Agency), Vol. 22, No. 11, Jan. 1942; Vol. 23, No. 2, Apr. 1942; Vol. 24, No. 10, Oct.–Nov.–Dec. 1946.

[18] Computed from data given in Table 28.

Ristori, the Soviet Union used 81% of its petroleum production.[19] The latter figure is only approximate, as the compilers have recognized, and other authorities have suggested a much higher percentage. Even for these two types of countries, five-year or ten-year averages would be desirable.

Long-term averages are even more desirable in the case of countries that produce relatively large amounts of petroleum and export most of it. In such countries there may from time to time be wide discrepancies between the amounts retained and those actually used. A year of good markets and large exports may be followed by a year of poor markets and large additions to the stock pile. Nevertheless, this method of computation, in spite of its defects, offers the best indication of petroleum consumption for countries where changes in stocks are unrecorded. These include the United States before 1918, and many parts of the world today.

For the United States, production, imports, and exports from 1871 on are summarized in the annual *Statistical Abstract of the United States,*[20] and the "consumption" indicated by these figures is presented in Table 28. For the sake of uniformity all the "consumption" figures in the table, except for the single years 1938, 1947, and 1948, were computed by the method under discussion, even though for recent years more exact estimates of consumption have been issued by the government. This overlapping offers an opportunity to check the validity of the method as applied to the United States. According to the government estimates published by the U.S. Bureau of Mines,[21] the average yearly consumption of petroleum during the decade 1931 to 1940 was 1,046,830,000 barrels, a figure not greatly different from the 999,760,000 barrels shown in Table 28. This difference may be largely accounted for by the fact that the production figures given in the *Statistical Abstract,* the source used in preparing the table, are for crude petroleum alone, whereas the data on domestic demand published by the Bureau of Mines include natural gasoline and benzol production. During the decade 1931 to 1940 production of these two together averaged 46,000,000 barrels yearly.[22]

For countries other than the United States the most convenient summary of data on petroleum production, imports, and exports is contained in the *Statistical Summary* of the mineral industries published annually by the Imperial Institute of Great Britain, covering each year since 1919.[23]

[19] Computed from figures of production and consumption compiled by Garfias *et al.* as given in *Petroleum Facts and Figures,* Amer. Petrol. Inst., N.Y., 7th ed., 1941, pp. 16 and 18.

[20] Published by the U.S. Department of Commerce.

[21] *Minerals Yearbook, 1940,* p. 935; *1941,* p. 1022.

[22] For annual figures see *Mineral Resources of the United States, 1931,* and *Minerals Yearbook, 1932* on.

[23] Imperial Institute (prior to 1926, Imperial Mineral Resources Bureau), *The Mineral Industry of the British Empire and Foreign Countries: Statistical Summary (Production, Imports, and Exports).* The first volume to include trade figures as well as production figures for petroleum was published in 1923, covering the years 1919–1921.

Similar data may be obtained for many countries in the *Commerce Year-book of the United States,*[24] in *International Petroleum Trade,*[25] and, for the years 1933 to 1946, in the four numbers of the *Statistical Year-book of the World Power Conference.*[26]

In compiling Fig. 59, data from the *Statistical Summary* of the Imperial Institute were used in the computation of the 1938 average per capita "consumption" (actually the amount retained) in countries for which there was no other source of information. Since the petroleum statistics are given in long tons by the Imperial Institute, amounts consumed have been converted into barrels.

COMPILATIONS COMBINING GOVERNMENT STATISTICS WITH ESTIMATES BASED ON ALL AVAILABLE DATA

The most comprehensive compilations are those published in the transactions of the American Institute of Mining and Metallurgical Engineers from 1934 to 1943, as part of its annual review of the petroleum industry.[27] These tabulations, prepared by V. R. Garfias and R. V. Whetsel, joined in 1939 by J. W. Ristori,[28] cover the years 1931 through 1942, figures for the latter year being rough estimates. Total consumption of all petroleum products and the consumption of the major refined products—motor fuel, kerosene, gas and fuel oil, and lubricants—are given for each country that used a million or more barrels a year. From 1939 (Vol. 132) on, data are given also for those countries that consumed between 100,000 and 1,000,000 barrels per year.

As the authors explained in the second paper,[29] "statistics from practically every country have been studied, and where actual figures were not available, the estimates have been based on all the data available." No distinction is made in the tables, however, between figures based on more or less official statistics and those estimated by the authors. The relative reliability of the figures is, therefore, unknown to the reader, but, as the only comprehensive set of estimates of petroleum consumption by country, they fill a distinct need for that period. Resumption of the publication of comprehensive compilations, under the same or other auspices, is most desirable. The Garfias and Whetsel estimates were used wherever possible in the computation of the per capita consumption figures for individual countries in 1938, shown in Fig. 59 and in Table 26.

[24] Published annually by the U.S. Department of Commerce since 1922.

[25] Published monthly by the U.S. Bureau of Mines since 1932. Contains data on imports and exports and on production and refining gathered from the U.S. Foreign Service and other sources.

[26] Published by the World Power Conference, London, in 1936, 1937, 1938, and 1948.

[27] *Petroleum Development and Technology. A.I.M.M.E. Trans.,* Vols. 107, 114, 118, 123, 127, 132, 136, 142, 146, 151, for the years 1934 to 1943.

[28] All with the Foreign Oil Department, Cities Service Company. Prior to 1936, Messrs. Garfias and Whetsel were with the Foreign Oil Department of Henry L. Doherty and Company.

[29] *Op. cit.,* Vol. 114, 1935, p. 245.

For no single year were these compilers able to account for the entire available supply of petroleum: the world's annual consumption of petroleum products, as tabulated, does not equal the production. At first it was thought that the difference represented additions to stocks, but when it persisted, some other explanation became necessary. In 1938[30] the compilers suggested as reasons for the discrepancy: (1) unreported consumption or stockpiling for military purposes; (2) errors in converting weight to volume units, and *vice versa;* (3) lack of authentic statistics on production and consumption for many countries and especially for the U.S.S.R.

The compilations and estimates prepared by Garfias, Whetsel, and Ristori have been reproduced in publications of the American Petroleum Institute[31] and the National Industrial Conference Board.[32] For the United States only, these publications include government statistics on demand for petroleum products by states and by uses, and related data from governmental and commercial sources on such topics as the consumption of petroleum products by aircraft, by the Army, Navy, and Coast Guard, and by farm tractors.

Since 1945 yearly compilations of government and private data on United States consumption of all oils and of individual refined products have been issued by E. DeGolyer and L. W. MacNaughton, under the title of *Twentieth Century Petroleum Statistics.*[33]

Data on petroleum consumption are published from time to time in the *Foreign Commerce Weekly.*[34] During 1947 the countries considered included Afghanistan, Austria, Bolivia, Brazil, Canada, Denmark, Eire, Ethiopia, France, Guatemala, India, Madagascar, Mozambique, the Netherlands, New Zealand, Peru, the Philippines, Switzerland, and Tunisia. In some cases the figures represent sales of petroleum products within the country during 1946, while in others, usually those countries that are wholly dependent on imported petroleum, the figures represent the 1946 imports. The sources from which the information was obtained are not given.

In the publications discussed, information is usually available in totals for individual countries: rarely is it given for smaller administrative units. Even in the United States figures for quantities of petroleum utilized are not compiled for units smaller than states, and only supplementary data —as on automobiles and on dwelling units using petroleum products— are available for cities or counties. Detailed studies of the use of petroleum within small areas, and comparisons of its use from area to area, in

[30] *Op. cit.,* Vol. 127, 1938, p. 304.

[31] *Petroleum Facts and Figures,* Amer. Petrol. Inst., N.Y., 5th ed., 1937; 6th ed., 1939; 7th ed., 1941; 8th ed., 1947.

[32] *The Petroleum Almanac,* The National Industrial Conference Board, N.Y., 1946,

[33] First prepared in 1945 in the Office of the Director of Naval Petroleum and Oil Shale Reserves, United States Navy Department, this compilation is now published by DeGolyer and MacNaughton, Continental Building, Dallas, Tex.

[34] Published by the U.S. Department of Commerce.

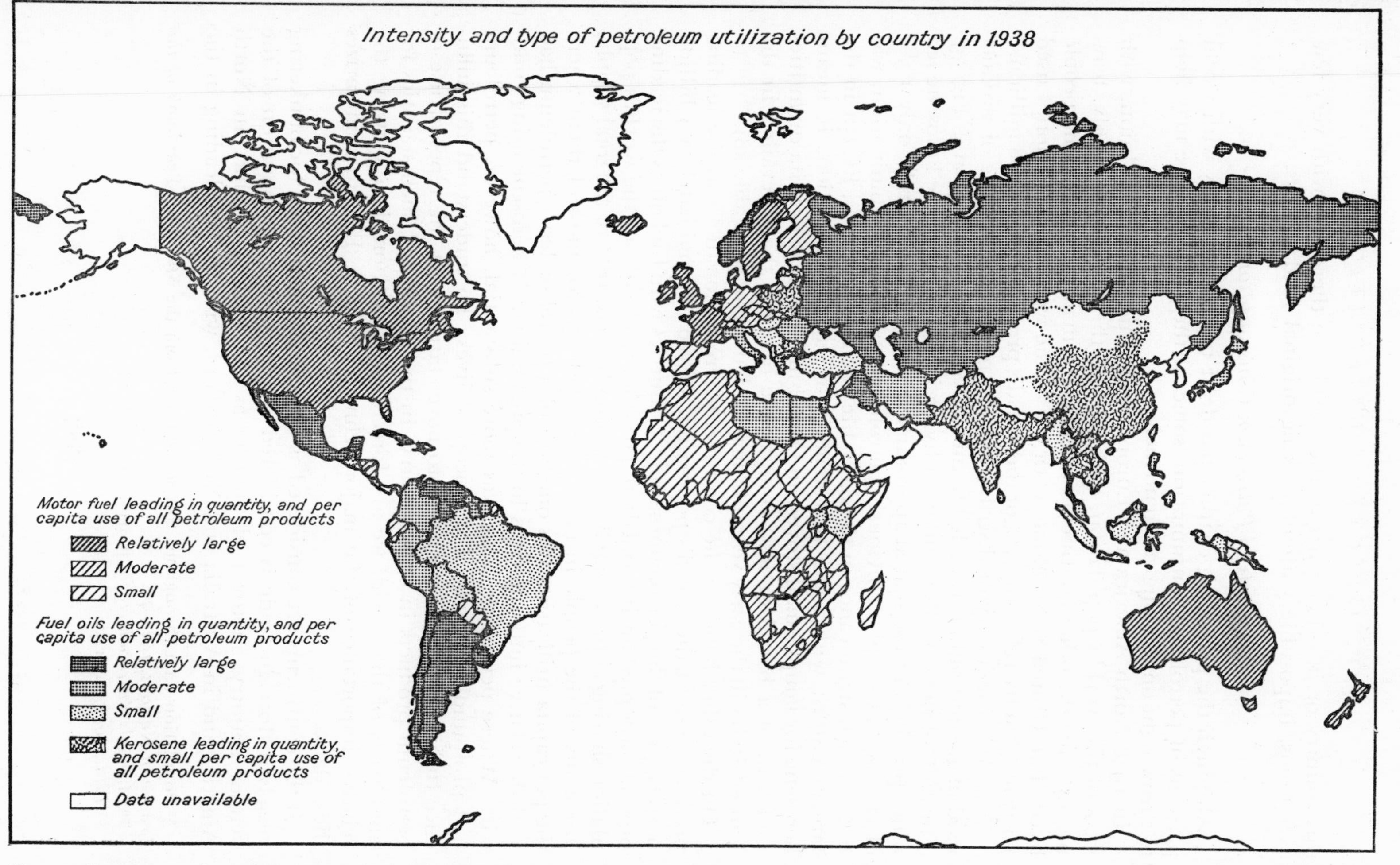

FIG. 59. Cartogram of petroleum utilization in 1938. The principal political units are classified according to their rates of per capita usage of petroleum products as a whole, and according to the type of product used in greatest quantity. (Bunker fuels were eliminated whenever the data permitted.) The method of computing the rates is described in the note at the end of this chapter.

this country or in other parts of the world, are therefore, with very few exceptions, impossible without making original field surveys.

Trends in Petroleum Consumption

Although the preceding chapter has dealt more thoroughly with world patterns of petroleum consumption, some further analysis here may help to throw light on the long-run tendencies.

In 1938 North America and Europe, including the Soviet Union, with 35% of the world's population, took 86% of the petroleum supply, leaving 14% for the other continents where the remaining 65% of the people lived. The United States alone, with 6% of the world's population,[35] used 1,137 million barrels,[36] or 57% of the world's production of 1,988 million.[37]

That this inequality of distribution was related to the type of product used in greatest bulk is revealed by an analysis of the data shown in Fig. 59. The 40% of the world's population living in countries where kerosene led other products in use shared approximately 29 million barrels, only 1.5% of the world's supply. Obviously petroleum, except for illumination, was not of much significance in the daily life of these countries. Except in the larger centers, with some petroleum-powered transportation, it meant kerosene for lamps, small stoves, and small motors, in per-capita quantities (one-fifth of a barrel or less) comparable to those that prevailed in the United States in the early years of the petroleum era (see Table 28).

On the other hand, in the countries where motor fuel was the leading product, 25% of the world's population shared approximately 1.4 billion barrels, 71% of the petroleum supply. Countries with fuel oil as the leading product occupied an intermediate position, with 30% of the world's population sharing approximately 375 million barrels, 19% of the total supply. For many of the people in these areas—perhaps for most of them where the per-capita utilization was one-fifth of a barrel or less—petroleum had as little relative importance as in the predominantly kerosene-using countries. Where per-capita usage was more substantial, however, petroleum had profoundly influenced the peoples' lives. To understand this influence fully it would be necessary to have statistics on the types of use of petroleum products in each country. In the absence of such statistics for most parts of the world, we are forced to depend on data showing the relative importance of the major refined products. Fig. 59 summarizes these data.

If the continents are considered as units, and fuel oil sold for bunkering vessels in foreign trade is considered as part of the consumption of the vendor country, motor fuel appears as the leading product in North America and in Australia and New Zealand, with fuel oil leading in the

[35] Population data throughout this section are from the *Statistical Year-book of the League of Nations, 1939–40.*

[36] *Minerals Yearbook, 1940,* p. 946.

[37] *Ibid., 1941,* p. 1110.

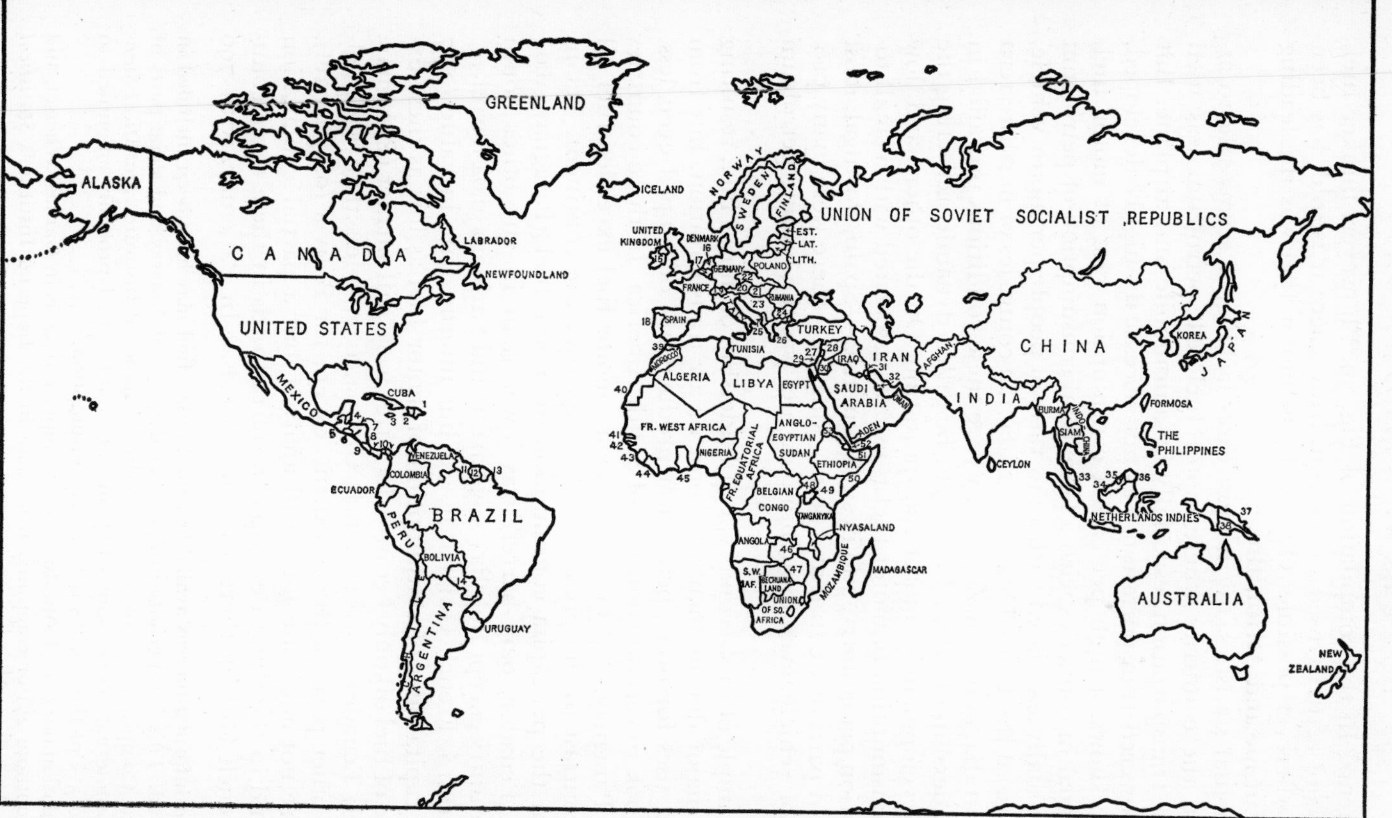

FIG. 60. Index map of the world's principal political units with boundaries as of January 1938. The names represented by numbers are as follows: 1, Dominican Republic; 2, Haiti; 3, Jamaica; 4, British Honduras; 5, Guatemala; 6, El Salvador; 7, Honduras; 8, Nicaragua; 9, Costa Rica; 10, Panama; 11, British Guiana; 12, Surinam; 13, French Guiana; 14, Paraguay; 15, Eire; 16, Netherlands; 17, Belgium; 18, Portugal; 19, Switzerland; 20, Austria; 21, Hungary; 22, Czechoslovakia; 23, Yugoslavia; 24, Bulgaria; 25, Albania; 26, Greece; 27, Lebanon; 28, Syria; 29, Palestine; 30, Transjordan; 31, Kuwait; 32, Bahrein; 33, Malayan Union; 34, Sarawak; 35, Brunei; 36, British North Borneo; 37, Northeast New Guinea; 38, Papua; 39, Spanish Morocco; 40, Spanish Western Sahara; 41, Gambia; 42, Portuguese Guinea; 43, Sierra Leone; 44, Liberia; 45, Gold Coast; 46, N. Rhodesia; 47, S. Rhodesia; 48, Uganda; 49, Kenya; 50, Italian Somaliland; 51, British Somaliland; 52, French Somaliland; 53, Eritrea.

other continents. In the compilation of Fig. 59,[38] however, bunker fuel, because it is not actually used in the countries where it is sold, has been eliminated, wherever possible. On this basis, motor fuel was the leading product in Europe and Africa, also.

Though actual statistics are in most cases lacking, it seems safe to assume that in the countries where motor fuel led, petroleum was used primarily for transportation, especially by automobile and airplane. But petroleum-powered transportation was not necessarily highly developed. As Table 26 shows, a high per capita consumption rate of motor fuel, which is normal in countries with a large total consumption of petroleum products, is usually associated with a low ratio of people per motor vehicle, and in the case of the United States, Canada, the countries of northwestern Europe, Australia, and New Zealand, where these conditions prevailed in 1938, the leadership of motor fuel might be called dynamic, reflecting the large-scale development of motor transportation. On the other hand, low per capita consumption of motor fuel is usually associated with a small total consumption per country, and in such countries—Spain, Portugal, most of Africa, and parts of Central and South America—there were many people per motor vehicle, even though motor fuel was the leading petroleum product.

For an example of this second type of leadership of motor fuel, resulting not from a major development of motor or air transportation, but from the fact that uses for other petroleum products are developed even less, one might look to certain countries of South America. In three countries —Ecuador, Paraguay, and French Guiana—motor fuel exceeded fuel oil in use, in contrast to the predominant trend over the continent. Yet in two of them the per-capita consumption of motor fuel (Paraguay, 0.04 barrel, and Ecuador, 0.05 barrel) was lower than in any other South American country except Bolivia (0.03 barrel) and Surinam (0.04 barrel); [39] and in the third, French Guiana, it was but intermediate in value (0.17 barrel). Per-capita rates of consumption of motor fuel as low as these exceeded those of fuel oil only because the latter were still smaller (Paraguay, 0.005 barrel; Ecuador, 0.017; French Guiana, 0.09)—less than the fuel-oil rates in other parts of the continent, except for Bolivia (0.07 barrel). Our data are not adequate for a full analysis, but a partial explanation may be found in the slight development of railroads in these three countries. In French Guiana there are no railroads.[40] In Paraguay, with 790

[38] Where no information was available some bunker fuel may have been included in the figures used. This is especially probable in the case of Norway and some parts of South America. Computation from data given in the Imperial Institute's *Statistical Summary* (see footnote 23, above) shows that in 1938 bunker fuel probably amounted to between 4 and 5% of the world's petroleum consumption.

[39] For population data, see footnote 35. Consumption data as in Garfias, Whetsel, and Ristori (see footnote 27) or computed from data in the Imperial Institute's *Statistical Summary* (see footnote 23).

[40] *The Pan American Yearbook, 1945,* p. 534.

Table 26. The use of petroleum products, by countries, in relation to (1) number of persons per motor vehicle in each country, and (2) petroleum product most used in each country.

1	2	3	4	5	6	7	8	9	10	11
Number of people per motor vehicle	Number of countries	Number of countries, within each class of column 2, where product most used in 1938 was:								
		motor fuel, with all consumption (per capita):			fuel oils, with all consumption (per capita):			kerosene, with all consumption (per capita):		
		large	moderate	small	large	moderate	small	large	moderate	small
Under 10	4	4[a]								
11–20	2	2[b]								
21–30	3	2[c]	1[d]							
31–50	6	1[e]	3[f]		2[g]					
51–100	9		5[h]		4[i]					
101–200	11		2[j]	2[k]	3[l]	4[m]				
201–300	8		3[n]		1[o]	2[p]	1[q]			1[r]
301–500	6			2[s]		3[t]	1[u]			
501–1,000	17		2[v]	4[w]	1[x]	5[y]	4[z]			1[aa]
Over 1,000	16			8[bb]		1[cc]	2[dd]			5[ee]
Total	82	9	16	16	11	15	8	—	—	7

These countries and the number of people per motor vehicle in each were:

[a] United States, 4; New Zealand, 6; Canada, 8; Australia, 9.

[b] United Kingdom, 18; France, 19.

[c] Denmark, 25; Sweden, 29.

[d] Union of South Africa, 28.

[e] Ireland, 47.

[f] Belgium, 37; Germany, 43; Switzerland, 44.

[g] Norway, 32; Argentina, 44.

[h] Newfoundland, 58; French Guiana, 75; Finland, 79; Southwest Africa, 84; French Morocco, 89.

[i] Netherlands, 56; Uruguay, 70; Cuba, 95; Chile, 99.

[j] Algeria, 105; Tunis, 134.

[k] Portugal, 151; Spain, 199.

[l] Venezuela, 131; Palestine, 139; Mexico, 190.

[m] British Malaya, 105; Italy, 106; Costa Rica, 164; British Guiana, 193.

[n] French Somaliland, 215; British Honduras, 224; Czechoslovakia, 234.

[o] U.S.S.R., 251.

[p] Ceylon, 202; Philippine Islands, 265.

[q] Brazil, 260.

[r] Latvia, 285.

[s] Syria, 304; Paraguay, 475.

[t] Peru, 334; Colombia, 351; Egypt, 458.

[u] Hungary, 408.

[v] Honduras, 785; Nicaragua, 986.

[w] Madagascar, 510; Ecuador, 660; French West Africa, 930; Angola, 953.

[x] Iraq, 553.

[y] Greece, 502; Japan, 532; Dominican Republic, 583; Rumania, 700; Surinam, 830.

[z] Salvador, 527; Guatemala, 580; Yugoslavia, 820; Netherlands East Indies, 874.

[aa] Poland, 768.

[bb] Haiti, 1,092; Nyasaland, 1,098; Anglo-Egyptian Sudan, 1,282; British Somaliland, 1,364; Belgian Congo, 1,480; New Guinea, 1,583; French Equatorial Africa, 2,662; Liberia, 20,000.

[cc] Iran, 1,041.

[dd] Bolivia, 1,468; Turkey, 1,708.

[ee] Indochina, 1,122; Siam, 1,244; Bulgaria, 1,411; British India, 2,080; China, 10,431.

Note: Countries and territories not listed above have been omitted because of insufficient data.

Sources: People per motor vehicle as cited in *Petroleum Facts and Figures*, 6th ed., 1939, pp. 18–20. Motor vehicles include passenger cars, trucks, and buses. See Note on Fig. 59, at end of this chapter, for sources of data on consumption and definition of terms.

miles of line, many locomotives burn wood.[41] In Ecuador many burn oil, but the country has only 800 miles of railroad, compared with well over 2,000 miles each in Colombia, Peru, and Bolivia.[42]

In the countries where fuel oil is the leading petroleum product it is more difficult, in the absence of statistics, to assume how it is used, since fuel oil, unlike motor fuel, does not have one dominant function. Yet certain probabilities can be stated. Home-burners, which in 1938 accounted for about one-fifth of the fuel oil utilized in the United States (Table 27), can be largely disregarded in other parts of the world, because of mild climatic conditions or because home heating by fuel oil is less common than in the United States. Except in the U.S.S.R., where, early in 1939, agriculture was reported to consume 80% of the fuel oil,[43] agricultural utilization was probably small in the countries where fuel oil was the most used product. In most of them, therefore, the supplying of power for industry and for transportation—railroads, diesel-burning trucks and buses, and ships—must have been the chief work of fuel oil; but whether industry or transportation took the greater quantity is not easy to judge. The industrial development of most of these countries is minor, but it cannot therefore be assumed that their industries necessarily used only a small proportion of their total fuel oil consumption; even a relatively small quantity of fuel oil used for industrial power can form a significant percentage of the total demand for the product when the latter is also relatively small.

It seems reasonable, therefore, to conclude that the furnishing of power for modern rapid transportation facilities is the most significant function of petroleum, except in the U.S.S.R. and in the predominantly kerosene-using countries of eastern Europe and southeastern Asia. In the case of most countries we do not have statistics to prove this. In the United States, however, relatively accurate figures are available. These, summarized for 1938 in Table 27, indicate that more than half of the petroleum consumed in that year was used in transportation, and that motor vehicles were the largest single user.[44] Class I steam railroads derived about 17% of their energy requirements from petroleum.[45] In addition, 73% of the tonnage of the United States merchant fleet (and 55% of the world's) was powered by petroleum in 1938.[46] It has been estimated that, for all the kinds of transportation in the United States in 1938, about 60% of the energy was supplied by gasoline or fuel oil.[47]

[41] *Op. cit.,* p. 394.

[42] *Op. cit.,* pp. 288, 225, 412, and 112 respectively.

[43] Lazar Volin, "Machine-Tractor Stations in the Soviet Union," *Foreign Agriculture,* Vol. 12, No. 4, April 1948, pp. 80–86 (ref. on p. 85), quoting K. Chebatarev. Of all petroleum products consumed in the Soviet Union at that time, agriculture took 60%. This included, in addition to the fuel oil, more than 60% of the kerosene supply.

[44] *Minerals Yearbook, 1939,* p. 977.

[45] Computed from data in the *Statistical Abstract of the United States, 1942,* p. 499.

[46] Computed from data in *Petroleum Facts and Figures,* 7th ed., 1941, p. 56.

[47] *Energy Resources and National Policy,* National Resources Committee, 1939, p. 20.

4. A STATISTICAL SURVEY

In spite of the incomplete data, certain characteristics of petroleum utilization in 1938 are clearly recognizable. (1) National averages of per-capita usage varied over an enormous range, from 0.007 barrel in Liberia to 8.47 barrels in the United States (omitting bunker fuel for vessels in foreign trade). (2) Wherever per capita consumption was high, either motor fuel or fuel oil was the product used in greatest quantity. (3) The chief function of these two classes of product was to furnish power for transportation.

Table 27. Consumption of petroleum products in the United States, 1938.

	Motor Fuel	Fuel Oils	Kerosene	Lubricants	Miscellaneous	All Petroleum Products
			(In percentages)			
Distribution by product	46.0	36.0	4.9	1.9	11.2	100.0
Distribution by use						
Transportation	90.0	35.3	—	49.8	—	55.0
House Heating	—	20.9	54.7	—	—	10.2
Industrial Heat & Power	—	41.0	—	50.2	—	15.8
Miscellaneous	10.0	2.8	45.3	—	100.0	19.0
	100.0	100.0	100.0	100.0	100.0	100.0

Source: Computed from data on consumption published by U.S. Bureau of Mines, in *Minerals Yearbook 1939,* p. 977 (motor fuel); *Ibid., 1941,* p. 1093 (fuel oils); *Ibid., Review of 1940,* p. 995 (kerosene), and p. 1009 (lubricants).

How long had this situation persisted? As our survey has shown, 1931 is the earliest year for which a world-wide compilation of consumption by countries (and that based in part on estimates) has been published. From production and trade figures, however, the utilization of petroleum by countries can be calculated back to 1919. For earlier years, world utilization can be estimated by assuming that each year it was approximately equal to world production. Estimates of the latter have been made by the U.S. Geological Survey and by the U.S. Bureau of Mines [48] for each year since 1857. These figures, together with those for the United States (computed by the production plus imports minus exports method) are given in Table 28, averaged by decades.

While the United States has always been far ahead in per-capita consumption of petroleum, at times the difference has been less than now. During the 1860's the United States, using nearly half of the world production, had a per-capita utilization roughly 25 times as great as the rest of the world's, which is about the same ratio as now. In the 1870's and 1880's, however, because of increasing production elsewhere and increasing exports from the United States, the ratio decreased; in the last decade of the century the United States, using some 30% of world production,

[48] See *Mineral Resources of the United States, 1930,* part 2, pp. 824, 825, for a summary through 1930; and yearly volumes of *Mineral Resources of the United States* and *Minerals Yearbook* thereafter.

had a per-capita consumption approximately 10 times that of the rest of the world.

Table 28. Comparative growth of production and consumption of petroleum in the United States and in the rest of the world, 1857 through 1948.

	Yearly Averages for the Periods								
	1857–1863	1864–1870	1871–1880	1881–1890	1891–1900	1901–1910	1911–1920	1921–1930	1931–1940
					(In million barrels)				
Production									
U.S.A.	1.18	3.53	12.52	29.25	55.30	137.42	305.20	771.86	1,065.79
Rest of World	0.05	0.34	1.90	16.35	55.34	99.19	157.84	358.54	668.63
Total	1.23	3.87	14.42	45.60	110.64	236.61	463.04	1,130.40	1,734.42
Consumption									
U.S.A.[a]	. .	1.79	5.96	15.45	34.04	107.80	280.39	756.64	999.76
Rest of World	. .	2.08	8.46	30.15	76.60	128.81	182.65	373.76	734.66
Total	. .	3.87	14.42	45.60	110.64	236.61	463.04	1,130.40	1,734.42
Per-Capita Consumption					(In barrels)				
U.S.A. only	. .	0.05	0.13	0.27	0.49	1.28	2.84	6.62	7.86

	Yearly Totals for the Years		
	1938	1947	1948
		(In million barrels)	
Production			
U.S.A.	1,267.47	1,988.80	2,161.32[b]
Rest of World	720.57	1,033.23	1,243.25[b]
Total	1,988.04	3,022.03	3,404.57[b]
Consumption			
U.S.A.	1,137.12	1,988.97	2,106.54[b]
Rest of World	850.92	1,033.06	1,298.03[b]
Total	1,988.04	3,022.03	3,404.57[b]
Per-Capita Consumption		(In barrels)	
U.S.A. only	8.76	13.68	14.41[b]

[a] Averages for 1864–1900 derived from data on production and exports; averages for 1901–1940 from data on production, imports, and exports. Figures for 1938, 1947, and 1948 include allowances for changes in primary stocks, and the production figures for these years include natural gasoline and benzol, which are omitted in the other years.

[b] Provisional figures.

Sources: Production figures through 1930 from *Mineral Resources of the United States, 1930*, Part II, pp. 824, 825; production figures 1931 to 1940, and data on petroleum imports and exports and on population, from various issues of the *Statistical Abstract of the United States*, except for data on petroleum exports 1864–1870 from *U.S. 10th Census, 1880*, Vol. 10, p. 278; all data for 1938, 1947, and 1948 from DeGolyer and MacNaughton, *Twentieth Century Petroleum Statistics, 1949*.

Up to the turn of the century kerosene had been the most important refined product, constituting 75% of the refinery yields in the United States in 1880; 66% in 1889; and 58% in 1899.[49] During the next decade, as the rapidly growing use of automobiles in the United States increased

[49] *Petroleum Facts and Figures*, 5th ed., 1937, p. 116.

the demand for gasoline, total production rose sharply, and a much smaller percentage was exported. As a result the United States, using more than three times as much as it had in the previous decade, took about 45% of the world total; and per capita utilization was more than 15 times as great as the rest of the world's. This was the beginning of the present era of large-scale consumption of petroleum, with the United States using well over half of the world total, and 20 to 30 times as much per capita as the average of the rest of the world.

In the twentieth century, United States petroleum production has in general exceeded the rapidly growing domestic demand. In recent years, however, this excess has been relatively small, and in 1947, for the first time since the depression years of 1931 and 1932, domestic consumption exceeded production (see Table 28). This slight deficit was, however, provided for by a withdrawal from stocks, and exports remained a little larger than imports. It was not until 1948 that the United States became a net importer, although in that year, because of an addition to stocks, consumption did not quite equal production.[50]

Information on postwar petroleum utilization is still incomplete. The world total for 1948, however, has been estimated at 3,446,988,000 barrels, of which the United States used 2,106,537,000 barrels,[51] or about 61%. Apparently the pattern of distribution is essentially the same as in 1938, with the United States using about three-fifths of the total. Forecasts indicate that during the next decade the utilization of petroleum by the rest of the world will increase about twice as rapidly as will that by the United States.[52] Even so, the United States will be using between 50 and 60% of the world total.

Although, in terms of relative amounts used in different parts of the world, the pattern has remained the same for several decades, and will probably not change appreciably in the near future, the weighting of demand among the various products is undergoing a change, as is shown in Table 29. In parts of Europe there has been a drastic, and probably temporary, reduction in the actual and relative importance of motor fuel

[50] Figures on production, consumption, trade, and stocks for these years are conveniently tabulated in *Twentieth Century Petroleum Statistics, 1949*.

[51] Computed from figures for daily demand given in *World Oil, Review-Forecast Issue,* February 15, 1949, pp. 47, 60.

[52] See, for example, *World Oil,* February, 1948, p. 268, where Dr. Joseph E. Pogue, economist for the Chase National Bank, is quoted as forecasting that from 1947 to 1956 there would be an expansion of 37 per cent in United States consumption and of 76 per cent in consumption by the rest of the world. Elsewhere in this same publication (p. 263), the yearly petroleum demand is forecast through 1951, assuming increases of 5 per cent per annum for the United States and 10 per cent per annum for the rest of the world. The forecasts made by the Committee for European Economic Co-operation of the 1951 demand by the co-operating countries of Europe assume a slightly higher rate of increase, 12 to 13 per cent per year (see Table XXIII, p. 161, of the United Nations publication, *A Survey of the Economic Situation and Prospects of Europe,* prepared by the Research and Planning Division of the Economic Commission for Europe, Geneva, 1948).

—partly because of rationing—and an increase in the relative importance of the fuel oils, especially of residual fuel oil, because of large-scale conversion from coal to oil for heat and power, and because of the planned mechanization of agriculture.[53] These developments are features of the European Recovery Program, and a continuation of this trend is forecast through 1951. Whether the leadership of fuel oil will be maintained, at

Table 29. Annual requirements of petroleum products in the United States and in the C.E.E.C. countries, 1938 and 1947.

	Motor Fuel	Distillate Fuel Oils	Residual Fuel Oils	Other Products and Crude	All Oils
United States [a]			(In million barrels)		
1938	523.0	117.4	291.8	204.9	1,137.1
1947	794.8	298.2	518.4	377.6	1,989.0
C.E.E.C. Countries [b]			(In thousand metric tons)		
1938	14,793	6,198	9,068	6,165	36,224
1947	12,407	9,792	17,997	7,257	47,453
United States [a]			(In percentages)		
1938	46.0	10.3	25.7	18.0	100.0
1947	39.9	14.9	26.1	19.1	100.0
C.E.E.C. Countries [b]					
1938	40.8	17.1	25.0	17.1	100.0
1947	26.1	20.6	37.9	15.4	100.0

[a] Domestic demand.

[b] Austria, Belgium, Denmark, France, Greece, Iceland, Ireland, Italy, Luxembourg, Netherlands, Norway, Portugal, Sweden, Switzerland, Turkey, United Kingdom, Western Germany; and dependent overseas territories of the above.

Sources: Actual requirements in the United States for 1938 from U.S. Bureau of Mines, *Minerals Yearbook 1940,* p. 946, and for 1947 from *Idem, Monthly Petroleum Statements 286–297* (February 1947 through January 1948). Actual requirements for C.E.E.C. Countries from U.N. Economic Commission for Europe, *A Survey of the Economic Situation and Prospects of Europe* (Geneva, 1948), p. 161. Percentages computed from the quantities.

least to as great a degree, after the economic recovery of Europe has been accomplished is doubtful. However, a slight change in the same direction has taken place in the United States, where the utilization of distillate and residual fuel oils together now almost equals (as in 1946 and 1948) or slightly exceeds (as in 1942, 1945, and 1947) that of motor fuel.[54] While the latter is still the leading single petroleum product, it accounts for a smaller percentage of total consumption than it did a decade ago. Apparently paralleling the great increase in motor-fuel demand, there has been and may continue to be a proportionately greater increase in demand for the fuel oils, once the little-desired by-products of motor fuel.

[53] U.N. Economic Commission for Europe, *op. cit.,* p. 160.

[54] DeGolyer and MacNaughton, *op. cit.,* p. 47.

4. A STATISTICAL SURVEY

Note on the Method of Computation Used in the Preparation of Fig. 59, "Intensity and Type of Petroleum Utilization by Country in 1938."

The data on annual consumption of all petroleum products and of the four principal types (motor fuel, fuel oils, kerosene, and lubricants) by each political unit in 1938 were taken wherever possible from the Garfias and Whetsel compilation in the *A.I.M.M.E. Transactions, 1939,* Vol. 132, cited in footnote 27, above. Where these were incomplete, consumption rates were computed from data on production and trade in the [British] *Imperial Institute's Statistical Summary of the Mineral Industry . . . 1936–1938,* cited in footnote 23. Bunker fuel was eliminated whenever possible, because it is not used in the country where it is sold. Distillate and residual fuel oils were treated as one, because they are so listed in the principal source.

From these sources consumption data for 105 political units in 1938 were obtained. The quantities used in each political unit were divided by the respective populations given in the *League of Nations Statistical Yearbook 1939–40,* to yield the per capita ratios.

The classification of the amounts of per capita use of all petroleum products into relatively large, moderate, and small was made after inspecting them arranged in order of magnitude. It then appeared that in about two-tenths of the cases the amounts exceeded 0.80 barrel (33½ gallons) of petroleum products used per person, and this was regarded as constituting relatively large consumption. About three-tenths of the cases fell within the range from 0.79 barrel (33 gallons) to 0.22 barrel (9¼ gallons) per person, and this was regarded as moderate consumption. In some five-tenths of the cases 0.21 barrel (9 gallons) or less was used per person, and that was regarded as the range of small general consumption.

5. THE EFFECT OF THE WORLD DISTRIBUTION OF PETROLEUM ON THE POWER AND POLICY OF NATIONS

BY HERBERT FEIS *

IN THE preceding chapters an assembly of experts have given an account of the location and development of the world's oil resources. These facts of geography, physical and human, are of political consequence. They figure in the position and conduct of each national state, and are among the matters which govern the dealings of national states with each other. Conversely, politics, national and international, determine the meaning of the facts revealed by geography.

This chapter is a brief note upon these crisscross connections. My aim is not to give a systematically complete analysis of the subject, but rather to seek, in the face of existing confusion and tension among nations, for policies which would turn the work of the geographer and geologist to lasting benefit. It will be understood that I am expressing only my own ideas and views.

The Importance of Oil

To a student of politics, the outstanding feature of oil geography is that, while the demand for oil is world-wide, the great sources of supply are few and separated.

Not so long ago a sufficient and regular supply of oil was vital to only a few countries—those which used much machinery in production and transport. Now, as oil is wanted for more and more purposes, it is becoming so to all nations. All are in large measure dependent upon it as a source of energy for power plants, factories, and farms; as fuel for trucks, ships, and trains; as lubricants for moving parts; and, in poorer parts of the world, for light. In many of these uses, and at many of the points of use, there is no good available substitute. If the supply fails, work is greatly hindered, and many parts of many countries suffer idleness, cold, hunger, and darkness.

Oil provides a better means for penetration throughout the world, for the quick development of previously unused areas and resources. Oil-fueled planes, trucks, and machines now enable men to travel to and live in many places previously out of reach, and to utilize resources previously inaccessible or unworkable. Thus, to cite a few familiar recent instances, the availability of oil has made it possible and economically practicable to transform the formerly barren Rio Grande valley into a

* Adviser on International Economic Affairs, U.S. Department of State.

most fertile farming country, to find and develop new mineral deposits and oil fields in the very cold areas of Canada, and to open up for farming and lumbering valleys in the Andes. In such ways an ample and available supply of oil changes economic geography, and, even more notably, military geography and strategy.

Its importance in war has become self-evident. Foreknowledge that the amount of oil supply could mean the difference between victory and defeat figured constantly in the diplomatic struggles which preceded United States' entry into the war. Japan hesitated for a long time to risk war, because it was uncertain whether it would have enough oil. Doubts about that vital point were among the restraints upon its restless wish for expansion. The American-British-Dutch embargo imposed in 1941 compelled a choice between retreat and quick attack, before carefully accumulated reserves were greatly reduced.

When war came, the possession of adequate supplies gave the United States and the Allies a great advantage over the Axis. Sufficiency of supply and superior means and systems of distribution made it possible for our side to secure the raw materials needed to make weapons from any and all sources, and also to send the weapons wherever they might best be used. Industries producing for war in the Allied countries could operate without interruption. We could supply our fighting forces with oil wherever we wanted them to be, and this made it possible for the Allies to choose and carry out the most favorable strategy.

Adequate oil supply made it possible for Britain to keep control of the seas, and for us later to secure the control of the Pacific; to fuel Britain's decisive battle in the air; to wage the telling bombing campaigns against Germany and Japan; to keep large British tank and plane forces in Africa and to conduct our own campaigns there; to keep China in the war; and to carry out the invasion of the western front.

In contrast, shortage of oil and of the means of transporting it greatly impeded Germany, Italy, and Japan. It caused interruptions in their war production, and it caused them to resort to inefficient production methods. It compelled them to economize in their training for combat aviation, in their tank operations, and in their naval actions.

Scarcity of oil also influenced both their military strategy and their operations. For example, one reason Germany persisted in its drive towards the Caucasus, despite the growing risk and burden, was the hope of getting more oil. Similarly, Japan dispersed its naval and land forces unwisely in order to get oil-producing spots in the Southwest Pacific. Reluctance to drain limited supplies influenced the German decision not to risk greater armed forces in the battles, first for the Suez Canal, and then later for Tunis. The wish to save oil sometimes kept the German warships in port instead of in the Atlantic. It affected the disposition of German fighting strength on both east and west fronts, hindering military movements that might have been advantageous.

In still another way, during this war period, oil showed itself to be an important factor in events. Ability to supply oil was an attraction; ability to deny oil won respect. Thus, the line-up of our friends and enemies was affected. For example, our ability to supply oil, and the enemies' failure to do so, influenced the Spanish decision to stay out of war in the dark months of 1940 when Axis victory seemed all but sure. That decision saved for us the Straits of Gibraltar, the entry into the Mediterranean, the use of Malta, and the later chance to land in North Africa.

The world has absorbed these lessons. Hence, each country now intently wishes to have an assured prospect of enough oil at all times and for all uses. The nations will continue to do so unless and until new types of fuel displace oil, or new weapons turn out to be so quickly and utterly destructive that nothing else counts.

The Contrasting Distribution of Supply

In contrast to the universal need for oil, the major sources of supply, as shown in the preceding chapters, are few and widely scattered.

The major sources are the United States, the Caribbean, the Soviet Union, and the fields of the Middle East. In 1949 about 95% of all the natural crude oil produced came from them. Currently, Canada is also emerging as a source of some importance; and production in the southwest Pacific is also likely to be increased before long. The chief change to be anticipated during, say, the next ten years, if there is no great war, is a growth in the relative importance of the Middle East and Canada.

It is not intended to suggest that the total potential world supply is limited to these well-known sources. Large new areas of supply will be located, new sections within presently known fields, and extensions of them, as well as new separate sources. In this connection it should be noted that, as total world consumption increases, the relative importance of any particular discovery declines. Formerly, the emergence of a single large field could quickly and greatly change the whole supply position; now this is less possible.

The present impressive fact—the fact which governments must take into account—is that the great regions of supply are few. And almost all the oil which countries will need for peace or war in the years immediately ahead must come from them, and nowhere else. Synthetic oil products can provide an important supplement or reserve; but very few countries have the natural resources and the capital necessary to establish a large synthetic industry.

Another way of appreciating the situation is to note how few countries produce within their own boundaries the oil that they need. There are the United States, which has done so up to now, but is beginning to import; possibly Canada after a while; Mexico, and a few other countries of the Caribbean, particularly Colombia and Venezuela; the Soviet Union, barely up to the present; Rumania; possibly Hungary and Austria; some few of the Arab states and Iran in the Middle East; Indonesia; and British Borneo. All other countries must draw all or most of their supplies from foreign sources, usually distant and overseas.

Some countries import mostly crude oil, and refine it themselves. Others import mainly products that have been refined elsewhere. Many are striving to enlarge both the total size of the refining industry within their borders and the range of grades and products refined in it. But the prospect is that many or most will remain dependent on imports, not only for much of their supply of crude, but also for important types of refined products, particularly for high-grade lubricants and motor fuels such as are used in aviation.

Leaving aside the fields of the Soviet Union and those within countries under its influence (Rumania, Hungary, and Poland), American and British companies (along with the Dutch) own the rights to develop all the great sources of supply. There are some smaller sources that have been reserved for owners of local nationality, as in Mexico, the Argentine, and Bolivia. There are others where ownership is divided between local and American interests, as in Canada, or where a share of the product goes to interests of other nationalities: France, for example, is entitled to one-fourth of the production of the Iraq Petroleum Company, and the Colombian government is entitled to a share of the production within that country. But at least 80% of all the oil produced outside of the Russian zone of control is supplied by enterprises owned by American or British (plus Dutch) groups. They are the great purveyors of oil for the world.

In past periods of peace countries have had no trouble in securing all the oil they could pay for. The international oil industry usually has sought new customers and markets, in the confident belief that all the oil that could be sold would be found. But the quantities demanded have increased remarkably rapidly.

So has the trade in oil. This has grown not only immensely in itself, but also in relation to the world's total commodity trade.

The International Prospect

These features of the situation lead to a few simple but vital conclusions. (1) The location of the great sources of supply means that most countries can get enough only through trade; (2) since they will want more and more, they will need to acquire an increasing sum of foreign funds for payment; (3) this they will be able to get only by increased exports of other products. (4) Thus, if the need for oil is to be satisfied, even in peacetime, general productivity and trade throughout the world will have to grow. (5) If it fails to grow, and if many countries find themselves unable to pay for the oil they need, their resentment is likely to be directed against the British and American companies, which control so large a part of the supply, and sell it only for dollars or pounds sterling. The British and American companies' position in the world's oil business will become vulnerable, unless all the other important branches of trade thrive as well.

One favorable point in the prospect is that much of the oil that countries import contributes, if properly used, to their greater productivity, their national income, and their ability to export. In these ways it usually tends ultimately to create its own means of payment. Thus, under favorable circumstances, there is a self-sustaining rhythm of growth in the oil trade.

But, in wartime, calculations of sale and purchase cease to govern oil's international movement. The supply available to any country is much affected by its location with reference to the areas of supply and the areas of warfare, and is subject to the will of belligerents who control sources of supply and trade routes. Therefore, as long as there is a serious chance that war will come again, command over oil supplies will be among the important aims of national diplomacy and military planning.

The Struggle for Control of Oil Resources

This being so, it is not surprising that competitive national efforts to acquire control of oil resources have resulted in some serious quarrels. What is surprising, rather, and to a degree requiring explanation, is that the disputes have not been more frequent and harsh.

The explanation lies in a variety of historical and economic circumstances. These it is useful to understand; for, as they change, as all such circumstances do, the points and prospects of conflict over oil will change, and any program devised to avoid or regulate conflict will have to be revised.

First among these circumstances is the geographical fact that three of the five great oil regions have been within the sphere of American or British (conjoined with Dutch) diplomatic control or influence—in the United States, the Caribbean, and the Southwest Pacific. Another great oil region, that within the Soviet Union, has been and remains in a sense a local field. Its outflow has been needed within the country, and, during the past twenty years, has been kept there. The fifth great area of supply, the Middle East, has long been and remains at the crossroads of dispute.

In the past the United States has been the greatest supplier, and the countries of the Caribbean have been next in order. There have been few serious diplomatic contests over opportunities in these areas. Geography, business economics, and technical knowledge, rather than international politics, have decided the disposition of control. Because of their financial resources, ability to provide markets, experience, and willingness to take risks, the American companies were in the main able to out-bid and out-perform others. True, some episodes occurred, not without importance at the time; for example, the rivalry between American and British interests in Mexico during the Diaz regime; and a few attempts by German and Japanese interests to secure rights in the same country before the outbreak of World War II.

The situation was roughly similar in regard to the oil resources of the Southwest Pacific, until Japan entered it. The discovered sources were under Dutch and British control, but, at first grudgingly and later more freely, place was granted to American interests offering attractive terms. It was on this area of supply, however, that Japan centered its hopes of independence in oil. Determined attempts were made by that country to secure concessions, which were just as persistently denied. The Dutch feared that if Japanese interests gained entry into large land areas, they would use the chance to disturb Dutch control over the colonies, to prepare the way for invasion. This denial of oil concessions was one of the reasons by which Japan sought to justify its resort to war.

Such was the diplomatic situation that in the past kept within bounds the international contest for control of the great sources of oil supply other than the Middle East. In that region historical events warded off serious trouble before World War II. Germany was eliminated as a contestant by its defeat in 1918, and after the Bolshevik Revolution, the new Soviet government withdrew Russia's claims.

Before 1914 German oil companies, supported by the centralized banking syndicates of Imperial Germany, had claim to an important share of the oil resources of Mesopotamia, and, with the vigorous support of the German government, they were reaching for more. Germany was planning to establish a continuous system of land transport to the Persian Gulf, at the same time as it was engaged in a program of naval expansion designed to challenge British control of the sea routes to and from the Middle East. Defeat disposed of these designs.

Imperial Russia held oil rights in Iran, and might have entered the contest over the resources of the other Arab lands. But the early revolutionists renounced the purpose, and only recently has it been taken up again. The elimination of Germany and the Soviet Union left Great Britain and the United States the only serious rivals for the oil of the Middle East during the inter-war period. The needs of France, the other eager seeker, were met by conceding to her a share in one of the major enterprises, the Iraq Petroleum Company.

A second important circumstance that lessened disputes over control of oil reserves was the long prevailing British control of the seas, which gradually has been shared with the United States. Other governments, weighing the advisability of diplomatic effort and financial outlay to acquire rights in foreign sources of oil supply, customarily take into account the chance of having access to this oil in wartime. Up to now the only two countries who thought they could count upon getting oil from overseas in wartime were the British Empire and the United States; and the British estimate turned out, it may be noted, to be wrong in some respects. In both wars Britain was forced to rely mainly on oil from the Western Hemisphere. The plans of other countries were laid along lines that, it was thought, would circumvent this Anglo-American control of the seas. Germany, as remarked, planned to get its oil overland. Japan counted upon its power to dominate the sea routes to the oil regions of the Southwest Pacific.

In the future the factor of control of the seas will be of much less importance. The emergence of new weapons—the great bombing planes, the guided missiles, and the vastly more powerful kinds of explosives, especially atomic and incendiary bombs—will greatly change the geography of power in the event of war. To risk a bold surmise, it seems probable that the existence of these weapons will further deter all but the few most powerful countries from seeking control over distant sources of oil supply.

Still a third circumstance that explains the past pattern is the economic dominance of American and British oil enterprise. They have had greater capital, experience, and technical skill in the search for and extraction of oil than their competitors. They could usually offer greater financial inducements and more assured markets. In addition, American oil enterprise has benefited by the belief that the United States was not seeking control over foreign peoples or governments. It was sometimes the more welcome for that reason; for example, in Saudi Arabia.

Rivalries of the kind discussed up to this point, it should be noted, have not been the only cause of international disputes over oil. With the exception of the United States and the Soviet Union, all countries whose lands contain much oil have found it necessary to rely, wholly or partly, upon foreign enterprise for development. Most of them have accepted that necessity, and tried to turn it to advantage. But many have disliked it, and some have preferred to exclude all foreign interests, even though their national efforts to find and extract oil could not match what foreign interests offered. Their right so to decide has been recognized; while provoking criticism, it has not been a serious cause of controversy. But sharp disputes have occurred when foreign interests have been invited to search for and develop oil, and then, after discovery, have either been expelled or gravely hindered. The Soviet Union disposed of such a dispute ruthlessly, by confiscation without payment. The private interests concerned, and the governments to which they looked for protection, decided they could do nothing about it. But the resentment persists. Mexico and Bolivia expropriated foreign oil interests within their territories. In time they reached a settlement; but ill feeling has lingered, and oil development in these two countries has been almost halted ever since.

The large international oil companies, American and British with an admixture of Dutch and French, have in general adjusted their operations with skill and flexibility to these facts of geography, politics, and economics. They have kept their place, both in total oil production and trade, to a greater extent than might have been anticipated. In the main they have succeeded in adjusting their relations smoothly with the governments of the producing areas. They appear to have learned to avoid excessive competition with one another for new opportunities. They have made huge profits, but they have used a great part of them to extend their operations within the countries where they have been earned.

Until World War II, all countries during peacetime could get all the oil products they could pay for, and at prices that tended to decline. Oil, as fuel, lubricant, and illuminant, became cheaper among the world's products, and its use steadily expanded. The economic achievement was, until the advent of the war, only slightly retarded by the frictions of international politics. But will this continue in the future?

The Problem in the Middle East

Anxiety hovers particularly over one of the great producing areas, the Middle East. On that most rich supply the world expects to draw more and more. It is an enormous economic stake, which will yield great wealth to its proprietors, and a great military stake to nations if they are at war. It lies under the territory of a number of relatively small, weak, and backward Arab states and protectorates: Iran, Iraq, Saudi Arabia, Bahrein, and Kuwait contain the large proven fields, but there are many other good locations.

Throughout the whole area politics and oil are mixed together in a dangerous whirl; politics of the great foreign states and of the small local ones. Fear exists that the Soviet Union may seek to gain control, first over Turkey and Iran, then further south along the Persian Gulf. Britain retains an important military base in Jordan, and seems determined to try to maintain its influence among the Arab rulers. The United States has committed itself to oppose the extension of Russian influence, and shown a will to preserve, against all forms of attack, American oil concessions throughout the area. At any time these national rivalries and suspicions among the great powers could result in an armed clash between them. This might arise over some political or strategic issue, or come about as a direct dispute over the control of the oil resources.

Then, too, between and within the local oil-containing Arab states there are many quarrels. During the struggle of the newly arisen state of Israel for independence, there was fear that the Arab states might cancel the concessions owned by American or British oil enterprises if their governments consented to the partition of Palestine. That fear has shown itself unfounded; but until peace within Palestine is firmly established, boundaries drawn, and decisions reached as to who shall govern the parts of Palestine remaining to the Arabs, the oil fields will remain subject to disturbance. Equally unfounded, I believe, is the fear that, in the event of a war with the Soviet Union, the Arab states might willingly turn towards the U.S.S.R., and transfer control of the oil resources. But the chance always exists, and may grow greater, that local anger against some oppressive government, or a revolutionary plot, or an economic depression will bring upsetting changes in one or several of the Arab states.

Such are some of the main items in the catalogue of possible troubles which must be avoided or wisely adjusted if the oil resources of the Middle East are to be developed without dangerous mischance. They create a worrisome political situation, in which the foreign oil interests concerned would do well to avoid upsetting each other. Until recently there was reason to fear that the great American and British oil companies which had, or desired to obtain, a place in the region might so upset each other that their rivalry for concessions, pipe-line privileges, and markets might stimulate political disorder. It may still do so at some point or other, at some time or other. But in a series of recent agreements they appear to have reached an adjustment of interests, which apparently consists in some ways of a division of opportunity and in other ways of a partnership. These arrangements merit support, as long as they do not operate as monopoly or exclude other qualified participants. They lessen one cause of anxiety, but leave other stubborn ones.

Among them is the Russian quest for a concession in northern Iran, which seems associated with a Russian design to acquire domination over part or all of that country. It is impossible to predict what means may be employed at some later time to advance either the acquisition of this concession, or the establishment of other Russian influence. This is only one of the points where the shadow of possible Russian attack rests over the existing arrangements for the oil development of the area. No one knows where or when the shadow may appear next. It compels constant and alert attention, and is another reason why those entrusted with the concessions must give no just cause of discontent. The anxiety will remain unless and until relations between the Soviet Union and the United States become tranquillized. In that event, objections to Russian entry as concession-seeker and oil-producer in the Arab lands of the Middle East would fade. But, meanwhile, the tension will continue; the retarding element of risk will be present; the work of production will lag behind the economic opportunity.

Proposals for an International Oil Policy

The record of discovery encourages the belief that the oil resources of the world are ample to provide all countries with this means of bettering their material condition of life. Will they do so? Or will selfishly national economic policies, antagonisms, and wars (perhaps, themselves, centered on oil) reduce and waste them? An attempt to sketch out the lines of a beneficial policy is in order. The exertions of the geographer and the geologist, of the engineer and the driller, call for an effort to formulate the terms of agreement which will turn their work to best account.

But the task is not a simple one. Prescription of what is desirable must be governed by measurement of what is achievable. Suggestions of policy cannot be based on geographic and economic facts alone; they can be useful only if the many disturbing features of the social, political, and international life of the day are taken fully into account.

Although these have been mentioned above in one form or another, it is well to reiterate them. First, that almost all countries have a bent toward becoming self-sufficient in their oil supplies. Second, that trade between nations is so severely controlled that many countries find it difficult or impossible to get the means of paying for all the oil they need. Third, that some countries are determined that all branches of oil production in their lands shall be carried on only by nationals of the country, and exclude all foreign interests. Fourth, that some countries reserve all chance to develop oil resources to the government, and exclude all private interests, foreign or domestic. Fifth, that there is the recurrent rivalry between the oil interests of different countries, sometimes supported by their governments, to secure control over sources of supply. Sixth, that there is the world-wide struggle between the Soviet Union and the countries devoted either to the traditional forms of democracy and private enterprise, or to democratically controlled socialized enterprise. This now reaches close not only to the oil regions of the Middle East, as noted, but also to those of the Southwest Pacific, and could extend itself to this hemisphere.

Bearing in mind, then, all these discordant features of the contemporary world, what suggestions regarding international oil policy may be offered?

The rules that could best serve to satisfy peacetime and peaceful needs are simple and few: production should be encouraged; supplies should be made ample; the product should become available to all nations on equal and fair terms. To bring about these results governments are called upon to encourage the search for oil both within and without their national territories. It would be well also if all could be persuaded to grant qualified foreign interests fair opportunity, and to provide reasonable security of operation. Experience shows that if there is mutual trust and experienced judgment, the urge for profit is reconcilable with the welfare and independence of the countries in whose domains the oil is found.

Most governments would also be well advised not to interfere with the import or export of oil, except for good reasons of national development, security, or solvency. It must be recognized that national governments may often be justified, if their judgment of actual circumstances is valid, in using their trade-regulatory powers for several purposes: (*a*) to foster the establishment of a refining industry; (*b*) to encourage development of promising domestic sources of supply; (*c*) to raise revenues (though this source of revenue is being excessively used in some countries, to the detriment of their economic life); (*d*) to economize in the use of scarce foreign currencies.

Further, if countries, whether as sellers or buyers of oil, want the supply of oil to be ample and cheap, they would be well advised to abstain from all but clearly necessary hindrances to other branches of international trade. If oil-importing countries are to be able to pay for the quantities they need, they will have to be competent and industrious in the production of goods for export, while the countries who receive income from the sale of oil must be ready to admit imports into their markets.

For their part, the companies which do an international oil business, and the governments to which they give primary allegiance, should commit themselves to serve all markets on equal terms in peacetime, to avoid monopolies, and to recognize their obligation to advance the economic development of the countries in which they operate.

These rules are propounded with peacetime purposes in mind. But it will be found, I think, that they will also best serve the wish of countries to have an ample and assured supply in case of war—not all countries, but most of them; and not under all circumstances, but under most.

The more ample the total world supply of oil, and the more numerous and scattered the sources of supply, the better the chance of each country to obtain enough in the event of war. It is to be expected that the initiative of different countries will be directed towards different sources of production. The interest of each will be directed towards sources within its territories, or near its boundaries, or accessible by protected water-haul; and safety from enemy action from the air will, above all else, determine each focus of interest. But the combined result of these separate aims should benefit all, if peace lasts, by enlarging total supply.

Thus the possibility of war would not seem to nullify the general soundness for most countries of the policies outlined above. But in two respects it does: it becomes prudent to oppose efforts of any clearly identified possible enemy to improve its oil-supply position; and it becomes justified to use strong measures to defend one's own sources of supply. General principles become qualified by special dangers, and ordinary methods must be supplemented to serve special purposes.

The adaptation of policy to the danger of war will differ for each country, and be made to fit its own situation. United States' policy during this present phase between peace and possible war might well include the following components:

(*a*) Continuation of the search for new sources of supply of natural crude within the United States.

(*b*) Encouragement of the foreign oil ventures of the American companies. This should be world-wide, but attention should be preferably directed to nearby sources.

The gradual growth of production in the Middle East is also desirable for several reasons: to supplement supplies available to the United States, as during the recent period of near shortage; to lessen the present and prospective rate of draft on resources of this hemisphere; to provide cheaper oil for Europe, Africa, and Asia and the Middle East itself. But it might be well to defer construction of contemplated new pipe lines in the area until several dangerous international tensions are adjusted. In the event of war, the pipe-line equipment might prove of greater use to an enemy than to ourselves.

The most effective forms of encouragement for the oil ventures carried on by American enterprises in foreign lands are well recognized, particularly (1) policies (including our present European reconstruction program) which enable foreign countries to buy oil for dollars, and (2) a moderate United States tariff policy.

(*c*) Proper diplomatic protection of United States oil enterprises abroad.

(*d*) Should it prove necessary, and in such ways as at the time may be judged feasible, armed protection should be given in the event of an attempt to dispossess these United States enterprises by force. Such a contingency is possible in the near future in the Middle East.

This protection should be associated with such guidance of these oil enterprises by the government as may prove necessary, to avoid either just grounds for foreign opposition to them, or actions that might be advantageous to an enemy in the event of war.

(*e*) Willingness to have United States oil interests associate themselves with foreign oil interests for the division of opportunity and of product.

(*f*) Adequate arrangements quickly to increase supply in time of emergency, to transport it where it may be needed, and to protect it. This matter should receive urgent attention, and for these purposes we should set aside reliable reserves (including synthetic plants) at different locations, maintain a reserve of equipment for production and transport, and build safe storage facilities.

Some of these measures are, of course, recommended for military reasons. They are made necessary because of the chance that we may again become engaged in war. It must be recognized that some of them, like all acts of military preparation, may arouse fears and countermeasures by possible enemies, and hence increase the chance that war will come. This cannot wholly be avoided. But it can and should be offset. We must do our utmost to prove in deed and word that we wish, not war, but peace, if it is a roughly just and secure peace; that we will recognize and protect the independence and safety of other countries if they will do the same. Accordingly, while we may take steps to assure ourselves of enough oil in the event of war, we should strive to achieve that purpose in ways that do not antagonize, or give just reason for complaint.

Thus the United States oil companies should operate the properties they control as an international trusteeship; and the United States government should see that they do so. The wish for profit must be disciplined, preferably self-disciplined.

This means that the product must be made freely available on equal and fair terms to all countries for peaceful purposes. It means also that the United States should continue to aid other friendly countries to improve their economic position, and to accord them a good chance to sell their products in foreign markets, and thereby obtain the funds to pay for oil and other essentials. Further, it means that United States oil interests and the United States government should be disposed to share new opportunities in oil production with other peaceful countries.

Oil as an Instrument for Keeping the Peace

The geography of oil suggests one other positive idea: oil, to repeat, has become essential in peace and war. Only a very few countries could long live satisfactorily or fight effectively if deprived of outside sources of supply. In fact only two could possibly do so: the United States and the Soviet Union.

Thus if governments could enter into a reliable pact for keeping the peace (preferably by making the United Nations Organization effective), the joint control of many nations over oil supply would condemn any single aggressor or small combination of aggressors (except possibly the Soviet Union or the United States) to quick and disastrous defeat. In other words, the geographic distribution of oil places in the hands of governments, if ever they learn to work together, a good means for enforcing international judgment against any defiant country, with the two possible exceptions noted.

An examination of the bearing of oil on the prospects of maintaining a peaceful international order comes to the same conclusion as similar inquiry begun from any other point. The United States and the Soviet Union have the power to decide the issue. If the possibility of war between them can be eliminated or greatly reduced, the chances of preserving international peace indefinitely are good. Then the course of oil production or trade could be left to those practices and arrangements that would bring the greatest benefits in peaceful living. The question of who controlled any source of supply, or any part of the total supply, would become just an ordinary economic one. Then, there is reason to be confident, nations would find the terms on which to deal fairly with one another. Then the facts of the geography of oil could be turned to best account, and knowledge of these facts could be fully and solely employed to improve the condition of man.

APPENDIX

WORLD REGIONS: PETROLEUM PRODUCTION AND EXPORTS, 1938 AND 1947

Table 30. World Regions: production of crude petroleum and exports of crude petroleum and of petroleum products, 1938 and 1947.

A. Production of Crude Petroleum and Total Inter-regional Exports

	1938		1947	
	Crude Production	*Total Exports*	*Crude Production*	*Total Exports*
	(In thousand b/d)			
WESTERN HEMISPHERE				
Alaska, Newfoundland	—	—	—	—
Canada	19	—	21	5
United States	3,327	531	5,085	451
Mexico	105	38	154	40
Caribbean, except U.S.A. & Mexico	623	569	1,319	1,140
Other South America [a]	97	23	102	—
Total, Western Hemisphere	4,171	1,161	6,681	1,636
Of which to Eastern Hemisphere		*427*		*947*
EASTERN HEMISPHERE				
Europe, except U.S.S.R.	160	21	133	44
U.S.S.R.	562	23	535 [b]	8
West Africa	—	—	—	—
North Africa	4	6	24	3
South & East Africa	—	—	—	—
Middle East	329	255	837	666
Other Asia [c]	35	16	11 [b]	—
East Indies & Oceania	176	93	58	3
Total, Eastern Hemisphere	1,266	414	1,598	724
Of which to Western Hemisphere		2		*12*

B. Analysis of Exports

	1938		1947	
	Crude	*Products*	*Crude*	*Products*
	(In thousand b/d)			
WESTERN HEMISPHERE				
Alaska & Newfoundland	—	—	—	0.3
To: Canada	—	—	—	0.3
Canada	—	0.2	—	4.5
To: Alaska, Newfoundland	—	0.1	—	2.8
United States	—	0.1	—	1.2
Europe, except U.S.S.R.	—	—	—	0.5
United States	211.3	319.5	126.4	324.2
To: Alaska, Newfoundland	—	3.8	—	7.8
Canada	68.1	16.3	106.4	55.1

	1938		1947	
	Crude	*Products*	*Crude*	*Products*
	(In thousand b/d)			
Mexico	0.4	4.6	0.3	13.5
Caribbean, except U.S.A. & Mexico	0.2	31.0	—	3.9
Other South America a	6.8	34.3	6.8	27.5
Europe, except U.S.S.R.	74.2	147.6	12.9	141.8
U.S.S.R.	—	4.2	—	2.7
West Africa	—	1.8	—	0.9
North Africa	0.3	3.8	—	4.3
South & East Africa	—	3.9	—	3.2
Middle East	—	1.7	—	1.5
Other Asia c	61.3	36.1	—	15.8
East Indies & Oceania	—	30.4	—	46.2
Mexico	13.5	24.4	18.9	21.2
To: Canada	—	—	—	0.6
United States	7.0	1.6	15.3	12.5
Caribbean, except U.S.A. & Mexico	0.2	3.9	—	1.6
Other South America a	—	0.2	—	5.7
Europe, except U.S.S.R.	6.1	17.5	3.6	0.6
North Africa	0.2	—	—	0.2
Other Asia c	—	0.7	—	—
East Indies & Oceania	—	0.5	—	—
Caribbean South America	170.0	398.8	451.2	689.0
To: Alaska, Newfoundland	—	0.5	—	0.5
Canada	21.0	—	84.9	7.7
Mexico	—	—	—	0.6
United States	65.4	74.3	250.9	149.4
Other South America a	0.9	77.3	16.4	176.1
Europe, except U.S.S.R.	81.5	202.9	92.1	306.1
West Africa	—	24.2	—	22.0
North Africa	1.2	15.9	6.9	25.8
South & East Africa	—	3.0	—	0.2
Middle East	—	—	—	0.3
Other Asia c	—	0.7	—	0.2
East Indies & Oceania	—	—	—	0.1
Other South America a	18.9	4.1	—	1.6
To: Canada	3.7	1.2	—	—
Caribbean, except U.S.A. & Mexico	2.8	—	—	1.3
Europe, except U.S.S.R.	12.4	2.4	—	—
Other Asia c	—	0.5	—	—
East Indies & Oceania	—	—	—	0.3
EASTERN HEMISPHERE				
Europe, except U.S.S.R.	0.1	21.2	—	44.5
To: United States	—	0.1	—	0.3
Caribbean, except U.S.A. & Mexico	—	0.1	—	—
Other South America a	—	0.2	—	—
U.S.S.R.	—	—	—	35.8
West Africa	—	0.2	—	0.2
North Africa	0.1	13.8	—	7.3
South & East Africa	—	0.6	—	0.1
Middle East	—	4.9	—	0.4
Other Asia c	—	0.9	—	0.4
East Indies & Oceania	—	0.4	—	—

	1938		1947	
	Crude	*Products*	*Crude*	*Products*
		(In thousand b/d)		
U.S.S.R.	—	22.7	—	8.5
To: Europe	—	16.7	—	8.5
North Africa	—	3.3	—	—
Middle East	—	2.2	—	—
Other Asia [c]	—	0.5	—	—
North Africa	—	6.5	—	3.1
To: Europe, except U.S.S.R.	—	4.4	—	3.1
South & East Africa	—	0.3	—	—
Middle East	—	0.4	—	—
Other Asia [c]	—	0.8	—	—
East Indies & Oceania	—	0.6	—	—
South & East Africa	—	0.3	—	—
To: West Africa	—	0.3	—	—
Middle East	93.5	161.6	124.4	541.2
To: United States	—	—	1.0	6.6
Other South America [a]	—	—	4.5	—
Europe, except U.S.S.R.	87.2	81.7	105.3	171.0
West Africa	—	—	—	2.0
North Africa	5.8	19.7	1.4	40.2
South & East Africa	—	17.2	—	43.0
Other Asia [c]	—	34.2	7.1	179.9
East Indies & Oceania	0.5	8.8	5.1	98.5
Other Asia [c]	3.9	12.6	—	—
To: Europe, except U.S.S.R.	—	1.4	—	—
U.S.S.R.	3.9	—	—	—
North Africa	—	1.5	—	—
South & East Africa	—	1.2	—	—
Middle East	—	1.3	—	—
East Indies & Oceania	—	7.2	—	—
East Indies & Oceania	0.7	92.3	—	2.9
To: United States	—	0.3	—	—
Caribbean, except U.S.A. & Mexico	—	1.7	—	—
Other South America [a]	—	0.1	—	—
Europe, except U.S.S.R.	0.1	14.5	—	—
North Africa	0.6	5.6	—	—
South & East Africa	—	7.4	—	—
Other Asia [c]	—	62.7	—	2.9

[a] Other than Caribbean countries.

[b] Sakhalin included in "Other Asia" in 1938 and in U.S.S.R. in 1947.

[c] Other than Middle East, U.S.S.R., East Indies & Oceania.

Sources: Crude production computed from figures of U.S. Bureau of Mines by countries, cited in *Petroleum Data Book 1947*. Basic data on exports furnished by the Coordination and Economics Department, Standard Oil Company (New Jersey).

BIBLIOGRAPHY

THIS short bibliography, based primarily on the choices of the contributors to this volume, includes a number of works that, for each main topic, will supply the inquirer with more numerous and specialized references. By far the most ample source in one volume is the 730-page *Bibliography on the Petroleum Industry* by E. L. DeGolyer and Harold Vance (Bulletin No. 83, School of Engineering, Texas Engineering Experiment Station, College Station, Tex., 1944); but the reader will find useful references in all the entries below that have the note "bibl." or "footnote refs."

The great majority of the references given are to works in English; however, some dealing with general topics in French, German, and Spanish, as well as works in these languages and in Russian, Rumanian, Dutch, and Portuguese relating to particular regions, have been included.

The topical arrangement follows the sequence of the chapters and their sections in this book; under each topic the titles are arranged in reverse chronological order by year of publication and those appearing in the same year, in alphabetical order by author. In the descriptions of illustrative matter, the abbreviation "ill." or "ills." indicates photographs, "fold maps" indicates insert maps, whether in black and white or in color, "pocket maps" indicates separate maps, "other maps" indicates text maps. Similar indications are used for diagrams.

Periodicals, their titles in italic, are listed on much the same principles as the books and articles. The monthly *Bulletin* of the American Association of Petroleum Geologists (Tulsa, Okla.) is the most frequently quoted periodical source; and the monthly *World Oil,* with its annual Forecast and Review (February) and World Oil Atlas (midsummer) issues is also a necessary and constant reference.

The periodicals listed below do not constitute an exhaustive list, but have been selected on the basis of the contributors' or editors' use of them. (Annual reports of colonial governments and trusteeships, and of oil companies and other commercial enterprises, are not here listed as periodicals.) Lists of official United States serial publications should also be consulted by the inquirer; a number of wartime and other discontinued series, not listed below, may provide exactly what the reader is looking for. The article on "Petroleum Periodicals," by Rocq, M., Nutting, E., and Karpenstein, K., in *Special Libraries* (monthly), N.Y., Vol. 36, No. 8, Oct. 1945, pp. 376–391, is comprehensive.

The Map Department of the American Geographical Society has over seven hundred maps catalogued under "Petroleum"; these may be consulted in the Society's reading room, but lack of space has prevented a listing of them in this volume. The principal published maps used for reference in the preparation of maps for this book are, however, noted in the legends under them.

The abbreviations used in the references given in this bibliography should explain themselves, except A.A.P.G. (American Association of Petroleum Geologists), A.I.M.M.E. (American Institute of Mining and Metallurgical Engineers), and A.P.I. (American Petroleum Institute).

PERIODICALS IN ENGLISH

American Association of Petroleum Geologists: *Bulletin*. Tulsa, Okla. Monthly.
American Institute of Mining and Metallurgical Engineers, Petroleum Branch: *Journal of Petroleum Technology*. N.Y. Monthly.
American Journal of Science. New Haven, Conn. Monthly.
American Petroleum Institute (N.Y.): *Statistical Bulletin*. Weekly. *Quarterly*.
Economic Geology. Society of Economic Geologists, Lancaster, Pa. Monthly.
Geographical Journal. Royal Geographical Society, London. Monthly.
Geographical Review. American Geographical Society, New York. Quarterly.
Geological Magazine. London. Monthly.
Geological Society of America (N.Y.): *Bulletin*. Monthly. *Memoirs*.
Geological Society of London: *Quarterly Journal* (including abstracts of proceedings).
Geophysics. Society of Petroleum Geophysicists, Houston, Tex. Quarterly.
Hebrew University: *Bulletin*. Jerusalem. Monthly.
Independent Petroleum Association of America: *Monthly*. Tulsa, Okla.
Indian Association for the Cultivation of Science: *Special Publications Series*. Calcutta.
Indian Minerals. Geological Survey of India, Calcutta. Quarterly.
Institute of Fuel: *Journal*. London. Monthly.
Institute of Petroleum, London: *Journal*. Monthly. Also, *Annual Review of Petroleum Technology*.
National Petroleum News. Cleveland, Ohio. Weekly.
Nature. London. Weekly.
Oil and Gas Journal. Tulsa, Okla. Weekly.
Petroleum Press Service. London. Monthly.
Petroleum Engineer. Dallas, Tex. Monthly.
Petroleum Times. London. Fortnightly.
Philippines, University of: *Bulletin of Natural and Applied Science*. Manila.
Royal Central Asian Society: *Journal*. London. Monthly.
Scientific Monthly. Washington, D.C.
School of Mines (Golden, Colo.): *Mines Magazine*. Monthly. *Quarterly*.
U.S. Geological Survey: *Bulletin*. Washington, D.C.
World Oil. Houston, Tex. Monthly.
World Petroleum. New York. Monthly.

PERIODICALS IN FRENCH

Chronique des Mines Coloniales. Bureau d'Études Géologiques et Minières Coloniales, Paris. Monthly.
Institut Français du Pétrole: *Révue*. Paris. Monthly.

BIBLIOGRAPHY

PERIODICALS IN GERMAN

Erdöl und Kohle. Hamburg and Berlin. Monthly.
Zeitschrift der Deutschen Geologischen Gesellschaft. Berlin. Quarterly.

PERIODICALS IN PORTUGUESE

Brasil, Ministério da Agricultura, Departamento Nacional da Produção Mineral (Rio de Janeiro): *Avulso* and *Boletim.*
Revista Brasileira de Geografia. Rio de Janeiro. Quarterly.

PERIODICALS IN SPANISH

Boletín de la Secretaría de Industria y Comercio. Buenos Aires.
Boletín de la Sociedad Geológico del Perú. Lima.
Boletín de Minas. Habana, Cuba.
Boletín de Minas y Petróleo. La Paz, Bolivia.
Boletín de Minas y Petróleo. Mexico, D.F.
Boletín de Minas y Petróleo. Santiago do Chile.
Boletín de Minas y Petróleos. Bogotá, Colombia.
Boletín del Cuerpo de Ingenieros de Minas del Perú, Dept. del Petróleo. Lima.
Boletín del Instituto Sudamericano del Petróleo. Montevideo, Uruguay.
Boletín Minero, Sociedad Nacional de Minería. Santiago de Chile.
Boletín Oficial de Minas e Industrias. Lima.
Estudio Tecnico, Instituto Colombiano de Petróleos. Bogotá.
Gaea, Boletín de la Sociedad Argentina de Estudios Geográficos. Buenos Aires.
Instituto de Geología, Anales. Mexico, D.F.
Instituto Geológico y Minero de España, Memorias. Madrid.
Petróleo. Mexico, D.F.
Petróleo del Mundo. New York.
Petróleo Interamericano. Tulsa, Okla.
Revista del Museo de la Plata, Sección Geológica. La Plata, Argentina.
Revista Geográfica Americana. Buenos Aires.
Yacimientos Petrolíferos Fiscales, Memoria correspondiente al año—. Buenos Aires.

PART I. PETROLEUM IN THE GROUND

GENERAL

Lalicker, C. G. *Principles of Petroleum Geology.* N.Y., Appleton-Century-Crofts, 1949. Pp. 377, ills., maps, diagrs., tables, footnote refs.
Structure of Typical American Oilfields. . . . 3 vols. Tulsa, Okla., A.A.P.G., 1929 (Vols. 1 and 2) and 1948 (Vol. 3). Vol. 1, pp. 510; Vol. 2, pp. 780; Vol. 3, pp. 516. Each with fold map and others, diagrs., tables, footnote refs.
Levorsen, A. I. "Trends in Petroleum Geology," *Smithsonian Inst. Ann. Rep., 1942,* pp. 227–234. Washington, Gov't Printing Off., 1943.
Heroy, W. B. "Petroleum Geology," *Geology, 1888–1938,* pp. 511–548. N.Y., Geological Soc. of Amer., 50th Anniversary Vol., 1941. Bibl. Also reprinted in *Smithsonian Inst. Ann. Rep., 1943,* pp. 161–198. Washington, Gov't Printing Off., 1944.
American Petroleum Institute, Division of Production. *Finding and Producing*

Oil: A Comprehensive Manual. Dallas, Tex., A.P.I., 1939. Pp. 338, ills., map, diagrs., tables, bibls.

Dunstan, A. E.; Nash, A. W.; Brooks, B. T.; Tizard, Sir H., Eds. *The Science of Petroleum. . . .* 4 vols. N.Y., London, Oxford Univ. Press, 1938. Vol. I: *Origin and Production . . .* (with 28 regional studies), pp. 836. Vols. II and III: *Chemical and Physical Principles of Refining . . .*, pp. 839–1670 and 1671–2388. Vol. IV: *Utilization . . .*, pp. 2389–3193. Each vol. has ills., maps, diagrs., tables, bibls.

Steinman, G.; Wilckens, O., Eds. *Handbuch der regionalen Geologie.* Issued in 29 parts. Heidelberg, Carl Winters, 1910–1937. Maps, diagrs., tables, bibls.

Wrather, W. E.; Lahee, F. H., Eds. *Problems of Petroleum Geology. . . .* Tulsa, Okla., A.A.P.G.; London, Murby, 1934. Pp. 1073, ills., maps, diagrs., tables, bibls.

Emmons, W. H. *Geology of Petroleum.* N.Y., London, McGraw-Hill, 1931. 2nd ed. Pp. 736, ills., maps, diagrs., bibl.

Redwood, Sir B. *A Treatise on Petroleum.* 3 vols. London, Griffin, 1913. 3rd ed., rev. Vol. I (pp. 367, fold and other maps, diagrs.) contains section on geographic and geologic distribution; Vol. II (pp. 417) on refining; Vol. III (pp. 383) world-wide state and municipal regulations.

GEOLOGICAL PRINCIPLES GOVERNING THE OCCURRENCE OF PETROLEUM

Zobell, C. E. "Bacterial Release of Oil from Oil-bearing Materials," *Fundamental Research on the Occurrence and Recovery of Petroleum,* pp. 168–181. N.Y., A.P.I., 1949. Ills., tables, bibl.

Umbgrove, J. H. F. *The Pulse of the Earth.* The Hague, Martinus Nijhoff, 1947. 2nd ed. Pp. 358, fold and other maps, diagrs., tables, bibls.

Pratt, W. E. "Distribution of Petroleum in the Earth's Crust," *Bull. A.A.P.G.,* Vol. 28, No. 10, Oct. 1944. Pp. 1506–1509.

Sheppard, C. W. "Radio Activity and Petroleum Genesis," *Bull. A.A.P.G.,* Vol. 28, No. 7, July 1944. Pp. 924–952.

Zobell, C. E.; Grant, C. W.; Haas, H. F. "Marine Micro-organisms Which Oxidize Petroleum Hydrocarbons," *Bull. A.A.P.G.,* Vol. 27, No. 9, Sept. 1943. Pp. 1175–1193, tables, bibl.

Sedimentation, Report of a Conference on. Tulsa, Okla., Research Committee, A.A.P.G., 1942. Pp. 68, refs. in text. Mimeo.

Trask, P. D.; Patnode, H. W. *Source Beds of Petroleum: Rept. of Investigation Supported Jointly by Amer. Petroleum Inst. and U.S. Geol. Survey.* Tulsa, Okla., A.A.P.G., 1942. Pp. 566, maps, diagrs., footnote refs.

Levorsen, A. I., Ed. *Stratigraphic Type Oil Fields: A Symposium.* Tulsa, Okla., A.A.P.G., 1941. Pp. 902, ills., fold and other maps, diagrs., tables, bibl.

Origin of Oil, Report of a Conference on. Tulsa, Okla., Research Committee, A.A.P.G., 1941. Pp. 81, bibl. Mimeo.

Trask, P. D., Ed. *Recent Marine Sediments: A Symposium.* Tulsa, Okla., A.A.P.G.; London, Murby, 1939. Pp. 736, maps, diagrs., tables, bibl.

Twenhofel, W. H. *Principles of Sedimentation.* N.Y., London, McGraw-Hill, 1939. Pp. 610, ills., diagrs., bibl.

Levorsen, A. I. "Stratigraphic versus Structural Accumulation," *Bull. A.A.P.G.,* Vol. 20, No. 5, May 1936. Pp. 521–530, table.

National Research Council. *Report of Committee on Sedimentation.* Washington, Nov. 1932. Pp. 229, ills., diagrs., tables, bibls. (*Nat. Res. Counc. Bull. No. 89.*) Also *Idem.* Washington, July 1935. Pp. 246, bibls., footnote refs. (*Nat. Res. Counc. Bull. No. 98.*)

White, David. "Metamorphism of Organic Sediments and Derived Oils," *Bull. A.A.P.G.,* Vol. 19, No. 5, May 1935. Pp. 589–617, footnote refs.

Trask, P. D.; Hammar, H. E.; Wu, C. C. *Origin and Environment of Source Sediments of Petroleum.* Houston, Tex., Gulf Publishing Co. and A.P.I., 1932. Pp. 323, fold map and others, diagrs., tables, bibls.

Twenhofel, W. H. *Treatise on Sedimentation.* Baltimore, Williams and Wilkins, 1932. Pp. 661, ills., maps, diagrs., tables, footnote refs.

Rich, J. L. "Function of Carrier Beds in Long-Distance Migration of Oil," *Bull. A.A.P.G.,* Vol. 15, No. 8, Aug. 1931. Pp. 911–924, diagrs., footnote refs.

Moore, R. C., Ed.; and Barton, D. C.; DeGolyer, E.; Deussen, A.; Hull, J. P. D.; Pratt, W. E., and others. *Geology of Salt Dome Oil Fields: A symposium on the origin, structure, and general geology of salt domes, with special reference to . . . North America.* Tulsa, Okla., A.A.P.G., 1926. Pp. 797, fold maps and others, diagrs., tables, footnote refs.

Grabau, A. W. *Principles of Stratigraphy.* N.Y., Seiler, 1924. 2nd ed. Pp. 1185, maps, ills., tables, diagrs., bibls.

THE SEARCH FOR PETROLEUM IN THE GROUND

Price, P. H. "Evolution of Geologic Thought in Prospecting for Oil and Natural Gas," *Bull. A.A.P.G.,* Vol. 31, No. 4, April 1947. Pp. 673–697, bibl.

Levorsen, A. I. "Discovery Thinking," *Bull. A.A.P.G.,* Vol. 27, No. 7, July 1943. Pp. 887–929.

Levorsen, A. I., Ed. *Petroleum Discovery Methods: Symposium by Research Committee of A.A.P.G.* Denver, Colo., and Tulsa, Okla., A.A.P.G., 1942. Pp. 164. Mimeo.

Pratt, W. E. *Oil in the Earth.* Lawrence, Kan., Univ. of Kansas Press, 1942. Pp. 105.

Levorsen, A. I., and others. "New Ideas in Petroleum Exploration," *Bull. A.A.P.G.,* Vol. 24, No. 8, Aug. 1940. Pp. 1335–1474, ills., maps, diagrs., tables, bibl.

Thompson, A. B. *Oil Field Exploration and Development. . . .* 2 vols. N.Y., Van Nostrand, and London, Crosby Lockwood, 1925. Pp. 1168, ills., fold and other maps, diagrs., tables.

Cunningham-Craig, E. H. *Oil Finding. . . .* Introduction by Sir B. Redwood. London, Arnold, 1912, and N.Y., Longmans Green, 1920. Pp. 324, ills., diagrs., tables.

GEOPHYSICAL EXPLORATION

Nettleton, L. L., Ed. *Geophysical Case Histories,* Vol. I (60 papers). Menasha, Wis., Society of Exploration Geophysicists, 1949. Pp. 671, fold and other maps, diagrs., tables.

Holmer, R. C. "The Stratigraphic Basis for Using a Seismograph for Mapping Geologic Structures," *Mines Mag.,* Golden, Colo., Vol. 38, No. 10, Oct. 1948. Pp. 12–15, tables, bibl.

Muffly, G. "The Airborne Magnetometer," *Geophysics,* Vol. 11, No. 3, July 1946. Pp. 321–334, diagrs.

Wantland, D. "Magnetic Interpretation," *Geophysics,* Vol. 9, No. 1, Jan. 1944. Pp. 47–59, diagrs., footnote refs.

Adler, J. L. "Geophysical Exploration for Stratigraphic Oil Traps," *Geophysics,* Vol. 8, No. 4, Oct. 1943. Pp. 337–347, footnote refs.

Eckhardt, E. A. "A Brief History of the Gravity Method of Prospecting for Oil," *Geophysics,* Vol. 5, No. 3, July 1940. Pp. 231–242, diagr., footnote refs.

Heiland, C. A. *Geophysical Exploration.* N.Y., Prentice-Hall, 1940. Pp. 1013, ills., maps, diagrs., footnote refs.

Jakosky, J. J. *Exploration Geophysics.* Los Angeles, Times-Mirror Press, 1940. Pp. 786, ills., diagrs., footnote refs.

Nettleton, L. L. *Geophysical Prospecting for Oil.* N.Y., London, McGraw-Hill, 1940. Pp. 444, ills., diagrs., tables, bibls.

Ricker, N. "The Form and Nature of Seismic Waves and the Structure of Seismograms," *Geophysics,* Vol. 5, No. 4, Oct. 1940. Pp. 348–366, ills., diagrs., formulae.

Weatherby, B. B. "History and Development of Seismic Prospecting," *Geophysics,* Vol. 5, No. 3, July 1940. Pp. 215–230, diagrs., bibl.

RESULTS OBTAINED IN THE SEARCH FOR PETROLEUM

Weeks, L. G. "Highlights on 1948 Developments in Foreign Petroleum Fields," *Bull. A.A.P.G.,* Vol. 33, No. 6, June 1949. Pp. 1029–1124, maps, tables.

Cadman, W. H. "The Oil Shale Deposits of the World and Recent Developments in their Exploitation and Utilization, Reviewed to May, 1947," *Jour. Inst. Petroleum,* London, Vol. 34, No. 290, Feb. 1948. Pp. 109–132, ills., tables, bibl.

Lees, G. M. "Oil Reserves of the World: Their Distribution and Future," *Jour. Inst. Fuel,* London, Vol. 21, No. 121, Aug. 1948. Pp. 299–300.

Weeks, L. G. "Highlights on Developments in Foreign Petroleum Fields, 1947," *Bull. A.A.P.G.,* Vol. 32, No. 6, June 1948. Pp. 1093–1160, maps, tables.

Baker, W. L.; Logan, L. J. "The Significance of World Petroleum Production Trends," *Petroleum Technology,* Vol. 10, No. 4, July 1947. Pp. 1–10.

Fanning, L. M. *American Oil Operations Abroad.* N.Y., London, McGraw-Hill, 1947. Pp. 270, ills., map, diagrs., tables, including 43 of foreign oil statistics.

Weeks, L. G. "Highlights on Developments in Foreign Petroleum Fields [1946]," *Bull. A.A.P.G.,* Vol. 31, No. 7, July 1947. Pp. 1135–1193, maps, tables.

Fuchs, W. M. *When the Oil Wells Run Dry.* Dover, N.H., Industrial Research Service, 1946. Pp. 447, ills., maps, tables, diagrs.

Fanning, L. M., Ed. *Our Oil Resources.* N.Y., London, McGraw-Hill, 1945. Pp. 331, ills., maps, diagrs., bibl.

PART II. THE FUNCTIONAL ORGANIZATION OF THE PETROLEUM INDUSTRY

Nelson, W. L. *Petroleum Refinery Engineering.* N.Y., London, McGraw-Hill, 1949. 3rd ed. Pp. 830, ills., diagrs., tables, bibls.

DeGolyer, E. L. "Seventy-five Years of Progress in Petroleum," *Seventy-five Years of Progress in the Mineral Industry,* pp. 270–302. N.Y., A.I.M.M.E., 1947. Tables, diagrs., bibl.

U.S. Dept. of Commerce. *Petroleum Refining, War and Postwar.* Washington,

Gov't Printing Off., 1947. (*Industrial Ser. No. 73.*) Pp. 137, ills., diagrs., tables.

Garner, F. H., Ed. *Modern Petroleum Technology.* London, Inst. of Petroleum, 1946. Pp. 466, ills., diagrs., tables, bibl. 34 articles on exploration, drilling, refining, utilization, distribution, standards.

Uren, L. C. *Petroleum Production Engineering: Oilfield Development.* N.Y., London, McGraw-Hill, 1946. 3rd ed. Pp. 764, ills., diagrs., tables, bibl.

U.S. Senate, 79th Cong., 2nd Sess. *Hearings before a Special Committee Investigating Petroleum Resources Pursuant to Sen. Res. 36.* 6 vols. Washington, Gov't Printing Office, 1946. ("New Sources in the United States," June 19–25, 1945, pp. 531, tables, diagrs., maps. "American Petroleum Interests in Foreign Countries," June 27, 28, 1945, pp. 462, tables, diagrs., maps. "Petroleum Requirements Postwar," Oct. 3, 4, 1945, pp. 119, tables, diagrs., map. "War Emergency Pipe Lines and Other Petroleum Facilities," Nov. 15–17, 1945, pp. 431, tables, diagrs., maps. "War-time Petroleum Policy," Nov. 28–30, 1945, pp. 280, diagrs. "The Independent Petroleum Company," Mar. 19–28, 1946, pp. 569, tables, diagrs.)

U.S. Tariff Commission. *Petroleum.* Washington, Gov't Printing Off., 1946. (*War Changes in Industry Ser., Rep. No. 17.*) Pp. 152, tables.

Kalichevsky, V. A. *The Amazing Petroleum Industry.* N.Y., Reinhold, 1943. Pp. 234, maps, ills., diagrs., bibl.

Kalichevsky, V. A.; Stagner, B. A. *The Chemical Refining of Petroleum. . . .* N.Y., Reinhold, 1942. 2nd ed. Pp. 550, ills., diagrs., tables, bibls. (*Amer. Chem. Soc. Monograph Ser.*)

Ball, M. W. *This Fascinating Oil Business.* Indianapolis and N.Y., Bobbs-Merrill, 1940. Pp. 444, maps, diagrs.

DeGolyer, E. L., Ed. *Elements of the Petroleum Industry.* N.Y., A.I.M.M.E., 1940. Pp. 519, ills., diagrs., tables, bibls.

Pogue, J. E. *The Economics of Petroleum.* N.Y., Wiley, 1921. Pp. 375, fold and other maps, diagrs., tables.

PART III. THE WORLD'S PETROLEUM REGIONS *

Western Hemisphere

THE CARIBBEAN AREA AS A WHOLE

Rich, J. L. "Oil Possibilities of South America in the Light of Regional Geology," *Bull. A.A.P.G.,* Vol. 29, No. 5, May 1945. Pp. 495–561, ills., maps, diagrs., bibl. on pp. 561–563.

James, P. E. *Latin America.* N.Y., Odyssey Press, 1942. Pp. 908, ills., maps, tables, bibl.

Schuchert, C. *Historical Geology of the Antillean-Caribbean Region.* N.Y., Wiley; London, Chapman & Hall, 1935. Pp. 811, ills., fold and other maps, diagrs., tables, bibls.

* In addition to the titles listed below, the reader is referred to the regional studies in *The Science of Petroleum,* Vol. 2, cited in the "General" section of the bibliography for Part I, above.

VENEZUELA

Caribbean Petroleum Co. staff members. "Oil Fields of Royal Dutch-Shell Group in Western Venezuela," *Bull. A.A.P.G.*, Vol. 32, No. 4, Apr. 1948. Pp. 517–628, fold and other maps, diagrs., tables, footnote refs.

Funkhouser, H. J.; Sass, L. C.; Hedberg, H. D. "Santa Ana, San Joaquín, Guario, and Santa Rosa Oil Fields (Anaco Fields), Central Anzoátegui, Venezuela," *Bull. A.A.P.G.*, Vol. 32, No. 10, Oct. 1948, pp. 1851–1908. Maps, tables, diagrs.

Schaub, H. P. "Outline of Sedimentation in the Maracaibo Basin," *Bull. A.A.P.G.*, Vol. 32, No. 2, Feb. 1948. Pp. 215–227, map, diagr.

Halse, G. W. "Oil Fields of West Buchivacoa, Venezuela," *Bull. A.A.P.G.*, Vol. 31, No. 12, Dec. 1947, pp. 2170–2192. Maps, diagrs.

Hedberg, H. D.; Sass, L. C.; Funkhouser, H. J. "Oil Fields of Greater Oficina Area, Central Anzoátegui, Venezuela," *Bull. A.A.P.G.*, Vol. 31, No. 12, Dec. 1947. Pp. 2089–2169, ills., fold and other maps and diagrs., tables.

Suter, H. H. "El Mene de Acosta Field," *Bull. A.A.P.G.*, Vol. 31, No. 12, Dec. 1947. Pp. 2193–2206, map, diagr., tables.

Liddle, R. A. *Geology of Venezuela and Trinidad*, pp. 1–679. Ithaca, N.Y., Paleontological Research Institute, 1946. 2nd ed., rev. Ills., fold maps and diagrs. and others, bibl.

Sutton, F. A. "Geology of the Maracaibo Basin," *Bull. A.A.P.G.*, Vol. 30, No. 10, Oct. 1946. Pp. 1621–1741, ills., fold and other maps, tables, bibl.

U.S. Tariff Commission. *Mining and Manufacturing Industries in Venezuela*. Washington, Gov't Printing Off., 1945. Pp. 31, maps, tables, footnote refs.

Uslar Pietri, A. *Sumario de economía venezolana. . . .* Caracas, Universidad Central, 1945. Pp. 316, maps, diagrs., tables, footnote refs. (*Ediciones del Centro de Estudiantes de Derecho.*)

Allen, H. J. *Venezuela*. N.Y., Doubleday, 1940. Pp. 283, ills., fold map.

Juana, C. G. de. "A Contribution to the Study of the Zulía-Falcón Sedimentary Basin," *Proc. 2nd Venezuelan Geol. Congr.*, Caracas, 1937. Pp. 123–140, map, diagrs. (Venezuela: Servicio Técnico de Minería y Geología, *Bol. de Geología y Minería*, Vol. 1.)

Wiedenmayer, C. "Comparison of the Maturin and Maracaibo Sedimentary Basins," *ibid.*, Caracas, 1937. Pp. 221–250, maps, diagrs.

TRINIDAD

Great Britain. *Colonial Reports—Annual* (on Trinidad and Tobago, 1947). London and Trinidad, H.M.S.O., 1949. Pp. 103, ills., fold map, tables, bibl.

Liddle, R. A. *Geology of Venezuela and Trinidad*, pp. 681–821. Ithaca, N.Y., Paleontological Inst., 1946. 2nd ed., rev. Ills., fold, pocket, and other maps and diagrs., bibl.

Wilson, C. C. "Los Bajos Fault of South Trinidad," *Bull. A.A.P.G.*, Vol. 24, No. 12, Dec. 1940. Pp. 2102–2125, ills., maps, tables, diagrs.

Kugler, H. G. "Summary Digest of the Geology of Trinidad," *Bull. A.A.P.G.*, Vol. 20, No. 11, Nov. 1936. Pp. 1439–1453, map, diagr., table.

Waring, G. A. *The Geology of the Island of Trinidad. . . .* Baltimore, Johns Hopkins Press, 1926. Pp. 178, ills., fold and other maps, tables, footnote refs. (*Johns Hopkins Univ. Studies in Geol., No. 7.*)

MEXICO

Beal, C. H. *Reconnaissance of the Geology and Oil Possibilities of Baja California. . . .* N.Y., Geological Society of America, 1948. Pp. 138, ills., pocket and fold maps, tables, bibl. (*Geol. Soc. of Amer. Memoir No. 31.*)

Ysita, E. "Mexican Petroleum," *Commercial Pan America,* Washington, D.C., Vol. 15, Nos. 3 and 4, Mar.–Apr. 1946. Pp. 1–47, map, diagrs., bibl.

Kellum, L. B. "Geologic History of Northern Mexico and Its Bearing on Petroleum Exploration," *Bull. A.A.P.G.,* Vol. 28, No. 3, Mar. 1944. Pp. 301–325, ills., maps.

"Informe de Petróleos Mexicanos: V aniversario de la nacionalización de la industria," *Petróleo,* Mexico, D.F., Vol. 2, segundo trimestre, No. 2, 1943. Pp. 12–18.

MacMahon, A. W.; Dittmar, W. R. "The Mexican Oil Industry since Expropriation," *Polit. Sci. Quart.* (N.Y.), Vol. 57, No. 1, Mar. 1942, pp. 28–50; No. 2, June 1942, pp. 161–188.

Muir, J. M. *Geology of the Tampico Region, Mexico.* Tulsa, Okla., A.A.P.G., 1936. Pp. 280, ills., pocket maps, diagrs., bibl.

Ordonez, E. "Principal Physiographic Provinces of Mexico," *Bull. A.A.P.G.,* Vol. 20, No. 10, Oct. 1936. Pp. 1277–1307, map.

Müllerried, F. K. G. "Geología petrolera de las zonas sur del estado de Tamaulipas y norte del estado de Veracruz," *Anales del Instituto Geológico de Mexico,* Mexico, D.F., Vol. 3, 1929. Pp. 55–66, table.

Vivar, G. "El petróleo en Aragon, Guadalupe, Hidalgo, D.F." *Anales del Instituto Geológico de Mexico,* Vol. 3, 1929. Pp. 85–92.

Ver Wiebe, W. A. "Salt Domes of Isthmus of Tehuantepec," *Pan-American Geologist,* Des Moines, Ia., Vol. 45, No. 5, June, 1926. Pp. 349–358, table.

Ver Wiebe, W. A. "Oil Fields of the Isthmus of Tehuantepec," *Pan-American Geologist,* Vol. 45, No. 3, April, 1926. Pp. 189–200, diagrs.

Ver Wiebe, W. A. "Tectonics of the Tehuantepec Isthmus," *Pan-American Geologist,* Vol. 45, No. 1, Feb. 1926. Pp. 15–28, map.

Ver Wiebe, W. A. "Geology and Oil Fields of the State of Tabasco," *Pan-American Geologist,* Vol. 44, No. 4, Nov. 1925. Pp. 273–284, map.

Ver Wiebe, W. A. "Geology of the Southern Mexico Oil Fields," *Pan-American Geologist,* Vol. 44, No. 2, Sept. 1925. Pp. 121–138, map, diagrs.

Ver Wiebe, W. A. "The Stratigraphy of the Petroliferous Area of Eastern Mexico," *Amer. Jour. Science,* New Haven, Conn., 5th ser., Vol. 8, No. 46, Oct. 1924, pp. 277–295; No. 47, Nov. 1924, pp. 385–394; No. 48, Dec. 1924, pp. 481–502. Tables, bibl.

DeGolyer, E. "Oil Industry of Mexico: An Historical Sketch," *Petroleum Review,* London, Vol. 30 (New Ser.), No. 613, Apr. 18, 1914, pp. 439–440; No. 614, Apr. 25, 1914, pp. 469–470. Ills.

Millward, R. H. "Petroleum in the Americas," *Bull. Pan-Amer. Union,* Washington, D.C., Vol. 31, No. 5, Nov. 1910. Pp. 756–777, ills.

COLOMBIA

Anderson, J. L. "Petroleum Geology of Colombia, South America," *Bull. A.A.P.G.,* Vol. 29, No. 8, Aug. 1945. Pp. 1065–1140, ills., diagrs., maps, bibl., pp. 1140–1142.

Wheeler, O. C. "Petroleum Developments in Colombia, 1942–1944," *Trans. A.I.M.M.E.*, N.Y., Vol. 160, *Petroleum Development and Technology, 1945*. Pp. 615–628. Maps, tables.

Notestein, F. B.; Hubman, C. W.; Bowler, J. W. "Geology of the Barco Concession. . . ." *Bull. Geol. Soc. Amer.*, Vol. 55, Oct. 1944. Pp. 1165–1216, ills., fold map and others, diagrs., bibl.

Ospina-Racines, E. *La Economía del Petróleo en Colómbia*. Bogotá, Editorial Antena, 1944. Pp. 107, ills., diagrs.

Mendoza, F.; Alvarado, B. *La Industria del Petróleo en Colómbia*. Bogotá, Ministerio de la Economía Nacional, Dept. de Petróleos, 1939. Pp. 112, ills., maps, diagrs., tables; followed by same text, ills., etc., in English.

South America Other than Caribbean

GENERAL

Weeks, L. G. "Paleogeography of South America," *Bull. Geol. Soc. Amer.*, Vol. 59, No. 1, Mar. 1948. Pp. 249–281, fold map and others, bibl.

Weeks, L. G. "Paleogeography of South America," *Bull. A.A.P.G.*, Vol. 31, No. 7, July 1947. Pp. 1194–1229, maps, bibl., pp. 1230–1241.

Rich, J. L. "Oil Possibilities of South America in the Light of Regional Geology," *Bull. A.A.P.G.*, Vol. 29, No. 5, May 1945. Pp. 495–563, ills., maps, bibl.

Singewald, J. T., Jr. *Bibliography of Economic Geology of South America*. N.Y., Geological Society of America, 1943. (*Spec. Paper No. 50.*) Pp. 159.

Hedberg, H. D. "Mesozoic Stratigraphy of Northern South America," *Proc. VIIIth Amer. Sci. Congr.*, Washington, May 1940, Vol. 4, pp. 195–227. Washington, Dept. of State, 1942. Bibl.

Weaver, C. E. "A General Summary of the Mesozoic of South America and Central America," *Proc. VIIIth Amer. Sci. Congr.*, Washington, May 1940, Vol. 4, pp. 149–193. Washington, Dept. of State, 1942. Fold map, tables, bibl.

ARGENTINA

Perón, Juan. *Nuestro Petróleo*. Buenos Aires, Peuser, 1948. Pp. 11.

Reed, L. C. "San Pedro Oil Field, Province of Salta, Northern Argentina," *Bull. A.A.P.G.*, Vol. 30, No. 4, Apr. 1946. Pp. 591–605, maps, diagrs.

Rozlosnik, A. "Antecedentes y perspectivas futuras de nuestra exploración petrolífera," *Bol. de la Secretaría de Industria y Comercio*, Buenos Aires, Año 1, Nos. 9–10, June–July 1945. Pp. 427–451, maps, diagrs.

Baldwin, H. L. "Tupungato Oilfield, Mendoza, Argentina," *Bull. A.A.P.G.*, Vol. 28, No. 10, Oct. 1944. Pp. 1455–1484, maps, diagrs.

Bracaccini, O. "La problema de la exploración petrolera en la República Argentina," *Bol. Sociedad Argentina de Estudios Geográficos Gaea*, Buenos Aires, No. 6, Sept. 1943. Pp. 3–6.

Rossbach, A. "Evolución de los estudios geológicos en el Golfo de San Jorge," *Bol. de Informaciones, Dir.-Gen. de Yacimientos Petrolíferos Fiscales*, Buenos Aires, Vol. 16, No. 183, Nov. 1939. Pp. 27–42, maps, diagrs.

Victoria, J. "La zona petrolífera Salteña," *Revista Geográfica Americana*, Buenos Aires, Vol. 12, No. 75, Dec. 1939. Pp. 425–450, ills., diagr.

Windhausen, A. *Geología Argentina, Part II.* Buenos Aires, Peuser, 1931. Pp. 645, ills., fold and other maps, bibls.

BOLIVIA

Ahlfeld, F. *Geología de Bolivia.* La Plata, Universidad Nacional, 1946. (Reprinted from *Revista del Museo de la Plata,* Nueva Série, Sec. Geología, Vol. 3.) Pp. 370, ills., pocket map and others, diagrs., tables, bibl.

De Paiva, G.; Muñoz Reyes, J.; Mariaca, G. *Geologia da faixa subandina da Bolívia.* Rio de Janeiro, 1939. Pp. 83, ills., fold maps and diagrs., tables, bibl. (*Departamento Nacional da Produção Mineral, Divisão de Geologia e Mineralogia, Boletim 101.*)

Lopez, P. N. "El petróleo en Bolivia," *Bol. Minero, Soc. Nacional de Minería,* Santiago de Chile, Año LII, Vol. 68, No. 440, Dec. 1936. Pp. 802–807.

Heald, K. C.; Mather, K. F. "Reconnaissance of the Eastern Andes between Cochabamba and Santa Cruz, Bolivia," *Bull. Geol. Soc. Amer.,* Vol. 33, No. 9, Sept. 1922. Pp. 553–570, ills., map.

Mather, K. F. "Front Ranges of the Andes between Santa Cruz, Bolivia, and Embarcación, Argentina," *Bull. Geol. Soc. Amer.,* Vol. 33, No. 11, Nov. 1922. Pp. 703–764, ills., maps, diagrs., bibl.

BRAZIL

Abreu, S. F. "Petróleo," *Bol. Geográfico de la Paz,* Vol. 6, No. 62, May 1948. Pp. 130–144, maps.

Abreu, S. F. "Aspectos geográficos, geológicos et políticos da questão do petróleo no Brasil," *Revista Brasileira de Geografia,* Rio de Janeiro, Vol. 8, No. 4, Oct.–Dec. 1946. Pp. 509–534. English summary.

Oliveira, A. I. de. *Pesquisas de petróleo no estado da Bahia.* Rio de Janeiro, Ministerio da Agricultura, Serviço de Documentação, 1946. Pp. 19, maps.

Hennessey, J. F., Jr. "Brazilian Fuel and Power," *Commercial Pan America,* Wash., D.C., Vol. 14, No. 5, May 1945. Pp. 1–21, tables. Mimeo.

Ribeiro Lamego, A. *A bacia de campos na geologia litorânea do petróleo.* Rio de Janeiro, 1944. Pp. 69, ills., fold maps and others, diagrs., tables, footnote refs. (*Dept. Nacional da Produção Mineral, Divisão de Geologia e Mineralogia, Bol. 113.*)

Iglesias, D. *Bibliografia e indice da geologia do Brasil, 1641–1940.* Rio de Janeiro, 1943. Pp. 323. (*Dept. Nacional da Produção Mineral, Divisão de Geologia e Mineralogia, Bol. 111.*)

Oliveira, A. I. de; Leonardos, O. H. *Geologia do Brasil.* Rio de Janeiro, 1943. Pp. 813, ills., fold map and others, diagrs., tables, footnote refs. (*Ministerio da Agricultura, Serviço de Informação Agricola, Série Didática No. 2.*)

Oliveira, E. P. de. *História da pesquisa de petróleo no Brasil.* Rio de Janeiro, Serviço de Publicidade Agricola, 1940. Pp. 208, bibl.

Abreu, S. F. "O recóncavo da Baía e o petróleo do Lobato: consideraçoes de caráter geográfico," *Revista Brasileira de Geografia,* Vol. 1, No. 2, Apr. 1939. Pp. 57–83. English summary.

Oddone, D. S. *Reconhecimento geomagnético nos arredores do planalto de Reserva, Estado do Parana.* Rio de Janeiro, 1939. Pp. 33, fold maps, diagrs., footnote refs. (*Dept. Nacional da Produção Mineral, Divisão de Fomento, Bol. No. 35.*)

Paiva, G. de. *Provincia petrolifera do Nordeste.* Rio de Janeiro, 1939. Pp. 127. (*Dept. Nacional da Produção Mineral, Div. de Fomento, Avulso No. 41.*)

Paiva, G. de; Carvalho do Amaral, I. *Justificativas para a locação de um poço para petróleo no Recôncavo, Baía.* Rio de Janeiro, 1939. Pp. 23, fold map and diagrs. (*Dept. Nacional da Produção Mineral, Divisão de Fomento, Avulso No. 40.*)

Paiva, G. de; Leinz, V. *Contribuição para a geologia do petróleo no Sudoeste de Matto Grosso.* Rio de Janeiro, 1939. Pp. 92, ills., fold map and others, diagrs., bibl. English summary. (*Dept. Nacional da Produção Mineral, Divisão de Fomento, Bol. No. 37.*)

Washburne, C. W., translated into Portuguese and edited by Pacheco, J. *Geologia do petróleo do estado de São Paulo.* Rio de Janeiro, Dept. Nacional da Produção Mineral, 1939. Pp. 228, ills., fold map and others, diagrs., tables, bibl.

Moura, P. de. *Possibilidades de petróleo no território do Acre.* Rio de Janeiro, 1937. Pp. 14, map. (*Dept. Nacional da Produção Mineral, Serviço de Fomento da Produção Mineral, Avulso No. 16.*)

Oppenheim, V. "Geology of Devonian Areas of the Paraná Basin in Brazil, Uruguay and Paraguay," *Bull. A.A.P.G.,* Vol. 20, No. 9, Sept. 1936. Pp. 1208–1236, ills., maps, table, bibl.

CHILE

Gallet, E. S. "Estudios y exploraciónes petrolíferas efectuados en Magallanes e Isla Grande de Tierra del Fuego," *Bol. Minero,* Santiago, Año LXII, Vol. 53, No. 549, Jan. 1946. Pp. 6–8.

Hemmer, A. "Geología de los terrenos petrolíferos de Magallanes y las exploraciónes realizadas," *Bol. Minero de la Sociedad Nacional de Minería,* Santiago, Año LI, Vol. 47, No. 419, Mar. 1935, pp. 139–149, and No. 420, Apr. 1935, pp. 181–188. Fold map and diagrs.

Bruggen, J. "Las formaciónes de sal y petróleo de la puna de Atacama," *Bol. de Minas y Petróleo,* Santiago, Vol. 4, No. 2, Mar. 1934. Pp. 105–122, maps.

Felsch, J. "Informe preliminar sobre las reconocimientos geológicos de los Yacimientos Petroleros en la cordillera de la provincia de Antofagasta," *Bol. de Minas y Petróleo,* Vol. 3, No. 29, Dec. 1933. Pp. 411–422, fold diagrs.

ECUADOR

Tschopp, H. J. "Bosquejos de la geología del Oriente Ecuatoriana," *Bol. del Inst. Sudamericano del Petróleo,* Montevidéo. Vol. 1, No. 5, Feb. 1945. Pp. 466–484, map, diagr.

Dirección General de Minería y Petróleos. *15 años de producción petrolera en el Ecuador, 1925–1939.* Quito, "El Comercio," 1940. Pp. 236; entire contents in tabular form.

Sheppard, G. *Geology of Southwest Ecuador.* London, Murby, 1937. Pp. 275, ills., map, diagrs., bibl.

Sheppard, G. "Observations on the Geology of the Santa Elena Peninsula," *Jour. Inst. Petroleum Technologists,* London. Vol. 13, No. 62, June 1927. Pp. 424–461, ills., maps, diagrs., bibl.

Wasson, T.; Sinclair, H. J. H. "Geological Explorations East of the Andes in

Ecuador," *Bull. A.A.P.G.,* Vol. 11, No. 12, Dec. 1927. Pp. 1253–1281, ills., maps, footnote refs.

PERU

Valverde, R. L. "La cuenca petrolera peruana del Amazonas y su correlación estratigráfica con las formaciónes petrolíferas del continente Sud Americano," *Bol. de la Sociedad Geológica del Perú,* Lima. Vol. 19, 1946. Pp. 81–110.

Petersen, G. "Historia de las refinerías de petróleo en el Departamento de Tumbes," *Bol. del Cuerpo de Ingenieros de Minas del Perú,* Lima. No. 129, 1944. Pp. 15–44, ills., map, diagrs., bibl.

Petersen, G. "Sobre el petróleo de Islaycocha," *Bol. de la Sociedad Geológica del Perú,* Lima. Vol. 12, Fasc. 2, 1943. Pp. 1–32, fold map, diagr., tables, bibl.

Cabrera La Rosa, A. "Descubrimiento de Petróleo en Pirín," *Bol. Inf. Petroleras,* Buenos Aires, Vol. 16, No. 180, Aug. 1939. Pp. 57–58.

Jochamowitz, A. *El problema petrolífero del Perú.* Lima, 1939. Pp. 131, ills., pocket maps, diagrs., tables. (*Bol. del Cuerpo de Ingenieros de Minas del Perú, No. 125.*)

Quiroga, O. "La industria del petróleo en el norte del Perú," *Revista de Marina,* Lima, Año 24, No. 5, Sept.–Oct. 1939. Pp. 647–671, tables.

Huntley, L. G. "El petróleo del nororiente del Perú," *Bol. de la Sociedad Geológica del Perú,* Lima, Vol. 9, 1937. Pp. 1–8.

Cabrera La Rosa, A.; Petersen, G. *Reconocimiento geológico de los yacimientos petrolíferos del Departamento de Puno.* Lima, 1936. Pp. 102, ills., pocket maps and others, tables, bibl. (*Bol. del Cuerpo de Ingenieros de Minas del Perú, Dept. de Petróleo, No. 115.*)

Moran, R. M.; Fyfe, D. "Geología de la región del Bajo Pachitea," *Bol. Ofic. de Minas e Industrias,* Lima, Año 12, No. 41, Dec. 1933. Pp. 43–54, fold map.

Salfeld, H. "Observaciónes geofísicas y geológicas en el despoblado de Sechura," *Bol. de la Sociedad Geológica del Perú,* Lima. Vol. 5, 1933. Pp. 67–78, fold map.

URUGUAY

Oppenheim, V. "Petroleum Geology of the Central Sedimentary Basin of Uruguay," *Bull. A.A.P.G.,* Vol. 19, No. 8, Aug. 1935. Pp. 1205–1218, map, diagrs., tables, bibl.

NORTH AMERICA *

Rister, C. C. *Oil! Titan of the Southwest.* Norman, Okla., Univ. of Okla. Press, 1949. Pp. 467, ills., map, tables, bibl.

Fanning, L. M. *Rise of American Oil.* N.Y., London, Harper, 1948. Pp. 169.

Giddens, P. H. *Early Days of Oil.* Princeton, Princeton Univ. Press, 1948. Pp. 160, ills.

Levorsen, A. I. "Our Petroleum Resources," *Bull. Geol. Soc. Amer.,* Vol. 59, No. 4, Apr. 1948. Pp. 283–299, maps, diagrs.

* In addition to the titles listed below, a number of the references given for Part II, "The Functional Organization of the Petroleum Industry," and Part IV, "Aspects of Utilization," apply particularly to North America, and Mexican references are listed under "Mexico" in the bibliography for Part III.

Structure of Typical American Oil Fields: Symposium on Relation of Oil Accumulation to Structure. 3 vols. Tulsa, Okla., A.A.P.G., 1929 (Vols. 1 and 2) and 1948 (Vol. 3). Vol. 1, pp. 510. Vol. 2, pp. 780. Vol. 3, pp. 516. Ills., pocket, fold, other maps, diagrs., tables, footnote refs.

"Appalachian Basin Ordovician Symposium." Papers by 10 authors, given before Pittsburgh Geol. Soc., May 16, 1947. *Bull. A.A.P.G.,* Vol. 32, No. 8, Aug. 1948. Pp. 1395–1657, maps, diagrs., tables, bibl.

Price, W. A. "Equilibrium of Form and Forces in Tidal Basins of Texas and Louisiana," *Bull. A.A.P.G.,* Vol. 31, No. 9, Sept. 1947. Pp. 1619–1663, maps, bibl.

Tait, S. W., Jr. *The Wildcatters: An Informal History of Oil Hunting in America.* Princeton, Princeton Univ. Press, 1946. Pp. 205, ills.

Lafferty, R. C. "Central Basin of Appalachian Geosyncline," *Bull. A.A.P.G.,* Vol. 25, No. 5, May 1941. Pp. 781–825.

Levorsen, A. I., Ed. *Possible Future Oil Provinces of the United States and Canada.* Tulsa, Okla., A.A.P.G., 1941. Pp. 154, maps, diagrs., bibl. Also in *Bull. A.A.P.G.,* Vol. 25, No. 8, Aug. 1941. Pp. 1433–1586.

Ruedemann, R.; Balk, R., Eds. *Geology of North America:* Vol. I, *Introductory Chapters; Geology of the Stable Areas.* Berlin, Borntraeger, 1939. (In series *Geologie der Erde*). Pp. 642, ills., fold and other maps, diagrs., tables, bibls.

Eastern Hemisphere

THE MIDDLE EAST AS A WHOLE

Boesch, H. *Erdöl im Mittleren Osten.* Zurich, 1949. Pp. 14, diagrs., tables. (*Geogr. Inst.,* Univ. of Zurich, *Ser. A., No. 36.*)

Barber, C. T., and others. *Petroleum Times Review of Middle East Oil: A Comprehensive Illustrated Review. . . .* London, 1948. Pp. 115, ills., maps, diagrs., tables, footnote refs. (*Petrol. Times Export No., June 1948.*)

Summary of Middle East Oil Developments. N.Y., Arabian American Oil Co., 1948. 2nd ed. Pp. 30, fold map and others, diagrs., tables.

Jones, J. H. "My Visit to the Persian Oil Fields," *Jour. Royal Central Asian Soc.,* London, Vol. 34, Pt. I, Jan. 1947. Pp. 56–68.

Longhurst, H. "Abadan: Britain's Largest Oil Refinery," *Great Britain and the East,* London, Vol. 63, No. 1778, Nov. 1947. Pp. 48–49 ME, ill.

Tromp, S. W. *Preliminary Compilation of the Stratigraphy, Structural Features, and Oil Possibilities of Southeastern Turkey. . . .* Ankara, 1941. Pp. 34, pocket maps and diagrs., English and Turkish text. (*Mining Research Inst. of Turkey, Ser. A., Brief Communications, No. 4.*)

Clapp, F. G. "The Geology of Eastern Iran," *Bull. Geol. Soc. Amer.,* Vol. 51, No. 1, Jan. 1940. Pp. 1–102, ills., fold map and diagrs. and others, tables, bibl.

Lockhart, L. "Iranian Petroleum in Ancient and Medieval Times," *Jour. Inst. Petrol.,* London, Vol. 25, No. 183, Jan. 1939. Pp. 1–18, maps, footnote refs.

Hume, W. F. *Geology of Egypt,* Vols. I and II. Cairo, Gov't Printing Press, 1925 and 1935. Vol. I, *Surface Features. . . .* Pp. 408, ills., pocket and fold maps, diagrs., tables, bibl. Vol. II: *Pt. I, Fundamental Pre-Cambrian Rocks of Egypt and the Sudan . . . ; Pt. II, Later Plutonic and Minor Intrusive Rocks.* Pp. 688 + 107 pp. of index; pocket maps.

Wilson, Sir A. T. *The Persian Gulf. . . .* Oxford, Clarendon Press, 1928. Pp. 327, ills., pocket map, bibl.

Palgrave, W. H. *Narrative of a Year's Journey through Central and Eastern Arabia, 1862–63.* 2 vols. 3rd ed. London and Cambridge, 1866.

SAUDI ARABIA AND BAHREIN

Longrigg, S. "Liquid Gold of Arabia," *Jour. Royal Central Asian Soc.,* London, Vol. 36, Pt. I, Jan. 1949. Pp. 20–33, map.

Mikesell, R. F.; Chenery, H. B. *Arabian Oil: America's Stake in the Middle East.* Chapel Hill, Univ. of N. Carolina Press, 1949. Pp. 216, map, tables.

Adolph, E. F., and associates. *Physiology of Man in the Desert.* N.Y., London, Interscience Publishers, 1947. Pp. 357, diagrs., bibl.

Twichell, K. S. *Saudi Arabia.* Princeton, Princeton Univ. Press, 1947. Pp. 204, ills., map.

Harrison, P. W. *Doctor in Arabia.* N.Y., John Day, 1940. Pp. 303, ills.

Lamare, P. *Structure géologique de l'Arabie.* Paris, Librairie Polytechnique Béranger, 1936. Pp. 63, maps, diagrs., bibl.

Philby, H. St. J. B. *The Empty Quarter.* N.Y., Holt, 1933. Pp. 432, ills., fold map.

Thomas, B. *Arabia Felix: Across the "Empty Quarter."* N.Y., Scribner, 1932. Pp. 397, ills., fold and other maps.

Musil, A. *Arabia Deserta: A Topographical Itinerary.* N.Y., Amer. Geogr. Soc., 1927. Pp. 631, ills., pocket and fold maps, diagrs., bibl. (*Amer. Geogr. Soc. Oriental Explorations and Studies, No.* 2.)

Doughty, C. M. *Wanderings in Arabia.* 2 vols. N.Y., Scribner, 1908. Pp. 308 and 292; ills.; map. (Abridgement of *Arabia Deserta,* ed. by Edward Garnett.)

Kunz, G. W.; Stevenson, C. H. *The Book of the Pearl.* N.Y., Century, 1908. Pp. 548, ills., maps, bibl.

Doughty, C. M. *Travels in Arabia Deserta.* 2 vols. Cambridge, Cambridge Univ. Press, 1888. Pp. 623 and 690, ills., fold and other maps, diagrs. And various later complete and abridged edns.

Forster, C. *Historical Geography of Arabia. . . .* 2 vols. London, Duncan and Malcom, 1844.

UNION OF SOVIET SOCIALIST REPUBLICS

Fohs, F. J. "Oil-Reserve Provinces of the Middle East and Southern Soviet Russia," *Bull. A.A.P.G.,* Vol. 31, No. 8, Aug. 1948. Pp. 1372–1381, maps, bibl. pp. 1381–1383.

Fohs, F. J. "Petroliferous Provinces of U.S.S.R.," *Bull. A.A.P.G.,* Vol. 32, No. 3, Mar. 1948. Pp. 317–350, map, bibl.

Heymann, H., Jr. "Oil in Soviet-Western Relations in the Interwar Years," *Amer. Slavic and East Europ. Rev.,* N.Y., Vol. 7, No. 4, Dec. 1948. Pp. 303–316, footnote refs.

Shanazarov, D. A. "Petroleum Problem of Siberia," *Bull. A.A.P.G.,* Vol. 32, No. 2, Feb. 1948. Pp. 153–196, maps, bibl.

Kovalev, I. V. "New Railroads in U.S.S.R.," *Amer. Rev. on Soviet Union,* N.Y., Vol. 8, No. 2, Mar. 1947. Pp. 36–42, maps.

Possony, S. T. "European Russia's Inland Waterways," *Proc. U.S. Naval Inst.*, Annapolis, Vol. 73, No. 534, Aug. 1947. Pp. 936–947, map, ills., tables.

Adams, E. "Estimates of Russian Oil Resources," *Petroleum Engineering*, Dallas, Vol. 18, No. 1, Oct. 1946. Pp. 178–182.

Perejda, A. D. "The Position of Russia in the Oil Age," *Amer. Rev. on Soviet Union*, N.Y., Vol. 7, No. 4, Aug. 1946. Pp. 3–19, maps, tables, bibl.

Schwarz, S. M. "How Much Oil Has Russia?" *Foreign Affairs*, N.Y., Vol. 24, No. 4, July 1946. Pp. 736–741.

Sujkowski, Z. "The Geologic Structure of East Poland and West Russia. . . ." *Quart. Jour. Geol. Soc. London*, Vol. 102, Pt. 2, July 1946. Pp. 189–201, fold map, diagrs., tables, bibl.

Khachaturov, T. S. "Organization and Development of Railway Transport in the U.S.S.R.," *Intern. Affairs*, London. Vol. 21, No. 2, Apr. 1945. Pp. 220–235.

Kolomitzeff, D. "Le pétrole sovietique," *Bull. de l'Université l'Aurore*, Shanghai, Ser. III, Vol. 6, No. 4, 1945. Pp. 589–725, fold map and others, tables, bibl.

Cressey, G. B. *Asia's Lands and Peoples*. N.Y., London, McGraw-Hill, 1944. Pp. 253–372, ills., maps.

Gester, G. C. "World Petroleum Reserves and Petroleum Statistics," *Bull. A.A.P.G.*, Vol. 28, No. 10, Oct. 1944. Pp. 1485–1505, map, diagrs.

Gregory, J. S.; Shave, D. W. *The U.S.S.R.: A Geographical Survey*. London, Harrap; N.Y., Wiley, 1944. Pp. 636, maps, diagrs., bibl.

Sanders, C. W. "Emba Salt-Dome Region, U.S.S.R., and Some Comparisons with Other Salt-Dome Regions," *Bull. A.A.P.G.*, Vol. 23, No. 4, Apr. 1939. Pp. 492–516, maps, diagrs.

U.S.S.R., The Scientific Editorial Institute. *Great Soviet World Atlas*, Vol. I. (*Pt. 1, Maps of the World; Pt. 2, Maps of the Union of Soviet Socialist Republics.*) Moscow, 1938. Pt. 1, 83 plates in color; Pt. 2, 168 plates in color. In Russian.

Dunbar, C. O.; Miller, A. K. "On the Carboniferous and Permian of the South Urals," *Amer. Jour. Sci.*, New Haven, Conn., 5th Ser., Vol. 33, No. 198, June 1937. Pp. 470–472.

Krems, A. J., Ed. *Petroleum Excursion of 17th International Geological Congress, Moscow, 1937*. In 5 fascicles. Moscow and Leningrad, Chief Editorial Office of Mining-Fuel and Geological-Prospecting Literature, 1937. *Fasc. 1; Permian Prikamye; Bashkirian A.S.S.R.; Samarskaya Luka*. Pp. 63, ills., fold maps, diagrs. tables, bibl. *Fasc. 2: Azerbaijanian S.S.R.* Pp. 105, ills., maps, diagrs., bibl. *Fasc. 3: Daghestanian and Chechenian-Ingushetian A.S.S.R.* Pp. 64, fold and other maps, diagrs., tables of deposits, bibl. *Fasc. 4: Georgian S.S.R.* Pp. 65, ills., fold maps, diagrs., tables of deposits, bibl. *Fasc. 5: Kuban-Black Sea Region*. Pp. 48, ills., fold map, diagrs., tables, bibl. In English.

Leuchs, K. *Geologie von Asien*, Book I, *Pts. 1 and 2*. Berlin, Borntraeger, 1935–1937. Pp. 553, ills., fold and other maps, diagrs., bibl. (In series, *Geologie der Erde*.)

Goubkin, I. M. "Tectonics of the Southeast Caucasus and Its Relation to Productive Oil Fields," *Bull. A.A.P.G.*, Vol. 18, No. 5, May 1934. Pp. 603–671, ills., maps, diagrs., tables, bibl.

Obruchev, V. A. *Geologie von Siberien.* Berlin, Borntraeger, 1926. Pp. 504, fold map, diagrs., bibl. pp. 505–523.

Europe West of the U.S.S.R.

GENERAL

Walters, R. P. "Oil Fields of the Carpathian Region," *Bull. A.A.P.G.,* Vol. 30, No. 3, Mar. 1946. Pp. 319–336, maps, diagrs.

Van der Gracht, W. A. J. M. Van W. "The Possibility of Oil and Gas Production from Paleozoic Formations in Europe," *Bull. A.A.P.G.,* Vol. 20, No. 11, Nov. 1936. Pp. 1476–1493, diagrs., footnote refs.

Stille, H.; Schlüter, H. "European Oil and Gas Occurrences and their Relationship to Structural Conditions," *Bull. A.A.P.G.,* Vol. 18, No. 6, June 1934. Pp. 736–745, map, bibl.

Zuber, S. "Ponto-Caspian and Mediterranean Types of Oil Deposits," *Bull. A.A.P.G.,* Vol. 18, No. 6, June 1934. Pp. 760–776, map, diagrs., bibl.

Zuber, S. "Paleogeography of Oil-bearing Deposits in Ponto-Caspian Countries," *Bull. A.A.P.G.,* Vol. 18, No. 6, June 1934. Pp. 777–785, map, table, footnote refs.

Bubnoff, S. von. *Geologie von Europa,* being Vols. I and II of *Geologie der Erde.* Berlin, Borntraeger, 1926. *Band I: Einfuhrung, Osteuropa, Baltischer Schild.* Pp. 330, ills., fold map and others, diagrs., bibls. *Band II: Der auszeralpina Westeuropa. Erster Teil: Kaledoniden und Variscíden.* Pp. 691, maps, diagrs., tables, bibls. *Zweiter Teil: Entwicklung des Oberbaues.* Pp. 694–1134, maps, diagrs., tables, bibls. *Dritter Teil: Struktur des Oberbaues und das Quartar nordeuropas.* Pp. 1135–1603, maps, diagrs., tables, bibls.

ALBANIA

Maddalena, L.; Zuber, S. "Sur la géologie des pétroles albanais," *Proc. 2e Cong. Mond. Pétrole,* Paris, 1937. Vol. I, pp. 559–579.

BELGIUM

Pirson, S. J. "Oil Possibilities of Belgium and the Belgian Congo," *Bull. A.A.P.G.,* Vol. 18, No. 9, Sept. 1934. Pp. 1160–1174, maps, diagrs., tables, footnote refs.

FRANCE

"La recherche de pétrole dans l'Union Française," *Chronique des Mines Coloniales,* Paris, Vol. 15, No. 129, Mar. 15, 1947. Pp. 128–134, bibl.

Durand, J. "La recherche du pétrole en France," *Houille, Minerais, Pétrole,* Paris, Vol. 1, No. 1, Jan.–Feb. 1946. Pp. 5–15, maps, footnote refs.

Eardley, A. J. "Petroleum Geology of the Aquitaine Basin, France," *Bull. A.A.P.G.,* Vol. 30, No. 9, Sept. 1946. Pp. 1517–1545, maps, tables, diagrs.

Macovei, G. *Les gisements de pétrole: géologie, statistique, économie.* Paris, Masson, 1938. Pp. 502, maps, diagrs.

Clapp, F. G. "Oil and Gas Possibilities of France," *Bull. A.A.P.G.,* Vol. 16, No. 11, Nov. 1932. Pp. 1092–1143, maps, tables, bibl.

De Launay, L. *Géologie de la France.* Paris, Armand Colin, 1921. Pp. 501, ills., fold maps, diagrs.

Lemoine, P. *Géologie du bassin de Paris.* Paris, A. Herrmann, 1911. Pp. 408, fold and other maps, diagrs., bibl.

GERMANY

British Intelligence Objectives Sub-Committee. "German Oil Industry Exploration," *Oil Fields Investigation: The War Development of the German Crude Oil Industry, 1939–1945, Pt. II, Sec. 1.* London, H.M.S.O., 1946. In 2 vols., pp. 88, maps, diagrs., tables. Mimeo.

Reeves, F. "Status of the German Oil Fields," *Bull. A.A.P.G.,* Vol. 30, No. 9, Sept. 1946. Pp. 1546–1582, maps, tables, diagrs., bibl.

Bubnoff, S. von. "Der Ostdeutsche Grenzraum: Struktur und Bodenschatze," *Geol. Rundschau,* Leipzig, Bd. 30, H. 7–8, 1939. Pp. 695–702, map.

Barton, D. C. "Magnetic and Torsion-Balance Survey of Munich Tertiary Basin, Bavaria," *Bull. A.A.P.G.,* Vol. 18, No. 1, Jan. 1934. Pp. 69–98, maps, diagrs., tables, footnote refs.

Van der Gracht, W. A. J. M. Van W. "Occurrence and Production of Petroleum in Germany," *Bull. A.A.P.G.,* Vol. 16, No. 11, Nov. 1932. Pp. 1144–1151, tables, footnote refs.

Kauenhowen, W. "Oil Fields of Germany," *Bull. A.A.P.G.,* Vol. 12, No. 5, May 1928. Pp. 463–498, maps, tables, bibl.

GREAT BRITAIN

Lees, G. M. *The Exploration for Oil in Great Britain and Its Economic Consequences.* Univ. College, Nottingham, 1946. Pp. 10. (*Abbott Memorial Lecture, 1946.*)

Lovely, H. R. "Geological Occurrence of Oil in the United Kingdom, with reference to present exploratory operations," *Bull. A.A.P.G.,* Vol. 30, No. 9, Sept. 1946. Pp. 1444–1516, maps, fold diagrs.

Phemister, J., and others. "Geophysical Prospecting and English Oilfields," *Nature,* London. Vol. 158, No. 4026, Dec. 28, 1946. Pp. 931–934.

Rumsby, P. L. "Recent Developments in the British Petroleum Drilling Campaign," *Bull. of the Imperial Inst.,* London. Vol. 44, No. 3, July–Sept. 1939. Pp. 252–259.

HUNGARY

Skeels, D. C.; Vajk, R. "Geophysical Exploration and Discovery of the Budafapuszta (Lispe) Oil Field in Hungary," *Geophysics,* Menasha, Wis., Vol. 12, No. 2, Apr. 1947. Pp. 208–220.

Lóczy, L. de. "Tectonics and Paleogeography of the Basin System of Hungary Elucidated by Drilling for Oil," *Bull. A.A.P.G.,* Vol. 18, No. 7, July 1934. Pp. 925–941, maps.

ITALY

Behrmann, R. B. "Geologie und Erschliessung der Erdölvorkommen italiens," *Oel und Kohle,* Berlin, Jg. 36, H. 46, 1940. Pp. 522–537.

Anelli, M. "Notice sur les resultats géologiques obtenus par l'exploration de la vallée du Po." *Congr. Intern. des Mines, de la Métallurgie et de la Géologie Appliquée,* Sec. de *Géologie Appliquée,* Paris, Vol. I, 1936. Pp. 449–452.

POLAND

Bohdanowicz, K. "Oil Fields in Poland: Geological and Structural Summary," *Bull. A.A.P.G.,* Vol. 17, No. 9, Sept. 1933. Pp. 1084–1097, tables.

Bohdanowicz, K. "Geology and Mining of Petroleum in Poland," *Bull. A.A.P.G.,* Vol. 16, No. 11, Nov. 1932. Pp. 1061–1091, map, diagrs., tables, footnote refs.

Cizancourt, H. de. "Geology of the Oil Fields of the Polish Carpathian Mountains," *Bull. A.A.P.G.,* Vol. 15, No. 1, Jan. 1931. Pp. 1–41, ills., map, tables, diagrs., bibl.

RUMANIA

Loczy, L. von. "Beiträge zur Ölgeologie des innerkarpathischen Beckensystems," *Petroleum,* Berlin and Vienna, Jg. 35, No. 27, July 1939. Pp. 461–468, ill., maps, fold diagrs.

Voitești, I. P. *Evoluția geologico-paleogeografică a Pamantului Romanesc.* Cluj, Institutul de Arte Grafige "Ardealul," 1936. Pp. 211, maps, diagrs, bibl.

Voitești, I. P. "Aperçu général sur la géologie de la Roumanie," *Ann. des Mines de Roumanie,* Bucharest, Vol. 4, Nos. 8–9, Aug.–Sept. 1921. Pp. 751–821, maps, diagrs., bibl. Text in French and Rumanian.

SPAIN

Sanz, D. R. *El petróleo en España,* followed by *Yacimientos de petróleo: terminología y clasificación.* Madrid, 1948. Pp. 177 and 61, fold and other maps and diagrs., tables. (*Memorias del Instituto Geológico y Minero de España.*)

INDONESIA

Vening Meinesz, F. A. *Gravity Expeditions at Sea.* 4 vols. Delft, 1932–1948. Ills., fold maps, diagrs., tables. (Publication of the Netherlands Geodetic Commission.)

Horton, C. W. "Interpretation of Isostatic Anomalies South of Java," *Geophysics,* Vol. 11, No. 2, Apr. 1946. Pp. 183–195, diagrs.

Schuppli, H. M. "Oil Basins of East Indian Archipelago," *Bull. A.A.P.G.,* Vol. 30, No. 1, Jan. 1946. Pp. 1–22, map, diagr., bibl.

Beltz, E. W. "Principal Sedimentary Basins in East Indies," *Bull. A.A.P.G.,* Vol. 28, No. 10, Oct. 1944. Pp. 1440–1454, map.

Umbgrove, J. H. F. "Geological History of East Indies," *Bull. A.A.P.G.,* Vol. 22, No. 1, Jan. 1938. Pp. 1–70, maps, fold and other diagrs., tables, footnote refs.

Vening Meinesz, F. A. "Gravity Anomalies in the East Indian Archipelago," *Geogr. Journ.,* London, Vol. 77, No. 4, April 1931. Pp. 323–337, diagrs.

BURMA

Evans, P.; Crompton, W. "Geological Factors in Gravity Interpretation, Illustrated by Evidence from India and Burma," *Quart. Jour. Geol. Soc. London,* Vol. 102, 1946. Pp. 211–249, fold map and others, diagrs., bibl.

Evans, P. "The Oilfields of India and Burma," *Jour. Royal Soc. Arts,* London. Vol. 94, No. 4717, May 10, 1946. Pp. 369–379, map.

Evans, P.; Sansom, C. A. "The Geology of British Oil Fields: Part 3, The Oil

Fields of Burma," *Geol. Mag.*, London, Vol. 78, No. 5, 1941. Pp. 321–350, map, diagrs.

Clegg, E. L. G. *The Geology of Parts of the Minbu and Thayetnayo Districts, Burma.* Calcutta, Gov't of India, 1938. Pp. 347, ills., fold maps, tables, footnote refs. (*Mem. Geol. Surv. India, Vol. 72, Pt. 2.*)

Cotter, G. de P. *The Geology of Parts of the Minbu, Myingyan, Pakokku, and Lower Chindwin Districts, Burma.* Calcutta, Gov't of India, 1938. Pp. 176, ills., fold maps, footnote refs. (*Mem. Geol. Surv. India, Vol. 72, Pt. 1.*)

Stamp, L. D. "Geology and Economic Significance of Burma Oil Fields," *World Petroleum,* N.Y., Vol. 7, No. 11, Nov. 1936. Pp. 580–592, ills., maps, diagrs., tables.

Barber, C. T. *The Natural Gas Resources of Burma.* Calcutta, Gov't of India, 1935. Pp. 200, ills., fold and other maps, diagrs., tables, footnote refs. (*Mem. Geol. Surv. India, Vol. 46, Pt. 1.*)

Stamp, L. D. "Natural Gas Fields of Burma," *Bull. A.A.P.G.*, Vol. 18, No. 3, Mar. 1934. Pp. 315–326, ill., maps, diagrs., footnote refs.

Other Parts of Asia, Africa, and Oceania

ASIA AS A WHOLE

Stach, L. W. "Petroleum Exploration and Production in the Western Pacific in World War II," *Bull. A.A.P.G.,* Vol. 31, No. 8, Aug. 1947. Pp. 1384–1403, maps, tables, diagrs.

Cressey, G. B. *Asia's Lands and Peoples.* N.Y., London, McGraw-Hill, 1944. Pp. 607, ills., maps, diagrs., bibl.

Stamp, L. D. *Asia: A Regional and Economic Geography.* N.Y., Dutton, 1936. Rev. ed. Pp. 704, ills., maps, diagrs., bibls.

Leuchs, K. *Geologie von Asien, Vol. I, Part I, Überblick über Asien, Nordasien.* Berlin, Borntraeger, 1935. Pp. 226, maps, diagrs., bibl. pp. 227–231. (In series *Geologie der Erde.*)

De Launay, L. *La géologie et les richesses minérales de l'Asie.* Paris, Librairie Polytechnique Béranger, 1911. Pp. 785, fold and other maps and diagrs.

CHINA

Weller, J. M. "Outline of Chinese Geology," *Bull. A.A.P.G.,* Vol. 28, No. 10, Oct. 1944. Pp. 1417–1429, map, bibl.

Weller, J. M. "Petroleum Possibilities of the Red Basin, in Szechuan Province, China," *Bull. A.A.P.G.,* Vol. 28, No. 10, Oct. 1944. Pp. 1430–1439, map, diagrs., bibl.

Pan, C. H. "Non-marine Origin of Petroleum in North Shensi, and the Cretaceous of Szechuan, China," *Bull. A.A.P.G.,* Vol. 25, No. 11, Nov. 1941. Pp. 2058–2068.

Lee, J. S. *The Geology of China.* London, Murby, 1939. Pp. 506, ills., maps, diagrs., bibl.

Grabau, A. W. *Stratigraphy of China.* 2 vols. Peking, Geol. Survey of China, 1923–1928. *Pt. I, Palaeozoic and Older.* Pp. 528, ills., fold maps, tables, diagrs. *Pt. II, Mesozoic.* Pp. 774, fold and other maps and diagrs.

Berkey, C. P.; Morris, F. K. *Geology of Mongolia.* N.Y., Amer. Mus. of Nat. History, 1927. Pp. 475, ills., pocket and other maps and diagrs., bibl. (*Natural History of Central Asia,* Vol. II.)

Fuller, M. L.; Clapp, F. G. "Geology of the North Shen-si Basin, China," *Bull.*

Geol. Soc. Amer., N.Y., Vol. 38, June 1927. Pp. 287–378, ills., fold and other maps and diagrs., footnote refs.

Fuller, M. L.; Clapp, F. G. "Oil Prospects in Northeastern China," *Bull. A.A.P.G.*, Vol. 10, No. 11, Nov. 1926. Pp. 1073–1117, ills., maps, tables.

INDIA AND PAKISTAN

Coates, J. "Petroleum in India," *Indian Minerals*, Vol. 1, No. 1, Jan. 1947. Pp. 13–21, ills., diagrs.

Wadia, D. N. *Petroleum Resources of India.* Calcutta, 1947. (*Special Publ. No. 11, Indian Assn. for Cultiv. of Sciences.*) Condensed in *Indian Minerals* (Calcutta), Vol. 1, No. 4, Oct. 1947, pp. 221–227, and Vol. 2, No. 1, Jan. 1948, pp. 5–10, maps, diagrs.

Sale, H. M.; Evans, P. "The Geology of British Oil Fields: Part I, The Geology of the Assam-Arakan Oil Region," *Geol. Mag.*, London, Vol. 77, No. 5, May 1940. Pp. 337–363, map, diagr.

Wadia, D. N. *Geology of India.* London, Macmillan, 1939. 2nd ed. Pp. 460, ills., fold and other maps, diagrs., footnote refs.

Condit, D. D. "Natural Gas and Oil in India," *Bull. A.A.P.G.*, Vol. 18, No. 3, Mar. 1934. Pp. 283–314, maps, diagrs., tables, bibl.

JAPAN

Cerkel, J. D. *Petroleum Resources and Production in Japan.* Tokyo, Allied Powers General Headquarters, 1947. Pp. 54, ill., map, diagrs. (*Nat. Res. Sec., Rept. No. 80.*)

Pollock, C. M.; Stach, L. W. "Production and Resources of Petroleum in Japan," *Bull. A.A.P.G.*, Vol. 31, No. 1, Jan. 1947. Pp. 156–158.

Willis, B. "Why the Japanese Islands?" *Scientific Monthly*, Vol. 51, No. 2, Aug. 1940. Pp. 99–111, map, footnote refs.

Imperial Geological Survey of Japan. *Outlines of the Geology of Japan.* Tokyo, 1902. Pp. 252, tables, descriptive text (in English) to accompany geological map of the Empire, 1:1,000,000.

THE PHILIPPINES

Willis, B. "The Philippine Archipelago: an illustration of continental growth," *Proc. VIth Pacific Sci. Congr.*, Berkeley, Calif., 1939, Vol. I. Pp. 185–200.

Willis, B. *Geologic Observations in the Philippine Archipelago.* Manila, 1937. Pp. 129, ills., pocket maps. (*Nat. Research Council of the Philippines, Bull. No. 13.*)

Africa

GENERAL

"La recherche de pétrole dans l'Union Française," *Chronique des Mines Coloniales*, Paris, Bureau d'Études Géologiques et Minières Coloniales, Mar. 1947. And see other issues.

Krenkel, E. *Geologie Afrikas.* In 3 parts. Berlin, Borntraeger, 1925–1938. Teil I, pp. 461; Teil II, pp. 527; Teil III, pp. 930. Each with ills., maps, diagrs., tables, bibl. (In series *Geologie der Erde.*)

Reed, F. R. C. *Geology of the British Empire.* London, Arnold, 1921. Pp. 480, fold maps and others, diagrs., bibl.

NORTH AFRICA

De Cizancourt, H. "Les recherches de pétrole en Algérie et en Tunisie," *Les ressources minérales de la France d'Outre-Mer,* Vol. 5. Paris, Société d'Éditions Géographiques, Maritimes et Coloniales, 1937. (*Publication du Bureau d'Études Géologiques et Minières Coloniales.*) Pp. 123–162, maps, diagrs., bibl.

Migaux, L. "Les recherches de pétrole au Maroc," *ibid.* Pp. 81–119, maps, diagrs., bibl.

Anderson, R. Van V. *Geology in the Coastal Atlas of Western Algeria.* N.Y., Geol. Soc. of Amer., 1936. Pp. 450, ills., fold map and diagrs., bibl. (*Geol. Soc. Amer. Mem.* 4.)

De Cizancourt, H. "Petroleum Research in North Africa," *Bull. A.A.P.G.,* Vol. 16, No. 5, May 1932. Pp. 443–467, ills., map, diagrs., tables, bibl.

WEST AFRICA

Robert, M. *Le Congo physique.* Liège, Vaillant-Carmanne, 1946. 3rd ed. Pp. 449, ills., fold and other maps, diagrs., bibl.

De Cizancourt, H. "Les recherches de pétrole en A.E.F.," *Les ressources minérales de la France d'Outre-Mer,* Vol. 5. Paris, Société d'Éditions Géographiques, Maritimes et Coloniales, 1937. Pp. 163–176, map, bibl.

Pirson, S. J. "Oil Possibilities of Belgium and the Belgian Congo," *Bull. A.A.P.G.,* Vol. 18, No. 9, Sept. 1934. Pp. 1168–1174, maps, diagrs., tables, footnote refs.

EAST AFRICA AND MADAGASCAR

De Cizancourt, H. "Les recherches de pétrole à Madagascar," *Les ressources minérales de la France d'Outre-Mer,* Vol. 5. Paris, Société d'Éditions Géographiques, Maritimes et Coloniales, 1937. Pp. 177–193, map, bibl.

Gregory, J. W. *The Rift Valleys and Geology of East Africa.* London, Seeley, Service, 1921. Pp. 479, ills., fold and other maps, diagrs., bibl.

SOUTH AFRICA

Du Toit, A. L. *The Geology of South Africa.* Edinburgh, Oliver & Boyd, 1926. Pp. 445, ills., fold map and diagrs., tables, footnote refs.

AUSTRALIA

David, T. W. E.; Browne, W. P. *The Geology of the Commonwealth of Australia.* In 3 vols. London, Arnold, 1950. Vol. I, pp. 725, ills., maps, sections. Vol. II, pp. 600, ills., maps, sections. Vol. III, color maps of Australia and New Guinea.

Reeves, F. "Geology of Roma District, Queensland, Australia," *Bull. A.A.P.G.,* Vol. 31, No. 8, Aug. 1947. Pp. 1341–1371, fold maps and others, diagrs., tables, bibl.

Condit, D. D.; Raggatt, H. G.; Rudd, E. A. "Geology of Northwestern Basin, West Australia," *Bull. A.A.P.G.,* Vol. 20, No. 8, Aug. 1936. Pp. 1028–1070, map, diagrs.

Condit, D. D. "Oil Possibilities in Northwest District, Western Australia," *Eco-*

nomic Geology, Vol. 30, No. 8, Dec. 1935. Pp. 860–878, map, footnote refs.

Clapp, F. G. "A Few Characteristics of the Geology and Geography of the Northwestern Basin, West Australia," *Proc. Linnean Soc., New So. Wales,* Sidney, Vol. 50, Pt. 2, No. 201, 1925. Pp. 47–66, ills., map, diagrs., footnote refs.

NEW ZEALAND

Henderson, J. *Petroleum in New Zealand.* Wellington, N.Z. Dept. of Scientific and Industrial Research, *Bull. No. 60.*

The World as a Whole

THE POLAR AREAS

Hubschmann, E. W. "Arktisches Erdöl," *Polarforschung,* Hamburg, Band II/1947, Heft 1/2, Aug. 1948. Pp. 173–175.

Reed, J. C. "Recent Investigations by the U.S. Geological Survey of Petroleum Possibilities in Alaska," *Bull. A.A.P.G.,* Vol. 30, No. 9, Oct. 1946. Pp. 1433–1443, ills., maps.

Lloyd, T. "Oil in the Mackenzie Valley," *Geogr. Rev.,* Vol. 34, No. 2, Apr. 1944. Pp. 275–307, ills., map, footnote refs.

Pratt, W. E. "Petroleum in the North," *Compass of the World* (edited by H. W. Weigert and V. Stefansson). N.Y., Macmillan, 1944. Pp. 336–347, map.

Nansen, F. "The Oceanographic Problems of the Still Unknown Arctic Regions," *Problems of Polar Research.* N.Y., Amer. Geographical Society, 1928. (*Amer. Geogr. Soc. Spec. Publ. No. 7.*) Pp. 3–13, fold and other map, diagrs.

Mackenzie, A. *Voyages . . . through the Continent of North America to the Frozen and Pacific Oceans, in the years 1789 and 1793. . . .* Toronto, Radisson Society, 1927. Pp. 498, ills., map, bibl. (*Masterworks of Canadian Authors, Vol. 3.*)

Paige, S.; Foran, W. T.; Gilluly, J. *A Reconnaissance of the Point Barrow Region, Alaska.* Washington, Gov't Printing Office, 1925. Pp. 33, ills., fold map and diagrs. (*U.S. Geol. Surv. Bull. No. 772.*)

THE CONTINENTAL SHELVES

Shepard, F. P. *Submarine Geology.* N.Y., Harper, 1948. Pp. 348, ills., fold and other maps and diagrs., bibls.

Pratt, W. E. "Petroleum on the Continental Shelves," *Bull. A.A.P.G.,* Vol. 31, No. 4, Apr. 1947. Pp. 657–672, footnote refs.

Umbgrove, J. H. F. *The Pulse of the Earth.* The Hague, Martinus Nijhoff, 1947. 2nd ed. Pp. 358, fold and other maps, diagrs., tables, bibls.

Umbgrove, J. H. F. "Origin of the Continental Shelves," *Bull. A.A.P.G.,* Vol. 30, No. 2, Feb. 1946. Pp. 249–253, bibl.

Sverdrup, H. U.; Johnson, M. W.; Fleming, R. H. *The Oceans: Their Physics, Chemistry, and General Biology.* N.Y., Prentice Hall, 1942. Pp. 1087, ills., fold and other maps, diagrs., tables, bibls.

Jones, O. T. "Continental Slopes and Shelves," *Geogr. Jour.,* London, Vol. 97, No. 2, Feb. 1941. Pp. 80–96, map and charts.

Ewing, M.; Woollard, G. P.; Vine, A. C. "Geophysical Investigations in Emerged and Submerged Atlantic Coastal Plain, Part IV. . . ." *Bull. Geol. Soc.*

Amer., N.Y., Vol. 51, No. 12, Dec. 1940. Pp. 1821–1840, fold and other diagrs., tables, bibl.

Tams, E. *Grundzüge der physikalischen Verhältnisse der festen Erde.* Berlin, Borntraeger, 1938. Pp. 377, ills., maps, diagrs., tables, bibl. (In series, *Geologie der Erde.*)

Ewing, M.; Crary, A. P.; Rutherford, H. M.; Miller, B. L. "Geophysical Investigations in the Emerged and Submerged Atlantic Coastal Plain, Parts I and II," *Bull. Geol. Soc. Amer.*, N.Y. Vol. 48, No. 6, June 1937. Pp. 753–812, maps, diagrs., tables, footnote refs.

Shepard, F. P.; Cohee, G. V. "Continental Shelf Sediments off the Mid-Atlantic States," *Bull. Geol. Soc. Amer.*, N.Y., Vol. 47, No. 3, Mar. 1936. Pp. 441–457, fold and other maps.

Daly, R. A. *The Changing World of the Ice Age.* New Haven, Yale Univ. Press, 1934. Pp. 271, ills., maps, diagrs., tables.

Bucher, W. H. *The Deformation of the Earth's Crust. . . .* Princeton, Princeton Univ. Press, 1933. Pp. 518, fold map and others, diagrs., tables, footnote refs.

Kossinna, E. v. "Die Erdoberfläche," *Handbuch der Geophysik*, Vol. 2, *Aufbau der Erde.* Berlin, Borntraeger, 1933. Pp. 869–954, ills., maps, diagrs., tables, footnote refs.

Shepard, F. P. "Sediments of the Continental Shelves," *Bull. Geol. Soc. Amer.*, N.Y., Vol. 43, No. 12, Dec. 1932. Pp. 1017–1040, fold maps and others, footnote refs.

Twenhofel, W. H., and others. *Treatise on Sedimentation.* Baltimore, Williams and Wilkins, 1932. Pp. 661, ills., diagrs., tables, footnote refs.

Van der Gracht, W. A. J. M. Van W., and others. *Theory of Continental Drift.* Tulsa, Okla., A.A.P.G., 1928. Pp. 240, maps, diagrs.

Johnson, D. W. *Shore Processes and Shoreline Development.* N.Y., Wiley, 1919. Pp. 584, ills., diagrs., bibl.

PART IV. THE UTILIZATION OF PETROLEUM *

Wilson, R. E.; Roberts, J. K. "Petroleum and Natural Gas: Uses and Possible Replacements," *Seventy-five Years' Progress in the Mineral Industry 1871–1946*, pp. 722–744. N.Y., A.I.M.M.E., 1947.

Frey, J. W.; Ide, H. C., Eds. *A History of the Petroleum Administration for War.* Washington, Gov't Printing Office, 1946. Pp. 463, ills., maps, diagrs., statistical tables.

Voskuil, W. H. *Postwar Issues in the Petroleum Industry.* Urbana, Univ. of Ill., 1946. Pp. 32, diagrs., tables, bibl. (*Univ. of Ill. Bull., Vol. 43, No. 30*).

Abraham, H. *Asphalts and Allied Substances. . . .* N.Y., Van Nostrand, 1945. 5th ed. 2 vols. Pp. 2142, ills., diagrs., bibl.

Brooks, B. T. *Peace, Plenty and Petroleum.* Lancaster, Pa., Cattell Press, 1944. Pp. 197, maps, tables, diagrs.

Mather, K. F. *Enough and to Spare. . . .* N.Y., London, Harper, 1944. Pp. 186, maps, diagrs., tables, bibl.

* In addition to the titles listed below, many references are given in the text and footnotes of Part IV, Chapter Four, "A Statistical Survey," and in the bibliography for Part II, "The Functional Organization of the Petroleum Industry."

ABBREVIATIONS

b/d = barrels per day (barrels of 42 United States gallons).

°API = measurement of specific gravity of petroleum at a temperature of 60° F. according to an arbitrary scale adopted by the American Petroleum Institute and ranging from 5 degrees at the heavy end to 65 degrees at the light end. A gallon of 5° API oil has a specific gravity of 1.037 and weighs 8.636 pounds. A gallon of 65° API oil has a specific gravity of 0.7201 and weighs 5.994 pounds.

. . . = information not available.

— = quantity nil or very small in amount.

Other abbreviations used in the text in this volume are listed in their alphabetical places in the Index and are explained in the text at the place of their first occurrence.

NOTE ON GLOSSARIES

A number of the works listed in the Bibliography contain useful glossaries. In particular, Kalichevsky, V. A., *The Amazing Petroleum Industry* (N.Y., Reinhold Publishing Corp., 1943) gives some two hundred items, and Dunstan, A. E., and others, *The Science of Petroleum* (N.Y., London, Oxford Univ. Press, 1938, 4 vols.) gives over two hundred items in the section, "The Nomenclature of Petroleum Products" by Nash, A. W., and Hall, F. C., in Volume I.

Definitions of a considerable number of more or less specialized petroleum terms are to be found in recent editions of Webster and other standard dictionaries. The three works listed below may be consulted for other references.

Fay, A. H. *A Glossary of the Mining and Mineral Industry*. Washington, D.C., Gov't Printing Office, 1948. (U.S. Bureau of Mines Bulletin 95.) Pp. 754, bibl.

Irizarry, O. B. *Petroleo Interamericano's Glossary of the Petroleum Industry*. (Pt. I, English-Spanish. Pt. II, Spanish-English.) Tulsa, Okla., Petroleum Publishing Co., 1947. Pp. 316.

Porter, H. P. *Petroleum Dictionary for Office, Field, and Factory*. 4th ed. Houston, Tex., Gulf Publishing Co., 1948. Pp. 326.

STANDARD DEFINITIONS OF SOME TERMS RELATING TO PETROLEUM *

Crude Petroleum. A naturally occurring mixture, consisting predominantly of hydrocarbons, and/or of sulfur, nitrogen and/or oxygen derivatives of hydrocarbons, which is removed from the earth in liquid state or is capable of being so removed. NOTE: Crude petroleum is commonly accompanied by varying quantities of extraneous substances such as water, inorganic matter and gas. The removal of such extraneous substances alone does not change the status of the mixture as crude petroleum. If such removal appreciably affects the composition of the oil mixture, then the resulting product is no longer crude petroleum.

Crude Shale Oil. The oil obtained as a distillate by the destructive distillation of oil-shale.

Gas Oil. A liquid petroleum distillate having a viscosity intermediate between that of kerosine and lubricating oil. NOTE: It should be understood that oils, other than gas oil as defined above, may be and are used in the manufacture of gas.

Gasoline. A refined petroleum naphtha which by its composition is suitable for use as a carburant in internal combustion engines.

Kerosine. A refined petroleum distillate having a flash point not below 73°F (23° C), as determined by the Abel Tester (which is approximately equivalent to 73°F as determined by the Tag Closed Tester, A.S.T.M. Standard Method D 56) and suitable for use as an illuminant when burned in a wick

* From American Society for Testing Materials Standards, 1944. See: *A.S.T.M. Standards, including Tentative Standards, Part III, Nonmetallic Materials—General*. American Society for Testing Materials, Philadelphia, 1945. Pp. 300, 301.

lamp. NOTE: In the United States of America local ordinances or insurance regulations require flash points higher than 73°F (23°C), Tag Closed Tester.

Oil Shale. A compact rock of sedimentary origin, with an ash content of more than 33 per cent and containing organic matter that yields oil when destructively distilled but not appreciably when extracted with the ordinary solvents for petroleum.

Topped Crude Petroleum. A residual product remaining after the removal, by distillation, or other artificial means, of an appreciable quantity of the more volatile components of crude petroleum.

Weathered Crude Petroleum. The product resulting from crude petroleum through loss, due to natural causes, during storage and handling, of an appreciable quantity of the more volatile components.

Fuel Oil. Any liquid or liquefiable petroleum product burned for the generation of heat in a furnace or firebox, or for the generation of power in an engine, exclusive of oils with a flash point below 100°F (38°C), Tag Closed Tester, and oils burned in cotton or woolwick burners. Fuel oils in common use fall into one of four classes: (1) residual fuel oils, which are topped crude petroleums or viscous residuums obtained in refinery operations; (2) distillate fuel oils, which are distillates derived directly or indirectly from crude petroleum; (3) crude petroleums and weathered crude petroleums of relatively low commercial value; (4) blended fuels, which are mixtures of two or more of the three preceding classes.

CONVERSION FACTORS *

Barrels (U.S.) multiplied by:

9702	equals	cubic inches
5.6146	"	cubic feet
0.15898	"	cubic meters
34.9726	"	Imperial gallons
42	"	U.S. gallons
158.984	"	liters
0.1588	"	metric tons (water 60°F.)
. . . .	"	tons of crude oil (see crude-oil conversion factors table below.)

Crude Oil Conversion Factors

Averages for Principal Producing Countries

Country	Barrels per Ton Long Tons	Metric Tons	Short Tons
North America			
Canada	7.891	7.766	7.046
Mexico	7.281	7.166	6.501
United States	7.459	7.341	6.660
South America			
Argentine	7.134	7.021	6.370
Bolivia	7.841	7.717	7.001
Colombia	7.143	7.030	6.378
Ecuador	7.707	7.585	6.881
Peru	7.661	7.540	6.840
Trinidad	7.193	7.079	6.422
Venezuela	6.958	6.848	6.212
Europe			
Albania	6.706	6.600	5.988
Austria	6.897	6.788	6.158
Czechoslovakia	6.891	6.782	6.153
England	7.440	7.322	6.643
France	7.177	7.064	6.408
Germany	7.363	7.247	6.574
Hungary	7.754	7.631	6.923
Italy	7.800	7.677	6.964
Poland	7.526	7.407	6.720
Rumania	7.525	7.406	6.719

* All the equivalents cited in this section are from the table of conversion factors in *World Oil Atlas*, July 1948, p. 55.

CONVERSION FACTORS

Country	Barrels per Ton Long Tons	Metric Tons	Short Tons
U.S.S.R. (except Sakhalin)	7.390	7.273	6.598
Africa			
Egypt	7.116	7.003	6.354
Asia (Middle East)			
Bahrein	7.446	7.328	6.648
Iran (Persia)	7.669	7.548	6.847
Iraq	7.600	7.480	6.786
Kuwait	7.482	7.364	6.680
Saudi Arabia	7.537	7.418	6.729
Asia (Far East)			
Borneo	7.345	7.229	6.558
India	7.355	7.239	6.567
Japan	7.143	7.030	6.378
Indonesia	7.911	7.786	7.063
Sakhalin (U.S.S.R.)	6.930	6.820	6.188
Sarawak	6.762	6.655	6.038

A.P.I. Gravity Conversion Factors

Gravity	Barrels per Ton Long Tons	Metric Tons	Short Tons
28° Crude Oil	7.220	7.106	6.446
32° " "	7.401	7.284	6.608
36° " "	7.582	7.462	6.770
40° " "	7.763	7.641	6.931
44° " "	7.945	7.819	7.094

INDEX

Abadan, 163, 168-176, 188; Figs. 32-34; Plate 16
Abednego, 161, 181
Abha, population, 221
Abqaiq field, 176, 204, 209, 215-219; pipe line to Ras Tanura, 32; Figs. 32, 34, 36; Plates, 65, 67
Abu Dhabi, oasis, 169-170; shaikhdom, 200
Abu Durba, 187
Abu Hadriya field, 176, 204, 215
Abyssal zone, 5, 6
Abyssinia (*see also* Ethiopia), 194
Acre Territory, Brazil, 129
Adana basin, Turkey, 185
A.D.A.R.O., government corporation, Spain, 269
Aden, 206
Adriatic Sea, 246, 265; refineries, 266; trough, 248
Aegean Sea, 191
Afghanistan, 301-302
Africa (*see also* Egypt), 301, 302; central, 194; east, 171, 302; motor fuel leading product used, 384-385; northern, 301, 367-368; petroliferous areas, 301, 302; products utilization, 367-368: civil aviation, 367, fuel oil, 368, kerosene, 367-368, lubricants, 368, per capita consumption by regions, 367; refining, 368; west, 301, 302
African shield, 5, 326, 328
African-Arabian crystalline shield, 171
Agha Jari field, 173, 175
A.G.I.P., *see* Azienda Generale Italiana Petroli
Agriculture, fuel oil use in U.S.S.R., 386; postwar mechanization in Europe, 390; tractors on U.S. farms, 362
Agua Caliente field, Peru, 120, 124-125, Plate 1
Agua Dulce discovery, Mexico, 98
Ahmad Shah, 197
Ahmadi, Kuwait, 184, 185; Figs. 32, 35; Plate 61
Ahwaz, Persia, 159, 188, 190
A.I.M.M.E., *Transactions,* consumption data, 379
Ain Dar field, Saudi Arabia, 162, 204, 216
Ain Zaleh field, Iraq, 182-183
A.I.O.C., *see* Anglo-Iranian Oil Company, Ltd.
Air mapping, in geological exploration, 15, 16; Indonesia, 289, 291-293; Fig. 45; Plate 85
Air transportation in pipe-line construction, Colombia, 112
Aislaby field, 262
Ajman shaikhdom, 200
Akyab, 297
Alabama, 45, 154, 158, Fig. 19, 134, Fig. 20, 135, Fig. 31, 157, Table 11, 139, Table 12, 142
Alagôas State, Brazil, 120
Alaska, Alaskan Theater, World War II, 350-351; northern, 314; Peninsula, 316; plains, 154-156; U.S. Dept. of Interior, exploration, 313-318; U.S. Naval Petroleum Reserve No. 4, 313-314 and Fig. 48; work feasibility chart, 315; Figs. 27, 29, 47-49; Plates 89, 95
Albania, 246, 249, 254, 271; refinery, 250, 253; Table 20, 250; Fig. 38, 242
Alberta, 150, 153-154, 156-158, 314; Fig. 29; Plates 48, 49
Alcan Highway, 317, 351
Aleppo, 186, 190
Aleutians, 350
Alexander the Great, 160-161
Alexandretta, 188
Alexandria, 188
Algae, *see* plankton
Algeria, 301
Al Hasa province, Saudi Arabia, 223
Al Kharj, 218, 222, 228; Fig. 32, Plates 68, 69
Al Khobar, 217
Alkylate, 36
Alkylation, 36, Fig. 55, 358-359
All Saints' Bay, Brazil, 128
Aller River, 257
Allier basin, 259
Allies and Axis, oil supplies, 393-394 (*and see* World War II)
Alpine geosynclinal system, Europe, 246
Alpine-Himalayan system, 164
Alps, 240, 245-246
Al Qurna, Iraq, 169
Alternative sources of petroleum, 324, 342
Alwand, Iraq, 181
Amara (Amarah), Iraq, 169, 194
Amazon River, 120; drainage system, 102, 124; upper basin, 124-125; lower basin, 128-129
Ambal, Persia, 168
American Independent Oil Co., operations in Mexico, 99; area and period of Middle East concessions, Table 15, 177-179
American Institute of Mining and Metallurgical Engineers, *see* A.I.M.M.E.
American Mission Hospital, Manama, 225
American Petroleum Institute, consumption data, 380
American-British-Dutch embargo, 1941, 393
Amir Abdul-Illah, Regent of Iraq, 198
Amiranian Oil Company, 175

Amotape Mts., Peru, 122
Amur River, 237
Anabar shield, 232
Anaiza, population, 221
Anatolian highlands, 194
Anatolian railway, 190
Anchorage, Alaska, 350, 351
Andes, Argentina, 126, 127; Bolivia, 125; Chile, 127; Colombia, 102, 123; Ecuador, 122, 124; Peru, 124; Trinidad, 82; Venezuela, 52-54, 73, 77, 102; opening up for farming, lumbering, 393
Andian National Corporation, Colombia, 111, 113
Andian trunk-line system, Colombia, 111-113
Andrews, E. B., 300
Angara shield, 232
Anglo-Egyptian Oilfields, Ltd., 188; concession in Egypt, Table 15, 179
Anglo-Iranian Oil Company, Ltd., 159, 163, 172-174, 181-182, 188-190, 204, 262-263; Table 15, area and period of concessions, 177-179; Technical College and Apprentice school, 195
Anglo-Portuguese Oil Company, 268
Anglo-Saxon Petroleum Company, 295
Angola, 302
Ankara, 185
Annual requirements of petroleum products, U.S.A. and C.E.E.C. countries, 1938 and 1947, Table 29, 390
Antarctica, 308, 310, 327
Antarctic shield, 327
Anticlinal structure, first observed, 300
Anticline, definition, 12
Anti-Lebanon Mountains, 190
Antioquia railroad, Colombia, 113
Antwerp, refineries, 268
Anzoátegui, Venezuela, 66-67, 73
Apennine Mts., 240, 246; foothills, 248, 265
Appalachian and north-central states, natural gas fields, Fig. 18, 133; oil fields, Fig. 17, 132
Appalachian Arch, 158
Appalachian Mts., overthrust belts, 154
Apsheron Peninsula, 233
Apure Basin, 57, 71; Table 3, 60; Fig. 8
Aqaba, Gulf of, 164, 171
Aquila S.A. Tecnico-Industriale, 266
Aquitaine Basin, 241, 248, 259
Arab League, 199
Arab nomadic tribes, 195
Arab states and Middle East Oil, 399-400
Arab-Jewish troubles, 186
Arabia (*see also* Aden, Bahrein, Kuwait, Muscat, Oman, Qatar, Saudi Arabia, Trucial Coast, Yemen), 162-172; Fig. 32, 165-167; Fig. 34, 176; Table 15, 177-179; 182-183; Fig. 35, 184; 189-196; Table 16, 196; 199-200; 203-223; 227-229
Arabian American Oil Company (Aramco), agriculture, 221; air transport, 218; concessions, area and period, Table 15, 177-179; educational facilities, 220-221; financial arrangements with Saudi Gov't., 222; formation, 213; harbor facilities, 219; hospitals, 220; industries, 219; King Ibn Saud, statement, 222-223; oil-field development, 213-217, Fig. 36, 214; opportunity schools, 220; personnel, 215; pipe lines, 32, 163, Table 17, 218; production and reserves, 216, 340-341; public utilities, 219; radio network, 220; sanitation, 220; trade schools, 220; transport operations, 227
Arabian Sea, 205, 212
Arafat, 229
Arakan coast, 299
Arakan Yoma, 275, 298
Aramco, *see* Arabian American Oil Company
Aratu field, Bahia, 129
Arbadil, 161
Arctic, and Antarctic, 308; eastern hemisphere, petroleum, 311-313; international crossroads, 308, 318; shields, 232; western hemisphere, petroleum, 313-318
Arctic mediterranean, 308
Arctic Sea, 326
Ardericca, Persia, 159
Argentina, continental shelf area, 328, 329; northern, accessible to Bolivian oil, 125; oil development, 126-127; petroleum products, use, 364, 365; Figs. 16, 49; Plates 40-42; Table 1
Aricesti field, Rumania, 251
Aripero, Trinidad, 85
Ark, the, 159
Arkansas, 45, 158, Fig. 19, 134, Fig. 20, 135, Fig. 31, 157, Table 11, 139, Table 12, 142
Arkhangel'sk, 311
Armenia, Colombia, 113
Aru Bay, 284
Aruba, 37, 49, 50; Figs. 6, 8, 9; Plates 13, 18, 19
Aruma escarpment, 207-208
Asbestos, 360
Asia, central, complex of shields in, 5, 232, basin, 237; products utilization, 368-369; southeastern, oil field areas, facilities, refineries, Fig. 40, 275
Asia, other than Middle East, Burma, and Soviet Union, 302-305
Asir, 205, 229
Asl field, Egypt, 187
Asphalts, deposits, Trinidad, 85, 89, Venezuela, 71-72; petroleum, varieties, uses, 360; shales, Morraro, Italy, 266; ship calking by pirates, Venezuela, 71; shipment, 14
Assam, 297-298, 302-304, 352-353; Plates 86, 87
Asterabad province, 175
Astrakhan', 235
Athabaska, oil sands, 156, 317
Atjeh, 278, 282

Atlantic Petroleum Company, 260
Australia, 301; South, 305-306, motor fuel leading product used, 382, 384; Table 1
Australian shield, 5
Austria, 241, 246, 249, 252-253, 271; refineries, 250, 253; Table 19, 241, Table 20, 250; Fig. 38, 242-243, Fig. 39, 247
Automobile routes in Saudi Arabia and neighboring areas, 227-229
Autun, shales, 260
Auyán-tepuí, Mt., 56
Availability of petroleum, the search, 335, 336; proved reserves and total resources, 336-338; making the most of resources, 338-340; U.S. and world's problem of supply, 340-341; world problem, 341-342; mineral interdependence, 342; alternative sources, 342
Aviation, commercial, Colombia, 113, 114
Aviation gasoline, components and how made, Fig. 55, 358-359
Awali, 215; hospital at, 226
Axis and Allies, oil supplies, 393-394
Axle grease, 360
Azerbaijan, 175
Azienda Generale Italiana Petroli (A.G.I.P.), 265

Baba Gurgur, Iraq, 181; Figs. 32, 34; Plate 58
Babalan River, 284
Babat field, 285
Babo, 293
Babylon, 160
Babylonian canals, 194
Badarpur, 297
Bacau fields, 251
Bachaquero field, Venezuela, 60; Fig. 9, 58; Fig. 10, 59
Bacteria, 7, 10
Baden, 259
Badra, Iraq, 165
Baghdad, 169, 174, 181, 189-190
Bahamas, magnetometer survey, gravimetric mapping, 48
Bahia, Brazil, 120, 128
Bahra, Kuwait, 182
Bahrein, 163, 169, 170, 176, 189, 195, 200, 203-204, 213, 214, 217, 223-229, 326, 329; refinery, 163, 215, 217; Figs. 32, 34, 36, 49; Table 1
Bahrein Petroleum Company, Ltd. (Bapco), 203, 213, 215, 217, 223-226; area and period of concessions, Table 15, 177-179
Baiji, Iraq, 190
Bajunglentir refinery, 284
Bakhtiari Mts., 192
Baku, 146, 160, 233-234, 238; Figs. 37, 38; Plate 73
Baku, Second, 234
Balakhany field, 233
Balaton Lake, 245
Balik Papan, Borneo, 275, 276, 281, 283, 287, Fig. 43, 288, 294; Figs. 40, 43; Plate 81
Balkassar field, Pakistan, 303
Baltic Sea, 236, 256, 267; Figs. 37, 38
Baluchistan, 164
Bandar Abbas, Persia, 168, 170, 189
Bandar Mashur, Persia, 174, 175, 189
Bandar Shahpur, Persia, 189
Bandar Shuwaikh, Kuwait, 185
Bandjermasin, 283
Baniyas, 176
Bapco, *see* Bahrein Petroleum Company
Baquba, Iraq, 195
Barbosa, Colombia, 113
Barcelona, Venezuela, 66
Barcelona Gap, Venezuela, 54
Barco, General Virgilio, 115, 116, 117
Barco concession, Colombia, 100, 107-110, 112, 115-116
Barents Sea, 311
Barger, T. C., 203
Bari, 254, 265
Barito (Tandjung) field, Borneo, 276, 280
Barito River, 283
Barquisimeto, 54
Barrancabermeja, Colombia, 101, 105, 111-113, 115
Barranquilla, Colombia, 112-113
Base stock, for aviation gasoline, 358
Basement rocks, 18
Basra, Iraq, 162, 169, 182, 188, 190, 194, 200, 225
Basra-Baghdad railway, 200
Basrah Petroleum Co., Ltd., 180, 182; area and period of concessions, Table 15, 177, 178
Bataafsche Petroleum Maatschappij (BPM), 261, 273, 278-282, 288, 290-291, 294
Batavia (Jakarta), 282
Bathyal zone, 5, 6
Batumi (Batum), 185, 236
Baudó valley, Colombia, 120
Bauer, C. J., 369
Bavaria, 259
Bear Lake, 311
Beckwith and Co., Ltd., concessions in Egypt, Table 15, 179
Beirut, 188, 190, 192
Belém, Brazil, 129
Belgian Congo, 301-302
Belgium, 245, 261, 268; refineries, 268; Fig. 38; Plate 92; Table 1
Belgo Petroleum Company, 268
Beme petroliferous tract, 298, reserve, 299
Bengal, 298
Bengal-Assam basin, 304
Bentheim gas field, 256
Berat skimming plant, 254
Bering Sea, 312
Bering Strait, 311
Berlin-Baghdad railway, 189-190, 200

Besano shales, 266
Bethlehem Steel Company, Venezuelan operations, 56
Béziers, 259
Big Delta, Alaska, 351
"Big Inch" pipe line, 33
Bisha, population, 221
Bissell, George H., 131
Bit, *see* drilling
Bizone of Germany, 255
Black Sea, geological exploration, 185; wind tracks, 191; ports, 236, 252
Blending agents, methods of producing, Fig. 55, 358-359
Blending, of high octane aviation gasoline components, 358
B.O.C., *see* Burmah Oil Company
B.O.D., Ltd., *see* British Oil Development, Ltd.
Bogotá, Colombia, 102, 112-113, 123
Bogotá marketing area, 113
Bohemia, western, 264
Boldesti field, 251
Bolívar, Simon, decree of Quito, 1829, 76
Bolívar coastal field, Venezuela, 58-62; Fig. 9, 58; Fig. 10, 59; Table 3, 60
Bolívar District, Venezuela, 72
Bolivia, oil fields, pipe lines, refineries, 125, Fig. 16, 121; fuel use, 384-386
Bordeaux region, refineries, 260, Fig. 38
Borneo (*see also* British Borneo, Barito, Balik Papan, Tarakan), 273, 277-283, 287-289, 293-296; refineries, 281, 296, Figs. 40, 43; Plates 80, 81
Boryslaw field, 255
Bottoms, *see* sea bottoms
B.P.M., *see* Bataafsche Petroleum Maatschappij
Brabant massif, 261
Brahmaputra Valley, 294, 298, 303, 353
Bratislava, refinery, 264
Brazil, exploration and oil development, 128-129; petroleum products, use, 364-365; refineries, 129; sedimentary areas, 120; Figs. 16, 49; Plates 43, 44; Table 1
Brazilian shield, 5, 326-327
Brea pits, southern California, 130
Bremen, 257-258
Brest, refineries, 260
Brighton pitch lake, Trinidad, 85, 89
Brines, fossil, 4
Britain, *see* Great Britain
British anti-locust mission, Oman, 1943, 228-229
British Borneo, 295-296; refinery, 296; Fig. 40, 275, Fig. 43, 288
British Columbia, overthrust belts, 154
British Commonwealth, sedimentary areas, 232
British Government, holdings in Middle East concessions, Table 15, 178, 179; interest in Anglo-Iranian Oil Company, 172-173
British-India Steam Navigation Company, 189
British Malayan Petroleum Company, 296
British Oil Development, Ltd. (B.O.D., Ltd.), 178; concession in Iraq, 180
British Overseas Airways Corporation, 225
British Petroleum Production Act, 1934, 262
Brooks Range, 313-315
Brunei, 295-296; continental shelf area, 328; Figs. 40, 43
"Bubble tower," 35
Bucaramanga, Colombia, 113
Bucharest, 252
Bucsani field, 251
Budafapuszta fields, 253-254
Budapest, 254
Buenaventura, Colombia, 101, 113
Bükkszék field, 253
Bulgaria, 269-270, 271; refineries, 270
Bulk stations, bulk terminals, *see* Part II, *Storage, Distribution*
Bunju Island, 280
Bunker fuel oil, European ships and, 365; in U.S. foreign trade, 362
Buqqa sub-field, 204
Buraida, population, 221
Burgan field, Kuwait, 182-183, Fig. 35, 184, 189, 204
Burgo de Osma, 269
Burgos province, Spain, 269
Burma, 274-276, 296-300, 297, 303, 352-353, 369; refineries, 297, 300; Figs. 40, 46, 49; Plate 82; Table 1
Burmah Oil Company, 172, Table 15, 178, 179, 182, 299-300
Bushire, Persia, 165, 168, 170, 172
Butadiene, 37, 42
Butane, from natural gas, 30; in LPG, 37; shipment, 42
Butler, Pa., to Brilliant, Pa., first oil pipe line, 143
Byrd Expedition, second, 310

Cable-tool drilling, *see* drilling
Cadiz, 269
Cadiz-Seville basin, 269
Cadman, W. H., 258
Calcutta, 352-353
Cali, Colombia, 112-113
California, 29, 33, 133, 141, 146, 154, 158, 233, 235, 238, 324, 328-329, 333, Fig. 21, 136, Fig. 22, 137, Table 11, 139, Table 12, 142, Pls. 50, 51
California Arabian Standard Oil Company, 213
Callao mine, Venezuela, 56
Caltex (*see also* Far Eastern Investment, Inc.), 175
Cambrian period, time, 3, 4, 5, 9
Cambyses, King of Persia, 182
Campeche, Mexico, 99

Campina field, 251
C.A.M.P.S.A. (Compañía Arrendataria del Monopolio de Petróleos S.A.), 269
Canada, climate compared with U.S.S.R., 230; current developments, 142, 146, 148, 150; eastern, 316; exploration, 130, 138; Fort Norman, U.S. Army development in World War II, 317; Mackenzie River, 316; motor fuel leading product used, 384, 385; northern, 316; northwestern, 318; plains, 154-156; potential production, 153; refineries, 148; reserves, 152, 154, 156; southern, 317; tanker fleet, 146; uranium ore, 316-317; utilization, 362; Figs. 2-5, 25-29, 56, 57, 59; Plates 45-49, 88; Table 1
Canadian shield, 5, 326-327
Canary Islands, refinery, 269
Candeias field, Brazil, 129; Plates 43, 44
Canol pipe line, 351
Cantagallo concession, Colombia, 106; link to Andian pipe line, 112; Figs. 14, 15
Cape Monze, 170
Cape Simpson, 313
Caquetá River, Colombia, 124
Caracas, Venezuela, 54, 74
Carare river, Colombia, 102-103, 115
Carboniferous period, time, 4
Carcross, Canada, 351
Caribbean Petroleum Company, 72
Caribbean petroleum province, oil field areas and facilities, Fig. 6, 47; proved reserves, 46, 326; compared with Middle East, 326
Caribbean Sea, 325, 327
Caroni River, Trinidad, 80, 82
Carpathian basin, 241, 246, 249, 255, 264, 266, 270
Carpathian fields, 255
Carpathians, 240-241, 244-246, 249, 251, 312
Cartagena, Colombia, 106, 111
Cartagena, Spain, refinery, 269
Casabe field, Colombia, 102, 106, 111-112
Casing, 28-29
Caspian Sea, 164, 190, 233, 235-237, 312, 326, 329
Catalysts in oil formation, 7, 10
Catalytic cracking, 34, 35-36; Houdry, thermafor, and fluid catalyst methods, Fig. 55, 358-359; Plates 17, 19
Catatumbo River, Colombia, 107; upper basin, 116; Fig. 9; Plate 33
Cauca river valley, Colombia, 102; Fig. 14
Caucasus, fields, region, 233-236, 326; Fig. 37
Caucasus Mts., 233; Fig. 37
Caucasus trend, 232-233; Fig. 37
Caunton field, 262-263
Cedros Point, Trinidad, 82
Celebes, 289, 294
Central Asian shield, 326
Central Cordillera, Colombia, 102
Ceptura field, 251
Ceram fields, 276, 281
Cerro Manantiales field, Chile, 127; Plates 38, 39
Cheb (Eger), 264
Cheleken Islands fields, 233
Cheribon, 277, 282
Chia-Surkh field, Iraq, 172
Chicago, pipe-line terminus, 146
Chiclana de la Frontera, 269
Chile, oil development, 127; use of lubricants, industrialization, 365; Fig. 16; Plates 38, 39
China, north, 232; petroliferous areas, 301; and Manchuria, 303-304; China-Burma-India theater, 352-353; Figs. 49, 56, 57, 59; Plate 93
China-Burma-India Theater, 352-353; Plate 93
Chindwin river basin, 274, 298-299
Chira River, Peru, 123
Chittagong, 352-353
Chubut Territory, Argentina, 126
Chungking, 304
Churchill, Winston, 172
Churchill-Roosevelt Highway, Trinidad, 89
Chuvash autonomous republic, 237
C.I.E.P.S.A. (Compañía de Industrializaciónes y Explotaciónes Petrolíferos S.A.), 269
Cieszyn (Tečin), 256
C.I.M.A. (Mexican American Independent Company), 99
Cities Service Oil Company, operations in Mexico, 99; Colombia, 111
Ciudad del Carmen, Mexico, 99
Ciudad Juarez, Mexico, 96
Clapp, F. G., 175
Climate, Trinidad, 82
Climate and vegetation, Middle East, 191-194
Climatic divisions, U.S.S.R., 230
Coal, synthesis of petroleum products from, 324, 342-343
"Coal oil," 34
Coast Range, Venezuela, 52-54, 64
Coatzacoalcos, Mexico, 99
Cochabamba, Bolivia, 125
Codigo de Minas of 1907, Venezuela, 73
Codimer, *see* iso-octane
Coevorden anticline, 261
Cojedes, Venezuela, 64
Coke, 360, 361
Colombia, as part of mediterranean basin, 325; climate and health, 114-115; concessions and pipe lines, Fig. 14, 101; effects of oil industry on national economy, 113-114; fuel use, 385-386; geography, geology, 100-111; major oil companies, Table 10, 114; mid-Magdalena fields, Fig. 15, 104; motor vehicles, imports, 114; national lands, 117-118; National Oil Co., 119; Pacific coast, 120-122; petroleum legislation, 115-119; production, consumption, exports, 100; re-

fining and transportation, 111-113; Figs. 2-5, 6, 14-16, 56, 57, 59; Plates 6, 7, 32-34
Colombian Andes, 52, 102
Colombian Petroleum Company, 107-108, 112, 116-117
Colón Development Company, 108
Colón District, Venezuela, 72
Colorado, U.S.A., Fig. 19, 134, Fig. 20, 135, Table 11, 139, Table 12, 142
Colorado field, Colombia, 105
Colorado River, Colombia, 103
Columbus, 1498 voyage, 82
Colville River, 314
Commerce Yearbook of the United States, 379
Comodoro Rivadavia fields, Argentina, 126-127, 328; Fig. 16; Plate 40
Comodoro Rivadavia-Buenos Aires natural-gas pipe line, 127
Compaction, 8, 10, 11
Compagnie Française des Pétroles, 180, Table 15, 178
Companhia Portugueza do Petróleo, 268
Compañía Colombiana de Petróleos, 116-117
Compañía Petróleo La Estrella de Colombia, 106
Concessionaire companies in Middle East, Table 15, 177-179
Concessions in the Middle East, April 1949, Table 15, 177-179
Condensates, 36, 139, 342, 356, 358, Fig. 55, 359
Conselho Nacional do Petróleo, Brazil, 128-129
Constanta, 252
Consumption (*see also* utilization), and production in U.S. and rest of world, 382-391, Table 28, 388; by Europe, North America, and rest of world, 1938, 382, 384; chief statistical sources, 375-382: government reports on petroleum utilized, 375-377; government statistics of annual production, imports, and exports as basis for estimating retained petroleum, 377-379; compilations combining government statistics with estimates based on all available data, 379-382; estimated, of 4 leading products, by continents in 1938, Table 24, 361; forecasts, U.S. and world, 389-390; world-wide, 1918, 1948, and estimated, 1951-52, 369-372
Continental shelves, 6, 45, 90, 152, 319-325, Fig. 49, 320, 328-329, 339, world map facing p. 14, Plates 90, 91
Continental slope, 6, 323-324
Conucos (small holdings), Venezuela, 55
Copenhagen, 268
Cordillera Central, Colombia, 102, 105
Cordillera de los Andes, 52, Fig. 8, 53
Cordillera de Mérida, Colombia, 102, 107
Cordillera Oriental, Colombia, 102-103
Cordilleras, North and South American, 327
Core drill, in exploration, 16
Cores, coring, 16, 19, 20
Cork, refinery, 270
Coro Ranges, Venezuela, 54
Corporación de Fomento, of Chile, 127
Cortemaggiore field, Italy, 265
Corumbá, Brazil, 125
Corys River, Arabia, 82
Costa Rica, 48, Fig. 6
Coveñas, Colombia, 102, 111-112
Cracking, 34-37; Fig. 55, 358-359; Plates 17, 19
Creole Petroleum Corporation, 49, 60, 62, 66, 70, 73-74, 80, 95
Cretaceous period, time, 4, 5
Creveny shales, 260
Cross section of central portion of an oil basin, Fig. 1, 13
Cuba, 45, 48
Cúcuta, Colombia, 112, 116
Cumarebo field, Venezuela, 57
Cumene, production, Fig. 55, 358-359
Cundinamarca railroad, Colombia, 113
Cunningham Craig, E. H., 85
Curaçao, 37, 49, 50
Cuts, *see* fractionator
Cyprus, 191
Cyrenaica, 191
Czechoslovakia, 241, 246, 254-256, 264, 271; refineries, 264; Fig. 38; Table 19

Dahana, Arabia, 207-208, 211-212, 218, 228
Dalaki, Persia, 172
Dalen, Netherlands, 257
Dalkeith field, Scotland, 262-263
Damascus, 190, 192, 229
Dammam dome, 209, 213
Dammam field, 204, 213-216
Dammam town, 219-220
Danish American Prospecting Company, 267, 268
Danube River, 249, 252-254, 264, 266
D'Arcy, William Knox, 161, 172, 175, 180, 181
D'Arcy Exploration Company, Ltd., 263
Darb Zubaida, 227, 228
Dar-i-Khazineh, Persia, 190
Darin, 221
Darwent, Walter, 85
Dead Sea, 164, 186, 187, 196
Debrecen, 253
Deformation, of strata, *see* Part I, *Traps* and *Geophysical Exploration*
DeGolyer, E. L., 161, 181, 450
De Mares, Roberto, 115, 116
De Mares concession, Colombia, 100, 102-106, 111, 115, 116
Denmark, 256-257, 267-268; refineries, 268, Fig. 38
Depositional zones, 5, 8
Depth of wells, 27, 29
Desert basin, Western Australia, 305, 306
Devoli field, Albania, 254
De Wijk' gas field, 261

Dhahran, Saudi Arabia, 204, 209, 213, 215-220, 225, 227, 228
Dhahran-Jeddah road, 228
Dhahran-Kuwait road, 227
Dhufar, 206, 212
Dhulian field, 303
Dibai, oasis, 169; shaikhdom, 200
Dibdiba, 209
Diesel oil, in U.S., 363
Difícil concession, field, Colombia, 106-107, 112
Difícil-Plato pipe line, Colombia, 112
Digboi field, 298, 303; Plates 86, 87
Dikaka, 210; Plate 64
Distillate fuel oil, in U.S., 390, Table 25
Distillation, 34-35; 355, 358; of gasoline-range crude petroleum in pressure still, Fig. 55, 358, 359
Diyala River, *see* Sirwan River
Diz River, 165
Diz-Karkheh River embayment, 165
Dizful, Persia, 159
Djambi fields, 276, 279-282, 287
Djirak field, 276
Doha, Saudi Arabia, 169, 212
Doherty, Henry L., and Company, 108, 116, 379
Dohuk, Iraq, 192
Dome, salt dome, Fig. 1, 13, 14
Domestic oil burners, 363
Dominican Republic, 48, Fig. 6
Don River, 236
Donbas, 235
Donets basin, 236
Dordtsche Petroleum Maatschappij, 278
Doughty, C. M., 207
Dragon's Mouths, 80
Drake, Colonel, *see* Drake, Edwin L.
Drake, Edwin L., 131
Drill collar, *see* drilling
Drill pipe, *see* drilling
Drilling, 27-29; Plates 8-11, 29, 41, 90
Drilling and Exploration Company, Dallas, Tex., 185
"Dry holes," definition, 20; index of exploratory effort, 21, 24; U.S., 138
Dudinka, 312
Duero-Ebro basin, 241, 248, 269
Duitama, Colombia, 113
Duke's Wood field, 262-263
Dukhan field (*and see* Qatar), 204, 215, Fig. 36
Dunkerque, refineries, 260
Dutch East Indies, *see* Netherlands Indies
Duwadami, 228
Duwaid, 227

Eakring field, 262-263
Earth movements, 4, 6, 7
East African rift system, 171
East Indies, continental shelf area, 329
East Texas field, 45, 141, 143, 145, 156; Fig. 30, 156; structure, 158; compared to Middle East, 162
Eastern and General Syndicate, 224
Eastern Cordillera, Colombia, 102
Eastern Gulf Oil Company, 224
Ebensee refinery, 253
Ecuador, coastal shelf, 122; fuel use, 384-386; oil prospects, 122, Oriente, 124; refineries, 122, Fig. 16
Edmonton, 153
Eire, *see* Ireland
Egypt, agriculture, 194; climate, 191; concessions, Table 15, 177-179; consumption of products, 368; oil fields, 172, 187, 188; political background, 201, 202; production of crude, 368; railways, 190; refineries, 187-188, 368; structure and topography, 171-172; Saudi Arabian treaty, 199; Figs. 32, 34, 56, 57, 59
El Baúl, Venezuela, 64
El Callao, Venezuela, 56
El Difícil concession, Colombia, *see* Difícil
El Dorado, Venezuela, 55
El Jauf, 211, 227
El Plan field, Mexico, 98
El Qantara, Egypt, 190
El Roble field, Venezuela, 66-68, 73
El Tablón concession, Colombia, 110
Elbe River, 258, 264
Eldorado Mines, 316
Elmendorf Airfield, 351
Emba basin fields, U.S.S.R., 237; Fig. 37; Plate 74
Emba-Ural'sk district, U.S.S.R., 234; Fig. 37
Embargo, American-British-Dutch, 1941, 393
Emir Faisal of Iraq, 198
Ems River, 257
Emsland field, 257
Energy per capita produced in each country in 1938: man-power years from 5 main sources of utilized energy, 371-374, Fig. 56, 370, Fig. 57, 371; U.S. and world, 1938 and 1947, Fig. 58, 372
Eocene period, time, 4
Epeirogenic movements, 4, 6
Erbil, Iraq, 190
Erin River, Trinidad, 82
Eritrea, 302
Ermelo, 302
Escarpment region, Arabia, 207
Esperanza field, Colorado, 107
Estonia, 237
Ethiopia, 302; concession, 178
Euphrates river, 160, 169-170, 182, 186, 189, 194-195, 203, 211
Europe, chief sources of imports, 271, 371; consumption, compared to U.S., 271; countries producing more oil than consumed, 249-254, with insufficient commercial production, 254-266, without commercial pro-

duction, 267-271; crude oil, imports, 249, production, refining capacity, Table 20, 250; Middle East chief source of imports, 371; motor fuel, fuel oils after World War II, 366, 367; motor fuel leading product used, 384-385; northern countries, 240; petroleum consumption, 1938, 1947, 367; petroleum development, by countries, 248-271; production of crude and refining capacity, Table 20, 250; Recovery Program (E.C.A.), 367, 390; refineries, 272; requirements and capacities, 271-272; sedimentary basins favorable for petroleum, 240-248, Table 19, principal basins, 241; southeastern countries, 240; utilization of petroleum products, 365-367: per capita differences between regions, 365, fuel oil and motor fuel, 365, bunker fuel, 365, kerosene, 365, U.S.S.R., 365-366, lubricants, 366, Germany during World War II, 366, 367; Figs. 38, 39, 49; Table 1
European Recovery Program, 367, 390
Exports, *see* trade, separate countries and regional divisions, utilization, Figs. 2-5, 40-41; Table 30, 406-408

Fadhili field, Saudi Arabia, 162, 204, 216
Fahaheel, *see* Mena al Ahmadi
Fairbanks, Alaska, 314-317, 350, 351
Faisal II of Iraq, 198
Falcon, N.I., 192
Falcón, state of, Venezuela, 54, 57
Falcón-Lara hills, Venezuela, 52, 54
Falknov, Czechoslovakia, 264
Fao, Iraq, 169, 195
Far Eastern Investment, Inc., 280
Faulting, fault traps, Fig. 1, 13, 14
F.E.A., agricultural mission in Arabia, 222
Fenno-Scandian Shield, 232
Finland, 267, 270; Fig. 38
Firdausi, *Shah-Nama,* 160
Firth of Forth refineries, 263
Fisher, W. B., 159
Fiume, refining center, 266
Fixed-bed catalytic cracking, 35, 36, Fig. 55, 358-359
Floresanto concession, Colombia, 110
Florida, 158, 324, Fig. 6, 47, Fig. 31, 157, Table 11, 139, Table 12, 142
Fluid-catalyst cracking, Fig. 55, 358, 359
Foran, Lieut.-Com. William T., 314
Forbes Magazine, quoted in footnote, 138
Forecasts, U.S. and world consumption, 389-390
Foreign Commerce Weekly, consumption data, 380
Forest Reserve field, Trinidad, 88
Formby field, 262, 263
Formosa, 301, 305; Figs. 40, 49; Table 1
Forst-Weiher field, 257
Fort Norman field, *see* Norman Wells field
Fort Richardson, 351
Fossil brines, 10
Fossils, 16
Fractionator, 35, 36, 37, 355; Fig. 55
France, 248, 254, 255, 259-260; refineries, 259-260; Fig. 38
Fraser, Sir William, 174
Fredonia, N.Y., first gas pipe line, 143
French Equatorial Africa, 302
French Guiana, fuel use, 384-385
French possessions, sedimentary areas, 232
French West Africa, 301, 302; fuel oil, 368
Frommern, 259
Frontignan, 259
Fuel oil (*see also* distillate fuel oil, residual fuel oil), 435; Africa, 368; Asia, 368; Europe, 365, 390; North America, 363, 386, 390; South America, 364-365; per cent used where it is most used product, 382; leading product in Asia, South America, 382-384
Fur, 160
Fushun, 304
Fyzabad field, Trinidad, 86, 88

Gabian field, 259-260
Gach Saran field, Persia, 168, 173-175; Figs. 32, 33; Plate 57
Galan field, Colombia, 105
Galeota Point, Trinidad, 82
Garfias, V.R., 361, 377, 379-380, 391
Gas oil, definition, uses, 355, 358, 434
Gasoline (*see also* motor fuel, World War II), 355-434; output, U.S., 362; production by cracking, 35, Fig. 55, 358-359; use in Australia, New Zealand, 369, North America, 382, 384; South America, 384, 386; U.S., 362, 386, 390; Tables 4, 5, 13, 22, 23, 24, 26, 27, 29
Gassum field, 268
Gathering lines, 31, 32, Plate 12
Gaza, Palestine, 186, 190
Gemsa, Egypt, 171, 187
Genesis, Book of, 159
Geneva, 269
Genoa, 266
Geological periods, geologic time-scale, 3-4
Geological Survey of India, 300
Geologists and geophysicists in oil discovery, 15-19, 153, 158, 338-340, *and see* all regional sub-divisions
Geophysics in geological exploration, 16-19, 338-340, *and see* regional sub-divisions
Georgia, U.S.A., 45; Fig. 6, Fig. 31
Germany, 241, 244-246, 249, 254-259, 271; hydrogenation, 258; Middle East oil before 1914, 397; occupation zones, 250, 258; oil shale, 258, 259; production and refineries, 258; World War II, consumption, rationing, utilization of petroleum products, 366, 367,

Germany (*continued*)
oil supplies and strategy, 393-394; Figs. 38, 39, 56, 57, 59; Table 1
Ghent, refineries, 268
Gilan province, Persia, 175
Giurgiu, 252
Gobi Desert, 304
Golden Lane field, Mexico, 156
Golfo de Paria, 52, Table 3, 60, 80, 82, 90-91
Golfo de Venezuela, 52
Gorgeteg gas field, 253
Gor'kiy, 235, 237
Graham Land, 310
Gran Chaco, Union Oil exploration concession, 125
Gran Colombia, 76
Gran Sabana, Venezuela, 56
Gravimeter, 16-18, 48
Great Artesian basin, Australia, 305, 306
Great Bear Lake, Canada, Fig. 47, 309, 311, 316
Great Britain, basins, Table 19, 241, 244, 254; fields, installations, 242-243; exploration, development, 261-264; Middle East, 399-400; refineries, 263; synthetic products, quantity producible, 343; World War II supplies, 345; Figs. 38, 49, Table 1
Great Lakes, tanker fleet, 146
Great Liberator, *see* Bolívar, Simon
Great Nefud, 207-208, 211, 212, 218
Great Powers, 1938, per-capita energy ratios, 373
Greater Dnepr scheme, 236
Greater Jusepín area, Venezuela, 64, 66
Greater Oficina fields, Venezuela, 64, 70-71, 73
Greater Zab River, 169
Greece, 270; refineries, 270, Fig. 38
Greenland, 310, 316; Figs. 47, 49; Plate 94
Greenman, Commodore, 315
Grinnell Land, 310
Grozny anticline, 233
Guadalajara, Mexico, 96
Guajira Peninsula, Colombia, 102
Guanoco asphalt lake, Venezuela, 72, 77
Guara field, Venezuela, 73
Guatemala, gravimetric survey, 48; plains, 154, Fig. 29, 155
Guavinita field, Venezuela, 71
Guayaguayare, Trinidad, 85
Guayana shield, 70
Guayanan hinterland, Venezuela, 55-57; Fig. 8
Guayaquil, Ecuador, 122
Gulbenkian, C. S., Table 15, 178, 180
Gulf Coast of Mexico, 325
Gulf Coast, U.S., *see* U.S., Gulf Coast
Gulf Exploration Company (Gulf Oil Corporation), Table 15, 178, 182, 185, 264, 265
Gulf of Aden, 205-206
Gulf of Mexico, 152, 325-326, 339
Gulf of Morrosquillo, Colombia, 112
Gulf of Oman, 205
Gulf of Paria (Golfo de Paria), 52, Table 3, 60, 80, 82, 90-91
Gulf of Selwa, 210
Gulf Oil Company, *see* Gulf Oil Corporation
Gulf Oil Corporation, 73, 74, 117, Table 15, 178, 182, 185, 260, 264, 265, 268

Habbaniya, Iraq, air base, 199
Hadhramaut, 206, 212, 229
Haft Kel field, 165, 168, 173, 175
Hahot anticline, 253
Haifa, 163, 176, 181, 188, 190, 196; Figs. 32, 34
Hail, population, 221
Haiti, 48
Halsey, Admiral William F., 352
Hamburg, 257, 258
Hanover basin, 245, 256, 257, 258
Haradh field, 162, 204, 216
Hardstoft field, 262, 263
Harun al Rashid, 228
Harvard University, 333
Harz Mts., 245, 256, 261
Hasa, annexation by Turks, 1871, 200
Hasa oasis, Saudi Arabia, 169
Hebron, Palestine, 186
Hearst Land, 310
Health measures in oil-field areas, Bahrein, 226; Colombia, 114, 115; Indonesia, 283; Saudi Arabia, 220
Heavy distillates, definition and uses, 358, 360
Heide fields, 245, 257
Hejaz kingdom, 199, 205
Hejaz Mts., 206, 207, 218
Herodotus, 159, 160, 161, 182, 270
Hesse, 259
Hewetson, H. H., 150
Himalayan-Malayan arc, 298
Hindiya Barrage, 194
Hindukush Range, 302
Hit, 160, 182
Hofuf, 209, 210, 228; population, 221
Hofuf-Sarar escarpment, 208
Hokkaido, 305
Holstein, Duke of, 161
Honduras, Fig. 6, 46; tanker fleet, 146
Honshu Island, 305
Hornum dome, 268
Houdry catalytic cracking, Fig. 55, 358, 359
Huallaga River, Peru, 124
Humboldt current, 122
Hume, W. F., 171
Hungarian basin, 241, 245, 246, 249, 252, 253, 266
Hungary, 245, 249, 252, 253, 254, 266, 271; refineries, 254; Figs. 38, 39; Tables 19, 20
Hunt, T. Sterry, 300
Hunter College, 375
Hurghada field, Egypt, 171, 172, 187

Husain, King of the Hejaz, 198, 199
Hydrogenation, 36, 254, 258, 358, Fig. 55, 357

Ibague, 113
Ibn Rashid, 199
Ibn Saud, King, 199, 204, 213, 219, 222-223
I.B.P., *see* Indo-Burma Petroleum Company
Iceland, 267, 270, 316
Illinois, 143, 146, Fig. 17, 132, Fig. 18, 133; Table 11, 139, Table 12, 142
Imperial Chemical Industries, 263
Imperial Institute of Great Britain, *Statistical Summary*, 378-379, 391
Imperial Oil Company, Ltd., 150
Inciarte asphalt deposits, Venezuela, 77
Indaw, 299
India, 196, 225, 227, 298; China-Burma-India Theater, World War II, 352-353; climate, 191; consumption, gasoline, kerosene, lubricants, 368-369; consumption and production compared, 369; oil fields, 301-304; Figs. 49, 56, 57, 59; Plates 86, 87; Table 1
Indian Ocean, 170, 205
Indiana, 143, 156-158, Fig. 17, 132, Fig. 18, 133, Table 11, 139, Table 12, 142
Indians, oil spring known to, 130
Indo-Burma Petroleum Company, 299, 300
Indochina, 303
Indonesia (*see also* Borneo, Java, New Guinea, Sumatra, Tarakan, and names of fields), 273-296; climate, 277; definition, 273; development of oil industry, 277-282; general development, 276-295; home and export markets, 294-295; Japanese occupation, 296; mapping by the oil companies, 289, 291-293; Mines Act, 278; population, 276, 277, 293-294; production, 278-282, Table 21, 274; schools established by oil companies, 294; Figs. 40-45, 49; Plates 77-81, 83-85; Table 1
Indonesian archipelago (*see also* Indonesia), 273; international communications, 289; Figs. 40-45, 49; Plates 77-81, 83-85
Indus Valley, 303, 304
Industria Matarazzo de Energia, S.A., Brazil, 129
Infantas field, 103, 105, 111
Inke gas field, 253
Inorganic soaps for grease making, 360
Insecticides, 42
Inspectoría Técnica de Hidrocarburos, Venezuela, 74
Instituto Nacional de Geofísica (Spain), 269
International oil policy, some desiderata, 401-404
International Petroleum Co., Ltd., 100, 123
I.P.C., *see* Iraq Petroleum Company
Ipireango S.A. Companhia Brasileira do Petróleos, 129
Iquitos, Peru, 124
Iran, *see* Persia
Iraq, 160, 161, 164, 183, 203, 211, 326; adjacent marine areas, 329; climate, 191, 192; communications, 188, 189, 197, 200, 227; concessions, Fig. 34, Table 15, 177-179, 180, 181; general development, 194-196; political background, 180, 198-199; oil-field development, 174, 180-182, 203, map, Fig. 33, 173; proved reserves, 340, 341; topography, 169, 183, 192, 193; Figs. 32-34; Plate 58; Tables 14, 15
Iraq-Persian border, 174
Iraq Petroleum Co., Ltd., 163, 180, 183, 188; area and period of concessions, Table 15, 177, 178
Ireland, 267, 270; refinery, 270; Fig. 38
Iron Gates, 249
Irrawaddy River, 273, 298, 299, 300, 353; *see also* Chindwin-Irrawaddy
Is (Hit), 160
Isfahan, 1637, 161
Iso-octane or codimer, production, Fig. 55, 358, 359
Israel, *see* Palestine
Israeli-Arab war, 181
Isthmus of Tehuantepec, 98, 99
Italy, 254, 255; crude production and refineries, 265, 266; oil shale, 266; oil supplies and World War II, 393-394; Fig. 38

Jabrin, 228
Jaffa, 188
Jafura dune area, 210, 212, 218
Japan, 305; consumption of products, 368-369; consumption, production compared, 369; oil shale, 369; petroliferous areas, 301; supplies in World War II, 369, 393-394, 397; Figs. 56, 57, 59, 60; Plate I
Japanese forces in Aleutians, 350; Burma, 300; Indonesia, 281; Manchuria, 304
Jask, Persia, 168
Jaszlo fields, 255, 256
Java, 293, 294; communications, 283, 289; geology, 273, 274; oil field development, 277-280, 282; refining and transport, 283-284
Javanese rivers, 282
Jebel Akhdar, 170
Jebel Aneiza, 169, 211
Jebel al Loz, 205
Jebel as Shafa, 205
Jebel Daka, 205
Jebel Dukhan, 183, 223
Jebel Maeth, 205
Jebel Usdum, 187
Jeddah, 205, 213, 218; population, 221
Jerusalem, 190
Jidda (*see also* Jeddah), 199
Johannesburg, 302
Jordan, 164, 181, 211, 229; automobile routes, 227; oil concessions, 177, 187; political background, 199, 201; Figs. 32, 34; Table 15

Jordan Exploration Co., Ltd., 187; area and period of concession, Table 15, 177
Jordan Valley, 164
Joya Mair field, 303
Jubail, 210, 221; Fig. 36; Plate 63
Jumaima, 227
Jura Mts., 269
Jurassic period, time, 4
Jusepín field, *see* Greater Jusepín
Jutland, 268

Kalba shaikhdom, 200
Kalundborg refinery, 268
Kamchatka basin, 237, 312
Kansas, 156-158, Fig. 19, 134, Fig. 20, 135, Fig. 31, 157, Table 11, 139, Table 12, 142
Kansu Province, 304
Kapuas Murung, 283
Kapuas River, 283
Karachi, Pakistan, 189, 352
Karatash, Turkey, 185
Karkheh River, 169
Karroo basin, 302
Karun River, Persia, 168, 169, 188, 189
Katalla, 316
Katalla-Yakataga region, 316
Kazan, 235
Kazerun, Persia, 165
Kelham Hills field, 262, 263
Kelly, *see* drilling
Kentucky, Fig. 17, 132, Fig. 18, 133, Fig. 31, 157, Table 11, 139, Table 12, 142
Kenya, 302
Kerosene, as illuminant where it is most used product, 382, consumption, in Africa, 367, 368, in China, India, Iran, 368, in Europe, 365, 366, in South America, 364, in U.S., 362; definition and uses, 355, 434-435; leading product used, eastern Europe, southeastern Asia, 386; leading refined product in U.S. to 1900, 388
Khanaqin, 181, 195
Khanaqin Oil Co., Ltd., 181; area and period of concession, Table 15, 178
Khatanga Gulf, 312
Khatanga River basin, 237
Khaur field, 303
Khor Musa, 189
Khorramshahr, Persia, 188
Khosrowabad, Persia, 189
Khurasan province, Persia, 175
Khuzistan, irrigation, 194
Kier, S. M., *Kier's Petroleum or Rock Oil,* 130
Kifri, Iraq, 165
King David's miners, 205
King of the Arabians, 525 B.C., 182
King Solomon's mines, 205
Kirkuk field, Iraq, 162, 163, 165, 176, 181, 182, 188, 190, 197, 198; Figs. 32, 34; Plate 58
Kladno coal basin, 264
Klamono field, 276, 280, 282, 283
K.L.M., *see* Koninklijke Luchtvaart Maatschappij
Kluang field, 276
K.N.I.L.M., *see* Koninklijke Nederlandsch-Indische Luchtvaart Maatschappij
Kolín refinery, 264
Komárom refinery, 254
Koninklijke Luchtvaart Maatschappij, 289; air mapping, 291, 293
Koninklijke Nederlandsch-Indische Luchtvaart Mattschappij (K.N.I.L.M.), 289
Koninklijke Paketvaart Maatschappij, 289
Koryak Mts., 312
K.P.M., *see* Koninklijke Paketvaart Maatschappij
Kraków, 256
Krasnovodsk, 233
Krenkel, E., 171
Kuh-i-Namak salt plug, Persia, Plate 59
Kuh-i-Seh Qalehtun syncline, Persia, 168
Kunming, 352, 353
Kurds, 195, 198
Kurdistan mountain zone, snowfall, 192
Kurnub, Palestine, 186
Kuskokwim River, 316
Kut, Iraq, 194
Kutaradja, 283
Kut-al-Imara, 169
Kuybyshev, 234, 237, Fig. 38, Plates 70, 75
Kuwait, adjacent marine areas, 329; communications, 176, 189, 227; concessions, Fig. 34, 176, Table 15, 177-179, 178, 204, 224; oilfield development, 182-185, map, Fig. 35, 184; political background, 199-200; refinery (topping plant), 184; reserves, 162, 326, 340, 341; reservoir rocks, 182; shipbuilding, 196; topography, 209-210; water supply, 170, 192; Figs. 32, 34, 35; Plates 60, 61; Table 1
Kuwait town, 169, 182, 185, 189, 196; Fig. 32; Plate 60
Kuwait-Mediterranean pipe line, 163
Kuwait Oil Co., Ltd., 182, 185, 204; Area and period of concession, Table 15, 178; Fig. 34
Kuwait-Saudi Arabian Neutral Zone, 210
Kyaukpyu, 297

La Brea asphalt lake, Venezuela, 77
La Brea well drilled, 1866, Trinidad, 85
La Cira field, 103, 105, 111, 113
La Dorada, Colombia, 113
Lago de Maracaibo, *see* Lake Maracaibo
Lago Petroleum Corporation, 72, 73
La Guaira, Venezuela, 52
Laguna de Los Terminos, Mexico, 99
Lagunillas field, Venez., 61, 72
Laila, population, 221
La Junta Petroleum Company, 110
Lake Albert graben, 301
Lake Lugano, 266

Lake Maracaibo, 50, 52, 54, 57, 58, 60, 71, 72, 73, 107, 328, 339; Figs. 9, 10, 16; Plates 24-29
Lake Titicaca, 123
Lakes Entrance, Australia, 306
Lalang River, 284
Lali field, Persia, 173, 175
Lalongue well, 259
Land Registry Offices, Venezuela, 78
Langkat, Sultan of, 277
La Paz field, 62, 72, 73
La Petrólea base camp and refinery, 116
Lapworth, Miss P. B., 159
Lara, state of, 52, 54
La Represalia, Venezuela, 57
La Risa concession, Colombia, 110
La Rosa field, 61, 62, 72
La Silla, Mt., 54
La Spezia refinery, 265
Las Mercedes field, 64, 71
Las Mercedes-Tucupido trend, 64, Table 3, 60
Latin-American countries, ownership of subsoil minerals, 76
Lauqa, 227
Lawrence, T. E., 190
Laws of the Indies, 1602, 76
League of Nations, 175, 198
Learning how to live in a world community, 343
Lebanon Mts., 190
Lebanon Republic, concessions and development, 177, 186; pipe line terminals and ports, 181, 188, 217; political background, 201; topography, 211; Figs. 32, 34; Table 15
Lebanon Petroleum Co., Ltd., area and period of concession, Table 15, 177
Lebkicher, Roy, 203
Ledo Road, 303, 353
Leduc field, Alberta, 150; Fig. 29; Plate 49
Lees, G. M., 262
Leghorn refineries, 254, 266
Le Havre refineries, 260
Lembeye, 259
Lena River, 230, 232, 233, 237, 312
Lendva field, 266
Leningrad, 235-237
Leona field, 73
Lesser Zab River, 169
Levant, air currents, 191
Ley de Hidrocarburos of 1943, Venezuela, 73
Liberator, *see* Bolívar
Liberia, per-capita use of petroleum products, 1938, 387
Libya, 301
Libyan Desert, 171
Lincoln, coal field, 263
Lingeh, Persia, 168
Link, T. A., 317
Liquid-phase thermal cracking, 35
Liquified petroleum gas (LPG), 37; in U.S., 139, 364; shipment, 42
Lisan Peninsula, 187
Lisbon, 268, 269
Littoral zone, 5, 8
Lizard Springs field, 86
Ljunström method, 267
Llanos, Colombia, 55, 123, 124; Venezuela, 55, 64, 73; Fig. 8; Plate 31
Lobato oil field, Brazil, 128
Lockhart, L., 150
Lodi field, 265
Loftus, W. K., 172
Long Beach field, Calif., 141; Plate 50
Los Angeles Basin, 130, 321, 322
Los Roques, islands, 55
Lot's wife, 161
Louisiana, 37, 152, 158, 324, 328, 329, 339, Fig. 6, 47, Fig. 31, 157, Fig. 19, 134, Fig. 20, 135, Table 11, 139, Table 12, 142
Lovászi field, 253
Lower Thames refineries, 263
LPG, *see* liquified petroleum gas
Lubricants, 364; consumption in Africa, 368, in Asia, 368-369, in South America, 365, in Europe, 366; from heavy distillates, range of uses, 358, 360; oils for greasemaking, 360
Lurs, nomadic tribes, 195
Lutong refinery, 296
Luxembourg, 267, 271
Luxor, 190
Lydda, 190

Ma'an, 190
Ma'aqala-Riyadh road, 208
Mackenzie, Alexander, 311
Mackenzie River, 316, 318
Madagascar, 302
Madura, 276
Magdalena Basin, 100, 102-106, 111-113, 115; Figs. 14, 15; Plates 32, 35
Magellan, Straits of, oil exploration, 127; Figs. 16, 49; Plate 39
Magnetometer, 16-18, 48
Magyar Amerikai Olajipari, R. T., 253
Mahad, Saudi Arabia, 205
Mahad Dhahab, 205
Mahakam River, 278, 287
Majlis (Persian Parliament), 180, 197
Major Areas of discovered and prospective oil, world as a whole, 325-329
Malamir, 224
Malaria, preventive measures, Colombia, 114-115; Indonesia, 283; Persian Gulf region, 220
Mamatain, Persia, 172
Mamberamo River, 283
Mamonal, 101, 111
Manama, 225, 227; population, 226
Manaos, Brazil, 124
M.A.N.A.T., former German-controlled company in Hungary, 253
Manchuria, 303-304

Manoa, Venezuela, 55, 56
Man-years of work, definition, 373
Manzanilla Point, 81
M.A.O.R.T., *see* Magyar Amerikai Olajipari, R. T.
Mapavri, Turkey, 185
Mara District fields, 57, 58, 62
Maracaibo Basin (*see also* Lake Maracaibo), Colombia, 100, 102, 107-110; Venezuela, 57-64; Figs. 9, 16; Table 3
Maracaibo District fields (*see also* Lake Maracaibo), 58, 62, 72; Fig. 9; Table 3, 60
Maracay, 54
Marañón River, Peru, 124
Margarita Island, 55
Margineni field, 251
Marie Byrd Land, 310
Marseille refineries, 260
Masandam Peninsula, 170
Masira Island, 170
Masjid-i-Sulaiman, Persia, 159, 160, 165, 168, 173-175, 180, 190; Figs. 32, 33; Plates 55, 56
Maturin basin, 64, 66, 68
Matzen field, 246, 252
Maykop, 233
Mazanderan province, Persia, 175
McCollum, L. F., estimate of Middle East reserves, 181
McKean County, Pa., first oil trunk line to seaboard (Philadelphia), 143
Mecca, 199, 205, 222, 227, 229; population, 221
Medan, 277
Medellín, 101, 112, 113
Medicinal oils, from heavy distillates, 358
Medina, 190, 207, 222, 227, 229; population, 221
Mediterranean basins, 162, 240, 325-326
Mediterranean Sea, communications, 185, 188, 189; eastern, 191-192, 326, 367, 368; pipe lines to, 163, 181, 182, 211; refineries, 260, 269
Melbourne, 306
Melville Island, 316
Mena al Ahmadi (formerly known as Fahaheel), 185, 189
Mendoza fields, Argentina, 126; Fig. 16; Plates 41, 42
Mene Grande field, 50, 57, 60, 72, 73, 76
Mene Grande Oil Company, C.A., 60, 70, 73, 74
Meneg, *see* Mene Grande Oil Company
Merchant shipping, world's, 1938, per cent powered by petroleum, 386
Merkwiller refinery, 259
Mesa de Guanipa, 73
Meshach, 161, 181
Mesopotamia, 161, 173, 194
Mesozoic era, time, 4
Mexico, 95-99; continental shelf, 328-329; eastern plains, 154-156; operations of American Independent Oil Co. and Cities Service Oil Co., 99; potential areas, 156-158; production, 95-98, Fig. 13, 96, Table 9, 98; refineries, 148; reserves, 152, 153; tanker fleet, 146; trade, 95; Figs. 6, 13, 25, 27, 28, 29; Table 1
Mexico City, refining capacity, 96
Mexican Petroleum Company of California, 97
Mica, 360
Michigan, Fig. 17, 132, Fig. 18, 133, Table 11, 139, Table 12, 142
Mid-Continent and western Gulf Coast states, natural gas fields, map, Fig. 20, 135; oil fields, map, Fig. 19, 134
Middle distillates: definition and uses, 355, 358; U.S. use, 362-363
Middle East (*see also* subdivisions and Chap. 9, Saudi Arabia and Bahrein) chief source of European imports, 271, 272; climate, 191-192; communications, 188-190; concessionaire companies, Table 15, 177-179; concessions map, Fig. 34, 176; definition, 159; estimated areas and populations, Table 16, 196; general development, 163, 194-196; location map, Fig. 32, 165-167; oil province, 161-164, 326; oil field development, 172-188; political background, 197-202, 396, 399-400, 403; potential reserves, 161, 162, 181; production and proved reserves, Table 4, 162; structure and topography, 164-172; vegetation, 192-194
Middle East pipe line, 163, 188
Middle East Pipe Line Company, 163
Middle Magdalena basin, fields, valley, 102, 103, 114, 115; fields, Fig. 15, 104
Midway Sunset field, California, Plate 51
Migration, of oil and gas, *see* Part I, *Role of Compaction and Migration,* 10-11, and *Traps,* 12-15
Mihály gas field, 253
Minatitlan, 96
Mining Ordinances of New Spain, 1783, 76
Ministry of Mines and Petroleum, Colombia, 118
Minneapolis, pipe-line terminus, 146
Minnesota, 146
Miocene period, time, 4
Miri field, Sarawak, 295, 296; refinery, 296
Mishmi Hills, 352
Misión field, 98
Mississippi, 45, 152, 158, Fig. 6, 47, Fig. 19, 134, Fig. 20, 135, Fig. 31, 157, Table 11, 139, Table 12, 142
Missouri, Fig. 17, 132, Fig. 18, 133, Fig. 19, 134, Fig. 20, 135, Table 11, 139, Table 12, 142
Molasse basin, 241, 246, 252, 257, 259, 264, 269
Molotov field, 234
Monagas, state of, 64, 68
Montaña, Peru, 120, 124, 125
Montana, U.S.A., 154, 314, Table 11, 139, Table 12, 142
Montería, Colombia, 110

Monterrey, Mexico, 96, 98
Montevideo, Uruguay, refinery, 126
Montserrat Hills, 81
Morava River, 266
Moreni-Baicoi trend, 251
Morocco, 301, Plate I, facing 14; Table 1
Morraro asphalt deposits, 266
Moruga River, Trinidad, 82
Moscow, 235, 311
Moscow-Volga canal, 235
Moses, 159
Mosul, 164, 165, 182, 189
Mosul Petroleum Co., Ltd., 180, 183; area and period of concessions, Table 15, 178
Motor fuel (*and see also* gasoline, World War II), and associated products, definition and uses, 355; 362, 382, 384-390; consumption in Africa, 367, Asia, 368, Europe, 365, 389-390, U.S., 362, 382, 384-390; leading product in North America, Australia, New Zealand, 382, in Europe, Africa, 383-384; per cent used where most used product, 382; Tables 4, 5, 13, 22, 23, 24, 26, 27, 29
Moving-bed catalytic cracking, 35, 36; Fig. 55, 358, 359
Mozambique, 302
Muara Enim fields, 278, 284
Muds and sands, 5, 6, 10, 11
Muhammad Riza, 198
Muharraq Island, 223, 225
Muharraq town, population, 226
Mukalla, 206
Mukden, 304
Munster Basin, 245, 257
Murbat, 206
Muscat, 183, 229
Musi River, 284, 285, 287
Musil, Alois, 211
Muwaih, 228
Myitkyina, 353

Nadir Shah, invasion of India, 1739, 160
Naft-i-Shah field, 165, 174
Naft Khaneh field, 165, 174, 181
Naft Safid field, 175
Nahr Umr field, 162, 182
Naiguatá, Mt., 54
N.A.M., *see* Nederlands Aardolie Maatschappij
Naples refineries, 266
Nariva swamp, Trinidad, 82
Nasiriya, Iraq, 169
National Industrial Conference Board, consumption data, 380
National Oil Company, Colombia, 1948, 119
National reserves, Venezuela, 78
Natural gas, *see* Part I, Petroleum in the Ground, Part II, *Production,* 29-30, *Transportation,* 33; United States, other separate countries; also condensates, natural gasoline, LPG
Natural-gas transmission lines (*see also* maps), Argentina, 127; North America, 143, 146, map, Fig. 25, 144; U.S., 33; U.S.S.R., 33
Natural gasoline, 37, 139, 358, Fig. 55, 359
Navy Special fuel, 352, 363
Nazi Government, oil policy, 257, 258
Near and Middle East, location map, Fig. 32, 165, 166-167
Near East Development Company, 180
Nebraska, Table 11, 139, Table 12, 142
Nederlands Aardolie Maatschappij (N.A.M.), 261
Nederlandsch-Indische Aardolie-Maatschappij (N.I.A.M.), 279, 280
Nederlandsche Koloniale Petroleum Maatschappij (N.K.P.M.), 279, 287
Nederlandsche Nieuw Guinea Petroleum Maatschappij (N.N.G.P.M.), 280, 283, 291, 294
Nederlandsche Pacific Petroleum Maatschappij (N.P.P.M.), 280
Neftedag fields, 233
Nejd, 170, 199, 207; Fig. 32
Nejran, 229; Population, 221
Neritic zone, neritic bottoms, 6, 7, 8
Netherlands, 244, 245, 254-256, 260-261, 271; refineries, 261; Figs. 38, 49, 56, 57, 59; Tables 1, 19, 20
Netherlands Indian Government, Air Force Dept., air mapping, 291; and Japanese attack, 281; partner in N.I.A.M., 279
Netherlands Indies (*see also* Indonesia), consumption and production compared, 369; effect of loss in World War II, 174; export trade and domestic consumption of petroleum products, 1913-1938, Table 22, 295
Netherlands New Guinea, *see* New Guinea, Netherlands
Netherlands West Indies, *see* Aruba and Curaçao
Neuquén fields, Argentina, 126
Nevada, U.S.A., 154
New Caledonia, 306-307
New Guinea, Australian, 306; Fig. 49; Table 1
New Guinea, Netherlands, 273, 274, 277-280; air mapping, 291-293; communications, 283; fields, 282; Figs. 40, 44, 49, 56, 57, 59; Plate 84
New Jersey, 33
New Mexico, Fig. 19, 134, Fig. 20, 135, Table 11, 139, Table 12, 142
New South Wales, 305, 306
New York and Bermudez Company, 72
New York City, gasoline price, 74
New York State, 33, 143, 154, Fig. 17, 132, Fig. 18, 133, Table 11, 139, Table 12, 142
New Zealand, 306-307, 369, 382, 384; Figs. 49, 56, 57, 59

N.I.A.M., *see* Nederlandsch-Indische Aardolie Maatschappij
Nicaragua, 48, Fig. 6
Nigeria, 302; Figs. 49, 56, 57, 59
Nile River, 159, 171, 189, 194
Niño current, 122
Nisibin, Turkey, 190
Nix, Dale, 227
N.K.P.M., *see* Nederlandsche Koloniale Petroleum Maatschappij
N.N.G.P.M., *see* Nederlandsche Nieuw Guinea Petroleum Maatschappij, 280
Noah, 159
Nordvik, 312, 313
Norman Wells field (Fort Norman Field), Fig. 29, 155; Fig. 47, 309; 311, 316-318; in World War II, 350-351; Plates 45-47, 88
Normandy-German operations, World War II, petroleum requirements, 344
North America (*see also* Canada, Caribbean Area, Mexico, United States, and lists of maps and tables at front of book), development of petroleum industry, 130-133; geological age range, 233; marketing, 147-151; petroleum records antedating Columbus, 130; production, 134-143; refining, 146-148; resources and reserves, 152-158; sedimentary areas, 326, 327; transportation, 143-146; utilization, 362-364, 382, 384; Figs. 25-29; Plates 1, 2-5, 8-11, 17, 20-22, 45-53, 88-91, 94, 95
North Burma area in World War II, 353
North Carolina, Fig. 31, 157
North-central and Appalachian states, natural gas fields, map, Fig. 18, 133; oil fields, map, Fig. 17, 132
North German basin, 240, 241, 244-245, 255-256, 261-262, 267-268
North Pole, 308
North Sea, 257
Northern Basin, Trinidad, 80
Northern District field, Mexico, 97
Northern Range, Trinidad, 80, 82
Northern sea route, 237
Northwest basin, Australia, 306
Northwest Territory, Canada, 311
Norway, 267, 271; refinery, 271; Fig. 38; Table 20
N.P.P.M., *see* Nederlandsche Pacific Petroleum Maatschappij
Nuevo Laredo, Mexico, 96
Nynäshamn, 267

Ob' River, 230, 235, 311
Oceania other than Indonesia, 305-307
Oceanographers, 5, 6
Odessa, 252
Oficina field, *see* Greater Oficina
Ohio, 131, 156-158, Fig. 17, 132, Fig. 18, 133, Fig. 31, 157, Table 11, 139, Table 12, 142
Oil accumulation, requisites for, 335
Oil as instrument of peace, 404
Oil burners, definition, 362
Oil Creek, Pa., 130, 131
Oilfield Employers Association, Trinidad, 90
Oilfield Workers Union of Trinidad, 89
Oil-forming process, *see* Part I
Oil occurrence, geological principles governing, 3-15
Oil Search, Ltd., Sydney, Australia, 301
Oil shale, 435; Czechoslovakia, 264; France, 260; Germany, 258-259; Great Britain, 263-264, 342; Italy, 266; Spain, 269; Sweden, 267, 343; U.S., 342, 343; U.S.S.R., 237; synthesis of petroleum from, 342
Okinawa campaign, 352
Oklahoma, 154, Fig. 19, 134; Fig. 20, 135; Fig. 23, 140; Fig. 24, 141; Table 11, 139; Table 12, 142; Plates 2, 4
Okhotsk Sea, 312
Oldham, Dr. Thomas, 300
Oligocene period, time, 4
Oman, 178, 183, 200; Figs. 32, 33, 49
Oman Mts., 170, 171, 206, 209, 212, 228
Ooze, *see* muds and sands
Opón River, Colombia, 103, 115
Oqair, 210, 228
Ordovician period, time, 4
Organic matter, role of, in oil formation, 7-10
Orinoco basin, 57, 64-71; Figs. 8, 11
Orinoco River, 52, 54, 55, 56, 64, 80, 102, 123
Orogenic movements, 4, 6
Oropuche River, 80
Orsk refinery, 237
Ortoire River, Trinidad, 82
Oslo Fjord, 271
Osnabrück nose, 245, 261
Ouachita Mts., Okla., overthrust belts, 154
Oviedo y Valdes, Gonzalo Fernandez de, 115
Oxidation, marine, 6, 8

Pachitea River, Peru, 124
Pacific, coastal shelf of South America, 102, 120, 122, 123; continental shelf of U.S., 152
Pacific Railway, Colombia, 113
Pacific Western Oil Corporation, 179; area and period of Middle East concession, Table 15, 177
Pahlavi dynasty, Persia, 197
Pahtoro fields, 254
Pakistan, petroliferous areas, 301-304; Figs. 49, 56, 57, 59; Table 1
Palacio field, 71
Palembang, 280-282, 284-287
Paleozic era, time, 4
Palestine, 163, 164, 171, 181, 186-187, 188, 190, 191, 192, 196, 201, 229, 399; refineries, 163, 181, 196; Figs. 32, 34; Tables 15, 16
Palestine Potash Syndicate, 187; Table 15, 177
Palgrave, W. H., 193

Palmyra, Syria, 186
Panama, test wells, 48; tanker fleet, 146; Fig. 6
Pangkalan Brandan, 278, 281, 284
Pangkalan Susu, 284
Pao, Venezuela, 56
Papantla, 97
Papua, 306
Paraffin, from heavy distillates, 358; industries, 360
Paraguay, 125, 384, 385; Fig. 16
Paraguay River, 125
Pardubice refinery, 264
Paris basin, 240-241, 248, 259, 261-262
Pascoe, Sir Edwin, 298
Patterns of world utilization, 361-369
Pau region, 259
Pearling, Persian Gulf, 195, 226-227
Pearson, S., *see* S. Pearson and Son, Ltd.
Péchelbronn field, 259
Pécs refinery, 254
Pegu Yoma, 298
Peklenica field, 266
Pemex, *see* Petróleos Mexicanos
Penang, 284
Pendopo field, 276
Península de Paraguaná, 54
Península de Paria, 54
Pennsylvania, 131, 143, Fig. 17, 132, Fig. 18, 133, Table 11, 139, Table 12, 142
Percussion drilling, *see* drilling
Perijá Mts., *see* Sierra de Perijá
Perlak field, 278, 284
Permafrost, 314-315
Permian period, time, 4
Pernis refinery, 261
Persia, 37, 159-198; central, 164, 191, 192, 194; climate, 191-192; communications, 188-190, 227; concessions, 175, Table 15, 177-179, 179-180, 328-329; consumption, 368, 369; crude production, 162, 173; geology, 164, 165, 168, 203; general development, 195-196; historical, 159-161; northeast and east, 164, 175, 190; northwest, 191; oil field development, 165, 172-180, 203; pipe lines, 163, 174, 175; political background, 197-198; refineries, 37, 163, 174; reserves, 162, 326, 340, 341; reservoir rocks, 183; vegetation, 192-193; western and southern, 163-165, 170, 172-175, 191-192, 209; Figs. 32-34; Plates 1, 16, 54-57, 59; Tables 14, 15
Persian Bank Mining Rights Corporation, 172
Persian Gulf (*and see* adjacent countries), 163-210, 213, 326-328, 340, 352, Figs. 32-36
Persian-Indian frontier, 168
Peru, 120, 122-125, 386; Fig. 16; Table 26; Plates 1, 12, 36, 37
Petrolatum, 358, 360
Petrólea field, 101, 107-109, 112, 116
Petróleos Mexicanos (Pemex), 96, 98, 99
Petroleum, and components, definition, 3; and products, standard definitions, American Society for Testing Materials, 434, 435
Petroleum asphalts (*see also* asphalt), 360
Petroleum coke, uses, 360, 361
Petroleum Concessions, Ltd., 186; area and period of concession, Table 15, 178
Petroleum Development companies associated with Iraq Petroleum Company, 183
Petroleum Development (Cyprus), Ltd., area and period of concession, Table 15, 177
Petroleum Development (Oman and Dhofar), Ltd., area and period of concession, Table 15, 178
Petroleum Development (Palestine), Ltd., 186, area and period of concession, Table 15, 177
Petroleum Development (Qatar), Ltd., 183, 204; area and period of concession, Table 15, 178
Petroleum Development (Trucial Coast), Ltd., area and period of concession, Table 15, 178
Petroleum legislation, Colombia, 115-119; Indonesia, 248-280; Trinidad, 91-94; Venezuela, 76-79
Petroleum products, chief types, Table 23, 356, 357; manufacture, *see* refining
Petroleum Press Service, 150
Petroleum province, definition, 153
Petroleum Times (London) "Review of Middle East Oil," 174, 185, 215
Petroleum warfare, 160; Plate 54
Pharaoh's daughter, 159
Philby, H. St. J. B., 212
Philippine Sea, second battle of the, 352
Philippines, 301, 305, 352; Figs. 40, 49, 56, 57, 59; Table I
Photography, aerial, in geological exploration (*see also* air mapping), 15
Piarco airport, 89
Piauhy-Maranhão basin, 120
Pico Bolívar, 53
Pilón field, 69
Pilote River, 82
Pipe lines, 31-32, 38-39, Arctic, Fig. 47, 309; Europe, Fig. 38, 242-243, Fig. 53, 346; Middle East, Fig. 34, 176; North America, Fig. 25, 144, Fig. 26, 145; South America, Fig. 16, 121; U.S.S.R., Fig. 37, 231; World War II, Africa, 345, Alaska, 351, China-India-Burma theater, 353, European theater, 345-347, Panama, 348; Fig. 35; Plates 12, 18, 23, 36, 44, 61, 62, 65, 75, 76, 81, 87, 93, 94, 97 (*and see* maps of facilities under separate countries)
Pipe still, 34, 35
Pitch lake, Brighton, Trinidad, 85, 89
Pladju refinery, 281, 284-285, 287
Plankton, 8, 9, 10
Plato, Colombia, 101, 112
Pliocene period, time, 4
Ploesti fields, 244, 252; refineries, 252

Po Basin, 246, 248, 254, 265-266
Pogue, Joseph E., 150, 389
Point Barrow, 313-316
Point Fortin, Trinidad, 85, 87, 88
Pointe-à-Pierre, Trinidad, 81, 87, 88
Poland, 241, 244, 254-256, 271
Policy, *see* power and policy of nations
Polish National Plan for Economic Reconstruction, 256
Polymerization, 36
Pontianak, 283
Porous rocks, porosity, *see* Part I, *Compaction and Migration,* 10-11, *Traps,* 11-15, *Estimation of Reserves,* 19-20; Part II, *Development,* 27-29
Port-of-Spain, Trinidad, 89
Port Said, 188
Port St. John, 351
Portugal, 241, 248, 268; refinery, 269; Fig. 38; Table 19
Post-Tertiary glacial and Recent time, 4
Power and policy of nations: importance of oil, 392; uneven distribution of supply, 394; the international prospect, 395; struggle for control of oil, 396; Middle East, 399; proposals for an international oil policy, 400; oil as an instrument for keeping the peace, 404
Poza Rica field, 96, 97, 98, 156
Pratt, Wallace E., 72, 73
Pre-Cambrian rocks, 5
Pressure in wells, 27, 29, 30
Pressure still, 35
Production statistics, see list of tables at front of book
Products pipe lines, in U.S., 38, 146
Propane, 30, 37, 42
Proved reserves, *see* reserves
Puerto Berrío, 101, 113
Puerto La Cruz, Venezuela, Fig. 11, 63, Plate 23
Puerto México, 96, 98, 99
Puerto Olaya, 113
Puerto Salgar, 101, 113
Pumps, pumping, in oil-field operations, 30-32
Punjab, 302-304
Punta Organos field, 123
Pusht-i-Kuh Mts., 165
Putumayo River, 124
Pyrenees, 240, 248, 259, 269

Qaiyarah, Iraq, 182
Qajar dynasty, Persia, 197
Qariyat al Milh, 227
Qatar, 183, 193-195, 200, 203-204, 209-212, 214, 215, 223; Figs. 32, 34, 36
Qatif field, 176, 204, 209, 214-216, 220, 221; Figs. 34, 36; Plate 65
Qatif, population, 221
Qatif Junction, 176
Qidam-Sarar escarpment, 210
Qishm Island, 172
Queensland, 305, 306
Quesada, Don Gonzalo Jimenez de, 115
Quindio road, Colombia, 113
Quirequire field, 64, 65, 66, 72, 73
Quirequire–Jusepín–San Joaquín trend, 64-68, Table 3, 60, Fig. 11, 63
Qurna, Iraq, 195

Ram Hormuz, 172
Ramadi, 182
Ramandag field, 183, 185
Ramree Island, 297
Rangoon River, 300
Rantau field, 276, 280
R.A.P., *see* Régie Autonome des Pétroles
Ras al Hadd, 170
Ras al Kaima Shaikhdom, 200
Ras al Mattala, 224
Ras Gharib field, 172, 187, 188
Ras Tanura, 32, 163, 176, 189, 215, 217, 220, 221; Figs. 32, 34, 36; Plate 62
Rawlinson, George, 160, 161
Reactor, 35, 36
Recent time, 4
Recovery factor, 19, 20
Red Ball Highway, World War II, 348
Red Sea, 164, 171, 205, 302
Red Sea coastal region of Arabia, 205
Red Sea hills, 171
Redwater field, 150
Reerink, Jan, 277
Refinaria do Petróleo Distrito Federal S.A., Brazil, 129
Refinaria e Exploração de Petróleo União S.A., 129
Refineries, *see* refining, *and see* under separate countries. Plates 16, 17, 19, 37, 42, 47, 62, 70, 77
Refining (*and see also* products' entries and refineries under separate countries), additives for greases, 360, alkylation, Fig. 55, 359, asphalts, 360; aviation gasoline, Fig. 55, 358-359; catalytic cracking, fixed-bed, fluid catalyst, Houdry, moving-bed, thermofor, Fig. 55, 358-359; Europe, capacity, 272; fuel oils, classification, 358; gas oil, 355, 358; greases, 360; heavy distillates, 358; kerosene, 355; lubricants, 358; medicinal oils, 358; middle distillates, 355, 358; motor fuel and associated products, 355; North America, 146, 148, 149, Fig. 28, 149; paraffin industry, 360; petrolatum, 358; petroleum coke, 360; polymerization, 36; principles and methods, 33-37, Fig. 55, 358-359; products, chief types, and their uses, 354-367, Table 23, 356-357; residual fuel oil, 360; residual products, 360; sludges, 360; synthesis, from natural gas, oil shale, coal, 342-343; technical oils, 358; thermal cracking, Fig. 55, 359;

thermofor catalytic cracking, Plate 17, Fig. 55, 359; toluene, 355; world capacity, 37-38; world's largest refineries, Abadan, Aruba, Curaçao, 37; Plates 16, 17, 19, 37, 42, 47, 62, 70, 77
Regenerator, 36
Régie Autonome des Pétroles, 260
Rembang, 283
Rengat, 282
Rentz, G. W., 203
Repressuring stations, 30
Reserves, proved (*and see* separate countries), definition, 19-20; Caribbean area and Middle East compared, 46, 326; world proved reserves, 20, 335, Table 1, 22-23
Reservoir pressure, 27, 29, 30
Reservoirs, reservoir rocks, *see* Part I, *Traps, Geophysical Exploration, Estimation of Reserves;* Part II, *Development, Production*
Residual fuel oil, 360, 363, 390
Residual products, definition, uses, 360
Reuter, Baron Julius De, Persian concessions, 172
Reynosa field, 98
Reza Pahlavi, Shah, 190
Rhadinace, 160
Rhine graben, *see* Upper Rhine graben
Rhineland, 258, 259
Rhône basin, 259, 260
Richfield Oil Company, Colombia, 111
Rieber Pass, 112
Rifle, Colorado, 342
Río Catatumbo, 108
Rio de Janeiro, Brazil, 129
Río de Oro field, 107, 108, 116
Rio Grande do Sul, Brazil, 129
Rio Grande Valley, transformed into fertile farms, 392-393
Río Meta, Colombia, 123
Rio Paují, 77
Ripalta field, 265
Ristori, J. W., 361, 378-380
Riyadh, 199, 218, 219-221, 228; population, 221
Riza Shah Pahlavi, 197, 198
Road oil, 360
Rocks, porosity, *see* Porosity
Rocky Mountain areas, oil discoveries, and population, 146
Rocky Mountain front, 317
Rhodessa field, 158
Roma gas field, 306
Roraima, Mt., 56
Rose Hill pool, Virginia, 154
Ross Sea, 310
Rotary drilling, *see* drilling
Rotterdam, 261
Rouen refineries, 260
Royal Crown of Spain, Decree in eleventh century, 76
Royal Dutch Airlines, *see* Koninklijke Luchtvaart Maatschappij
Royal Dutch Company for the Working of Petroleum Wells in the Netherlands Indies, 277
Royal Dutch Petroleum Company, 277, 278
Royal Dutch-Shell group, British Borneo, 295, 296; Colombia, 105, 106, 124; Curaçao, 112; Ecuador, 124; Egypt, 188; Haifa, 181; Indonesia, 277-280, 285; Iraq, 178, 180; Mexico, 97, 98; Netherlands, 261; Trinidad, 85; Venezuela, 50, 60-62, 72, 74
Royal Netherlands–Indian Airlines, 289
Royal Packet Navigation Company, 289
Rub' al Khali (Empty Quarter), 170, 203, 205, 206, 208, 210, 212, 213, 218, 221, 228
Ruhlertwist No. 2 discovery, 257
Ruiz field, 71
Rumania, 241, 244, 249, 251, 253, 254, 255, 266, 271, 312, 326
Runcu field, 251
Russia, *see* U.S.S.R.
Russian Government, influence in Persia, 1873, 172
Rutba fort, 169
Ruus al Jibal, 170

Sabán field, Venezuela, 71
Sabkha Mutti, 228
Sabkhas, 210
S.A.C.O.R., *see* Sociedade Anónima Concessionária de Refinação do Petróleos em Portugal
Sagoc pipe line, Colombia, 111
Sahagun concession, Colombia, 111
St. Hilaire, France, shales, 260
St. Lawrence River, 38
St. Marcet field, France, 259, 260
St. Nazaire, France, refineries, 260
Sakaka, 221, 227
Sakhalin, 235, 237, 312; Figs. 37, 39; Table 1
Salala, 212
Salt core, *see* salt dome
Salt dome, Fig. 1, 13, 14; Plate 59
Salta fields, Argentina, 126
Salvador, Bahia, Brazil, 128
San Andres test, Colombia, 111
San'a, Yemen, 205
Sands and muds, 5, 10, 11, 19
San Fernando, Trinidad, 85
San Joaquín field, Venezuela, Fig. 11, 63, 64, 66-68, 73
San Joaquín–Santa Rosa–Roble area, Fig. 11, 63, 66-68, 73
San Jorge Gulf, Argentina, 127
San Juan delta, Colombia, 122
San Remo Agreement, 180
Santa Ana field, Venezuela, 66
Santa Cruz, Bolivia, 125
Santa Elena Point, Ecuador, 122

Santa Rosa field, Venezuela, Fig. 11, 63, 64, 66-68
Santos, Brazil, 129
São Paulo, Brazil, 129
Saratov, U.S.S.R., 235, 237; Figs. 37, 38; Plate 76
Sarawak, 295-296; refinery, 296; Figs. 40, 43
Sarawak Oilfields, Ltd., 295
Sargasso Sea, 9, 10
Saudi Arabia (*see also* Arabia), and Aramco, *see* Arabian American Oil Company; concessions, Fig. 34, 176, Table 15, 177-179; continental shelf, 329; Dahana, 208, 212; deserts, 170, 211-213; east-central, 209; eastern, 208; eastern coastal, 209; escarpment region, 207-208; Great Nefud, 207, 211-212; Hejaz Mts., 205, 206-207; location map, Fig. 32, 165-167; Mining Syndicate, Ltd., 205; mountains, 212; northern, 211; oil fields, 213-216, Fig. 36, 214 and geographical adaptation, 217-220; oil industry and Arabian people, 220-223; Persian Gulf coastal region, 209-210; pipe lines, 32, 163, 188, 217, Table 17, 218; political background, 199; production rates, 162, 216-217; proved reserves, Table 14, 162, 340, 341; refineries, 217; remote, partially explored and unexplored regions, 211-213; Rub' al Khali, 212-213; Summan Plateau, 208-209; Tihama, 205, 206-207; topography and structure, 205-213; Figs. 32, 34, 36; Plates 62-69
Saudi royal family (*see also* Ibn Saud, King), 199
Sava River, 266
Scandinavian countries, 240
Scandinavian shield, 5, 326
Scheldt River, 38
Schleswig-Holstein fields, 245, 257
Schmidt, K., 88
Schoonebeek field, 256-257; 261
Schuchert, Charles, 6
Schuler field, 158
Scotland, 244, 261-262; shale, 263, shale industry, 343; Figs. 38, 49, 56, 57, 59
Sea power and command of oil, 397-398
Seaboard Oil Company of Delaware, 175
Seattle, 350
Secondary recovery, 30
Sedimentary areas of the various continents, Plate I, facing 14; 326-328
Sedimentary basins in the Arctic, Fig. 47, 309
Sedimentary basins, oil and gas fields, and facilities, Europe, Fig. 38, 242-243
Sedimentary formations as record of earth movements, 4
Sedimentation, 4-10
Seihun (Seyhan) River, Turkey, 185
Seine River, 260
Seismic equipment in geological exploration, 16-19
Selnica field, 266
Semarang, 282
Seneca Oil Company, 131
Sennacherib, King, 160
Separator, well-head, 30; catalytic, 36
Seria field, 296, 328
Serpent's Mouth, 80
Serranía de Imataca, 56
Serranía del Interior, 54
Sète, 259; refineries, 260
Seville basin, 241, 248, 269
Seward, Alaska, 350, 351
Shadrach, 161, 181
Shaikh Mubarak, 200
Shale oil, definition, 434, *and see* oil shale
Shan states, hills, 298
Shaqra, population, 221
Sharjah, 169, 200, 228, 229
Shatt, *see* Shatt-al-Arab
Shatt-al-Arab, 169, 185, 188, 190, 195
Shell Oil Company, *see* Royal Dutch Shell group
Shell still, 34, 35
Shell Transport and Trading Company, 278
Shensi, 304
Shields of the continents, 5
Shukkoku, 305
Siam and Indochina, 303
Siberia, 233, 237, 312-314
Sicilian central basin, 241, 248
Sicily, shales, 266
Sidon, Lebanon, 163, 176, 188, 211, 217
Sierra de Perijá, Colombia, 52, 62, 73, 102, 107, 108, 112
Sierra de San Luis, 54
Siérra Nevada de Mérida, 57
Silliman, Prof. Benjamin, 131
Silurian period, time, 4
Simplon-Oriente route, 190
Sinai Peninsula, 171, 172, 187, 190
Sinclair Oil Company of Colombia, Table 10, 114
Sinclair Oil Corporation, Table 15, 178
Sinclair Petroleum Co., area and period of concession in Ethiopia, Table 15, 178
Sind ranges, 170
Singapore, 284
Singu-Lanywa fields, Burma, 276, 296, 297, 299; Fig. 46
Sinkiang Province, China, 304
Sinú area, Colombia, 110, 111
Sinú Oil Company, Colombia, 110
Sinú River, Colombia, 110
Sinú valley, Colombia, 110
Sirwan River, 165, 169
Sirwan River embayment, 165
Sisak refinery, 266
Sitra Island, 223, 225
Skagway, Alaska, 351
Sludges, petroleum, 360, 361
Smackover field, 158
Smith, William, 131

S.N.P.A., *see* Société Nationale des Pétroles d'Aquitaine
S.N.P.L.M., *see* Société Nationale des Pétroles du Languedoc Méditerranéen
Sociedade Anónima Concessionária de Refinação do Petróleos em Portugal (S.A.C.O.R.) 269
Società Americana del Petrolio, 266
Société Nationale des Pétroles d'Aquitaine (S.N.P.A.), 259, 260
Société Nationale des Pétroles du Languedoc Méditerranéen (S.N.P.L.M.), 260
Socony-Vacuum Oil Company, Colombia, 106, 110, 111, 117; concessions in Middle East, Table 15, 177-179; Egypt, 188; Saudi Arabia, 213; Spain, 269; Venezuela, 70
Socuavo structure, Colombia, 108-110
Sogamoso River, Colombia, 102-103
Solvents, 355
Somaliland, British, 302
Somaliland, former Italian, 302
Sorgipe State, Brazil, 120
Soria province, Spain, 269
Sorong, 283
South Africa, products consumption, 367, 368
South America (*see also* under separate countries), Fig. 16, oil and gas fields and facilities, 121; foothills and great plains east of Andes, 123-127; northern Brazil, fields, 120, 128-129, refineries, 129; Pacific coastal zone, 120-123; utilization, 364-365
South Carolina, Fig. 31, 157
South Mountain pool, Ventura, California, 154
South Pole, 308
Southeast Asia oil province, 273-276
Southeast Trades, 82
Soviet Asia, polar areas discoveries, 308, 311
Soviet Union (*see also* U.S.S.R.), concession rights in Persia, 175, 176; per-capita energy output in man-power years, 371, 373; proved reserves, 22, 340, 341
Soviet zone, Austria, oil production, 250
Soviet zone, Germany, hydrogenation, 258
Space heaters, definition, 362
Spacing of wells, 28, 29
Spain, 267, 269, 394; refineries, 269; Fig. 38
Spanish Armada, 122
Spanish Colonies, mining laws, 1602 and 1783, 76-77
Spanish conquistadores, 71
S. Pearson and Son, Ltd., London, 97
Spitzbergen, 308, 310
Spring Hill field, Chile, *see* Cerro Manantiales field
Stalingrad, 235
Standard Française des Pétroles, 260
Standard Oil Company (N.J.), 74, 123, 124, 150, 163, Table 15, 177-179, 180, 186, 213, 240, 250, 253, 260, 261, 408
Standard Oil Company of Bolivia, 125
Standard Oil Co. of California, Table 15, 177, 204, 213, 224, 280
Standard Oil Co. of Egypt, concessions in Egypt, Table 15, 179
Standard Oil Company of Trinidad, 85
Standard Oil Company of Venezuela, 73
Standard Vacuum Oil Company, 180, 279, 280, 281, 291, 294, 301
Standard Vacuum Petroleum Maatschappij, 279, 280
Standard Vacuum Petroleum Mij, *see* Standard Vacuum Petroleum Maatschappij
Stanford University, School of Mineral Science, 130
Statistical Abstract of the United States, 378
Steel Brothers and Company, Ltd., 263
Stefansson, Vilhjalmur, 316
Steimbke field, Germany, 257
Sterlitamak, U.S.S.R., 234
Stills, *see* refining
Stockholm, 267
Stocks in storage, U.S. and world-wide, 31; Britain in World War II, 345
Storage (*see also* Tanks), 30, 31; Plates 14-16, 18
Strabo, on Dead Sea, 186
Straits of Malacca, 277
Strasbourg, 260
Styrian embayment, 252
Suban Djerigi field, Sumatra, 276
Sucre, Bolivia, 125
Sucre District, Venezuela, 72
Suderbruch field, Germany, 257
Sudr field, Egypt, 172, 187-188
Suez, Egypt, port and refineries, 187, 188
Suez Canal, 163, 190
Suez, Gulf of, 171-172, 187
Suldrup dome, Denmark, 268
Sumatra, 273-287, Fig. 41, 285, Fig. 42, 286, 289; air mapping, 291, Fig. 45, 292, 293-294; Plates 77, 78, 85; Table I
Summan Plateau, Arabia, 208, 209
Sumerian times, 159
Sumpal field, Sumatra, 284
Sungei Gerong, Sumatra, refinery, 281, 287; Fig. 42; Plates 77, 78
Surabaja, Java, 276, 278, 282-284
Susa (Shush), Persia, 159
Sweden, 255, 267; refineries, 267; Fig. 38
Swedish oil shale industry, 267, 343
Switzerland, 241, 246, 269; refineries, 269; Fig. 38
Synthesis, of liquid hydocarbons from natural gas, petroleum from oil shale, petroleum products from coal, 342
Synthetic gasoline and German war machine, 342
Synthetic rubber, *see* butadiene
"Syrabia," 171
Syria, 163-164; Fig. 32, 165-167; climate, 191-192; political background, 201; Saudi Ara-

Syria (*continued*)
bian treaty, 199; Syria Petroleum Company, Ltd., concession, Fig. 34, 176, Table 15, 177-179, 186; test drilling, 199; Fig. 32
Syria Petroleum Co., Ltd., area and period of concession, Fig. 34, 176, and *Table 15*, 177; 186
Syriam, Burma, refinery, 300
Syrian Desert, 163, Fig. 32, 165-167, 169, 205, 211
Szechuan province, China, 304

Tablazos, Peru, 122
Tablón concession, Colombia, 111
Tabuk, Arabia, 206
Táchira state, Venezuela, 52, 71
Tagus River, 268
Taima, Arabia, population, 221
Tajo basin, 241, 248, 269
Talang Akar field, Sumatra, 276
Talang Akar-Pendopo area, Sumatra, 279, 280
Talang Djimar field, Sumatra, 276, 280
Talara, Peru, 113; Plate 12, 37
Tamana Hill, Trinidad, 81
Tampico, Mexico, 96, 99
Tampico Embayment, Mexico, 97-98
Tanganyika, 302
Tank cars, ships, trucks, *see* Part II, *Transportation, Storage, Distribution;* Plates 20-22
Tankers, 38, 146, Fig. 54, 349; transportation by, costs in U.S., 146; U.S. and world fleets, 146; World War II, Pacific theater, floating tank farm, 352, pipe line across Pacific, 351, 352, Plates 96, 97, 98; shuttle service to Mediterranean, Britain, 345; Plates 96-98
Tanks, tank farms, *see* Part II, *Storage, Distribution;* Plates 13-15, 18, 26, 37, 62, 77, 88
Tapline, *see* Trans Arabian Pipe Line Company
Tarakan field, Borneo, 274, 276, 278-281, 287-289; Figs. 40, 43; Plate 80
Taranaka District, N.Z., 306
Targul Mures gas field, Rumania, 252
Tarentum, Pa., 130
Tartus, Fig. 34, 176, 188
Tarut Island, Arabia, 221
Tatar autonomous republic, 237
Technical oils, 358
Tečin, *see* Cieszyn
Tehran, Persia, 197
Telaga Said field, Sumatra, 277, 278, 284
Telbiti River, Iraq, 160
Telukbetung, Sumatra, 283
Tennessee, Fig. 17, 132, Fig. 18, 133, Fig. 31, 157, Table 11, 139
Temblador area, Venezuela, 64, 68, 69
Tepetzintla field, Mexico, 98
Teran-Guaguaqui fee property, Colombia, 106
Terdonck refinery, Belgium, 268
Tertiary era, time, 3, 4
Texas (*see also* Gulf Coast), 29, 37, continental shelf, 152, 324, 328-329, 339; East Texas field, 45, 141, 143, 145, 156, Fig. 30, 156; generation of oil in Paleozoic era, 333; leasing of oil, gas rights, 138; overthrust belts, 154, Fig. 29, 155; pipe lines, 33, natural gas, Fig. 25, 144, crude oil and products, Fig. 26, 145; potential trap areas, 156-158, Fig. 31, 157; production and reserves, oil, Table 11, 139, gas, Table 12, 142; refineries, 37, Fig. 28, 149; structure, 158, compared to Middle East, 162; plates 53, 91
Texas–Louisiana–Gulf Coast fields, 238
The Texas Company, Bahrein, 224, Fig. 34, 176, Table 15, 177-179; Colombia, 106, 110-111, 117; Netherlands Indies, 280; Saudi Arabia, 213, Fig. 34, 176, Table 15, 177-179; Venezuela, 62
Thames River area, refineries, 263; storage, 38
Thermal cracking, 35, Fig. 55, 358, 359
Thermofor catalytic cracking, Plate 17; Fig. 55, 358, 359
Thom, W. T., Jr., estimates of per-capita energy output in man-power years, 370-374
Thomas, Bertram, 170, 212
Thuringian basin, 241, 248
Tia Juana field, Venezuela, 60, 62, Fig. 9, 58, Fig. 10, 59
Tibet, 232
Tibú field and structure, Colombia, 108-110
Tidal and shore currents, 6
Tierra del Fuego, Chile, 127, 327
Tigris River, 164, 169, 180, 182, 183, 189, 194-195, 203
Tigris-Euphrates valley, 203
Tigris-Euphrates-Karun basin, 164
Tihama, Arabia, 205-206
Time-scale, geologic, 3, 4
Timor, 294
Tinsukia, Burma, 353
Titusville, Pa., 5½″ gas line, 1875, 143
Tjepu field, Java, 276, 278, 279, 282; refinery, 283
Tjerimai volcano, Java, 277
TNT, 355
Tobago (*see also* Trinidad), census, 83
Toenggal No. 1, Sumatra, 277, 278
Tolba River, U.S.S.R., 312
Toluene, production in 1939 and 1944, 355; as blending agent for high-octane aviation gasoline, 358
Torres, Gumersindo, 74
Torres Vedras, 268
Tortuguero, Mexico, 99
Trade (*and see* under separate countries), international, and payments for oil, 395-396, 401-404; Table 30, world's regions, production of crude petroleum and exports of crude petroleum and petroleum products, 1938 and 1947, 406-408; Figs. 2-5, world's inter-

regional, in crude petroleum and products, 1938 and 1947, 39-41
Transandine highway, 125
Trans-Arabian Pipe Line Company, 163, 188, 211, 216, 217, 219
Transdesert pipe lines (*see also* Trans-Arabian Pipe line), 163, 181-182
Trans-Jordan Petroleum Co., Ltd., area and period of concession, *Table 15*, 177-179, 187
Trans-Iranian railway, *see* Trans-Persian Railway
Trans-Persian railway, 189, 190, 197
Trans-Siberian railway, 237, 313
Trans-World Air Lines, 218
Transylvanian basin, 241, 248, 249, 251, 252
Traps—the closed pressure system of oil occurrence, 11-15, Fig. I, 13, 156-158, Fig. 31, 157
Treaty of Amiens, 83
Trebizond, Turkey, 185
Trees, Crawford and Benedum, Messrs., 116
Trentino, 266
Triassic period, time, 4
Trieste, 250, 265, 266; refineries, 266; Fig. 38
Trinidad, British, Dutch, and French influences, 83; Central Range, 81-82; climate, 82; continental shelf, 329; Crown Colony status, 83; crude oil production, by fields, Table 6, 86; economic conditions, 83-85, 90; geological data on fields, Table 7, 87; gov't., 83; imports in bond, 88; internat'l. airlines, 89; labor, 89-90; location and area, 80, Fig. 12, 81; oil fields and facilities, 86-88, Fig. 12, 81, Table 6, 86, Table 7, 87; pipe lines, 87-88; petroleum development, 85-89, 90; petroleum products exported, Table 4, 84; population, 83; refineries, 87-88; reserves, 90-91; Southern Basin, 82, 86; Southern Range, 82; tariff policy, 89; topography and geology, 80-82; trade unions, 89-90; transportation, 88-89; West Indies, shipping center for, 89; Figs. 6, 8, 11, 12, 16
Trinidad Leaseholds, Inc., 80, 85, 87, 88
Trinidad Petroleum Company, 85
Trinity Hills, Trinidad, 82
Tripet Company, Colombia, 111
Tripoli, Lebanon, 163, 176, 181, 186, 188
Triptane, 358
Tropical Oil Company, Colombia, 105, 110, 111, 112, 113, 115, 116, 124
Trucial Coast, 170, 183, 195, 200, 228
Trucial Oman, 212
Trujillo, Venezuela, 77
Truman, President, statement on continental shelf, 319
Tuapse, U.S.S.R., 236
Tucupido field, Venezuela, 64
Tucupita field, Venezuela, 64, 68
Tucupita-Temblador-Oficina trend, 68
Tunis, 301
Turbio, river, 54
Turin, 266
Turkey, 164, 172, 180, 181, 183, 185, 188, 191, 198, 199
Turkish Petroleum Company, Ltd., 180
Turkmen basin, U.S.S.R., 237
Turksib railroad, 237
Turner Valley pool, Alberta, 154; Fig. 29; Plate 48
Turnu Severin, Rumania, 252
Turukhansk, U.S.S.R., 312
Tuwaiq Mts., Arabia, 207, 208; Fig. 32, Plate 66
Tuxpan, Mexico, 96
Tuymazy field, U.S.S.R., 234
Twenhofel, W. H., 323
Twentieth Century Petroleum Statistics, consumption data, 380
Twingon petroliferous tract, Burma, 298, 299
Twinzayos, 298, 299
Twitchell, K. S., 205
Tzuliuching field, China, 304

Ucayali River, Peru, 124
Uganda, western, 301
Ujfalu field, 253
Umiat District, 314
Umm-al-Qaiwain shaikhdom, 200
Umm Nasan, 223
Umm Said, 215
Unconformity, *see* Part I, *Traps*, 11-15
Underwater surveys, 339
Union of South Africa, 302
Union of Soviet Socialist Republics, *see* U.S.S.R.
Union Oil Company of California, 125
Union Oil, Paraguay, 125
United British Oilfields, Trinidad, 87
United Geophysical Company, Houston, 185
United Kingdom (*see also* Great Britain), 255
United Nations Organization, 180, 198, 404
United States (*see also* North America, Caribbean Area, separate states, and lists of maps, tables, and photographs at front of book)
United States, consumption: totals, 150, 151, 271, 362, 386-389, diesel oil, 362, 363, distillate fuel oil, 362, 363, domestic fuel oil, 363, gasoline, 362, 386, kerosene, 362, middle distillates, 362, 363, residual fuel oil, 362, 363, 386
United States, exploration, 130, 131, 134-135, 138-143, foreign trade, 39-41, 151, 362, 405-407; international policy, 74, 340-342, 393-394, 399-404; marketing, 148-151
United States, output: crude oil, 139, 337; energy, five main sources, 371-374; natural gas, 142; petroleum, 238, 334, 340-341
United States, producing regions (*see also* continental shelves, separate states): Appalachian and north-central states, Fig. 17,

United States (*continued*)
132, Fig. 18, 133; Gulf Coast, 45-46, 132, 134, 135, 140, 141, 146, 152, 344, and Mid-Continent states, Fig. 19, 134, Fig. 20, 135; interior basin, 7, 238; northern plains, 154-156; Rocky Mt. area, 146; southeastern, 156-158, Fig. 31, 157; western Pacific, 153, Fig. 21, 136, Fig. 22, 137
United States, production, 132-143; refining, 37, 146-149, 354-361; reserves, 46, 139, 142, 152, 238-239, 337, 340-341; transportation of petroleum, 32, 33, 38, 143-146
Updip, *see* Part I, *Traps,* 11-15
Upper Rhine Graben basin, 241, 248, 256, 257, 259
Uracoa field, 69
Ural Mts., 230, 233-235, 237, 311, 312
Uranium-ore deposits, 316
Uruguay, 126; refinery, 126; Fig. 16, Plate I
U.S. Antarctic Service Expedition, 1939-1941, 310
U.S. armed forces in World War II, petroleum supplies, 344-353; Fig. 53; Plates 92-98
U.S. Army in World War II, petroleum supplies, *see* U.S. armed forces
U.S. Army-Navy Petroleum Board of the Joint Chiefs of Staff, 344-353
U.S. Bureau of Mines, consumption data, 376, 378; production, 361, 387; synthesis of petroleum from oil shale, 342
U.S.-Canadian Board of Defense, 350, 351
U.S. Census, consumption data, 376, 377
U.S. Department of State, 392
U.S. Geological Survey, Alaskan exploration, 313; world production data, 387
U.S. Govt. and Arabian agricultural project, 221
U.S. Naval Petroleum Reserve No. 4, 313, 314; Fig. 47; Plate 89
U.S. Navy in World War II: Alaska theater, 350-351; Caribbean theater, 348-349; Far East, 217; Mediterranean and European theaters, 344-348; Pacific theater, 351-352; Plates 89, 96, 97
U.S. Public Roads Administration, consumption data, 376
Use of petroleum products, *see* utilization
Uslar Pietri, A., 75
U.S.S.R. (Union of Soviet Socialist Republics), Ch. 10, 230-239: climate, 230; sedimentary areas and oil fields, 230-235; Fig. 37, oil field development and transport facilities, 231; sedimentary areas compared to Brit. Commonwealth, U.S., Brazil, French possessions, 232; inter-shield area, 232, 233; Caucasus trend, 232-233; Urals, 233-235; Siberia, eastern, 233; Caucasus region, 233-234; fields compared to California, 233; Volga drainage basin, 234; geology compared to West Texas, 234; Emba-Ural'sk district, 234-235; Ob' River drainage area, 235; Sakhalin Island, 235; oil field development and transportation, 235-237; Table 18, production of crude by areas, 236; oil shales, 237-238; comparison of probable ultimate yields, U.S.S.R. and U.S.A., 238-239. Other references:
U.S.S.R., agriculture, fuel oil in, 386; continental shelf, 302, 311-313, Fig. 49, 320, 321, 326, 329; Middle East, 399-400, 403; energy output, Figs. 56, 57, 59, 370, 371, 381; Persia (Iran), 175, 179-180, 400; pipe lines, oil, 32, gas, 33; products utilization, 365, 366; proved reserves, 22, 340, 341; Figs. 32, 37, 38, 49, 52; Plates 70-76
Utah, 154, Table 11, 139, Table 12, 142
Utilization (*see also* consumption; Figs. 56-60; tables 24-28), Africa, 367-368, 382, 384, 385; Asia, 368-369, 382, 384, 386; Australia, N.Z., 369, 382, 384; Europe, 271-272, 365-367, 386, 389-390; North America, 148-151, 362-364, 382, 384; South America, 364-365, 382, 384; United States, 148-151, Table 13, 151, 271, 355-374, 377-378, 382-390, U.S.S.R., 365, 371-372, 377-378, 386; world-wide, 38-42, 341, 354-374, 382-391

Vacuum distillation, 37
Vado, Italy, 266
Valencia, Lake of, Venezuela, 54
Valencia, Venezuela, 54
Vallo, Norway refinery, 271
Valona, 254
Vapor-phase thermal cracking, 35
Vejrum dome, 268
Velasquez Private Lease, Colombia, 106
Vendée shales, 260
Venezuela (*see also* names of companies, fields, districts), annual production of crude, 1920-1948, Fig. 7, 51; Apure basin, 71, Fig. 8, 63; as part of mediterranean basin, 325; Bolívar Coastal Field, 58-62, Fig. 10, 59; continental shelf, 329; eastern petroliferous basins, oil fields, and facilities, Fig. 11, 63; effect of oil industry on national economy, 74-76; European oil companies in, 74; exploration and exploitation laws, 77-79; exports, by types of crude, 49-50, by destination, 49-51, Table 2, 50; geographical background, 52-57, Fig. 8, 53; geology, 57-71; Gov't. and petroleum industry, 73-75; Maracaibo basin, 57-64, Figs. 8, 9, and 10, 53, 58, 59; oil companies, 74; oil fields, 57-71, Figs. 9, 10, and 11, 58, 59, 63; oil industry, history, 71-73; Orinoco basin, Fig. 8, 53, Fig. 11, 63, 64-71; production by basins and districts, cumulative and daily, Table 3, 60; refineries, 58, 63; reserves, 75, 340; Spanish conquistadores, 71; taxation, 79; Figs. 2-5, 6-11, 16, 56, 57, 59; Plates 23-31
Venezuela Gulf Oil Company, 73
Venezuelan Congress, 78
Venezuelan Oil Concessions, Ltd., 60, 72

Venice, refinery, 265
Ventura, California, overthrust belt, 154
Vera Cruz state, Mexico, 98, 99
Verkhoyansk, 230
Victoria, Australia, 306
Vienna, 252, 253
Vienna basin, 241, 246, 252, 264
Vinding, 268
Virginia, U.S.A., 154, Fig. 17, 132, Fig. 18, 133, Table 11, 139, Table 12, 142
Vogelkop area, 282
Volga drainage basin fields, 234, 235, 237
Volga River, oil-transport route, 235, 236, 237; oil-shale deposits, 237

Wade, Prof. F. Alton, 310
Wadi Hadhramaut, 203
Wadi Sirhan structural depression, 211
Wahhabi power, in Nejd, 199
Waring, G. A., 85
Warsaw, 256
Wasian-Mogoi fields, New Guinea, 276
Watson Lake Airfield, Canada, 351
Weddell Sea, 310
Weeks, L. G., 23, 98, 239, 323
Weingarten field, 257
Wels gas field, 247
Weser River, 257
West Guara field, 70
West Virginia, 131, Fig. 17, 132, Fig. 18, 133, Table 11, 139, Table 12, 142
Western Australia, 305-306
Western Baronga Island, 297
Western Cordillera, Colombia, 102
Western Desert, Egypt, 172, 187
Western Geophysical Company, 125
Whetsel, R. V., 361, 377, 379-380, 391
White Oil Springs field, Persia, 173
White oils, 360
White Sea, 236
Whitehall Petroleum Corporation, 301
Whitehorse refinery, 317, 351
Widyan region, 211
Wildcat drilling, U.S., 143; Venezuela, 72
Wildcat well, definition, 139
Wilkins, Sir Hubert, 310
Wilson, Sir A. T., 200
Wisconsin, Fig. 17, 132, Fig. 18, 133, Pl. 22
Wonokromo refinery, 283, 284
World maps: I, sedimentary basins and petroliferous areas, facing 14; Figs. 2-5, interregional trade in crude petroleum and in petroleum products, 1938, 1947, 40-41; Fig. 49, continents and continental shelves, 320; Fig. 56, chief sources of energy, 1938, 370; Fig. 59, petroleum utilization, 1938, 381; Fig. 60, index map, principal political units, Jan. 1938, 383
World patterns of civilian utilization, 354-374
World Power Conference, *Statistical Yearbook,* 379
World statistics (*see also* consumption, production, reserves), Table 1, production and reserves of crude petroleum, by principal producing countries, Jan. 1, 1949, 22-23; Table 24, production of crude petroleum and estimated consumption of 4 leading products, by continents, 1938, 361; Table 28, production and consumption in U.S. and in rest of world, 1857-1948, 388; Table 30, world regions production of crude petroleum and exports of crude petroleum and of products, 1938 and 1947, 406-408
World War II, geographical aspects of petroleum use in: Alaskan Theater, 350-351, Pl. 95; BT tanker, cross section, Fig. 54, 349; Caribbean Theater, 348-349; China-Burma-India Theater, 352-353, Pl. 93; Europe, bulk distribution of gasoline to forces, 347-348, Fig. 53, 346, Pls. 92, 98; Mediterranean and European Theaters, 344-348; Pacific Theater, 351-352, Pls. 96, 97; *and see* separate countries
Wu Su field, 304
Württemberg, 258
Wyllie, B. K. N., 159
Wyoming, 29, 141, 154, Table 11, 139, Table 12, 142, Pl. 52

Xicalango, Mexico, 99

Yacimientos Petrolíferos Fiscales, Argentina, (Y.P.F.), 127
Yakhutsk A.S.S.R., 230
Yana River, 312
Yaracuy River, 54
Yaracuy, state of, 54
Yaroslavl', 235
Yemen, 199, 205, 206, 212, 229
Yenangyaung field, 275, 296-300
Yenbu', population, 221
Yenisey River, 230, 232, 237, 312
Yondo concession, 100, 105-106
Y.P.F., *see* Yacimientos Petrolíferos Fiscales
Yucatan Peninsula, potential productiveness, 153
Yugoslavia, 245, 252, 254, 255, 266, 271
Yukon River, 316
Yumen field, 304
Yuruari River, 56

Zagros Mountains, 164, 165, 168, 170, 171, 191, 192
Zallaq, 224
Zanta, 270
Zhiguli Hills fields, U.S.S.R., Plates 71, 72
Zistersdorf field, 253
Zubaida, 228
Zubair field, 162, 182
Zulía, state of, 60, 62, 72, 77, 102
Zureh River, Persia, 168
Zylker, A. J., 277
Zylker concession, 278

THE AMERICAN GEOGRAPHICAL SOCIETY

Broadway at 156th Street
New York 32, N.Y.

The purpose of the American Geographical Society is to advance geographical science, to make it better known, and to bring home to people its bearing on the fundamental problems of human existence and human relations in all the regions of the earth.

Founded almost one hundred years ago, the Society seeks to achieve its purpose through a widely diversified program of activities. From the beginning, it has taken a lively interest in the progress of exploration, notably in the polar regions; it cooperated with Peary, Mikklesen, Stefansson, Byrd, Wilkins, Ellsworth, Mawson, and Ronne, and pioneered the use of techniques and instruments for exploratory surveying and mapping which have since become standard. It has rendered equally notable service in the cartographic field. Its Department of Hispanic American Research (organized by the late Dr. Isaiah Bowman, who took office as Director in 1915) was responsible for the production and revision, over a period of twenty-nine years, of the standard topographic map of Latin America.

The Society has also for many years sponsored and published studies relating to questions of wide public interest in world affairs. Agencies of the United States Government, private institutions and individuals frequently consult the Society and staff members for information, advice, and active cooperation.

The Library and Map Collections, growing for nearly a century, now contain representative materials for every branch of geography and every part of the world, in some 150,000 volumes, 200,000 maps, 2,800 atlases, and many regionally classified photographs and lantern slides. Parts of these materials are not to be found in any other American collection. A cross-indexed Research Catalogue provides the inquirer with clues to information he could hardly come upon elsewhere; and the curators regularly prepare and circulate bibliographies, book-lists, and map catalogues.

Publications

The Society's quarterly, the *Geographical Review,* deals with all phases of the subject—regional, physical, biological, social, political, economic, historical—and with the progress of exploration by land, sea, and air. Its articles, many of which are enduring contributions to human knowledge, are written by authorities in their respective fields, and are illustrated by original maps and diagrams and by photographs.

The several series of book publications—these, too, almost all illustrated by original maps and photographs—offer many titles of general interest and others of more specialized appeal. Space permits mention of only a few of the more recent titles.

Two transcontinental flight studies, *Focus on Africa* by Dr. Richard U. Light and *The Face of South America* by John L. Rich, based on extensive sequences of air photographs, demonstrate, with their accompanying texts, the precision and breadth of geographical knowledge to be gained from that medium. *Mirror for Americans,* by Ralph H. Brown, a study in historical geography, recreates the United States of 1810, region by region. The one-volume *Aids to Geographical Research,* by John K. Wright and Elizabeth T. Platt, provides a wealth of information about geographical references.

Japan: A Geographical View, by Guy-Harold Smith and Dorothy Good, is a useful short study prepared during the war. *Pioneer Settlement in the Asiatic Tropics: Studies in Land Utilization and Agricultural Colonization in Southeastern Asia,* by Karl J. Pelzer, deals with more specialized topics, as does *The Coast of Northeast Greenland, with Hydrographic Studies in the Greenland Sea: The Louise A. Boyd Arctic Expeditions of 1937 and 1938,* by Louise A. Boyd and others.

Among cartographic publications, *The Map of Hispanic America,* on the scale of 1:1,000,000 (about 16 miles to the inch), covers in its 107 sheets all parts of the Americas south of the United States. The arbiters in a number of international boundary disputes have used it in drawing up the lines of settlement, and during World War II Allied forces relied on the Millionth map, as it is commonly known, for cartographic information on the area as a whole. Government officials, scientists, business men concerned with Latin America, find it indispensable.

Other topographic maps in color prepared and published by the American Geographical Society are a *Map of the Americas, 1:5,000,000,* covering the entire land area in five large sheets, and a new *Map of the World,* prepared on the basis of work done during World War II for the Department of State, published in a single sheet. The Society has also prepared and published outline maps of North and South America, and many local and other research maps. The maps are sold at prices ranging from $0.25 to $2.50 a sheet.

Membership

Members are grouped in three classes: Honorary Members, Honorary Corresponding Members, and Fellows. The honorary classes are comprised of persons who have signally promoted the knowledge of geography or who are eminent in associated fields. Membership in the third category is open to all who take interest in exploration and travel, in the spread of geographical knowledge, and in the advancement of science. Annual dues for Fellows are $10, and for Sustaining Fellows, $25. Fellows receive the *Geo-*

graphical Review gratis, the right to one book or map publication gratis each year, and reduced rates for the purchase of other publications. Fellows may use the Society's facilities and attend the lecture meetings held each winter in New York City.

Full information about membership in the Society and a list of all publications in print will gladly be sent on request.